SYSTEMS APPROACHES FOR AGRICULTURAL DEVELOPMENT

Systems Approaches for Sustainable Agricultural Development

VOLUME 2

Scientific Editor
F.W.T. Penning de Vries, CABO-DLO, Wageningen, The Netherlands

International Steering Committee
An international steering committee will support the series

Aims and Scope
The book series *Systems Approaches for Sustainable Agricultural Development* is intended for readers ranging from advanced students and research leaders to research scientists in developed and developing countries. It will contribute to the development of sustainable and productive systems in the tropics, subtropics and temperate regions, consistent with changes in population, environment, technology and economic structure.
The series will bring together and integrate disciplines related to systems approaches for sustainable agricultural development, in particular from the technical and the socio-economic sciences, and presents new developments in these areas.
Furthermore, the series will generalize the integrated views, results and experiences to new geographical areas and will present alternative options for sustained agricultural development for specific situations.
The volumes to be published in the series will be, generally, multi-authored and result from multi-disciplinary projects, symposiums, or workshops, or are invited. All books will meet the highest possible scientific quality standards and will be up-to-date. The series aims to publish approximately three books per year, with a maximum of 500 pages each.

The titles published in this series are listed at the end of this volume.

Systems approaches for agricultural development

Proceedings of the International Symposium on Systems Approaches for Agricultural Development, 2–6 December 1991, Bangkok, Thailand

Edited by

FRITS PENNING DE VRIES
DLO Centre for Agrobiological Research, Wageningen, The Netherlands

PAUL TENG
International Rice Research Institute, Manila, Philippines

and

KLAAS METSELAAR
DLO Centre for Agrobiological Research, Wageningen, The Netherlands

Kluwer Academic Publishers
Dordrecht / Boston / London

Library of Congress Cataloging-in-Publication Data

Systems approaches for agricultural development / edited by F.W.T.
Penning de Vries, P.S. Teng, K. Metselaar.
p. cm. -- (Systems approaches for sustainable agricultural
development ; v. 2)
Includes index.
ISBN 0-7923-1880-3 (HB : alk. paper)
1. Agricultural systems--Congresses. 2. Crops--Congresses.
3. Crops--Mathematical models--Congresses. 4. Agricultural systems-
-Research--Congresses. 5. Rice--Congresses. I. Penning de Vries,
F. W. T. II. Teng, P. S. III. Metselaar, K. IV. Series.
S494.5.S95S95 1992b
338.1--dc20 92-20496

ISBN 0-7923-1881-1

Published by Kluwer Academic Publishers,
P.O. Box 17, 3300 AA Dordrecht, The Netherlands.

Kluwer Academic Publishers incorporates
the publishing programmes of Martinus Nijhoff,
Dr W. Junk, D. Reidel, and MTP Press.

Hardback edition sold and distributed in the U.S.A. and Canada
by Kluwer Academic Publishers,
101 Philip Drive, Norwell, MA 02061, U.S.A.

In all other countries, sold and distributed
by Kluwer Academic Publishers Group,
P.O. Box 322, 3300 AH Dordrecht, The Netherlands.

Paperback edition sold and distributed
by International Rice Research Institute,
P.O. Box 933, 1099 Manila, Philippines.

printed on acid-free paper

Printed in the Netherlands

Contents

SESSION 3. CROP PRODUCTION: SOIL CONSTRAINTS

SESSION 4. CROP PRODUCTION: BIOLOGICAL CONSTRAINTS

SESSION 5. FARMING SYSTEMS

Preface

The symposium

In the next decades, agriculture will have to cope with an ever-increasing demand for food and raw basic materials on the one hand, and with the necessity to use resources without further degrading or exhausting the environment on the other hand, and all this within a dynamic framework of social and economic conditions. Intensification, sustainability, optimizing scarce resources, and climate change are among the key issues. Organized thinking about future farming requires forecasting of consequences of alternative ways to farm and to develop agriculture. The complexity of the problems calls for a systematic approach in which many disciplines are integrated. Systems thinking and systems simulation are therefore indispensable tools for such endeavours.

About 150 scientists and senior research leaders participated in the symposium 'Systems Approaches for Agricultural Development' (SAAD) at the Asian Institute of Technology (AIT), Bangkok, Thailand, in December 1991. The symposium had the following objectives:
- to review the status of systems research and modeling in agriculture, with special reference to evaluating their efficacy and efficiency in achieving research goals, and to their application in developing countries;
- to promote international cooperation in modeling, and increase awareness of systems research and simulation.

The symposium consisted of plenary sessions with reviews of major areas in systems approaches in agriculture, plus presentations in two concurrent sessions on technical topics of systems research. Subjects of studies were from tropical and temperate countries.

The plenary sessions with eight invited presentations were aimed at senior research leaders. These broad papers dealt with (i) agricultural research, (ii) education, training and technology transfer, (iii) agricultural policy and (iv) practical applications. These contributions, plus five review papers from technical sessions, are published in three special issues of the journal Agricultural Systems (1992).

The second group of presentations was aimed at systems scientists. Half of the 36 contributions were invited papers, half were selected. In five sessions, they dealt with the following features: (i) crop production under genotypic constraints, (ii) crop production under weather-related constraints, (iii) crop production with soil constraints, (iv) crop production with biological constraints; farming systems (v). A separate session (vi) dealt with training and technology transfer. These contributions form the contents of this book. There are five papers in the first session, five in the second, five in the third, three in the fourth, nine in the fifth and five in the last session. All papers were reviewed.

The models presented in the first four of the technical sessions are generally mechanistic, explanatory and process-based, often with a one-day time interval of integration and simulate a single, cropped field. They are particularly suited for research and research applications. As there are many valid goals for systems research, many models are possible, and a discussion of which one is 'best' is futile. Several of the 'farming systems' papers explore the possibility of combining agrotechnical models with socio-economic considerations; they demonstrate that this integration is still in an early stage. In the 'training and technology transfer' session, one might note the contrast between the IBSNAT approach that emphasizes development of tools for predictive purposes, and the SARP-approach that emphasizes trainee's understanding of the mechanisms involved. They reflect the extremes of the range of modeling skills mentioned by Spedding in 'The study of Agricultural Systems': operate, repair, improve, construct models.

Systems research

Where systems research is often thought of as computer work, and personal computers have become a very visible tool for research and planning in all countries, a symposium on systems research may appear to result from that technological development. Indeed, much of the work presented here is carried on a wave of new technology. However important and fascinating these new toys are, their presentation was *not* the key issue of SAAD.

SAAD's key issues were the need to think about entire agricultural production systems, the observation that optimizing subsystems has lead to unsustainable agriculture, and the recognition that improving agricultural systems implies satisfying several, partially conflicting goals.

The real world is complex, and to think about it and handle it, we must divide it into parts. In our science, large parts are 'systems', that can often be divided into several 'subsystems'. We use the term 'system' for 'a part of the real world with many interacting components and processes, but with few interactions with the environment'.

Systems Research comprises an analytic and a synthetic phase. The tool Systems Analysis has been around for a while. It guides us to identify systems, subsystems and key processes. By taking systems analysis very far, one can dissect systems and identify relevant research problems at a disciplinary level. In this manner, systems analysis guides us to optimize our often traditionally approach of monodisciplinary research, and to derive conclusions and recommendations. Model building is the inverse of the analysis. Since systems are not just conglomerates of subsystems, special attention is given to the interactions of susbsystems. Only for subsystems without feedback, an optimum system is obtained by putting together optimized subsystems. But this is rare in agriculture. We must, therefore, look at the entire system and not stop at subsystems. Reaching an optimum for one or two subsystems, in fact, might be counterproductive for the total.

Two examples illustrate initially promising developments, based on optimized subsystems, that turned wrong at the systems level: (i) Elimination of insect pests with insecticides initially appeared attractive: all bugs could be killed so that yield loss would be history. We know now that this was short-sighted: natural enemies get killed as well, so that previously harmless insects can become new pests; persistent pesticides accumulate in soil, groundwater, ecological food chains, farmers, and are a serious problem. Killing all insect pests was an optimum solution in a too narrowly defined system. We are starting to learn that insect ecology is part of the agricultural system, as is the soil, the groundwater and the fauna. (ii) Too narrow a focus at fertilizer use efficiency has lead to practices of applying fertilizer at high levels where the response of the crop is small (but economically still positive), but with emissions of large quantities of nutrients to the environment. Pollution of ground and surface water has become a world-wide problem.

There are many examples of subsystems optimization leading to malfunctioning of the entire system of agricultural production. These subsystem optimizations make the entire production system look good for some time, but then the total becomes unsustainable. Indeed, sustainability requires systems research. We hope that this volume of SAAD papers will make you more aware of the point that one should not look too narrowly at problems, and that agricultural production and land use have many features that should be studied together.

There is also another emerging feature that has lead to SAAD: the possibility to handle quantitatively the common knowledge that yield is generally only one objective out of many. Farming can have several goals such as yield and yield stability, income, product diversity, an attractive landscape. Governments have goals regarding water and fertilizer use, absence of pollution, employment. In any system of crop production, all goals are reached at a certain level. We could not compare these goals and their implications objectively and quantitatively, or indicate how the degree of realization of goals can be traded, so that the full systems view was lacking. New developments make this possible, thereby opening ways to define and explore explicitly alternative options for the development of agriculture.

One world of systems research

We did not distinguish between systems research in agriculture in so-called developed and developing countries, because sustainability, a crucial issue in both worlds, always requires a systems approach.

In preparing SAAD, we were struck many times by the fact that so much of systems research is relevant at the same time to rich and poor countries, to extensive and intensive agriculture. The reason, we think, is that agricultural production has changed a lot almost anywhere in the past decades, will be changing even more in the next decades, and that planning for such changes cannot occur without systems approach.

Systems research uses its tools: computers, and basic knowledge in the form of models, data bases and geographic information systems. The importance of the fact that basic knowledge is valid and applicable all over the world cannot be underestimated. Though much was developed in rich countries, it is in principle applicable everywhere else. By compiling knowledge in the form of objective, verifiable models, poor countries can benefit from basic research performed elsewhere, and employ more effectively the large data bases they often possess already in research and planning.

SAAD would not have been a success without the constructive ideas of its International Steering Committee, its Steering Committee at IRRI, and particularly without AIT's efforts as organizer and excellent host. We thank them very much.

The initiative for SAAD was developed by the project SARP (Simulation and Systems Analysis for Rice Production, a collaborative project of the International Rice Research Institute, Los Banos, Philippines, the Centre for Agrobiological Research and the Department for Theoretical Production Ecology of the Agricultural University, both in Wageningen, The Netherlands, with funding from the Ministry of International Cooperation of The Netherlands) and the project IBSNAT (International Benchmark Site Network for Agrotechnology Transfer, with many international participants and US-research groups, coordinated by the University of Hawaii and with financial support from the United States Assistance for International Development).

Frits W.T. Penning de Vries
Paul S. Teng
Klaas Metselaar

SESSION 1

Crop production: genotypic constraints

Designing improved plant types: a breeder's viewpoint

L.A. HUNT
Department of Crop Science, University of Guelph,
Guelph, Ontario, Canada N1G 2W1

Key words: alfalfa, barley, breeding program, crop duration, design driven breeding, genotypic, traits, ideotype, maize, plant architecture, rice, sorghum, soybean, wheat, yield potential

Abstract
Plant breeders have traditionally devoted time to a consideration of the characteristics required for high performance in their target region. Some, over the last 30 years, have extended this traditional approach to encompass considerations of the basic processes of dry matter accumulation and distribution. In some cases this extension has involved a breeding approach based explicitly on the design and selection of model plants or ideotypes. This latter has not been particularly successful, but the concept that factors affecting dry matter accumulation and distribution should be taken into account is now widely accepted. Recently, the availability of computer simulation models of crop growth has made it possible to refine the original considerations, to examine traits in a quantitative rather than qualitative manner, and to examine environmental response traits as well as morphological characteristics. They thus constitute a useful addition to the array of tools used by plant breeders when deciding on parents and crosses. Their use in this arena, however, is likely to be limited, partly because of scepticism stemming from previous experience with 'ideotype' concepts. They may have greater utility in some related areas. First, to analyze trial data in terms of specific genotypic traits. Second, to calculate likely field performance over a 10 or 20 year period. And third, to calculate field scale performance from spaced plant data. Before any of these applications are possible, however, further model development to ensure that they account for all factors currently used in variety release/choice decisions may be necessary.

Introduction

The word 'design' can be used in at least two different contexts – in the context of engineering and in that of art. The inclusion of the word 'design' in the title of the paper is thus appropriate insofar as it emphasizes the dual personality of a plant breeder – that of an engineer (a genetic engineer!), and that of an artist. As an engineer, the plant breeder is concerned with the functional requirement of a cultivar for the target area(s) with which he/she is involved, while as an artist, the plant breeder is involved with the creation of a whole array of possible cultivars, with sorting out how these creations match up to the pre-set requirements, with eliminating those that do not match up, and perhaps even with starting all over again. In this role, the breeder is as 'choosey' as an artist who creates many paintings of a particular scene, rejects all but one, and perhaps even then modifies that before allowing it to be displayed to public view. Both design activities could possibly be aided by systems approaches and modelling, and this paper will explore some ways in which this could be accomplished. As a basis, however, some space will be devoted to a discussion of the plant breeding process, and to recent experiences with 'design driven breeding'.

F.W.T. Penning de Vries et al. (eds.), Systems Approaches for Agricultural Development, 3–17.

The plant breeding process

The goal of any breeding program is the development of new, improved cultivars or breeding lines/populations for particular target areas and for specific applications. Attainment of this goal often represents a time investment of from 10 to 15 or more years. Because of this, breeders are continually and acutely aware not only of the need for constant evaluation and refinement of procedures, but also of the cost of following procedures that may not be particularly efficient in the production of superior cultivars. Breeders are also aware that superior cultivars can be developed using different breeding pathways, that there may not be one 'best' method or even outcome, and that they will generally have to be satisfied with an improvement of a few percent. Because of this latter they have to use, at least in the final breeding phase, techniques that can detect small differences. Breeders are also acutely aware of time constraints, particularly in the period between harvesting and seeding, and of the need for data handling and analytical techniques that can not only be applied to large numbers, but that also work quickly and efficiently in a routine setting.

The situations and options that can lead to an efficient breeding process thus present fascinating challenges to the breeder. Many of these challenges resolve themselves into questions of resource distribution between different aspects of the program. For a self-pollinated cereal, these questions are best considered by dividing the breeding process into three phases, each of which involves the breeder in quite different functions. The entire process can be divided first into heterozygous and homozygous phases, with the heterozygous phase dealing with everything up to individual line selection and the homozygous phase dealing with all subsequent line evaluations and cultivar release. The heterozygous phase can be further subdivided into a planning and hybridization phase, and one concerned with the handling of the early generation progenies. The personal time of the breeder may be divided among these three phases in different ways – in one case with a wheat breeding program Jensen (1975) estimated that the breeder's time was distributed in a 40–10–50 percent fashion between the phases, while the division of the time of the technical 'crew' was quite different and in the order of 5–10–85 percent (Table 1).

For a cross-pollinated crop, the process is somewhat different, and an

Table 1. Division of time between the different phases of activity in a self-pollinated crop breeding program (after Jensen 1975).

Phase	Breeder's time	Technician's time
1. Planning and hybridization (heterozygous)	40%	5%
2. Segregation and stabilization	10%	10%
3. Line evaluation and release (homozygous)	50%	85%
	100%	100%

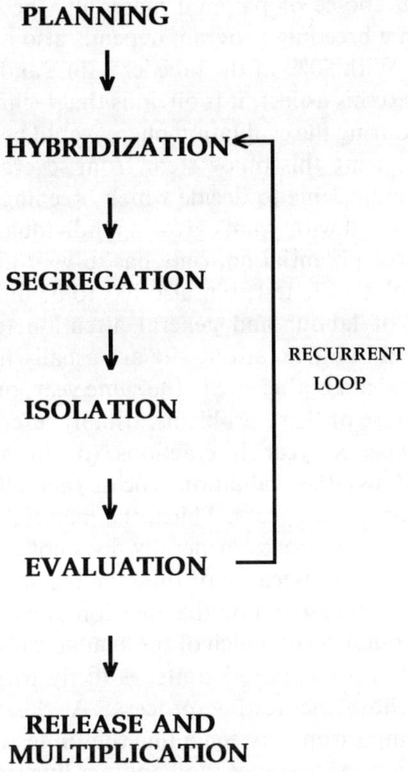

Figure 1. Phases involved in breeding a cross-pollinated crop.

element of recurrency is often introduced before material is suitable for release as a cultivar (Figure 1). The distribution of the breeder's time, however, may be quite similar to that for a self-pollinated crop, while the distribution of the technical crew's time would be increased for the hybridization and segregation phases.

Although the actual figures on distribution of effort vary between programs, all programs show a large percent of the breeder's time assigned to the planning and hybridization phase. Such a distribution reflects the fact that most breeders consider this phase of critical importance to the creation of a superior genotype. Indeed, most breeders have traditionally spent time on the design of the desired genotype for their target area. This design has not only involved consideration of the functional requirements for the area (e.g. appropriate maturity, lodging resistance, disease resistance, tolerance of mineral deficiency or excess, kernel size, grain yield and quality, absence of annoying 'rough' awns) but has also involved consideration of parents that could be used as a source of specific traits, and of the most appropriate methods for using parents that may only have one desirable attribute as well as those that are generally good performers in the target area.

with careful choice of parental materials, the fact remains that the f success in a breeding program depends also on the number of lines each year. With 50% of the breeder's time and 85% of the technical crew's time devoted to this aspect, it is obvious that techniques that could help reduce the demands during the evaluation phase would be of tremendous value. The large demands during this phase stem from several different problems. First, it is generally impossible to decide which breeding lines are superior on the basis of data obtained with plants growing individually (spaced plants), so evaluation of their true potential not only has to wait until there is sufficient seed available for a plot scale test, but also has to be undertaken in a fashion that demands a lot of labour and general attention to crop management. Second, data obtained in a trial at one site are usually not indicative of what would happen in a trial at another site in the same year, or even at the same site in another year. Because of these problems, usually referred to as genotype × location and genotype × year interactions (or in general, genotype × environment interactions), the evaluation process generally has to be continued at several sites and for several years. Third, the material that is produced as a result of the plant breeding process generally does not match up in all aspects to the breeding design ideal. Because of this, the breeder has to spend further time on design, but in this case not on the question of what would be ideal, i.e. the target, but on the question of which of the available breeding lines, each one a different combination of required traits, is likely to perform best for the seasons that will follow the testing process. A classical example of the conundrum is the comparison between a line that in testing proved to be high yielding but of low disease resistance, and another line that was lower yielding but of higher disease resistance. In such a case the breeder must determine whether the disease resistance requirements in the years following testing, and on the farm rather than in a research plot, are likely to be greater or less than during testing. In essence, the breeder must decide whether the line that was lower yielding during testing might, because of its disease resistance, be higher yielding on the farm and in the years subsequent to testing.

Design driven breeding – recent experiences

Although an element of design has been a part of the plant breeding process from the time of its inception as an activity, Donald (1962), in a classical paper entitled 'In search of yield', argued that too little attention had been paid to the basic processes governing dry matter production and its transformation into economic yield, and to characteristics governing these processes. Donald (1968a b) expanded these arguments and developed a breeding approach based much more explicitly on the design of model plants or ideotypes than traditional breeding. The basis of this approach was the use of known principles of physiology and agronomy to design a plant that was capable of greater production than existing types. Donald also suggested that such a model plant was likely to involve a combination of characters that would rarely, if ever, occur by chance in breeders plots.

As an example, Donald (1968a b) designed a small grain cereal ideotype for favourable environments. The ideotype features stressed by Donald were largely morphological: a short stem, a few small erect leaves, a high harvest index (ratio of seed yield to biological yield, the total above-ground dry matter), an erect ear or panicle, awns, and a single culm. Donald (1979) reported an attempt to develop such an ideotype in barley. The lines developed were uniculm, six-rowed awned barleys, with a high harvest index relative to the commercial cultivars available at that time. However they had lax medium-length leaves rather than the short erect leaves.

More recently Donald and Hamblin (1983) developed a general ideotype, which they argued was applicable to all annual grain crops including cereals, grain legumes and oilseeds. The principal characteristics of this general ideotype were: (i) annual habit, (ii) erect growth, (iii) dwarf stature, (iv) strong stems, (v) unbranched and non-tillered habit, (vi) reduced foliage, (vii) erect leaves, (viii) determinate habit, (ix) high harvest index and (x) early flowering in most circumstances.

The papers of Donald and his co-workers generated considerable interest and debate among crop improvement workers (e.g. Mock and Pearce 1975; Trenbath and Angus 1975; Atsmon and Jacobs 1977; Jones and Kirby 1977; Islam and Sedgley 1981; Adams 1982; Sedgley and Seaton 1987a b; Smith 1987; Rasmusson 1987). However, the impact of this approach on plant breeding programs is a matter of some debate and controversy. At the practical level, the formal design of model plants or ideotypes, and the selection of such types, has been adopted as a major breeding activity by relatively few programs.

A number of difficulties and disadvantages have hampered the greater acceptance and use of the concept of design driven breeding. Donald (1968a b) recognized several of these difficulties, but argued that they did not invalidate the utility of the approach. The difficulties and disadvantages fall into two main classes – (i) conceptual (whether the approach has validity), and (ii) practical (how could it be implemented). The first major difficulty is that the concept implies a single superior type for a given climatic or agricultural region. The second difficulty, once it is accepted that there is a single 'best' type, or perhaps a group of 'best' types, is that of defining the important characteristics. As noted by Rasmusson (1987), identifying individual traits that enhance yield universally, or even in a limited range of environments, is a difficult task. This difficulty was illustrated by Marshall (1991) with a consideration of the desirability of having awns on the florets of wheat and barley.

Marshall (1991) pointed out that Donald (1968a) had argued that awns were of proven value in terms of increasing yield, an argument supported by many studies. However, despite the fact that there have been substantial improvements in yield potential of modern European cultivars (Evans 1981; Austin 1987), many of them are still awnless. Marshall argued that the continued competitiveness of awnless cultivars could indicate that awns do not increase yields under northern European conditions as assumed by Donald (1968a) or, that the yield advantage offered by awns is more than offset by one or more disadvantages such as increased susceptibility of awned varieties to weather damage (King 1987).

A second example is that of restricted tillering and spike or panicle size. The value of reduced tillering, or of its complete elimination in a so-called uniculm, is still the subject of considerable debate. Some studies have shown that it is advantageous (Islam and Sedgley 1981; Donald 1979), others have not (Borojevic and Kraljevic–Balalic 1980).

Whatever the reasons, the examples illustrate the difficulty in establishing for any given environment the value of a particular character, even a well studied qualitative trait. Regardless of such difficulty, however, Rasmusson (1987, 1991) emphasized that a plant breeder interested in developing cultivars with greater yield potential would benefit from formally designing an appropriate model or ideotype for her/his target area(s). Doing this would aid in the process of sorting out what is known from what is not, would call attention to germplasm resources and needs, and promote goal setting for individual traits. Based on his own ideotype findings, other published research on small grains, and experience in traditional breeding, he proposed an ideotype for barley consisting of 14 traits (Table 2). In proposing the ideotype,

Table 2. Proposed ideotype for 14 traits of six-rowed spring barley to be grown in the upper midwestern USA (after Rasmusson 1987).

Ideotype trait		Proposed ideotype		
		Approximate phenotype	Present level preferred	Suggested change and breeding goal
Culm	Tiller number, m^{-2}	450	X	
	Culm diam., mm	5.0		Increase to 5.2 mm
	Culm length, cm	86		Shorten to 80 cm
Head[1]	Head number, m^{-2}	350	X	
	Kernels/head, no.	54		Increase to 60 kernels
	Kernel weight, mg	34		Increase to 40 mg
	Awn length, cm	13	X	
Leaf (penultimate)	Leaf length, cm	20	X	
	Leaf width, mm	16		Increase to 18 mm
	Leaf angle (upper leaves)	–		Semierect
Growth duration	Grain-filling period, days	30		Lengthen to 32 days
	Vegetative period, days	52		Lengthen to 54 days
Other[1]	Vegetative biomass (straw), Mg ha^{-1}	4.5		Increase to 5.2 Mg ha^{-1}
	Harvest index, %	47	X	

[1] The yield component and biomass figures are generalizations that may not be fully consistent with each other.

he identified traits believed worthy of attention in a breeding program, and specified the desired goal for each of the ideotype traits. Two of the 14 traits were phenological and the remainder morphological in nature. The strategy employed in designing the ideotype was to use two popular cultivars as the standard or base of reference. It was suggested that five traits be kept at the level of these cultivars, while departures from the standard were recommended for nine traits.

Morphogenetic compatibility or morpho-symmetry was an important feature of Rasmusson's ideotype. This feature was necessary because some trait combinations fall outside the realm of morphogenetic possibility, or at least are very difficult to obtain. For example, it may be impossible to obtain a large spike on a plant with small narrow leaves, as proposed for wheat by Donald (1968b). Accordingly, Rasmusson's model jointly concentrated on stem diameter, leaf size, number of kernels (spike size) and kernel size. A joint increase in each of these characteristics is consistent with morpho-symmetry, so making it easier to obtain the desired trait expressions in a single genotype.

The formal design driven breeding approach also suffers from a number of practical problems. One of the most serious is a lack of the appropriate genetic diversity for the traits in question. Marshall (1991) pointed out that a good example of this is the lack of diverse sources of the uniculm trait in small grained cereals. Naturally occurring uniculm mutants have been recognized and isolated in barley (from the cultivar Kindred) and wheat (from crosses with a north African local cultivar, Atsmon and Jacobs 1977), and mutagenic agents have also been used to generate uniculm variants artificially in Proctor barley (Kirby 1973). In all cases the uniculms were rather poor in traits other than the uniculm aspect, and would require substantial improvement through crossing and selection before being of value in a commercial breeding programme. A second problem relates to the substantial increase in the number of traits that must be selected for by the breeder. Marshall (1991) focussed on this problem, emphasizing that each increase in the number of desired traits, if they were controlled by a single gene, would require a doubling in the size of the selected population if the same progress for other criteria was to be maintained. If the trait were controlled by two genes then the size would have to be quadrupled, etc. Replacing one overall trait, yield, by a number of ideotype characters would thus soon make the breeders involvement with numbers even greater than with a conventional approach. This problem is magnified because the characters themselves are not the end point, which of course is harvested yield under commercial conditions, and that what is really of concern is the gain in yield following selection for the individual characters. In other words, what is of real interest is not the 'direct response' to character selection, but the 'correlated response' for yield. This response, which depends on the genetic correlations among characters as well as on their heritabilities and variances (see, for example, Kramer et al. 1982; Kramer 1983), is unlikely to be as great as for the direct responses.

Design driven breeding – newer approaches

The main thrust of ideotype breeding over the past 25 or so years has been morphological, with design being based largely on considerations of individual traits studied in isolation from each other. Knowledge of the correlations within a plant system has led to modification of the ideotype considered ideal, as witness the comments of Rasmusson (loc. cit.) on the need for what he termed morpho-symmetry. Such considerations, however, were dependent largely on an individual scientist's knowledge of the working of the plant system. They are now being complemented by more formal attempts to use systems analysis and simulation modelling approaches, to take a multi-disciplinary approach that encompasses persons knowledgeable about the climate, the soil, plant pathogens and deleterious insects, as well as the plant system per se, and that makes use of the power of the computer. To date, such an approach has been used in attempts to identify traits required in a high yielding rice of the future (Penning de Vries 1991; Dingkuhn et al. 1991), one that may be grown under different agronomic conditions (direct seeded) than in the past (transplanted). Traits identified have included: (i) enhanced leaf growth during crop establishment, (ii) reduced tillering, (iii) less foliar growth and enhanced assimilate export to stems during late vegetative and reproductive growth, (iv) sustained high foliar N concentration, (v) a steeper slope of N concentration from the upper to the lower leaf canopy layers, (vi) expanded capacity of stems to store assimilates, and (vii) a prolonged grain filling period (Dingkuhn et al. loc. cit.). Such a list, in which some aspects are not easily translated into plant characteristics with which a plant breeder can work, is reminiscent of the original ideotype lists of Donald and others. Its use in practical plant breeding would suffer from the same limitations identified for the ideotype concept from more than 20 years of effort. Nonetheless, the use of simulation models in this context provides a mechanism for quantitatively evaluating the impact of hypothetical changes. This has not been available to plant breeders in the past – it should become one more tool available for use in the overall plant breeding process.

The general ideotype of Donald and Hamblin (1983) included early flowering in addition to an array of morphological attributes. The inclusion of such an aspect, which is tantamount to specifying a particular response to environmental signals, represents a fundamentally different approach than one concerned largely with morphological attributes. It is one in which most attention has been placed on the flowering process, but it is by no means restricted to this aspect.

The timing of events during a crops life cycle (phenology) is the most important aspect of crop adaptation and yield determination. Phenology varies with genotype and with the environmental signals (basically daylength and temperature) to which the plant responds, and it determines the allocation of carbon to different plant organs. The first essential step in breeding higher yielding crops for a specific target area is thus to ensure that the phenology is matched to the seasonal cycle of temperature, rainfall and soil water balance.

Maximum yields are obtained when these events are timed to avoid extremes in drought and temperature, and when maturity occurs before drought, heat, temperature or frost. Information on requirements has traditionally been obtained from agronomic trials, or through trial and error. In many areas (e.g. the Canadian prairies) the earliest successes of plant breeding were in meeting these requirements, such that a change in phenology in large parts of these areas has ceased to be an important aspect of the breeding process. In other areas, despite adjustments in phenology in the past by breeding, scope may still exist for further modification that may increase yields. This is not only so for the lesser developed crops but also for highly bred species such as wheat. It may apply in cases where traditional agronomy (planting date) is practiced, but is particularly pertinent when a changed agronomy is being considered. Richards (1991) has emphasized that studies with wheat in the last five years in Australia have shown that yield increases are still possible by changing phenology and farming practice. These changes all involve sowing cereals earlier than practiced in the past, and using genotypes that have vernalization and photoperiod responses that delay floral initiation. Benefits presumably accrue from the warmer soil temperatures and therefore faster canopy growth and the increased water use efficiency associated with vegetative growth during the cooler months. Many of the genotypes currently recommended are inappropriate as they have no delaying mechanism for floral initiation. Such precocious development results in insufficient leaf area and therefore a low dry matter at anthesis and consequently a low grain yield. However, through the use of photoperiod or vernalization sensitivity genes to delay initiation, varieties can be developed that are more appropriate.

Another recent example where modification of phenology has made possible a changed agronomic practice has been the development in Australia of wheat cultivars with a more flexible sowing time. These cultivars can be sown between early April and June and yet still flower in the optimum week in September-October when the risk of frost has passed. As well as giving farmers more flexibility at sowing, these wheats also provide opportunities for grazing in mid-winter when animal feed is short. A variety with this flexibility was first proposed in 1937 (Macindoe 1937). Pugsley (1983) developed the concept, noting that it could only be achieved by combining vernalization sensitivity genes with photoperiod insensitivity. However, the anticipated benefits from early sowings are not always achieved (Batten and Khan 1987). This is possibly because of a different spectrum or intensity of weeds and diseases, and reasons for it need more investigation.

The fact that the anticipated benefits from early sowing are not always achieved emphasizes that a more traditional qualitative approach to design can easily result in the overlooking of important components of the system. The application of a systems approach to crop improvement, and the use of simulation models that account for all aspects of this system and that can rapidly generate yield data for many years, will help avoid this. Their application can be expected to aid in the design of plant types with the appropriate combination of phenological characteristics. The availability of such information may be of much greater utility in breeding than a mere consideration of morphological characteristics.

Design driven breeding – possible model uses

Examples of the use of simulation models to define desired planting dates and phenological characteristics have been provided by Stapper and Harris (1989) and Aggarwal (1991). Such examples reflect a move from morphological to phenological and environmental response characteristics. Further applications in this area can be foreseen. For example, studies of the impact of the recently documented long juvenile trait in soybeans (Kiihl and Garcia 1989; Hinson 1989) would help identify regions where use of the trait in breeding would be most advantageous. Other similar applications can be visualized as new traits are discovered through examination of material in germplasm banks, or become available to breeders through application of genetic engineering techniques.

Environmental response characteristics are of equal significance for growth as for development, but the same attention does not seem to have been paid to growth responses, especially if characteristics responsible for tolerance of specific extremes are eliminated from consideration. Perhaps this reflects the fact that a consideration of general response patterns, and the drawing of conclusions relative to a desired response pattern, are much more difficult than with morphological and developmental characteristics, and perhaps can only be tackled with a systems/simulation approach. Field (1974) and Field and Hunt (1974) reported on the application of such an approach to help determine a desirable temperature response pattern for alfalfa for the Eastern part of Canada. The reason for this research was that approximately fifty percent of the seasonal yield was produced in the first cut of the season with production declining from second to third cut. Results from controlled environment studies had suggested that increased ambient temperatures in the later part of the growing season may have been responsible for this decline (Pearson and Hunt 1972). However, because of the numerous combinations of day, night and soil temperatures experienced in the field, it was impossible to completely clarify the role that seasonal temperature changes played in determining alfalfa production patterns solely on the basis of controlled environment studies. Rather than arbitrarily choose different combinations of day–night–soil temperatures for extended experimental work, the authors used simulation to help analyze the possible mechanisms whereby temperature contributes to the productivity decline. In adopting this approach, the authors attempted to calculate the degree to which seasonal temperature changes controlled the pattern of production of alfalfa, and so used the system approach implicit in model construction to extend the results of phytotron measurements to the complex field situation. The results supported the hypothesis, and led to suggestions that breeding efforts should be directed at obtaining alfalfa clones with a more uniform performance at different temperatures, as subsequently explored in breeding work by McLaughlin and Christie (1980).

Simulation models that use weather and soil data and operate at a system level to predict crop growth and yields (e.g. Keefer 1986; Sinclair et al. 1987; Carberry et al. 1989; Hammer and Vanderlip 1989; Muchow et al. 1989) thus appear to have utility in making it possible to formalize the thinking that breeders have traditionally done, and that became extended in the ideotypes approach. Depending on their nature and structure, such models provide potentially powerful tools for formulating ideotypes that are not merely morphological, but that also account for physiological responses. They can provide insight into the impact of specific traits on crop performance under hypothetical conditions. Alternatively, the information on environmental variability embedded in historical weather records can be exploited to evaluate the probable long-term performance of 'novel' genotypes in target environments. The stage is set, therefore, for simulation models to be used as a tool to aid in the breeders thinking process. However, the specific results of such activities, as distinct from enhanced understanding, will likely be viewed from the perspective of previous experience with ideotype breeding. This has not been universally successful, perhaps because the understanding on which some ideotypic concepts were founded later proved to be less complete than originally recognized. Because of this, breeders are likely to view the designs that stem from simulation modelling as hypotheses, not proven facts, and to devote only as much effort to the breeding of ideotypes as they would to any other hypothesis testing activity. This is generally only a small part of a breeding program targeted on cultivar release.

The engineering design phase of a plant breeding program is, however, only part of the overall endeavour. A breeder also works in the artistic domain, evaluating the products of her/his efforts and endeavouring to judge which of the products would perform well on a farm scale, and in years or growing seasons subsequent to the one(s) for which data are available. This activity involves a considerable amount of judgement, largely because of the genotype × location and genotype × year (or growing season) interactions, but also because data is often only available from spaced plants or small plots. It is in this arena that simulation modelling could play an extremely valuable role in breeding. It could make it possible to extend results obtained at a restricted number of locations and over a limited number of years, or with spaced plants and small plots, to a wider time and area (i.e. field) base. To do this, though, the characteristics of the breeding lines would have to be known. These could be determined from the data generated in the traditional network of testing sites if crop models could be used as tools to resolve the data into underlying characteristics. With knowledge of such characteristics, the model could then be run with a historical weather data set to produce a profile of yielding ability over many years. Decisions to retain a breeding line or release a potential cultivar to the farming community could then be based on a comparison of cumulative probability curves of yield. Naturally, for such an approach to be effective, the models would have to account for those aspects that currently enter into variety retention and release decisions. This is not always the case with current models, and their further development to account for all variables currently recognized to be of significance in a production region is urgently required before widespread application will be possible. Nonetheless, attempts are being made to use models in this context (Dua et al. 1990).

Conclusions

A most significant development in recent years has been the transition from a qualitative to a quantitative approach, and from morphological to physiological traits in designing crop ideotypes. This change has stemmed from the integration of physiological understanding of crop growth and development into models to predict, quantitatively, crop performance in particular environmental conditions. These models have greatly facilitated the task of evaluating different phenologies and environmental responses for specific environments, particularly where competing risks have to be balanced. However, the experience of breeders in selecting for specific crop types has not been too favourable. Because of this past experience, and because of other constraints, the more sophisticated approaches to design are thus likely to be viewed with a degree of scepticism by plant breeders. As model development proceeds, and as they become capable of accounting for the whole array of factors affecting the system, this scepticism will likely erode. Breeders are accustomed to using a wide array of tools, and there is no reason to suspect that they would not add well validated models to their armoury of tools used in the formal designs phase of their work.

Simulation models could achieve greater everyday use in breeding programmes, however, if they could be used as tools that could be applied to field trial data to calculate specific genotypic characteristics, and then to make long time course predictions of the performance of specific genotypes for specific locations. If this could be done, software packages that are built around crop models could become as integral to breeding as software packages based on statistical theory. The challenges are formidable, but the potential gain in efficiency is considerable. There is urgent need, therefore, for further development work to ensure that models account for all aspects that are currently taken into account in decision making (e.g. diseases), for the continued development of appropriate software packages, and for case studies to validate the overall approach. Should the approach be fully validated, there would be further effort necessary not only to extend the concept to breeders but also to those individuals who determine whether varieties are to be released to the farming community. It should not be forgotten that breeders do not make variety release decisions in many countries. Those making decisions will have to be apprised of, and be prepared to accept, newer and more efficient approaches.

References

Adams M W (1982) Plant architecture and yield breeding. Iowa State J. Res. 56:225–254.

Aggarwal P K (1991) Estimation of the optimal duration of wheat crops in rice – wheat cropping systems by crop growth simulation. Pages 3–10 in Penning de Vries F W T, Van Laar H H, Kropff M J (Eds.), Simulation and systems analysis for rice production (SARP), PUDOC, Wageningen, The Netherlands.

Atsmon D, Jacobs E (1977) A newly bred 'Gigas' form of Wheat (*Triticum aestivum L.*): Morphological features and thermo-periodic responses. Crop Sci. 17:31–35.

Austin R B (1987) Some crop characteristics of wheat and their influence on yield and water-use. Pages 321–336 in Srivastava J P, Porceddu E, Acevedo E, Varma S (Eds.), Drought Tolerance in Winter Cereals. John Wiley & Sons, Chichester, UK.

Batten G D, Khan M A (1987) Effect of time of sowing on grain yield and nutrient uptake of wheats with contrasting phenology. Aust. J. Exp. Agric. 27:881–887.

Borojevic S, Kraljevic–Balalic M (1980) Productivity of single-culm and multi-culm plants of winter wheat cultivars in field conditions. Euphytica 29:705–713.

Carberry P S, Muchow R C, McCown R L (1989) Testing the CERES-Maize simulation model in a semi-arid tropical environment. Field Crops Res. 20:297–315.

Davidson J L (1979) Potential for high rainfall areas in Victoria. Pages 23–30 in White D H (Ed.), Agricultural Systems and Advances in Technology. Australian Institute of Agricultural Sciences, Melbourne, Australia.

Dingkuhn M, Penning de Vries F W T, De Datta S K, Van Laar H H (1992) New plant type concepts for direct seeded flooded rice. Proceedings International Rice Research Conference, Seoul 1990, International Rice Research Institute, Los Baños, Philippines. (in press).

Donald C M (1962) In search of yield. J. Aust. Inst. Agric. Sci. 28:171–178.

Donald C M (1968a) The breeding of crop ideotypes. Euphytica 17:385–403.

Donald C M (1968b) The design of a wheat ideotype. Pages 377–387 in Finlay K W, Shepherd K W (Eds.), Proc. Third Int. Wheat Genet. Symp., Canberra. Aust. Acad. Sci. Canberra, Australia.

Donald C M (1979) A barley breeding programme based on an ideotype. J. Agric. Sci., Camb. 93:261–268.

Donald C M, Hamblin J (1983) The convergent evolution of annual seed crops in agriculture. Adv. Agron. 36:97–143.

Dua A B, Penning de Vries F W T, Seshu D V (1990) Simulation to support evaluation of the potential production of rice varieties in tropical climates. Trans. ASAE 33:1185–1194.

Evans L T (1981) Yield improvement in wheat: empirical or analytical. Pages 203–222 in Evans L T, Peacock W J (Eds.) Wheat Science – Today and Tomorrow. Cambridge University Press, Cambridge, UK.

Field T R O, Hunt L A (1974) The use of simulation techniques in the analysis of seasonal changes in the productivity of alfalfa (*Medicago sativa* L.) stands. Pages 357–365 in Proc. Int. Grassl. Congr. XII, Moscow, USSR.

Field T R O (1974) Analysis and simulation of the effect of temperature on the growth and development of alfalfa (*Medicago sativa* L.). Ph.D. Thesis, University of Guelph, Canada. 154 p.

Hammer G L, Vanderlip R L (1989) Genotype-by-environment interaction in grain sorghum III. Modelling the impact in field environments. Crop Sci. 29:385–391.

Hinson K (1989) Use of a long juvenile trait in cultivar development. Pages 983–987 in Pascale A J (Ed.) Proc. World Soybean Res. Conf. IV. Asociacion Argentina de la Soja: Buenos Aires, Argentina.

Islam T M T, Sedgley R H (1981) Evidence for a 'uniculm effect' in spring wheat (Triticum aestivum L.) in a mediterranean environment. Euphytica 30:277–284.

Jensen N F (1975) Breeding strategies for winter wheat improvement. Pages 31–45 in Proceedings Winter Wheat Conference, Zagreb, Yugoslavia.

Jones H G, Kirby E J M (1977) Effects of manipulation of number of tillers and water supply on grain yield in barley. J. Agri. Sci., Camb. 88:391–397.

Keefer G D (1986a) Effect of sowing date on development, morphology and yield of two 'temperate' hybrids in Central Queensland. Pages 4.91–4.101 in Foale M A, Henzell R C (Eds.) Proc. First Aust. Sorghum Conf., Lawes. Organ. Comm., Aust. Sorghum Conf., Lawes, Queensland, Australia.

Keefer G D (1986b) Prediction of optimum sowing date for irrigated sorghum in Australia using a calibrated version of the SORGF model. Pages 4.73–4.83 in Foale M A, Henzell R C (Eds.) Proc. First Aust. Sorghum Conf., Lawes. Organ. Comm., Aust. Sorghum Conf. Lawes, Queensland, Australia.

Kiihl R A S, Garcia A (1989) The use of the long juvenile trait in breeding soybean cultivars. Pages 994–1000 in Pascale A J (Ed.). Proc., World Soybean Res. Conf. IV. Asociacion Argentina de la Soja: Buenos Aires, Argentina.

King R W (1987) Ear and grain wetting and pre-harvest sprouting. Pages 327–335 in Mares D J (Ed.) Fourth International Symposium on Pre-harvest Sprouting in Cereals. Westview Press, Boulder, Colorado, USA.

Kirby E J M (1973) Effect of temperature on ear abnormalities in uniculm barley. J. Exp. Bot. 24:935–947.

Kramer T (1983) Fundamental considerations on the density-dependence of the selection response to plant selection in wheat. Pages 719–724 in Proc. 6th Int. Wheat Genetics Symp., Kyoto, Japan.

Kramer Th, Ooijen J W, Spitters C J T (1982) Selection for yield in small plots of spring wheat. Euphytica 31:549–564.

Macindoe S L (1937) An Australian 'winter' wheat. J. Aust. Inst. Agric. Sci. 3:219–24.

Marshall D R (1991) Alternative approaches and perspectives in breeding for higher yields. Field Crops Res. 26:171–190.

Martin R H (1983) Register of cereal cultivars in Australia. J. Aust. Inst. Agric. Sci. 49:231–2.

McLaughlin R J, Christie B R (1980) Genetic variation for temperate response in alfalfa (*Medicago sativa* L.). Can. J. Plant Sci. 60:547–554.

Mock J J, Pearce R B (1975) An ideotype of maize. Euphytica 24:613--623.

Muchow R C, Sinclair T R, Bennet J M (1990) Temperature and solar radiation effects on potential maize yield across locations. Agron. J. 338–343.

Pearson C J, Hunt L A (1972) Effects of temperature on primary growth of alfalfa. Can. J. Plant Sci. 52:1007–1015.

Penning de Vries F W T (1991) Development and use of crop modelling in rice research: searching for higher yields. Proc. Symposium on Rice Research – New Frontiers, Hyderbad, India.

Pugsley A T (1983) The impact of plant physiology on Australian wheat breeding. Euphytica 32:743–8.

Rasmusson D C (1991) A plant breeder's experience with ideotype breeding. Field Crops Res. 26:191–200.

Rasmusson D C (1987) An evaluation of ideotype breeding. Crop Sci. 27:1140–1146.

Richards R A (1991) Crop improvement for temperate Australia: Future opportunities. Field Crops Res. 26:141–169.

Sedgley R H, Seaton K A (1987) Ideotypes for the central wheatbelt of Western Australia Part I – Approach in Agronomy 1987. Responding to Change. Proc. 4th Aust. Agron. Conf., Melbourne. Aust. Soc. Agron. Australia. 336 p.

Sedgley R H, Seaton K A (1987) Ideotypes for the central wheatbelt of Western Australia Part II – Wheat in Agronomy 1987. Responding to Change. Proc. 4th Aust. Agron. Conf., Melbourne. Aust. Soc. Agron. Australia. 336 p.

Sinclair T R, Muchow R C, Ludlow M M, Leach G J, Lawn R J, Foale M A (1987) Field and model analysis of the effects of water deficits on carbon and nitrogen accumulation by soybean, cowpea and black gram. Field Crops Res. 17:121–140.

Smith E L (1987) A review of plant breeding strategies for rainfed area. Pages 79–87 in Srivastava

J P, Porceddu E, Acevedo S, Varma S (Eds.) Drought Tolerance in Winter Cereals. John Wiley & Sons, Chichester, UK

Stapper M, Harris H C (1989) Assessing the productivity of wheat genotypes in a mediterranean climate using a crop-simulation model. Field Crops Res. 20:129–152.

Trenbath B R, Angus J F (1975) Leaf inclination and crop production. Field Crop Abstr. 28:231–244.

Woodruff D R, Tonks J (1983) Relationship between time of anthesis and grain yield of wheat genotypes with differing developmental patterns. Aust. J. Agric. Res. 34:1–11.

Improvement of rice plant type concepts: systems research enables interaction of physiology and breeding

M. DINGKUHN[1], F.W.T. PENNING DE VRIES[2] and K.M. MIEZAN[1]

[1] *West African Rice Development Association (WARDA/ADRAO), Irrigated Rice Program, B.P. 96 Saint Louis, Senegal.*

[2] *Center for Agrobiological Research (CABO–DLO), P.O. Box 14, 6700 AA Wageningen, The Netherlands.*

Key words: assimilate partitioning, crop duration, crop growth rate, crop leaf area, ideotype, MACROS, nitrogen-limited yield, rice, Sahel, senescence, simulation, temperature pattern, yield potential

Abstract

Varietal selection is mostly based on convention and breeders' intuition, with little of the knowledge of quantitative plant-environment interactions. Yet, growth models can predict the performance of existing rices. A recent application of models is the design of alternative plant types. Two studies are presented. One resulted in a new ideotype for direct seeded, flooded rice in favourable environments; the other addresses genotypic requirements for adverse thermal environments in the Sahel. Assimilate and nitrogen partitioning, gas exchange and yield of direct seeded and of transplanted IR64 rice were studied in the Philippines in 1986-1990. The growth model L3QT was developed, simulating crop growth and production under nitrogen limitation. Partitioning patterns were then modified to simulate hypothetical genotypes. In IR64, leaf area index limited growth prior to panicle initiation, but during later stages, the leaf area appeared to be excessive, the leaf nitrogen content low and growth limiting, particularly for direct seeded rice where a dense population and the absence of a transplanting shock led to a large vegetative biomass. Simulated yields increased if initial leaf growth was rapid at the expense of tillering, and leaf area was kept low later on. This required enhanced leaf assimilate export and storage in stems. The lower leaf area corresponded with a higher leaf nitrogen concentration, and hence a higher assimilation, and delayed monocarpic senescence. Transfer of more nitrogen to upper leaf strata, and a short vegetative and long ripening phase also improved yield.

The model L3QT is being adapted for ideotype development for Sahelian, irrigated rice environments. Focus is on seasonal and diurnal thermal variation and their effect on crop duration and yield potential.

Problem characterization

Oryza sativa occurs in a wide range of ecotypes of contrasting morpho-physiology. Such differences are less related to the basic physiological mechanisms governing growth and differentiation, but mainly to morphology and anatomy. A comparison of the tall, upland-adapted *japonica* cv IRAT-104 which has a superficial resemblance to maize, and the short, erect and thin-leafed lowland *indica* cv IR64 shows that differential growth and spatial orientation of organs account for much of the visible diversity of the species. Many of these differences can be described in quantitative terms of organ number (e.g. tillers), growth rate, surface and shape.

F.W.T. Penning de Vries et al. (eds.), Systems Approaches for Agricultural Development, 19–35.

Morpho-anatomic traits decide on whether a rice variety is a 'yielder' in favourable environments or 'survivor' under adverse physical conditions. Much research has been conducted on the effects on yield of leaf area index (LAI), canopy structure, tiller and spikelet number, the tissues' nitrogen (N) status and other determinants of assimilate source and sink capacity (Matsushima 1957; Tanaka et al. 1966; Potter and Jones 1977; Akita 1989; Dingkuhn et al. 1991a,b,c,d). In the presence of this mechanistic knowledge, crop simulation becomes useful in two ways: the prediction of the performance of a given cultivar in a hypothetical environment (e.g. climate change), and the performance of hypothetical plant types in specific target environments (Penning de Vries 1991, 1992). This study deals with the latter.

Breeders' selection criteria for segregating or introduced accessions are mostly based on intuition and convention, but increasingly also on mechanistic knowledge of how component traits contribute to the targeted performance. There are limits, however, to our grasp of a complex and dynamic system. Crop simulation can help to provide quantitative answers to 'what if..' questions in the search for alternative plant types.

Models are particularly accurate where the dynamics of growth and development, as influenced by the physical environment, can be described physiologically. This is the case for the weight relationships among plant organs (carbon assimilation and partitioning; Graf et al. 1990a,b and 1991; Penning de Vries et al. 1989 and 1990) and the photo-thermal responses of crop ontogeny (Angus et al. 1981; Summerfield et al. 1992).

Target environments

We address two target environments: the irrigated rice ecosystems of SE Asia, where the current shift from transplanting to broadcast seeding creates a need for new varieties, and the irrigated rice ecosystems of the Sahel where the currently available germplasm is ill-suited for rice-rice double cropping.

Both environments have in common that irrigation water is mostly available, and that high yields are essential to justify labour- and cost-intensive production. Cost is particularly crucial in the Sahel where rice production is considered uneconomical by many sources (e.g. USAID 1989), although local demand is rising rapidly (Nyanteng 1987; Martineau and Harre 1991). In SE Asia, it is also economic considerations that have caused the current, massive shift to direct seeding (De Datta 1986; Erguiza et al. 1990).

Both environments also share a preference for photoperiod insensitive, high-yielding, *indica*-type rice varieties. This and a geographical latitude well within the tropics (16 °N and less in the Sahel, the central plains of the Philippines and all of Indonesia) make temperature the sole environmental factor of importance governing crop duration.

Marked differences between the two environments and the resulting genotypic requirements for rice are related to weather. (Soils which are also

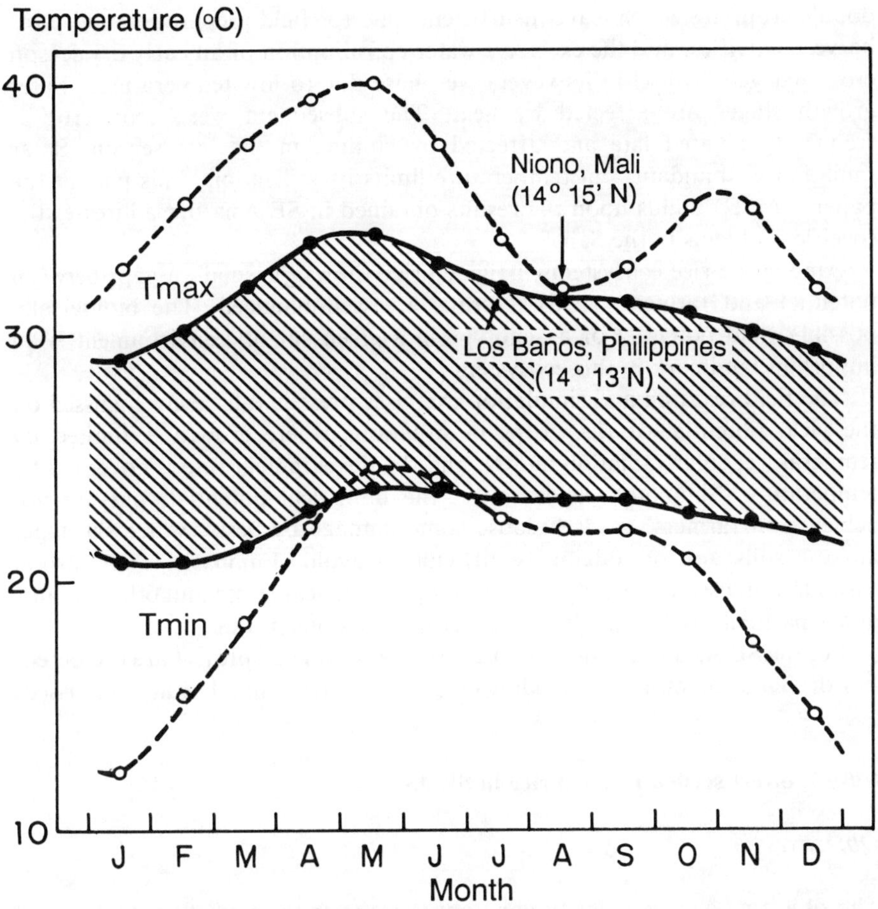

Figure 1. Mean monthly minimum and maximum temperatures for the IRRI lowland site in Los Baños, Philippines (1959–83; from IRRI) and the rice irrigation schemes in Niono, Mali (1950–80; from ICRISAT, Niamey).

diverse are not considered here, so that this paper addresses situations where no toxicities affect the crop and nutrient deficiencies are corrected.) Temperature patterns in Central Luzon (Philippines) are very balanced (Figure 1), resulting in an almost constant crop duration for a given cultivar. Temperatures are almost optimal for carbon assimilation (Yoshida 1981) and development rate of *indica* rices (Summerfield et al. 1992). Grain yields are much influenced by solar radiation and N nutrition. Consequently, the first part of the study (case I) focuses on the optimization of N and light use of direct seeded, irrigated rice in the Philippines.

In the Sahel, extreme seasonal and diurnal temperature differences (Figure 1) lead to a seasonal variability of crop duration of up to 50% (Figure 2). The two major problems resulting from this are acute time constraints in rice-rice double cropping which leave insufficient time for field preparation and post-harvest activities, and the excessive water consumption of an early dry season crop that gets 'stuck' in its vegetative phase due to low temperatures. Later growth stages are affected by heat. The subsequent wet-season crop is frequently planted late and affected by chilling in the late season. Solar radiation is abundant, but temperature limits its utilization. This part of the paper (case II) builds upon the results obtained in SE Asia but addresses the specific problems in the Sahel.

Among the rice ecosystems, irrigated rice shows the smallest gaps between potential and farmers' yields if crops are well-managed. The biologically possible yield potential of a genotype in a given climatic environment is an important landmark in this ecosystem.

The field research and simulation exercises presented here are focussed on the yield potential of rice crops in situations where growth is limited by temperature, solar radiation and N, and the development rate is governed by temperature. Simulated yields reflect the biological potential, but are not achieved in farmers' fields because some damage by pests, water shortage, adverse soils and/or lodging is difficult to avoid. Emphasis is on carbon assimilation, assimilate and N partitioning among leaf strata and other organs, and – particularly for the Sahel – on crop development rate.

For the SE Asian environment, results of a completed project are presented. For the Sahel, an on-going breeding program involving simulation is discussed.

Case I: direct seeded flooded rice in SE Asia

Objectives

The objective was to identify component traits for a plant type that would enable greater N use efficiency and yield potential for direct seeded, flooded, tropical rice.

Materials and methods

Crop growth experiment

The improved lowland-adapted, semidwarf, *indica* rices IR58, IR64, IR72, and IR29723 were grown in field experiments during the dry seasons (January–May) in 1986–1990 at two sites in the Philippines: the International Rice Research Institute (IRRI) lowland farm in Los Baños, Laguna and the Maligaya Rice Research and Training Center (MRRTC) farm in Muñoz, Nueva Ecija. The IRRI site had rich soil (Dingkuhn et al. 1990a). The MRRTC site had a soil poor in N, and higher solar radiation (Schnier et al. 1990a).

All experiments had in common the use of different plant establishment practices and sequential observations throughout the season of crop dry matter, N uptake and distribution within the plant, LAI, tiller number, tissue death and

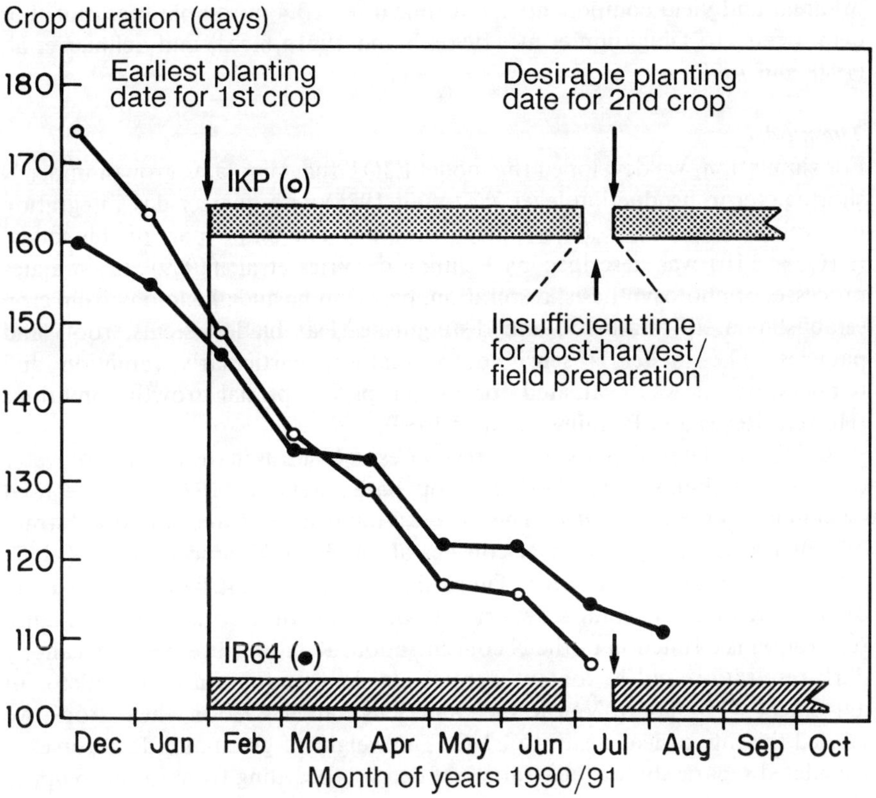

Figure 2. Crop duration as dependent on sowing date in 1990/91 for transplanted IR64 and locally popular I Kong Pao (IKP) rice at the WARDA farm site in Fanaye, Senegal, with a climate similar to Niono (Figure 1). Desirable planting dates for the 1st and 2nd crop are shown, indicating the need for shorter-duration cultivars.

canopy CO_2 exchange rate (CER). Plants were either transplanted at 20 d (TPR) or direct seeded (DSR) using pre-germinated seed, either as broadcast (BS), dibbled at an identical population as in TPR (DS), or row seeded (RS). A factorial, randomized complete block design with four replicates was used.

Plant populations averaged at 100 (TPR and DS), 180 (RS) and 250 (BS) m^{-2}, with four plants per hill in TPR and DS. Spacing was 0.2 m between hills in TPR and DS, and 0.25 m between rows in RS. Plots were kept flooded (0.05 m) until two weeks before maturity.

CER of the canopy was measured with a mobile depletion system, enclosing a 0.36 m^2 area for 60 s (Dingkuhn et al. 1990a). CER of single leaves was measured porometrically with an open system (Dingkuhn et al. 1991c). Sequential sampling for LAI, dry matter, N content and tiller number was based on a 0.6 or 0.4 $\times$ 0.4 m area per plot. Grain yield was taken from a 7.2

m^2 area, and yield components from four 0.2 × 0.4 m samples per plot. For details refer to Dingkuhn et al. (1990a,b and 1991a,b,c,d) and Schnier et al. (1988 and 1990a,b).

The model

For simulation, we developed the model L3QT that simulates growth under N shortage (crop production level 3; De Wit 1986) with quarter-day integration intervals. The model includes tiller formation. The major part of this model (L1Q + TIL) was described by Penning de Vries et al. (1989). It simulates processes of photosynthesis, respiration, partitioning and phenology from crop establishment till maturity, and distinguishes leaf blades, stems, roots and panicles. The model is sensitive to weather, particularly radiation and temperature. It was evaluated for rice crops in optimal growth conditions (Herrera-Reyes and Penning de Vries 1989).

L3QT simulates photosynthetic responses to N status in the absence of water deficits (Dingkuhn et al. 1991a). Crop parameters for IR64 rice are used (Penning de Vries et al. 1989). The foliage is divided into three horizontal strata of equal area that differ in N concentration. Leaf N concentration (LNC) affects both quantum efficiency (Dingkuhn et al. 1991a) and light saturated rate of photosynthesis (Penning de Vries et al. 1990). In all organs maintenance respiration is a function of the N concentration. Relationships between canopy dark respiration and N content were established by Ingram et al. (1989). In agreement with data of Van Keulen and Seligman (1987) for wheat crops, we related the rate of leaf death to LNC. The weight of green and dead leaves is calculated separately for the strata, leaf death proceeding from lower to upper strata.

The total N content in leaves was an input in the model because our focus was on crop responses to N present in the canopy, and not on N uptake. L3QT determines the mean LNC by dividing the total N content by total dry weight. The distribution of N in the foliage was observed only at heading stage in an independent study (Dingkuhn et al. 1991c), but the model demonstrated it to be essential to know its profile at all growth stages in order to understand growth dynamics. In L3QT, LNC of the central leaf stratum equals the mean, that of the bottom layer is lower, and that of the top layer higher by an equal value. This difference was 10 mg N g^{-1}, or less if it would decrease LNC in the lower layer below the minimum of 4 mg g^{-1} (Dingkuhn et al. 1991c). The gradient does not affect total amount of foliar N.

Leaf/stem assimilate partitioning followed pattern A (Figure 3) which approximates observations for IR64 across planting methods and N application rates. Patterns were reported to be independent of N level (Van Heemst 1986). The root/stem partitioning pattern was derived from observations by Diekmann (IRRI 1990, unpublished). The fraction of stem reserves mobilizable for grain growth was raised to 40% to simulate the significant weight loss of stems after flowering. Planting methods are not addressed by L3QT, but affect input values for initial crop weight and N content.

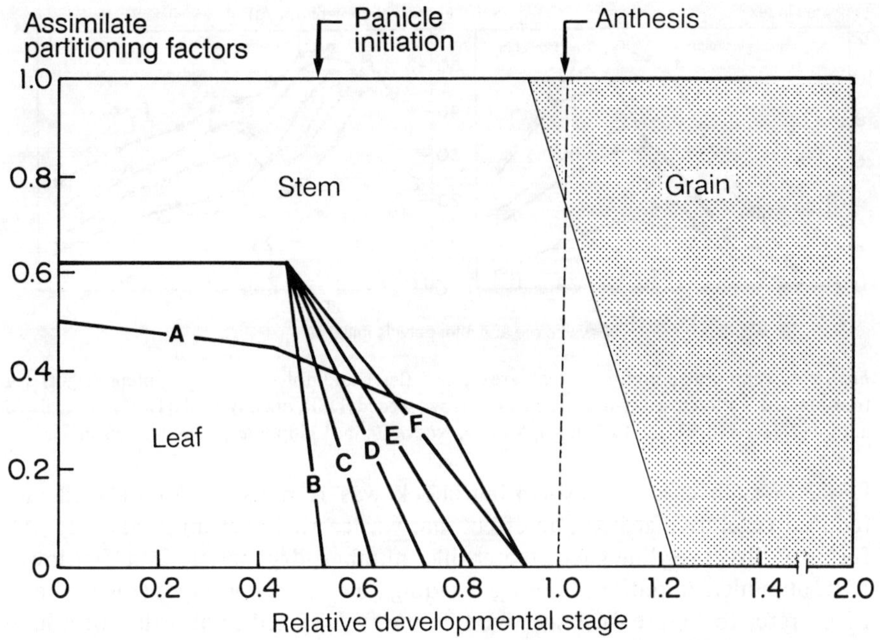

Figure 3. Assimilate partitioning diagram for leaf, stem and grain as used in the rice model L3QT. Pattern A describes observed patterns for IR64 rice. Patterns B to F stand for hypothetical rice plant types.

For model verification, data from Schnier et al. (1990b) were used, covering the response of DSR and TPR to six N application rates. Simulation started at 27 d after seeding for DSR and 21 d after transplanting for TPR. Meteorological data recorded at the experimental sites were used as inputs. For extrapolations of plant type performance, historic data of the IRRI lowland farm for 1979-1988 were used. Simulated plant types had modified leaf/stem assimilate partitioning, N distribution among leaf strata, or duration of growth phases.

Growth and yield of transplanted and direct seeded rice

Over a period of five years, the dynamics of N uptake, N utilization for carbon assimilation and assimilate partitioning were studied for semidwarf *indica* rices in the Philippines. The results were reported in detail by Dingkuhn et al. (1990a,b and 1991a,b,c,d) and Schnier et al. (1990a,b). Only principal findings are presented here.

Direct seeded rice profoundly differed from transplanted rice in the quantity of dry matter produced and allocated to plant organs, resulting in various degrees of dilution of absorbed N in the tissues. Generally, DSR produced more tillers, leaf area and assimilates per ground area than did TPR, and at an earlier time. This difference was mainly due to greater population in broadcast seeded rice but it materialized also when equal populations were planted in TPR and

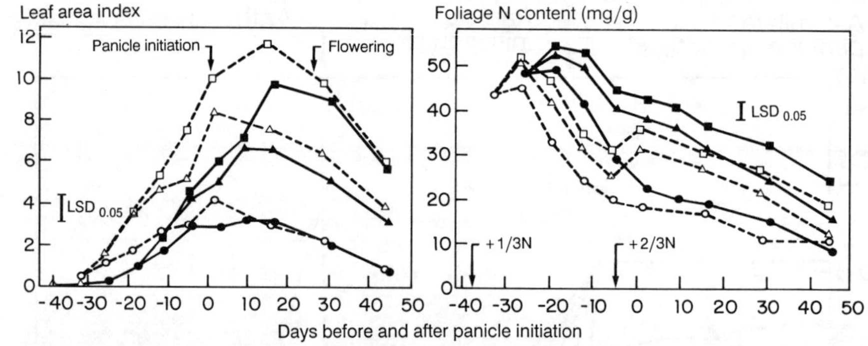

Figure 4. Temporal patterns of leaf area index (left) and foliar bulk N content (right) for transplanted (TPR, closed symbols) and direct (row) seeded (DSR, open symbols) IR64 rice fertilized at three N levels (0,90, 150 kg N ha^{-1}). Muñoz, Nueva Ecija, Philippines, 1988 dry season.

DSR. In such cases, transplanting shock was responsible for delayed and reduced vegetative growth, an effect that was carried over until maturity. All forms of direct seeding gave greater tiller number, dry matter, LAI, but lower LNC at panicle initiation, booting, heading and flowering stages (for LAI and LNC: refer to Figure 4). Excessive tillering in DSR led to abortion of a high percentage of tillers, incurring additional cost in terms of energy and assimilates.

Relative tillering rate (RTR) was uniformly correlated with relative growth rate (RGR) across treatments and years. Values of RGR above 5.5% d^{-1} were associated with tillering, whereas lower values correspond with tiller abortion. This relationship governed tiller dynamics until booting stage.

In DSR, the absence of transplanting shock increased initial RGR and thus, enhanced tillering. Dilution of N caused by rapid growth (Figure 4) resulted in a drop in RGR and consequently, in early tiller abortion. In such cases, N topdressing boosted RGR and sometimes initiated a second tillering phase, falling into the reproductive phase and producing infertile tillers (Schnier et al. 1990a).

Foliage expansion and death could similarly be related to basic growth parameters: whether the total leaf surface grew or was reduced by tissue death depended on LAI and the N status of the foliage. At low LAI, the foliage expanded even if LNC was low, whereas sustaining a high LAI required a high LNC. The amount of N needed to sustain a canopy therefore increased more than proportionally if LAI increased.

The fact that transplanting shock did not only delay the establishment of a closed canopy but also permanently changed its structure and nutritional status was due to differential effects of the shock on growth and development. Tillering and the onset of linear growth were delayed in TPR by up to three weeks, whereas reproductive development was delayed by one week. Consequently, DSR had greater dry matter and lower LNC than had TPR at any given phenological stage.

Under given weather conditions, CER of the rice canopy is largely a function of LNC and LAI (Dingkuhn et al. 1990b). Architectural differences between DSR and TPR no longer affected CER once a closed canopy was formed (Dingkuhn et al. 1991c). The response of CER to LAI followed an optimum curve, the optimal value of the LAI depending on LNC. The LAI of DSR exceeded the optimum during the reproductive growth stages, resulting in a low CER.

Typically, LAI limits CER during exponential growth when LNC is high. After panicle initiation and throughout linear growth, LNC is limiting. Fertilizer dose or timing can modify but not neutralize these limitations (Schnier et al. 1990a,b). During crop establishment, CER of DSR was clearly higher than that of TPR because of its greater LAI. At or after panicle initiation, this relationship was gradually inverted as foliar N deficiency developed in DSR. During ripening, the crop growth rate of DSR was inferior to that of TPR.

Late season N deficiency in DSR was not caused by inferior N uptake but by N being spread thinly over a large leaf area. Tanaka et al. (1966) observed that harvest index was negatively correlated with plant dry matter at heading. Consequently, a plant that has efficiently utilized solar radiation and has accumulated much assimilate before the activation of the reproductive sink will be handicapped by a vegetative burden. The handicap is twofold, particularly in tropical environments with short days and warm nights: high maintenance reduces the percentage of assimilates available for panicle growth, and low LNC limits assimilation. As a combined result, growth efficiency decreases rapidly in DSR.

The light interception by the great number of panicles in DSR, and the reduced flag leaf width related to the lower amount of N available per tiller (Dingkuhn et al. 1991c) further depress growth efficiency. Probably even more important is the early onset of monocarpic senescence in DSR which in turn is related to foliar N deficiency. Competition among neighbouring plants or tillers accelerates senescence and maturity (Debata and Murty 1986).

In spite of these limitations, DSR frequently produces the same and under certain conditions, greater grain yields than does TPR. Apparently, DSR compensates for its limited supply of 'fresh' assimilates during ripening with a greater mobilization of reserves (Dingkuhn et al. 1990b).

It is an important conceptional question why rice is incapable of efficiently carrying-over growth advantages from the vegetative into the ripening phase. An obvious but not necessarily correct answer is that the sink potential sets an upper limit to yield when vegetative growth is abundant. However, boosting the potential sink by either eliminating N deficiency at panicle initiation (Schnier et al. 1990a) or controlling spikelet degeneration through late season N application (Dingkuhn et al. 1991b,c) had only small effects on yield. An alternative hypothesis, is that rice, if given the resources, produces far more leaf area than required for assimilation and growth (Dingkuhn et al. 1990b) (Figure 5).

Modern, high yielding rices resulted from selection under transplanted culture. It led to a plant type that compensates for the transplanting shock by maximizing tillers and leaf area. This was necessary to achieve a structural basis

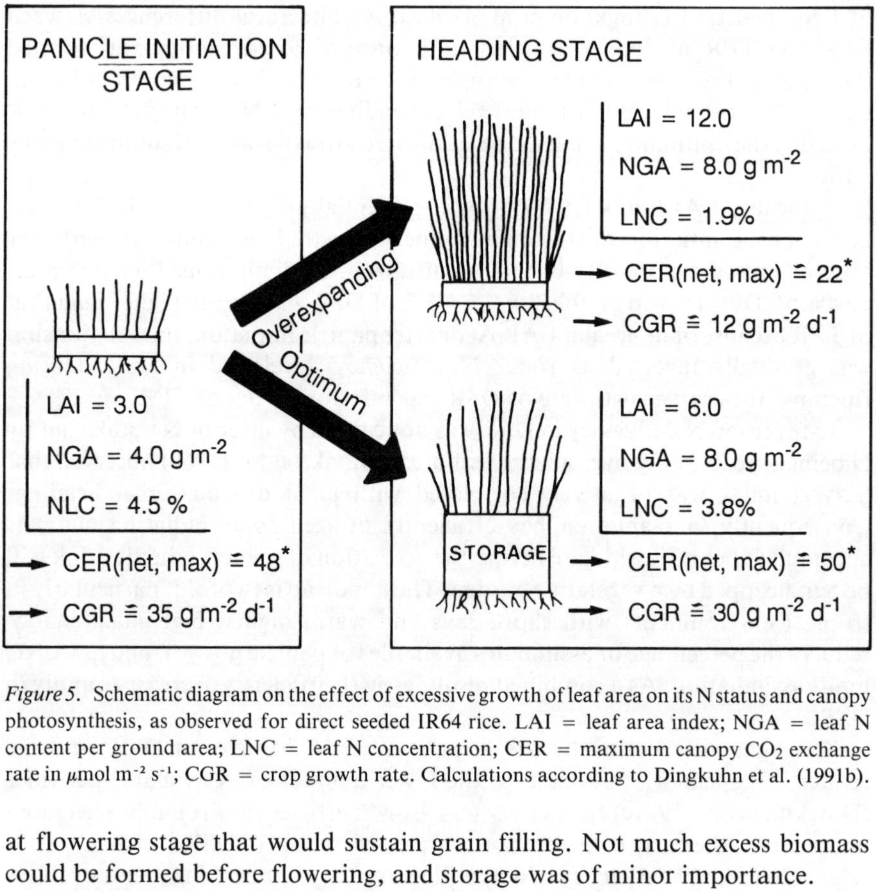

Figure 5. Schematic diagram on the effect of excessive growth of leaf area on its N status and canopy photosynthesis, as observed for direct seeded IR64 rice. LAI = leaf area index; NGA = leaf N content per ground area; LNC = leaf N concentration; CER = maximum canopy CO_2 exchange rate in μmol m^{-2} s^{-1}; CGR = crop growth rate. Calculations according to Dingkuhn et al. (1991b).

at flowering stage that would sustain grain filling. Not much excess biomass could be formed before flowering, and storage was of minor importance.

A crop that produces excess assimilates at an early stage is probably new in the evolution of cultivated rice. Modern varieties tend to use their foliage as a storage organ, the mobilization of which inevitably impedes assimilation. The creation or expansion of an alternative reserve pool with a low rate of maintenance respiration, would benefit grain filling and enable a reduction of the LAI and a higher LNC – a necessary precondition for delayed foliar self-destruction and improved growth efficiency during ripening.

This concept was derived from field studies. In the following, we simulate the yield-effects of plant type modifications.

Simulation

Approach

The model L3QT was chosen because it simulates in relative detail the processes of carbon assimilation. Having obtained satisfactory simulations of the response of DSR and TPR to N nutrition, gradual modifications were made to test hypothetical plant types. Modifications included alternative leaf/stem

Table 1. Simulated effects on yield of observed and optimized assimilate partitioning patterns for transplanted and direct (row) seeded IR64 rice. Partitioning patterns are detailed in Figure 3. GY = grain yield (0% moisture); BY = biological yield; HI = harvest index.

Treatment	Part.pattern	GY	BY	HI
TP 30 N	A (observed)	5.2	11.4	0.46
	C (optimum)	6.0	13.2	0.45
TP 90 N	A (observed)	7.8	14.4	0.54
	C (optimum)	8.5	16.9	0.50
TP 150 N	A (observed)	8.1	16.5	0.49
	C (optimum)	9.7	18.6	0.52
RS 30 N	A (observed)	4.4	11.3	0.39
	C (optimum)	5.3	13.1	0.40
RS 90 N	A (observed)	6.9	15.8	0.44
	B (optimum)	7.6	17.7	0.43
RS 150 N	A (observed)	6.7	17.1	0.39
	B (optimum)	8.1	19.9	0.41

assimilate partitioning patterns at unchanged total LNC, vertical distribution of N within the foliage, and the duration of phenological phases. Each of these simulated 'traits' corresponded to previous field observations: harvest index and yield were strongly affected by the degree of foliar N dilution through growth (Schnier et al. 1990a,b; Dingkuhn et al. 1990b), the LNC of the topmost leaves (Dingkuhn et al. 1991b,c) and crop duration (Dingkuhn et al. 1991d).

Results

Both dry matter and grain yield were sensitive to leaf/stem assimilate partitioning patterns. Effects were small if the relative pattern over time for IR64 (Pattern A in Figure 3) was retained and multiplied by a constant factor. This modification resulted in lower or greater assimilate export of leaves. For all N rates and in both TPR and DSR, dry matter and grain yield increased by 5% to 8% if the assimilate retention of leaves was reduced by 50% uniformly across growth stages (ergo, if export was enhanced). Effects were slightly greater in DSR than in TPR, in accordance with the observed excess production of leaves of DSR.

Effects were stronger if the pattern over time of assimilate partitioning was modified. If the initial leaf/stem partitioning ratio was increased from 0.5 (the initial value for IR64) to 0.6 but reduced rapidly after panicle initiation, the simulated grain yield increased by 10 to 25% or 0.8 to 1.6 t ha^{-1}, depending on the rate of N application (Table 1). Predicted gains in biological yield were even greater, indicating that a source-limited yield potential was probably still greater. Particularly for DSR, simulated harvest index was below the observed value, possibly due to an underestimation of mobilizable reserves.

The 'optimized' assimilate partitioning pattern led to the rapid establishment of a closed canopy. Thereafter, LAI remained roughly constant until the early ripening stage, resulting in a table-shaped LAI kinetics as opposed to the bell-shaped pattern in IR64. The plateau value for LAI depended on N input, and was similar to the optimal LAI predicted for IR64 in an earlier study (LAI 4 to 6; Dingkuhn et al. 1990b). Almost the same optima were suggested by Tanaka et al. (1966) but without taking into account N nutrition. The main advantages of the plateau-type pattern of LAI were prevention of foliar N dilution and the formation of more reserves in stems instead of in leaves.

Yield was also sensitive to modifications of the N gradient across strata of the canopy. The modifications did not affect LAI or total N present in leaves. Steep concentration gradients with greatest LNC at the top generally gave the greatest growth rates and yields. Yield differences were in the order of 3 t ha^{-1} for the absence of any gradient as compared with a N concentration difference of 20 mg g^{-1} between individual strata. The actual gradient was not known for IR64. A substantial gradient was observed for flowering IR72, a similar variety (Dingkuhn et al. 1991c). Best fit of simulated to observed IR64 yields was obtained with low gradients at low N rates, and steep ones at high N rates. More research is needed to adequately describe these relationships.

Strong interactions were observed between growth duration and foliar N dilution. An extension of the vegetative growth phase beyond that of IR64 increased dry matter, but not grain yield. The maintenance of a high LNC as opposed to a gradual decrease caused by dilution was without effect in this case. A previous field study had also shown that the additional biomass produced by varieties with a longer vegetative growth phase was not reflected in the yields. Long duration was associated with more N dilution in the tissues, followed by early senescence (Dingkuhn et al. 1991d).

A simulated extension of the ripening phase, however, improved both grain and straw yield – provided that LNC remained high enough to sustain photosynthesis. A delay of monocarpic senescence would therefore be essential for any yield benefits resulting from a longer ripening phase. This, in turn, would require sustained N uptake to satisfy the N demand of the grains. In the current high yielding cultivars, roots senescence rapidly during the ripening phase, particularly in DSR (Diekmann 1990, IRRI unpublished), which is detrimental to N uptake.

The course of monocarpic senescence in rice (Biswas and Choudhuri 1980), although triggered by the activation of the assimilate and nutrient sink in the panicle, depends on the N status of the plant (Dingkuhn et al. 1990b and 1991d). Rapid dilution of N in the foliage prior to flowering and a high LAI create a situation conducive to senescence, resulting in reduced production of 'fresh' assimilates and reduced ability of the root to absorb nutrients. It appears quite possible therefore that modified partitioning of N and assimilates would also affect the course of monocarpic senescence. These mechanisms can be simulated and might provide new ways to improve yield potential.

Simulation of the combined modifications was not yet attempted. We cannot judge at present whether the hypothetical 'traits' are mutually compatible.

Possible consequences for research and breeding

The results are encouraging, considering that a physiological yield barrier for tropical, irrigated rice has been perceived to exist (Yoshida 1983), but that appears now to be penetrable. But more research is required to verify the physiological compatibility of the hypothetical traits with agronomically essential plant characteristics. Our simulations were extrapolations into a hypothetical reality. They are suited as impulses, not as guidelines in breeding.

Some results require a physiological follow-up and might provide a better understanding the canopy structure of an annual cereal. The importance of the vertical distribution of N in the foliage was shown by both field research and simulation, but it is yet unclear to what extent different LNC among leaf positions translate into effective exposure of upper (high N) leaves to light and lower ones to shade. Leaf angles and the distribution of N within a leaf blade have to be considered. A refined model might be used to simulate sequential leaf appearance and death and the necessary transport of N to achieve and maintain the desired N gradient and kinetics of LAI. Reduced tillering as recommended for DSR also affects LAI and should be addressed by such a model.

More research is needed on the factors governing the onset and course of monocarpic senescence, including that of roots. Senescence is delayed if the assimilate sink does not match the source – an undesirable situation. The search for germplasm with late foliar senescence should therefore concentrate on genotypes with late senescence in spite of a large sink capacity. An expanded sink in terms of more fertile spikelets is also essential for the realization of greater source limited yield potential, regardless of whether it would be achieved with modified assimilate partitioning, foliar distribution of N or duration of the ripening phase.

Lastly, the availability of donor germplasm for larger storage capacity in stems and/or roots, the maintenance cost of such reserves and the amount of N immobilized by them need to be studied.

Case II: irrigated rice in the Sahel

Case I referred to an environment that is highly favourable to rice production. Consequently, the focus exclusively was on yield potential. In the following, we will present an ongoing research program addressing a favourable but more complex environment: the irrigated Sahel.

Objectives

The objective is to identify component traits that would enable high and stable yields in combination with a short and predictable crop duration for irrigated rice crops in the dry and wet seasons of the Sahel.

Plant-environment interactions
The Sahel represents a complex situation due to extreme seasonal and diurnal temperature fluctuations. But there are also similarities to the Philippine situation that justifies the use of the model L3QT for simulation: water stress is usually absent, similar varieties are used, and yield potential is of central importance. In both environments, farmers apply herbicides and fertilizers. Direct seeding is practiced in Sahelian Senegal, whereas transplanting is more common in other Sahelian countries.

Preliminary studies showed that varietal improvement will have to address different climatic constraints: (i) retardation of crop development due to low temperature; (ii) effects of adverse (high and low) temperatures on growth, yield and yield components; (iii) leaf area dynamics and vertical N distribution are affected by extreme weather conditions. As we will stick to the concept of potential yield, some other major but less predictable and regionally scattered constraints (such as spider mite, stemborer or nematode attacks, soil salinity) will be ignored.

Temperature and development rate
L3QT simulates the crop development rate as a function of temperature, using the actual ambient temperature and a genotype specific base temperature. A non-linear function accounts for the existence of a decrease in temperature sensitivity at high temperatures (Penning de Vries et al. 1989).

For the target environment, crop duration is crucial. In rice-rice double cropping, a delay of one to two weeks in maturity of the dry season (early) crop jeopardizes the schedule for the main (wet) season crop (which will then be stressed by low temperatures during flowering: Figure 2). The plant type required for the first season is a high yielding, long grain *indica* with a growth duration of 120–130 days in a thermal environment where a typical tropical, '120 day' *indica* variety matures in 140–180 days.

For L3QT, we will quantify the base (Tb), optimum temperature (To) and thermal units requirements (Tsum) of variety ideal for such an environment. Fifty genotypes are being characterized regarding their ontogenetic response to temperature. A hybridization program will provide information on the genetics of Tb, To and Tsum.

The temperature of the water influences crop development more than that of the atmosphere because the rice plant's growing point is submerged until booting stage. The evaporation caused by dry winds can depress the water temperature by 4°C (Rijks 1973). We have observed over extended periods mean water temperatures 5°C lower than those of the air. A submodel is being developed to simulate water and soil temperature on the basis of weather data and the light transmission of the plant canopy. This aspect will be incorporated in L3QT.

Growth under extreme thermal conditions
The first season crop experiences diurnal temperature fluctuations of about 20°C throughout the season, mean minimum temperatures of 10 to 15°C during

crop establishment and mean maximum temperatures of about 40°C during the reproductive phase. At pre-noon when crop gas exchange is normally greatest, the soil is frequently by 15°C colder than the air. Such thermal conditions affect ontogeny, fertility and basic growth processes.

A short duration variety was identified that produces over 9 t ha^{-1} grain and 22 t ha^{-1} dry matter under these conditions, due to intense solar radiation. Significantly greater yields were obtained with broadcast seeding than with transplanting. This confirms the earlier findings in the Philippines that a superior yield potential of DSR can be expected in environments with high solar radiation (Dingkuhn et al. 1990b).

The morpho-physiological basis of this performance is presently under investigation and simulation will aid in the identification of component traits that are responsible for high yield potential in the Sahel. Improvements of L3QT will be necessary regarding the response of tillering and leaf/stem assimilate partitioning to temperature. Temperature induced sterility, however, will be ignored because cultivars have already been identified with low sterility under Sahelian weather conditions.

Canopy structure vs. wind and atmospheric drought

The results from the SE Asian environment indicated a sensitivity of potential production to the distribution of foliar N among canopy strata. In the Sahel, however, we observed a neutralized or even inverted N (chlorophyll) gradient across leaf positions in some varieties, probably due to desiccation by dry winds.

The relationship between chlorophyll distribution in the canopy and weather parameters is being studied for different varieties, and simulation will quantify the impact of this factor in terms of yield. The effect of reduced flag leaf assimilation as a function of atmospheric drought and wind is being studied in the field, and will be simulated with an adapted version of L3QT.

Outlook

The concept of a close integration of eco-physiological field research, crop simulation and breeding for a given crop species and ecosystem has been contemplated, but not yet effectively implemented. The potential superiority of this approach over mere performance related selection evolves from a more rational identification of desirable component traits. Traits are characterized not only morpho-physiologically and genetically, but also regarding the role they play in the plant community and production system.

The authors are aware that their approach is ambitious and that, the project must remain attached to a conventional breeding program. The first project reported here (SE Asia) lacked the integration into such a breeding program and therefore produced results that could not be readily evaluated from a geneticist's and breeder's point of view. The active participation of all essential disciplines is being realized in the study for the Sahelian environment.

References

Akita S (1989) Physiological aspects for improving yield potential in tropical rice culture. Pages 41–76 in Progress in Irrigated Rice Research. International Rice Research Institute, P.O. Box 933, Manila, Philippines.

Angus J F, Mackenzie D H, Morton R, Schafer C A (1981) Phasic development in field crops. II. Thermal and photoperiodic responses of spring wheat. Field Crops Res. 4:269–283.

Biswas A K, Choudhuri M A (1980) Mechanism of monocarpic senescence in rice. Plant Physiol. 65:340–345.

Debata A, Murty K S (1986) Influence of population density on leaf and panicle senescence in rice. Indian J. Plant Physiol. 29:281–285.

De Datta S K (1986) Technology development and the spread of direct seeded flooded rice in Southeast Asia. Exp. Agric. 22:417–426.

De Wit C T (1986) Introduction. Pages 3–10 in Van Keulen H, Wolf J (Eds.) Modelling of agricultural production: Weather, soils and crops. Simulation Monograph Series, PUDOC, Wageningen, The Netherlands.

Dingkuhn M, Schnier H F, De Datta S K, Wijangco E, Dörffling K (1990a) Diurnal and developmental changes in canopy gas exchange in relation to growth in transplanted and direct seeded flooded rice. Aust. J. Plant Physiol. 17:119–134.

Dingkuhn M, Schnier H F, De Datta S K, Dörffling K, Javellana C (1990b) Response of direct seeded and transplanted flooded rice to nitrogen fertilization: II. Effects of leaf area and nitrogen status on foliage expansion, senescence, canopy photosynthesis and growth. Crop Sci. 30:1284–1292.

Dingkuhn M, Penning de Vries F W T, De Datta S K, Van Laar H H (1991a) Concepts for a new plant type for direct seeded flooded tropical rice. Pages 17–38 in Direct Seeded Flooded Rice in the Tropics. International Rice Research Institute, P.O. Box 933, Manila, Philippines.

Dingkuhn M, De Datta S K, Javellana C, Pamplona R, Schnier H F (1991b) Effect of late season nitrogen application on canopy photosynthesis and yield of transplanted and direct seeded tropical lowland rice. I. Growth dynamics. Field Crops Res. 28:223–234.

Dingkuhn M, De Datta S K, Pamplona R, Javellana C, Schnier H F (1991c) Effect of late season nitrogen application on canopy photosynthesis and yield of transplanted and direct seeded tropical lowland rice. II. Canopy stratification study. Field Crops Res. 28:235–249.

Dingkuhn M, Schnier H F, De Datta S K, Dörffling K, Javellana C (1991d) Relationship between ripening phase productivity and crop duration, canopy photosynthesis and senescence in transplanted and direct seeded lowland rice. Field Crops Res. 27 (in press)

Erguiza A, Duff B, Khan C (1990) Choice of rice crop establishment technique: Transplanting vs. wet seeding. IRRI Research Paper Series 139. International Rice Research Institute, P.O. Box 933, Manila, Philippines.

Graf B, Rakotobe O, Zahner P, Delucchi V, Gutierrez A P (1990a) A simulation model for the dynamics of rice growth and development: Part I. The carbon balance. Agric. Syst. 32:341–365.

Graf B, Gutierrez A P, Rakotobe O, Zahner P, Delucchi V (1990b) A simulation model for the dynamics of rice growth and development: II. The competition with weeds for nitrogen and light. Agric. Syst. 32:367–392.

Graf B, Dingkuhn M, Schnier F, Coronel V (1991) A simulation model for the dynamics of rice growth and development: III. Validation of the model with high-yielding varieties. Agric. Syst. 36:329–349.

Herrera-Reyes C G, Penning de Vries F W T (1989) Evaluation of a model for simulating the potential production of rice. Philipp. J. Crop Sci. 14(1):21–32.

Ingram K T, Dingkuhn M, Novero R P (1989) Growth and CO_2 assimilation of lowland rice in response to timing and method of N fertilization. Plant and Soil 132:113–125.

Martineau J C, Harre D (1991) L'impossible auto-suffisance. Jeune Afrique Economie 139(1):66–73.

Matsushima S (1957) Analysis of development factors determining yield and yield prediction in lowland rice. Bull. Nat. Inst. Agr. Sci. Japan Ser. A5:1–271.

Nyanteng V K (1987) Rice in West Africa: Consumption, imports and production with projections to the year 2000. West African Rice Development Association (WARDA), Monrovia, Liberia (new address: B.P.2551, Bouake, Ivory Coast), 40 p.

Penning de Vries F W T (1992) Rice production and climate change. Pages 177–191 in Penning de Vries F W T, Teng P S, Metselaar K (This volume) (Eds.) Systems Approaches for Agricultural Development, Proceedings of the International Symposium on Systems Approaches for Agricultural Development, 2-6 December 1991, Bangkok, Thailand.

Penning de Vries F W T, Van Keulen H, Alagos J C (1990) Nitrogen redistribution and potential production in rice. Pages 513–520 in Sinha S K, Lane P V, Bhargava S C, Agrawal R K (Eds.), International Congress Plant Physiology, India.

Penning de Vries F W T, Kropff M J, Teng P S, Kirk G J D (Eds.) (1991) Systems simulation at IRRI. IRRI Research Paper Series, No. 151. International Rice Research Institute, P.O. Box 933, 1099 Manila, Philippines.

Penning de Vries F W T, Jansen D M, Ten Berge H F M, Bakema A H (1989) Simulation of ecophysiological processes of growth in several annual crops. Simulation Monograph 29, PUDOC, Wageningen, The Netherlands and IRRI, Los Baños, Philippines.

Potter J R, Jones J W (1977) Leaf area partitioning as an important factor in growth. Plant Physiol. 59:10–14.

Rijks D (1973) Donnees meteorologiques recueillies a Guede, Kaedi et Same. Juin 1972 – Mai 1973. Projet pour le Developpement de la Recherche Agronomique et de ses Applications dans le bassin du Fleuve Senegal. Organisation pour la Mise en Valeur du Fleuve Senegal (Mali, Mauritanie, Senegal) (OMVS), Saint Louis, Senegal, 12 p.

Schnier H F, De Datta S K, Mengel K, Marqueses E P, Faronilo J E (1988) Nitrogen use efficiency, floodwater properties and nitrogen-15 balance in transplanted lowland rice as affected by liquid urea band placement. Fert. Res. 16:241–255.

Schnier H F, Dingkuhn M, De Datta S K, Mengel K, Wijangco E, Javellana C (1990a) Nitrogen economy and canopy CO_2 assimilation in tropical lowland rice. Agron. J. 82(3):451–459.

Schnier H F, Dingkuhn M, De Datta S K, Mengel K, Faronilo J E (1990b) Nitrogen fertilization of direct-seeded flooded vs. transplanted rice: I. Nitrogen uptake, photosynthesis, growth and yield. Crop Sci. 30:1276–1284.

Summerfield R J, Collinson S T, Ellis R H, Roberts E H, Penning de Vries F W T (1992) Photothermal responses of flowering in rice. Ann. Bot. 69:101–112.

Tanaka A, Kawano K, Yamaguchi J (1966) Photosynthesis, respiration and plant type of the tropical rice plant. Technical Bulletin 7. Int. Rice Res. Inst., P.O. Box 933, Manila, Philippines.

USAID (1989) Niger: Politiques du riz et du cotton. Agricultural Policy Analysis Project, Phase II. Rapport Technique 106. Bethesda, Maryland, USA, 86 p.

Van Heemst H D J (1986) The distribution of dry matter during growth of a potato crop. Potato Res. 29:55–66.

Van Keulen H, Seligman N G (1987) Simulation of water use, nitrogen nutrition and growth of a spring wheat crop. Simulation Monograph 28, PUDOC, Wageningen, The Netherlands.

Yoshida S (1981) Fundamentals of rice crop science. International Rice Research Institute, P.O. Box 933, Manila, Philippines.

Yoshida S (1983) Rice. Pages 103–128 in Potential Productivity of Field Crops Under Different Environments. International Rice Research Institute, P.O. Box 933, Manila, Philippines.

Designing improved plant types for the semiarid tropics: agronomists' viewpoints

R.C. MUCHOW[1] and P.S. CARBERRY[2]

[1] *CSIRO Division of Tropical Crops and Pastures, Cunningham Laboratory, 306 Carmody Road, St. Lucia, Queensland 4067, Australia*

[2] *CSIRO/QDPI Agricultural Production Systems Research Unit, P.O. Box 102 Toowoomba, Queensland 4350, Australia*

Key words: Australia, CERES-Maize, climate variability, drought resistance, genotypic traits, groundnut, ideotype, kenat, maize, radiation use efficiency, rice, risk, semi-arid zone, simulation, sorghum, transpiration efficiency, water-limited yield, wheat, yield potential

Abstract

The unpredictable and serious water constraint to crop production in the semiarid tropics poses a considerable challenge to the design of improved plant types. In this paper, simulation models of the contrasting crops, sorghum, maize and kenaf, and long-term climatic data were used to assess the value, in terms of both yield and risk, of different plant designs at a semiarid site in northern Australia.

The impact of three crop improvement strategies, modified phenology, improved yield potential and enhanced drought resistance, on crop production was quantified. There was no clear yield advantage of the modified traits in all years, and the choice of plant type would depend on attitude to risk. Using a mean-standard deviation space, risk-efficient cultivars were selected as dominating other cultivars either by their higher mean yield or lower standard deviation.

Relative to our standard cultivars and site, simulated earlier maturity tended to improve yield in the lower yielding years and increased yield stability, in sorghum and maize, but not in kenaf. In all species, higher yield potential was advantageous, both in terms of yield and risk, despite the severe water limitation. However, there were species differences in the value of improved designs based on drought resistance traits.

Introduction

The availability of water is a serious constraint to crop productivity in the semiarid tropics (SAT). Much has been written about the scope for genetic improvement to overcome this constraint (e.g. Baker 1989; Ludlow and Muchow 1990; Lawn and Imrie 1991; Richards 1991). In designing improved plant types for water-limited environments, an agronomist's viewpoint may be to better match crop phenology to the expected water supply, or to improve crop yield potential to capitalize on the good seasons, or to incorporate crop traits which impart improved drought resistance to minimize risk in the poor seasons. Whatever the strategy, traditional processes of identifying those traits with both high benefit relative to cost and high achievability, can often prove little better than trial and error in highly variable environments. The primary purpose of this paper is to describe one approach which can assist in assessing the value of different plant designs in environments as demanding as the semiarid tropics.

F.W.T. Penning de Vries et al. (eds.), Systems Approaches for Agricultural Development, 37–61.

The variability in the amount and temporal distribution of rainfall, both among and within seasons and across locations, poses special problems for the design (selection) of improved plant types and the assessment of the comparative value of different genotypes. Firstly, it is difficult to assess the expected value of different plant attributes, as final yield is an integral of the growth over the whole season, and a trait that influences the ability of a crop to grow under certain environmental conditions may be relatively unimportant in the context of the total life of the crop. Furthermore, the relative importance of different growth processes in determining final yield, and consequently the value of different traits, may differ among crop species and environments. Secondly, plant characters that influence performance have differing opportunities for expression in different seasons, and it would be expensive, if not impractical, to assess the value of different plant types using conventional multi-site, multi-season cultivar trials. Consequently, it would be difficult to obtain sufficient information to evaluate different plant traits on a probabilistic basis so that the risk preferences of farmers can be considered. In this context, the ability of improved plant types to yield in the poorer years is more important to subsistence farmers than the average response over all years which is appropriate to the risk-neutral person (plant breeder?).

When coupled to long-term climatic data, crop growth simulation offers the opportunity to quantitatively assess the value of different crop traits at different locations in climatically variable environments. Crop growth simulation is a valuable tool for resolving conflicts associated with different plant designs. For example, earlier maturity may improve yields in poorer years and result in greater yield stability, but in the better years, yield may be sacrificed. By generating probabilistic estimates of yield using simulation models, the appropriate phenology at different levels of risk aversion can be determined. However, this approach depends on having an adequate simulation model for the particular crop, and sufficient understanding of cultivar traits and their mode of operation.

Bailey and Boisvert (1989) used this approach to evaluate the performance of a range of groundnut cultivars at several locations in drought-prone India by incorporating economic concepts of risk efficiency. The crucial result from their analysis was that the ranking of cultivars depended on the simulation of yields, and therefore on the ability of the model to adequately simulate the response to available water. The consequences of different phenology on grain yield have been assessed using this approach for sorghum (Jordan et al. 1983; Muchow et al. 1991), rice (O'Toole and Jones 1987), and wheat (Stapper and Harris 1989). Hammer and Vanderlip (1989) simulated the impact on grain yield of differences in phenology and radiation use efficiency of old and new sorghum cultivars. Jordan et al. (1983) and Jones and Zur (1984) have also assessed the value of osmotic adjustment and deeper rooting, but Loomis (1985) has criticised this work, because the assimilate cost of deeper rooting was not incorporated into the model. Ludlow and Muchow (1990) question the magnitude of this cost, but it behoves the point that more research is required to understand the mode of action of some traits and the possible interaction among traits, so that their impact can be assessed using simulation models.

In this paper, we assess the scope for crop improvement for three contrasting crops in a semi-arid tropical environment using three selection strategies: modified crop phenology, improved yield potential and enhanced drought resistance. The two cereal crops, sorghum and maize, have grain as their economic yield, whilst in the third, the fibre crop kenaf, the economic yield is the vegetative stem material which can be used as a source for paper pulp (Muchow et al. 1990). Firstly, we briefly describe the development and validation of simulation models for these three crops in semi-arid tropical Australia, and highlight traits that are adequately conceptualized in the models. Secondly, we describe the methodology appropriate to assess the value of modified traits in terms of both yield and risk. Finally, we compare the responses to changed traits among the three crop species for our selected semiarid site in northern Australia.

Description and validation of the crop models

To assess the potential for dryland cropping in the semi-arid tropics of northern Australia, growth simulation models of maize, sorghum and kenaf have been developed and validated. All three models were initially based on the framework of CERES-Maize (Jones and Kiniry 1986), but calibrated to the genotypes and the environment of the study region (Carberry et al. 1989; Birch et al. 1990). Subsequently, however, the three models have progressed from the CERES framework with the incorporation of several innovative and alternative routines, which are described in detail elsewhere (Carberry and Abrecht 1991; Carberry and Muchow 1992).

A generic framework was used to develop the maize, sorghum and kenaf models. Briefly, all use daily maximum and minimum temperature, solar radiation and rainfall as climatic inputs. Crop phenology is divided into similar stages for the three models, except that the kenaf model does not incorporate the grain-filling stage of the cereal models. The duration of each phenological stage is predicted based on daily temperature and photoperiod. The duration from sowing to flowering in the three models is simulated as four stages, sowing to emergence, emergence to the end of the basic vegetative period (BVP), which is a photoperiod-insensitive stage, a photoperiod-induced stage which depends on the cultivar's photoperiod sensitivity and which ends at floral initiation, and a floral development period (FDP) which ends at 50% flowering. Leaf area development of maize and sorghum is described using functions for the initiation, appearance, expansion and senescence of leaves. Similarly, leaf area development of kenaf is a function of node production, leaf area per node and leaf area senescence. Potential above-ground biomass production is predicted from leaf area index (L), a light extinction coefficient and the crop's radiation use efficiency (RUE). Accumulated biomass is partitioned to leaf and stem components in each of the crop models, and specifically, to grain for the cereal crops and to immature stem for kenaf.

Plant water use and the soil water balance are treated identically in the three crop models. Soil evaporation, rainfall infiltration and runoff, and soil water drainage are predicted using the CERES soil water balance routines (Jones and Kiniry 1986). Potential transpiration is predicted as a function of daily biomass accumulation, a transpiration efficiency coefficient (TE, units Pa) and predicted daily vapour pressure deficit. Actual daily transpiration equals potential if the fraction of plant extractable soil water in the rooting zone is greater than a threshold value. On a given day, the extraction of soil water (WE) by the crop is dependent on the available soil water range for each depth increment, the extraction front velocity which corresponds to the rate of vertical soil penetration by roots, and the extent of exploration of each soil layer by the crop. The WE increases as an exponential function of days after sowing, to a maximum value at anthesis. Rate of phenological development, daily biomass accumulation, transpiration and leaf growth are decreased below potential values once the fraction of available soil water declines below a threshold value.

Oven-dry yields simulated by the models are compared with observed yields for crops grown at Katherine in Figure 1. The majority of data are independent of those used in model development, and were collected from experiments where optimum nutrition and control of pests and disease were maintained, but sowing date, plant population and water regime varied.

Table 1. Standard conditions used for simulations of crops grown at Katherine.

Condition	Units	Value		
General				
Sowing criterion:				
Start search	day of year		350	
Accumulated rainfall	mm		30.0	
Rainfall window	d		5	
Sowing depth	cm		5	
Latitude	°S		14.5	
Initial soil water	%		10.0	
Crop specific		Maize	Sorghum	Kenaf
Cultivar name	–	XL82	DK55	G4
Sown population	plants m^{-2}	7.0	16.0	18.0
Soil profile depth	m	1.8	1.8	2.7

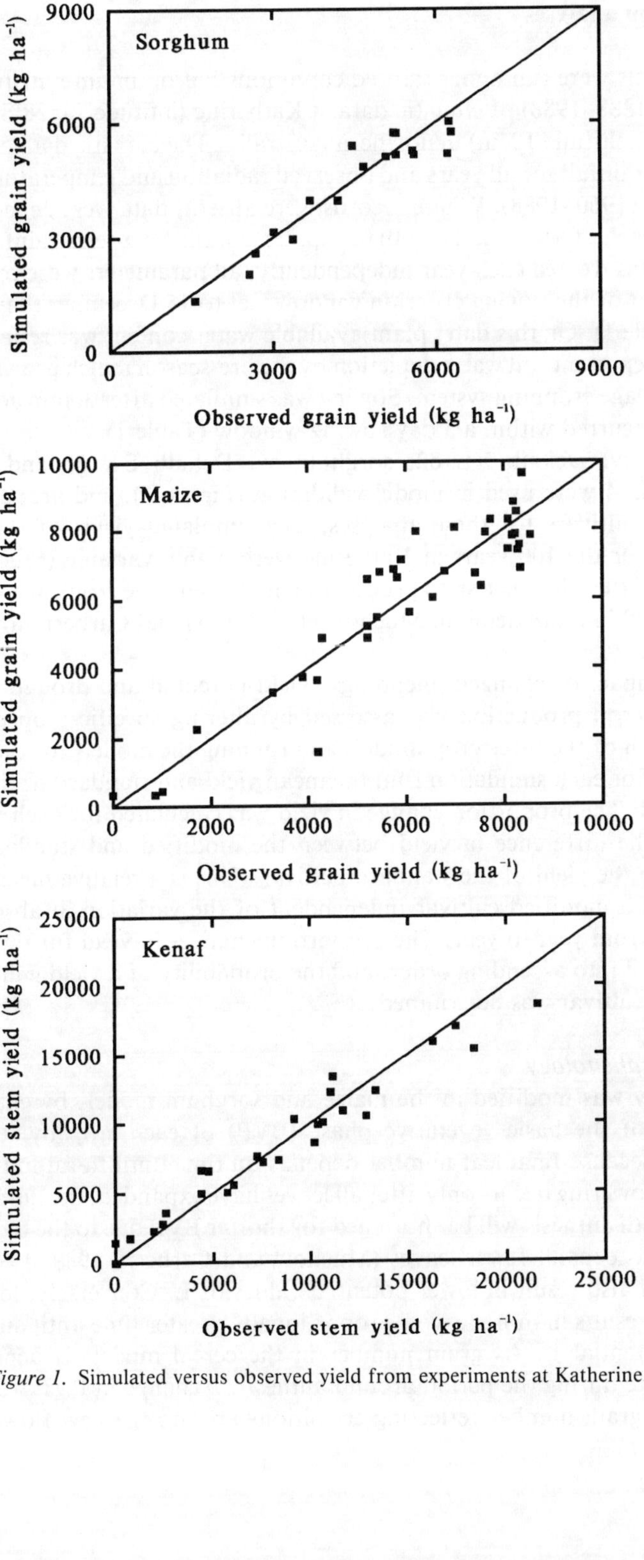

Figure 1. Simulated versus observed yield from experiments at Katherine.

Simulation analyses

The models were run under rainfed conditions but optimum nutrition for 100 seasons (1888–1988) of climatic data at Katherine (latitude 14°28'S, longitude 132°16'E, altitude 107m) in northern Australia. The climatic data consisted of observed rainfall for all years and observed radiation and temperature for years of record (1960–1988). Where records were absent, data were generated using the approach of McCaskill (1990) to supplement the long-term rainfall records. Simulations treated each year independently; all parameters were re-initialized at the start of the sowing criterion window, set to 15 December (day 350) each year (Table 1). On this date, plant available water content was re-initialized to 10%, to represent soil water depletion by the pre-season mulch growth required in a no-tillage cropping system. Sowing was simulated after 30mm accumulated rainfall occurred within a 5 day sowing window (Table 1).

Maize cv. Dekalb XL 82, sorghum cv. Dekalb DK 55 and kenaf cv. Guatemala 4 were used in model validation (Figure 1), and are taken as the standard cultivars for these analyses. The simulated yields of the standard cultivars for the 100 years at Katherine were highly variable (Muchow et al. 1992), and the climatic risk to production in this semiarid tropical environment is discussed in more detail in Muchow et al. (1991) and Carberry and Abrecht (1991).

The impact of changed phenology, yield potential and drought resistance traits on crop production was assessed by altering specific crop parameters within each of the three crop models and running the models for the modified cultivar. For each simulation run, the mean yield and standard deviation were calculated. The proportion change in yield was calculated for each year as the ratio of the difference in yield between the modified and standard cultivar relative to the yield of the standard cultivar. This is a relative measure of the impact of a modified cultivar independent of the variation in absolute yield obtained from year to year. The proportion changes in yield for the 100 years were sorted into ascending order, and the probability of a yield gain using the modified cultivar was determined.

Modified phenology

Phenology was modified in the maize and sorghum models by changing the duration of the basic vegetative phase (BVP) of each crop by ±20% and ±40%. Because final leaf number depends on time until floral initiation and because flowering occurs only after all leaves have expanded, the duration from sowing until anthesis will be shortened for shorter BVP due to the reduced time required to expand fewer leaves (Muchow and Carberry 1989, 1990). Fewer leaves will also result in lower potential values of L. Conversely, longer BVP generally results in more leaves, proportionally greater time until anthesis and higher potential L. As grain number in the cereal models is dependent on growth rate during the period around anthesis, a change in BVP will result in predicted grain number reflecting conditions around the new flowering date which is simulated. At Katherine, where the radiation and temperature regime remains relatively constant during the season, this effect on grain number will be minimal in most years, except where altered anthesis date and water deficit coincide, or where L is sufficiently affected to, in turn, affect biomass accumulation. Genetic yield potential was kept unchanged in the cereal models with the change in BVP.

Most of the genetic variation in the duration to flowering in kenaf is associated with the duration of the floral development period (Carberry et al. 1992). For these analyses the thermal time for FDP of kenaf was adjusted by ±20%. The sole effect of modified FDP in kenaf is on crop duration, and stem yield potential either increases or decreases as a consequence of longer or shorter durations, respectively.

Modified yield potential

Experience with CIMMYT semi-dwarf wheats has shown that high yield potential under optimal conditions has given superior performance under dry conditions (Edmeades et al. 1989). We assessed the impact of higher yield potential by increasing RUE. Higher RUE would result in higher daily dry matter production under non-limiting conditions, and could lead to higher grain number, greater ability to maintain or even increase grain size, and consequently higher yield. However, as water use is linked to biomass production, higher RUE also results in higher rates of water use and therefore, greater risk of soil water deficit under low rainfall conditions. The standard RUE in sorghum (1.25 g MJ^{-1}), maize (1.6 g MJ^{-1}), and kenaf (1.2 g MJ^{-1}) was increased by 20%.

Modified drought resistance

Four drought resistance traits were considered: higher WE potential, higher TE, lower RUE, and lower RUE combined with higher TE.

Higher WE increases the potential amount of plant available water which can be transpired. Likely yield gains would depend on the frequency and severity of water deficit in different years, and on whether rainfall was sufficient to recharge the soil water profile after enhanced depletion. The maximum WE of sorghum (191 mm), maize (154 mm), and kenaf (277 mm) was increased by 20%.

By increasing biomass production per unit of water transpired, higher TE should increase yield, but the likely yield gain would depend on the occurrence of water deficits and the relative proportions of soil evaporation and transpiration contributing to total water use. The two cereals crops have the same high TE (9 Pa), characteristic of C_4 plants, but sorghum has a lower RUE than maize and consequently, sorghum produces less daily biomass and uses less water than maize. Kenaf has lower RUE and, being a C_3 plant, lower TE and so its simulated daily transpiration rate is higher than in the two cereals. A 20% increase in TE was simulated for the modified cultivars.

Lower RUE would result in lower daily biomass production under non-limiting conditions, and could lead to lower grain number, less ability to maintain grain size, and consequently lower yield. However, as water use is linked to biomass production, lower RUE would result in a slower rate of soil water depletion and could prevent the occurrence of severe water deficit, particularly at critical growth stages, and thereby improve yield. Again, the likely yield gain would depend on the occurrence and severity of water deficit in different years. If stomatal closure was the sole cause of a decrease in RUE, a decrease in the ratio of internal to external CO_2 concentration would be expected, with a concommitant increase in TE (Farquhar and Richards 1984). A combined lower RUE with higher TE would slow the rate of water depletion even further, thereby lessening the occurrence of severe water deficit. Modified cultivars were simulated as having either a 20% lower RUE or a combined 20% lower RUE and 20% higher TE.

Simulation results

Modified phenology

Simulated yields of the standard cultivars varied greatly from year to year (Muchow et al. 1991), and there was similar year to year variation in the proportional change in sorghum yield with modified phenology relative to the standard cultivar (Figure 2a, b). This highlights the limited value of conventional experimentation, conducted over a limited number of seasons, to select the appropriate phenology in this climatically variable environment. The 100 years of data in Figure 2 (a, b) are plotted as cumulative distribution functions in Figure 2c. There was a 49% probability of obtaining a yield gain with the earlier sorghum, and only a 22% probability of obtaining a yield gain with the later cultivar (Figure 2c and Table 2). The relative advantage of earlier or later maturity was mostly independent of sowing day, even with a 47 day range in sowing day (Figure 3). Averaged over the 100 years, a 20% decrease in the BVP resulted in 3 days earlier sorghum maturity, but there was little change in mean yield (Table 2). However most of the substantial yield gains with earlier maturity occurred in the lower yielding years (Figure 4) and the coefficient of variation was reduced (Table 2). Conversely, five days average later maturity decreased mean yield by 6.9% and increased the coefficient of variation (Table 2), and substantial yield losses occurred in a number of the lower yielding years (Figure 4). Surprisingly, it was more likely for both the earlier and later maturing cultivars to yield less than the standard cultivar in the higher yielding years (Figure 4). Cultivars of even earlier or later maturity (±40% change in the BVP) resulted in lower mean yield and a lower probability of obtaining a yield gain, although the coefficient of variation was least with the earliest maturing and greatest with the latest maturing cultivar (Table 2).

In maize, an average 3 day earlier or later maturity (±20% change in BVP) had little effect on mean yield, with a 45 and 49% respectively, probability of obtaining a yield gain relative to the standard cultivar maturity (Table 2). Even

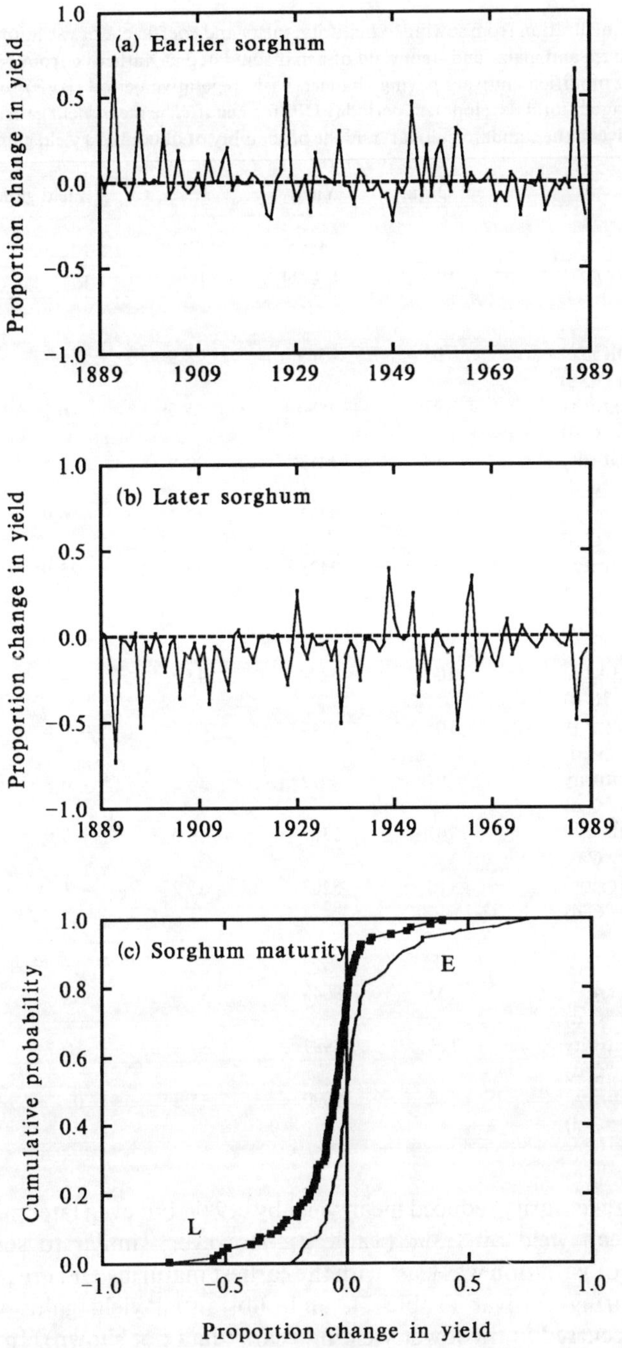

Figure 2. Proportion change in yield relative to standard sorghum yield for (a) earlier (BVP 306 °C d) and (b) later (BVP 460 °C d) maturity at Katherine from 1889 to 1988, and (c) cumulative distribution function for earlier (E) and later (L) maturity.

Table 2. Mean duration from sowing to maturity, mean and coefficient of variation (CV) for grain yield of sorghum and maize and stem yield of kenaf simulated at Katherine from 1889 to 1988, for standard and modified cultivars having changed basic vegetative period (BVP) for sorghum and maize or changed floral development period (FDP) for kenaf. The mean yield gain of the modified cultivar relative to the standard cultivar, and the probability of obtaining a yield gain are also given.

Cultivar	Duration	Yield		Yield gain	
		Mean	CV	Mean	Probability
	(d.)	(kg ha^{-1})	(%)	(%)	(%)
Sorghum					
A. Dekalb DK55 (BVP 383 °C d)	95	4569	28.9	–	–
B. Earlier maturity (BVP 306 °C d)	92	4580	25.5	0.2	49
C. Earlier maturity (BVP 230 °C d)	89	4453	22.9	– 2.5	32
D. Later maturity (BVP 460 °C d)	100	4253	33.7	– 6.9	22
E. Later maturity (BVP 536 °C d)	107	3428	46.5	–25.0	7
Maize					
A. Dekalb XL82 (BVP 171 °C d)	104	5274	43.8	–	–
B. Earlier maturity (BVP 137 °C d)	101	5244	40.6	– 0.6	45
C. Earlier maturity (BVP 103 °C d)	97	4912	36.3	– 6.9	39
D. Later maturity (BVP 205 °C d)	107	5307	46.5	0.6	49
E. Later maturity (BVP 239 °C d)	110	5203	51.2	– 1.3	47
Kenaf					
A. Guatemala 4 (FDP 528 °C d)	133	9048	37.4	–	–
B. Earlier maturity (FDP 422 °C d)	125	8987	37.5	– 0.7	0
C. Later maturity (FDP 634 °C d)	140	9090	37.3	0.5	100

earlier maize maturity reduced mean yield by 6.9%, but even later maturity only reduced mean yield by 1.3% (Table 2). However, similar to sorghum, the coefficient of variation was least with the earliest maturing and greatest with the latest maturing cultivar (Table 2), and substantial yield gains with earlier maturity occurred in the lower yielding years (data not shown). In contrast to maize and sorghum, earlier or later maturity in kenaf always decreased or increased stem yield respectively, but stem yield and the coefficient of variation were relatively unresponsive to changed maturity, with a 7–8 day average

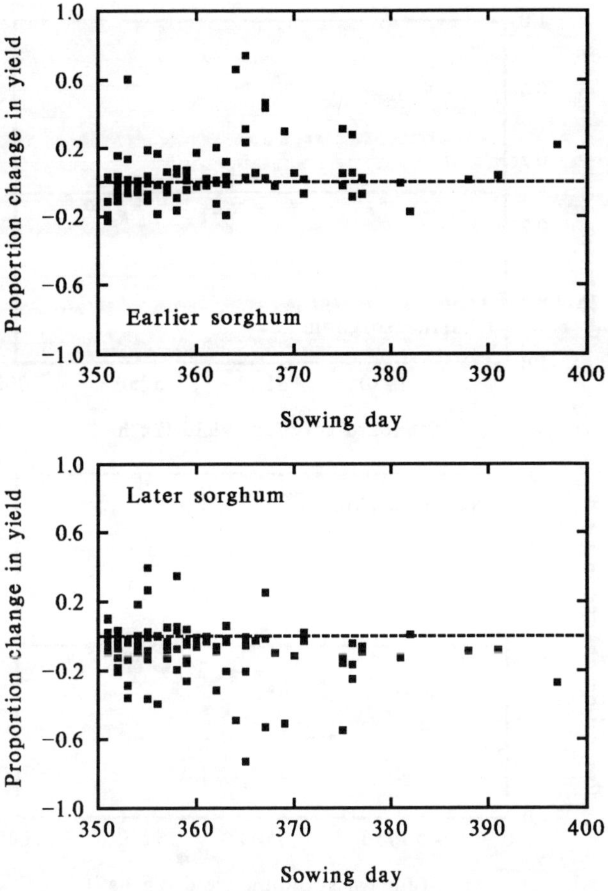

Figure 3. Effect of sowing day on proportion change in yield relative to standard sorghum yield for earlier (BVP 306 °C d) and later (BVP 460 °C d) maturity. Sowing day is day of year from 15 to 31 December; thereafter day of year plus 365 or 366 (leap year).

change in maturity affecting mean yield by less than 1% (Table 2). Even a 21 day later maturity only increased mean stem yield by 2% in this semi-arid tropical environment (Muchow and Carberry 1992). This is associated with the already long duration (average 133 days) of the standard kenaf cultivar and the low occurrence of rainfall at the end of the wet season.

Since there was no clear advantage of earlier or later maturity in sorghum or maize in all years, and since substantial yield gains are likely in the lower yielding years, the choice of cultivar phenology would depend on attitude to risk. Anderson et al. (1977) discuss approaches for agricultural decision analysis under conditions where outcomes are uncertain. McCown et al. (1991) employed one such approach, the (E,V) analysis using a Mean-Standard deviation space, for identifying risk-efficient strategies based on simulated gross margin returns for fertiliser application in the climatically variable Kenyan SAT. We believe a

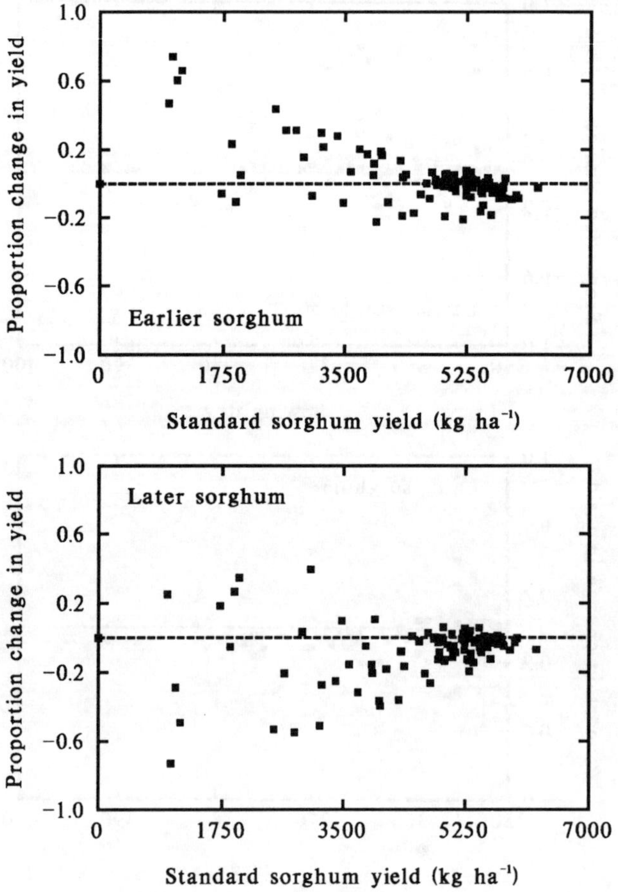

Figure 4. Proportion change in yield relative to standard sorghum yield for earlier (BVP 306 °C d) and later (BVP 460 °C d) maturity as a function of standard sorghum yield.

similar approach is appropriate here, as economic production costs will change little with cultivar selection within a species, and so modified cultivars can be compared using production yields directly rather than the gross margin return normally used. Using Mean-Standard deviation space, risk-efficient cultivars dominate other cultivars either by their higher mean yield or lower standard deviation. This analysis is shown in Figure 6 where the iso-utility or indifference line reflects the added risk associated with higher yield. A horizontal indifference line would represent the choice criterion of a risk-neutral farmer, whereas indifference lines of increasing slope depict increasing aversion to risk. Barah et al. (1981) used a 2:1 ratio of change in standard deviation to change in mean yield to rank sorghum cultivars for their yield and stability/adaptability from multi-location and multi-year trials in India. We used this 2:1 rule, with cultivars in Figure 5 above the 2:1 indifference line being more risk-efficient than the standard cultivar.

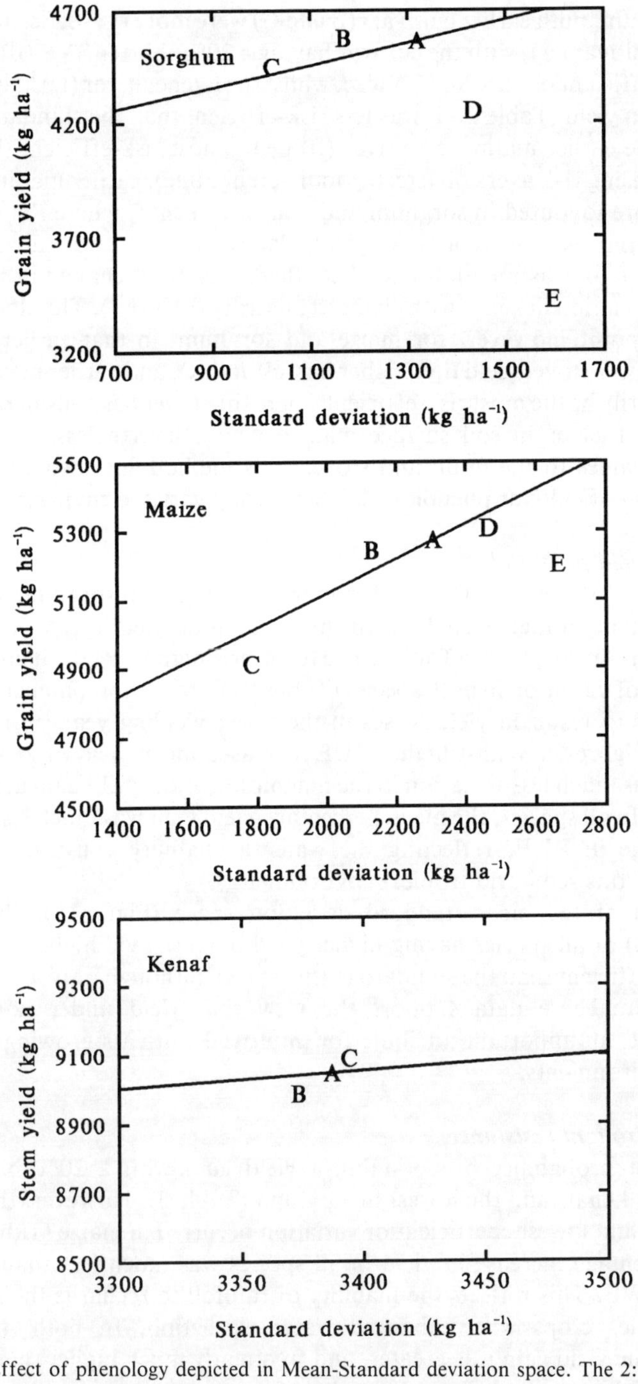

Figure 5. Effect of phenology depicted in Mean-Standard deviation space. The 2:1 indifference line is drawn through the standard cultivar values of sorghum, maize and kenaf, and symbols represent earlier or later maturity as shown in Table 2.

In sorghum, both earlier cultivars (B and C) were more risk-efficient than the standard cultivar (A), with the cultivar having a 20% shorter BVP (B) being the most risk-efficient (Figure 5). In maize, whilst the later cultivar (D) had a slightly higher mean yield (Table 2), it was less risk-efficient than the standard cultivar (A) with the earlier maturing cultivar (B) being most risk-efficient (Figure 5). With increasing risk aversion (steeper indifference line), earlier maturity would be even more favoured in sorghum and maize. In kenaf, where the vegetative stem material is the economic yield, later maturity was always more risk-efficient, but as mentioned earlier, there was little change in mean and standard deviation in kenaf with changed maturity (Figure 5). This does raise an important point, however, for maize and sorghum, in that earlier maturity results in lower stover yield (i.e. higher harvest index), and earlier maturity may not necessarily be the most risk-efficient where the stover has considerable value either as a fuel or in soil surface management. Nevertheless, by assigning economic values to the grain and stover, this methodology would allow the optimisation of cultivar phenology in climatically variable environments.

Modified yield potential

There was a 70% probability of obtaining a yield gain in sorghum and maize cultivars having a higher RUE, with the mean grain yield increasing more in maize than in sorghum (Table 3). However, higher RUE increased the coefficient of variation in both species (Table 3). Furthermore, higher RUE was more likely to result in yield losses in the lower yielding years in maize and sorghum (Figure 6). Whilst higher RUE increased mean yield of kenaf (Table 3), there was much less variation in the magnitude of the yield gain across years (Figure 7). In all species, the mean percentage yield gain was much less than the 20% change in RUE, reflecting the water availability constraint on yield potential in this semi-arid tropical environment.

Analysis of the mean-standard deviation space (Figure 7), shows that cultivars (B) of all species having higher yield potential via higher RUE, were more risk-efficient than the standard cultivars (A), although only marginally so in sorghum. These data support the view that yield under non-stressed conditions is an important attribute for improved cultivars growing in water-limited environments.

Modified drought resistance

The highest probability of obtaining a yield gain with a 20% higher WE occurred in kenaf, and the lowest in sorghum (Table 3). However, the highest mean yield and lowest coefficient of variation occurred in maize (Table 2). The mean percentage increase in yield in all species was much less than the 20% change in WE. This reflects the inability of rainfall to recharge the soil water profile under crops with enhanced water extraction. In both maize and sorghum, the yield gain was greater, and frequently much higher than 20%, in the lower yielding years (Figure 8). In kenaf, the variation in the magnitude of the yield gain was less, and the yield gain was independent of the standard

Table 3. Mean and coefficient of variation (CV) for grain yield of sorghum and maize and stem yield of kenaf simulated at Katherine from 1889 to 1988, for standard and modified cultivars having changed yield potential and drought resistance. A 20% increase in radiation use efficiency (RUE), water extraction (WE) and transpiration efficiency coefficient (TE), a 20 % decrease in RUE, and a combined 20% decrease in RUE and 20% increase in TE were simulated. The mean yield gain of the modified cultivar relative to the standard cultivar, and the probability of obtaining a yield gain are also given.

Cultivar	Yield		Yield gain	
	Mean (kg ha^{-1})	CV (%)	Mean (%)	Probability (%)
Sorghum				
A. Dekalb DK55	4569	28.9	–	–
B. Higher RUE (1.5 g MJ^{-1})	4780	35.2	4.6	70
C. Higher WE (229 mm)	4739	25.6	3.7	45
D. Higher TE (10.8 Pa)	4888	22.3	7.0	63
E. Lower RUE (1.0 g MJ^{-1})	4080	21.1	–10.7	26
F. Lower RUE higher TE (1.0 g MJ^{-1} 10.8 Pa)	4262	15.9	– 6.7	28
Maize				
A. Dekalb XL82	5274	43.8	–	–
B. Higher RUE (1.92 g MJ^{-1})	5833	51.3	10.6	70
C. Higher WE (185 mm)	5726	38.4	8.6	81
D. Higher TE (10.8 Pa)	6081	34.4	15.3	87
E. Lower RUE (1.28 g MJ^{-1})	4217	33.7	–20.0	22
F. Lower RUE higher TE (1.28 g MJ^{-1} 10.8 Pa)	4653	24.7	–11.8	37
Kenaf				
A. Guatemala 4	9048	39.4	–	–
B. Higher RUE (1.44 g MJ^{-1})	9426	39.2	4.2	88
C. Higher WE (332 mm)	9395	38.1	3.8	91
D. Higher TE (6 Pa)	10651	34.3	17.7	100
E. Lower RUE (0.96 g MJ^{-1})	8497	34.9	– 6.1	12
F. Lower RUE higher TE (0.96 g MJ^{-1} 6 Pa)	9877	31.3	9.2	84

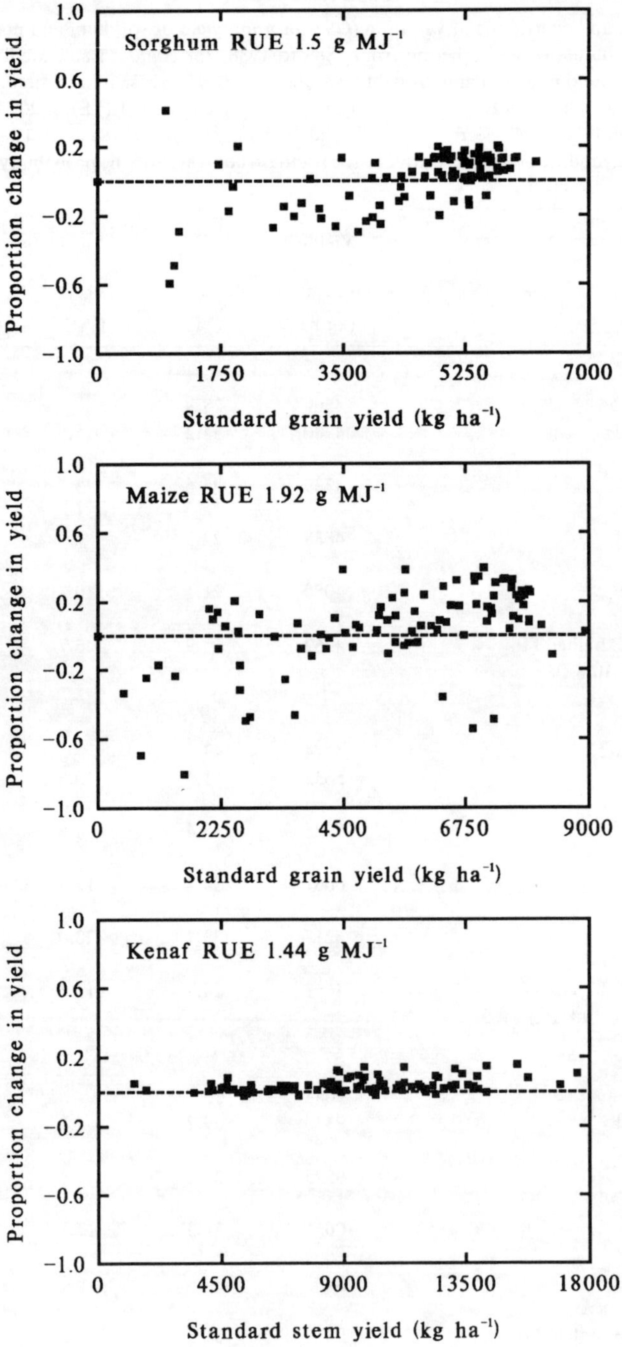

Figure 6. Effect of 20% higher radiation use efficiency on the proportion change in yield of sorghum, maize and kenaf relative to standard cultivar yield as a function of standard cultivar yield.

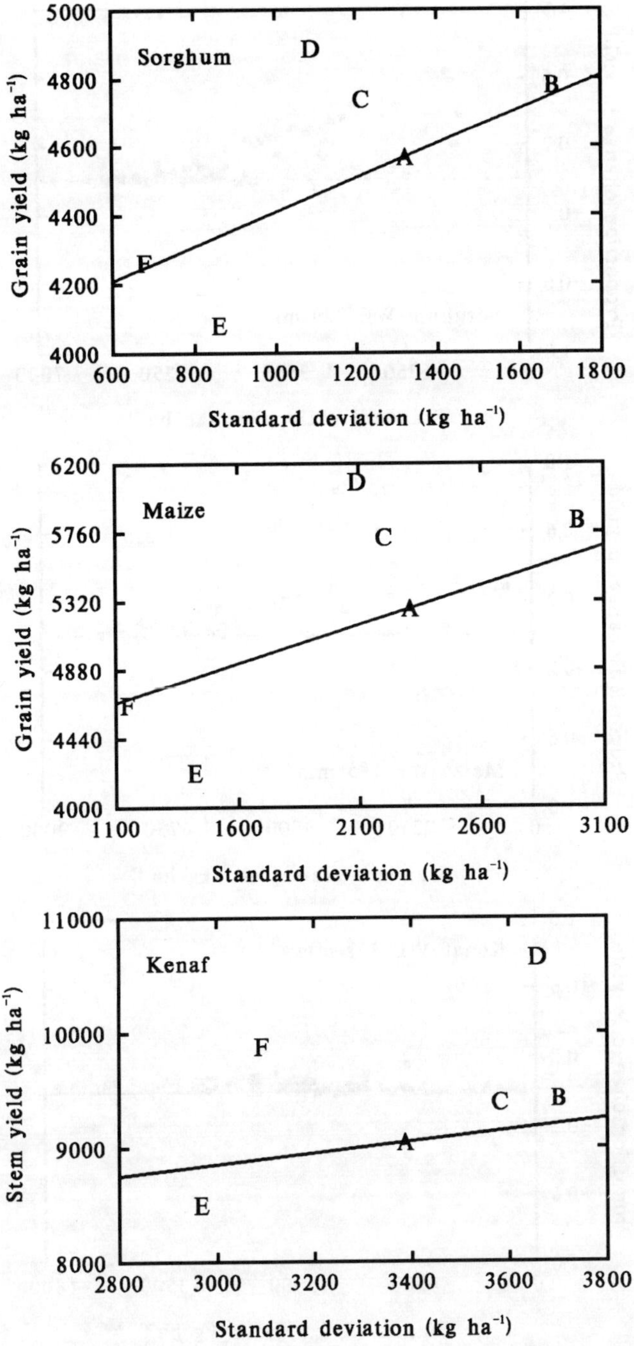

Figure 7. Effect of yield potential and drought resistance depicted in Mean-Standard deviation space. The 2:1 indifference line is drawn through the standard cultivar values of sorghum, maize and kenaf, and symbols represent modified yield potential or modified drought resistance shown in Table 3.

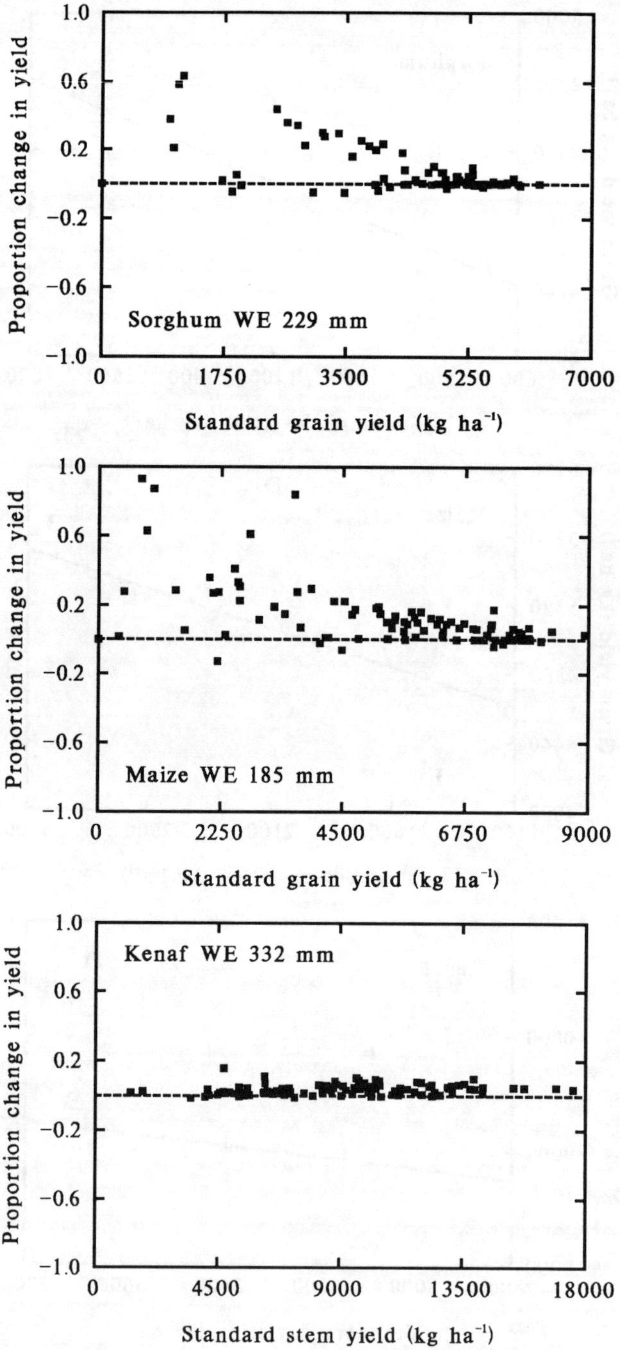

Figure 8. Effect of 20% higher soil water extraction on the proportion change in yield of sorghum, maize and kenaf relative to standard cultivar yield as a function of standard cultivar yield.

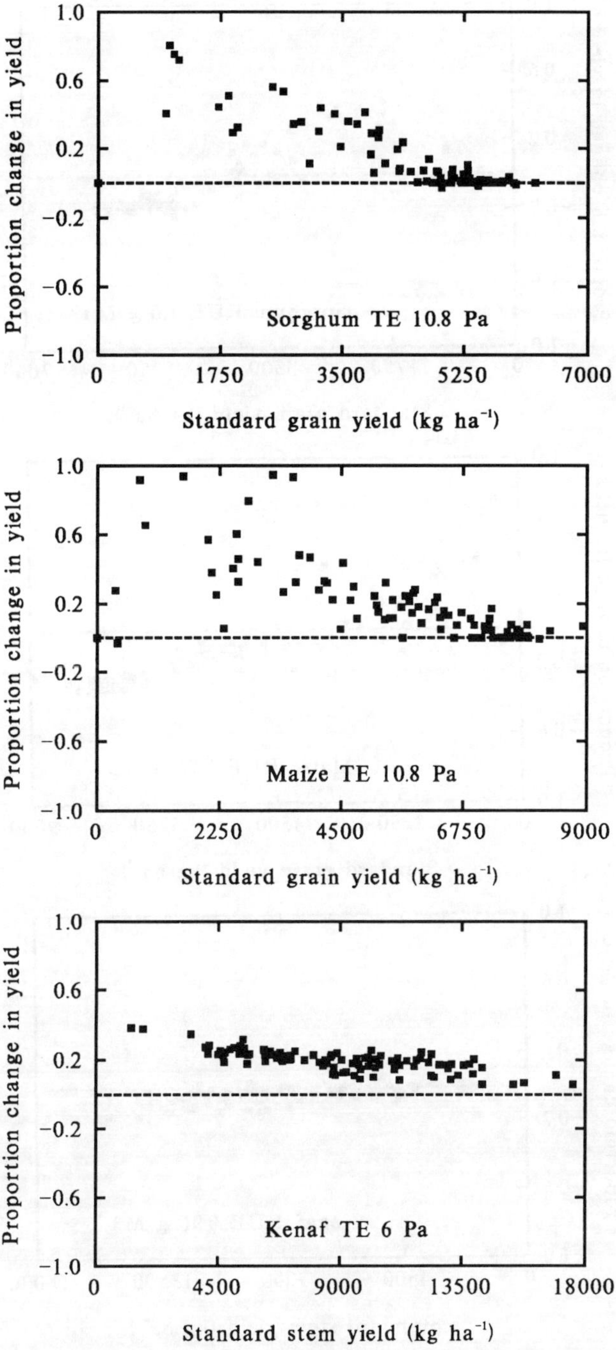

Figure 9. Effect of 20% higher transpiration efficiency coefficient on the proportion change in yield of sorghum, maize and kenaf relative to standard cultivar yield as a function of standard cultivar yield.

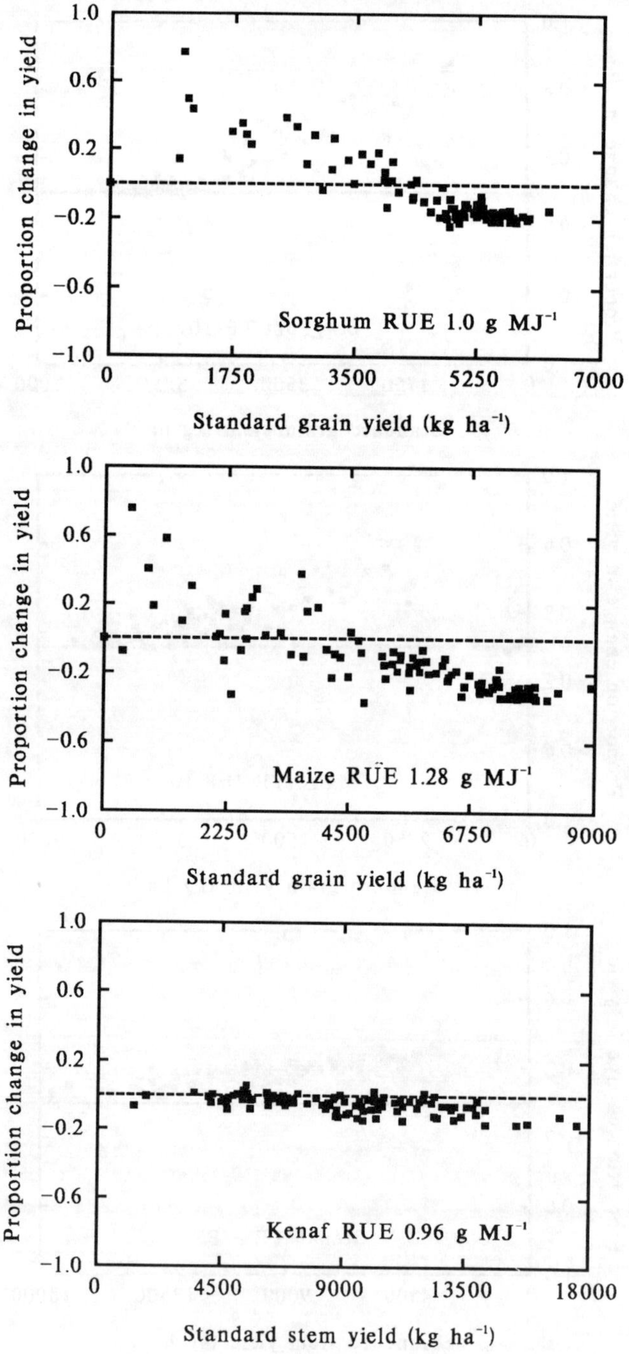

Figure 10. Effect of 20% lower radiation use efficiency on the proportion change in yield of sorghum, maize and kenaf relative to standard cultivar yield as a function of standard cultivar yield.

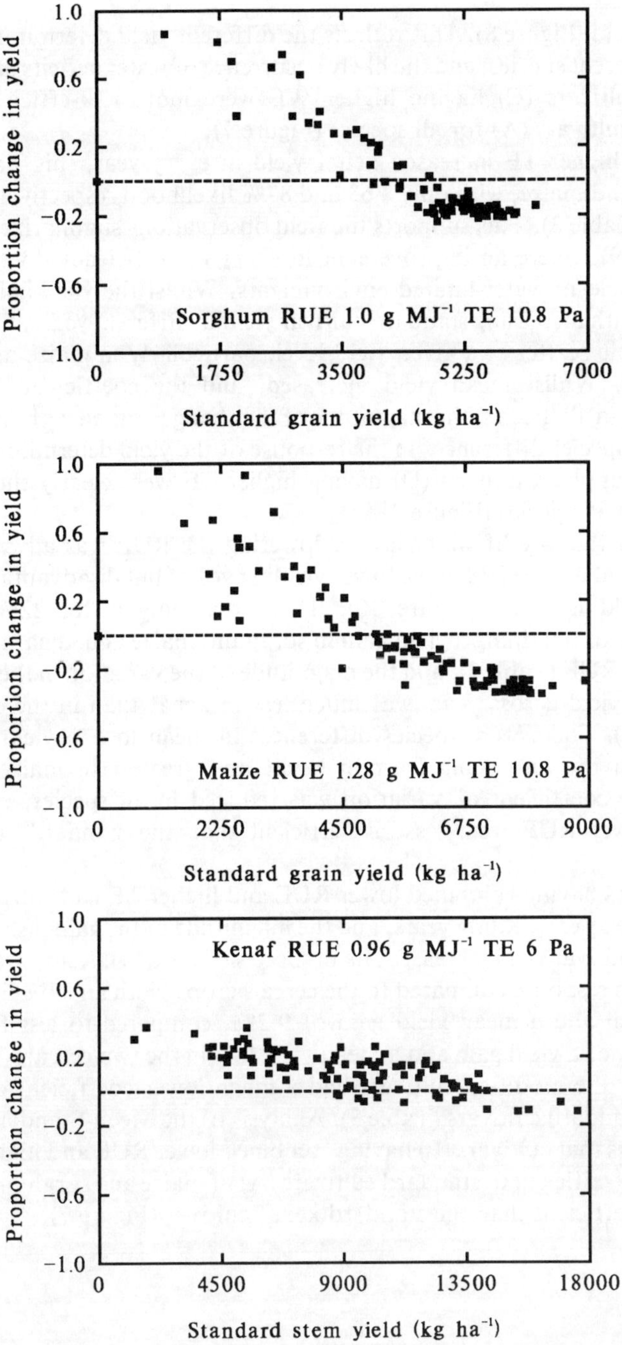

Figure 11. Effect of 20% lower radiation use efficiency combined with 20% higher transpiration efficiency coefficient on the proportion change in yield of sorghum, maize and kenaf relative to standard cultivar yield as a function of standard cultivar yield.

cultivar yield (Figure 8). This reflects the different yield determining processes of the different species, and the likely occurrence of water deficits at key growth stages. Cultivars (C) having higher WE were more risk-efficient than the standard cultivars (A) for all species (Figure 7).

Whilst higher TE increased kenaf yield in every year, this was not so in sorghum and maize, with only a 63 and 87% likelihood respectively, of a yield increase (Table 3). This supports the field observations summarized by Turner et al. (1989), where an improvement in water use efficiency does not always increase yield in water-limited environments. Whilst the yield gain tended to decrease with increasing standard cultivar yield in different years, there was still considerable scatter at a given yield level, particularly in maize and sorghum (Figure 9). Whilst mean yield increased, and the coefficient of variation decreased, in all species, the magnitude of the changes varied with species, again reflecting species differences in the response of the yield determining processes to water supply. Cultivars (D) having higher TE were clearly the most risk-efficient in all species (Figure 7).

Slowing the rate of water use by lowering the RUE was advantageous to sorghum and maize yield in the lower yielding years, but disadvantageous in the higher yielding years (Figure 10). There was only a 26, 22% and 12% probability of obtaining a yield gain in sorghum, maize and kenaf respectively, with lower RUE (Table 3), and the magnitude of the variation in the proportion change in yield across years was much less in kenaf than in the cereal crops (Figure 10). There were species differences in mean loss in yield with lower RUE, with the loss being least in kenaf and greatest in maize (Table 3). Whilst the coefficient of variation was reduced in all species, cultivars (E) having lower RUE were less risk-efficient than the standard cultivars (A) (Figure 7).

Cultivars having combined lower RUE and higher TE had substantial yield gains in the lower yielding years, and the magnitude of the gain decreased as the standard cultivar yield increased, in all species (Figure 11). Kenaf was markedly different in response compared to the cereals crops, with an 84% likelihood of a yield gain and a mean yield gain of 9.2%, compared to less than a 37% probability of a yield gain and a mean yield loss in the two cereals (Table 3). In all species, the modified cultivars had a much lower coefficient of variation than the standard cultivars (Table 3). Analysis of the Mean-Standard deviation space shows that cultivars (F) having combined lower RUE and higher TE were equally risk-efficient to standard cultivars (A) of maize and sorghum, but much more risk-efficient than the standard kenaf cultivar (Figure 7).

Conclusions

The unpredictable and serious water constraint to crop production in the semiarid tropics poses a considerable challenge to the design of improved plant types. We have outlined one approach using crop growth simulation coupled to long-term climatic data to assess the value of different plant designs both in terms of crop yield and production risk. Our simulation analysis for a semi-arid tropical environment has shown that crop species differ markedly in their response to changed plant design. The long-season fibre crop kenaf, in which the vegetative material comprises the economic yield, provided a contrast to the cereal crops maize and sorghum, where the reproductive organs comprise the economic yield. Whilst we have examined aspects related to changed phenology, yield potential and specific drought resistance traits where mode of operation is well understood and adequately encapsulated within the models, actual cultivars vary in many traits which may be linked, and there is a clear need to confirm favourable simulation outcomes in the field. We believe the results from crop growth simulation can focus this field testing using realistic designs, and make it more efficient and effective.

Our analysis has shown that the matching of crop phenology to the expected water supply is a difficult concept in the climatically variable semi-arid tropics, and is strongly dependent on attitude to risk aversion. Even a mean difference of three days in crop maturity caused considerable year to year variation in the proportion change in cereal yield relative to the standard cultivar. Earlier maturity tended to improve yield in the lower yielding years and increased yield stability in the cereals, but not in the fibre crop. The choice of appropriate phenology must balance yield with risk, and we have outlined a methodology using Mean-Standard deviation space to select the most risk-efficient phenology (Figure 5).

Designing improved plant types with higher yield potential was advantageous both in terms of yield and risk in all species, despite the severe water limitation in this environment. However, there were species differences in the value of improved designs based on drought resistance traits. In all species, the most risk-efficient drought resistance trait was higher transpiration efficiency. In kenaf, lower radiation use efficiency (to conserve water use) combined with higher transpiration efficiency was the next most risk-efficient trait, whilst in sorghum and maize, it was higher extent of soil water extraction.

The magnitude of the yield response to changed crop trait varied enormously from year to year, even when expressed as a relative response after accounting for the variation in absolute yield from year to year. This highlights the limited value of selecting improved plant designs using conventional experimentation conducted over a limited number of seasons in this climatically variable environment. Furthermore, whilst the proportion change in yield relative to the standard cultivar in individual years was often much greater than the 20% simulated change in crop trait, the mean proportion change, in all but one instance, was much less than 20%. This is due to the variability in the amount and occurrence of rainfall and the consequent impact on the yield determining processes of different crop species. Only with crop growth simulation, can the impact of different plant designs on crop yield be quantitatively assessed in the climatically variable semiarid tropics. Whilst our analysis has focussed on the water limitation, the approach also needs to be used to design improved plant types to overcome other limitations (e.g. nitrogen) as appropriate simulation models become available.

The question remains as to how feasible is the attainment of improved plant design and what is the likely cost-benefit. For example, our simulations have shown that higher TE is very beneficial in all our selected crop species. However, TE is a relatively conservative parameter, and there are few reports of TE higher than the standard values used here for C_4 and C_3 crops (Tanner and Sinclair 1983). Until higher values can be demonstrated, the prospects of improved design based on higher TE remain uncertain. Clearly, conventional field experimentation is required to quantify the expected variation in crop traits, and is a necessary adjunct to the simulation approach where the value of such genetic variation can be assessed.

Acknowledgments

We thank Drs M. R. Thomas and B.A. Keating for their contribution to the methodology used in this paper.

References

Anderson J R, Dillon J L, Hardaker J B (1977) Agricultural decision analysis. Iowa State University Press, Ames. 344 p.

Bailey E, Boisvert R N (1989) A comparison of risk efficiency criteria in evaluating groundnut performance in drought-prone areas. Aust. J. Agric. Econ. 33:153–169.

Baker F W G (ed.) (1989) Drought resistance in cereals. CAB International, Wallingford, UK. 222 p.

Barah B C, Binswanger H P, Rana B S, Rao N G P (1981) The use of a risk aversion in plant breeding: Concept and application. Euphytica 30:451–458.

Birch C J, Carberry P S, Muchow R C, McCown R L, Hargreaves J N G (1990) Development and evaluation of a sorghum model based on CERES-Maize in a semi-arid tropical environment. Field Crops Res. 24:87–104.

Carberry P S, Abrecht D G (1991) Tailoring crop models to the semi-arid tropics. Pages 157–182 in Muchow R C, Bellamy J A (Eds.) Climatic risk in crop production: Models and management in the semiarid tropics and subtropics. CAB International, Wallingford, UK.

Carberry P S, Muchow R C (1992) A simulation model of kenaf for assisting fibre industry planning in northern Australia. 3. Model description and validation. Aust. J. Agric. Res. (in press).

Carberry P S, Muchow R C, McCown R L (1989) Testing the CERES-Maize simulation model in a semi-arid tropical environment. Field Crop Res. 20:297–315.

Carberry P S, Muchow R C, Williams R, Sturtz J D, McCown R L (1992) A simulation model of kenaf for assisting fibre industry planning in northern Australia. 1. General introduction and phenological model. Aust. J. Agric. Res. (in press).

Edmeades G O, Bolanos J, Lafitte H R, Rajaram S, Pfeiffer W, Fischer R A (1989) Traditional approaches to breeding for drought resistance in cereals. Page 27–52 in Baker F W G (Ed.) Drought resistance in cereals. CAB International, Wallingford, UK.

Farquhar G D, Richards R A (1984) Isotopic composition of plant carbon correlates with water-use efficiency of wheat genotypes. Aust. J. Plant Physiol. 11:539–552.

Hammer G L, Vanderlip R L (1989) Genotype-by-environment interaction in grain sorghum. III. Modelling the impact in field environments. Crop Sci. 29:385–391.

Jones C A, Kiniry J R (1986) CERES-Maize: a simulation model of maize growth and development. Texas A&M University Press, College Station, USA. 194 p.

Jones J W, Zur B (1984) Simulation of possible adaptive mechanisms in crops subjected to water stress. Irrig. Sci. 5:251–264.

Jordan W R, Dugas W A, Shouse P J (1983) Strategies for crop improvement for drought-prone regions. Agric. Water Management 7:281–299.

Lawn R J, Imrie B C (1991) Crop improvement for tropical and subtropical Australia: Designing plants for difficult climates. Field Crops Res. 26:113–139.

Loomis R S (1985) Systems approaches for crop and pasture research. Pages 1–8 in Crops and Pasture Production – Science and Practice. Proceedings of the 3rd Australian Agronomy Conference, University of Tasmania, Hobart. Australian Society of Agronomy, Parkville, Australia.

Ludlow M M, Muchow R C (1990) A critical evaluation of traits for improving crop yields in water-limited environments. Adv. Agron. 43:107–153.

McCaskill M R (1990) TAMSIM – A program for preparing meteorological records for weather-driven models. CSIRO, Division of Tropical Crops and Pastures, Tropical Agronomy Technical Memorandum No. 65. Australia.

McCown R L, Wafula B M, Mohammed L, Ryan J G, Hargreaves J G N (1991) Assessing the value of a seasonal rainfall predictor to agronomic decisions: The case of response farming in Kenya. Pages 383–409 in Muchow R C, Bellamy J A (Eds.) Climatic risk in crop production: Models and management in the semiarid tropics and subtropics. CAB International, Wallingford, UK.

Muchow R C, Carberry P S (1989) Environmental control of phenology and leaf growth of maize in a semiarid tropical environment. Field Crops Res. 20:221–236.

Muchow R C, Carberry P S (1990) Phenology and leaf area development in a tropical grain sorghum. Field Crops Res. 23:221–237.

Muchow R C, Carberry P S (1992) A simulation model of kenaf for assisting fibre industry planning in northern Australia. 5. Impact of different crop traits. Aust. J. Agric. Res. (submitted).

Muchow R C, Hammer G L, Carberry P S (1991) Optimising crop and cultivar selection in response to climatic risk. Pages 235–262 in Muchow R C, Bellamy J A (Eds.) Climatic risk in crop production: Models and management in the semiarid tropics and subtropics. CAB International, Wallingford, UK.

Muchow R C, Sturtz J D, Spillman M F, Routley G E, Kaplin S, Martin C C, Bateman R J (1990) Agronomic studies on the productivity of kenaf (*Hibiscus cannabinus* L. cv. Guatemala 4) under rainfed and irrigated conditions in the Northern Territory. Aust. J. Exp. Agric. 30:395–403.

O'Toole J C, Jones C A (1987) Crop modelling: applications in directing and optimising rainfed rice research. Pages 255–269 in Weather and Rice. Proceedings of the International Workshop on the Impact of Weather Parameters on Growth and Yield of Rice. International Rice Research Institute, Los Banos, Philippines.

Richards R A (1991) Crop improvement for temperate Australia: Future opportunities. Field Crops Res. 26:139–167.

Stapper M, Harris H C (1989) Assessing the productivity of wheat genotypes in a Mediterranean climate, using a crop-simulation model. Field Crops Res. 20:129–152.

Tanner C B, Sinclair T R (1983) Efficient water use in crop production: Research or re search? Pages 1–27 in Taylor H M, Jordan W R, Sinclair T R (Eds.) Limitations to efficient water use in crop production. American Society of Agronomy, Madison, USA.

Turner N C, Nicolas M E, Hubick K T, Farquhar G D (1989) Evaluation of traits for the improvement of water use efficiency and harvest index. Pages 177–200 in Baker F W G (Ed.) Drought resistance in cereals. CAB International, Wallingford, UK.

Simulation in pre-testing of rice genotypes in Tamil Nadu

S. PALANISAMY[1], F.W.T. PENNING DE VRIES[2], S. MOHANDASS[1], T.M. THIYAGARAJAN[1] and A. ABDUL KAREEM[1]

[1] *Tamil Nadu Rice Research Institute, Aduthurai 612 101, Tamil Nadu, India*

[2] *Center for Agrobiological Research (CABO-DLO), P.O Box 14, 6700 AA Wageningen, The Netherlands.*

Key words: assimilate partitioning, breeding, climatic variability, India, MACROS, pre-release trials, release decisions, rice, selection trials, simulation, stem reserves, yield potential

Abstract

A long series of pre-testing trials of promising cultures at many locations is a prerequisite for the selection and release of a new rice variety. This involves considerable time and effort. To reduce time and expense, we sought to employ a crop growth simulation model that explains the interaction of environment and genotype.

Field experiments were conducted at Tamil Nadu Rice Research Institute (TNRRI), South India, during the 1990 wet season to establish the performance of six medium duration genotypes, in detail. Periodical sampling was done to establish the carbohydrate partitioning pattern of these cultivars, and the fraction stem reserves. Crop duration and yields of the 1990 TNRRI experiment were used to calibrate a few model parameters.

The calibrated model was then employed to predict the ranking of seven genotypes in 11 experiments at three locations (TNRRI, Coimbatore and Madurai) and four years (1987, 1988, 1989 and 1990). The ranking of the varieties was compared with that from field trials. In our first attempt, two genotypes were correctly predicted to be among the top three.

The paper discusses possibilities of using a model to compute the performance of pre-release rice genotypes at different locations of Tamil Nadu, using local weather data. We aim at reducing significantly the number of field experiments and the duration of the selection period.

Introduction

Growing rice has been given top priority in Tamil Nadu, South India, since it is the staple food for its people. In this state, rice was cultivated at 2.0 Mha, producing 6.25 Mt rice during 1989–'90. In the last ten years, average yield increased from 2.0 to more than 3.0 t ha^{-1}. The target for rice at the end of 8th plan period, in 1995, is 7.5 Mt. A rice production target of over 9.0 Mt by the year 2000 is aimed for, corresponding with an average yield of more than 4 t ha^{-1}. Most rice varieties bred or introduced yielded more than 10 t ha^{-1} grain in certain parts of the state where intensive cultivation was followed and with favorable weather, but is only 7–8 t ha^{-1} elsewhere. At fields of progressive farmers, such varieties are expected to produce at least 60 % of that yield level. Closing the current yield gap therefore provide sufficient opportunity for yield increases. Yet, plant breeding is an effort of long duration, and has to address yield stability, rice quality and resistance against various stresses in addition to

F.W.T. Penning de Vries et al. (eds.), Systems Approaches for Agricultural Development, 63–75.

yield itself, so that breeding remains important inspite of the yield gap that already exists.

In breeding rice varieties, variation in rice genotype is created by hybridization or by other means. High yielding genotypes are identified through a series of yield tests, including Progeny Row Trials (PRT), Preliminary Yield Trials (PYT), and Comparative Yield Trials (CYT), at different rice research stations of Tamil Nadu Agricultural University (TNAU). Selected cultures from the CYT's are pooled in various duration groups, and tested simultaneously at all the rice research stations in Multi Location Trials (MLT). The promising entries in MLT, which show consistently high yields over years and locations, are identified as pre-release cultures and are then tested at the farmers holdings in more than 80 locations throughout Tamil Nadu under Adaptive Research Trials (ART). Pre-release cultures with stable, high yields over many locations and seasons are released as new varieties.

The process of development of a rice variety from PRT to ART takes a minimum of 8–10 years, while the pre-testing of rice genotypes under MLT alone involves a three year period with more than 30 field experiments at 10–11 rice stations. It is estimated that an average of 40 man-days of scientists go into conduct of MLT every year at every station. The variety release process is summarized in Table 1. Tamil Nadu Rice Research Institute (TNRRI) at Aduthurai coordinates this rice breeding program, and shares responsibility for the official release of new varieties.

Table 1. The variety release process.

Trial	No. of entries	Period (years)	No. of expts.	No. of research stations
PRT	120	1	11	11
PYT	60	1	11	11
CYT	30	3	33	11
MLT	10	3	33	11
ART	2 or 3	2	200	100 (Locations)
Total		10	288	

Systems analysis and simulation have been used by engineers for more than 30 years and their success inspired agricultural scientists to apply similar techniques in agriculture. Crop growth simulation models are increasingly used to support field research and extension. A crop growth model has been used successfully to simulate potential growth rates and yield of rice under different weather situations (Herrera-Reyes and Penning de Vries 1989; Dua et al. 1990). The model simulates growth and yield for the best management possible, i.e. the crop has continuously ample water and nutrients, and is without pest and disease. In that case, crop growth depends only on the current state of crop and on current weather conditions, particularly radiation and temperature.

Although this situation is generally not attained at farmers fields, the potential production for a particular site and year is one of the important attributes of a genotype.

Results of crop simulation models are sometimes sensitive to values of biological parameters that are not well known, but respond generally properly to fluctuations in weather. Trends in simulated yield are therefore more reliable than absolute values. This may be a handicap for yield forecasting, but less so in cultivar selection, where ranking of genotype performance is the method to express superiority of a genotype, rather than establishing absolute yields.

If it would be possible to reduce the period of MLT from three years to one year by use of simulation model, then time, space, energy and money could be saved to a considerable extent. With the aim of reducing the period of pre-testing of rice cultures of the long duration genotypes (135–145 days) under MLT, the present study has the following specific objectives: (i) to predict the ranking for potential production of pre-release rice genotypes under MLT at TNRRI; (ii) to predict the ranking for potential production of pre-release rice genotypes under MLT at two other locations in Tamil Nadu; (iii) to predict the ranking for potential production of selected rice genotypes from MLT for advancing to ART testing.

Materials and methods

The sites

Aduthurai (11 °N, 79.5 °E, 19.5 m), Coimbatore (11 °N, 77 °E, 431 m) and Madurai (8.5 °N, 79 °E, 147 m) were selected from an aggregate of ten sites of rice research stations in Tamil Nadu to conduct the MLT's. The weather data of these sites (solar radiation, maximum and minimum temperature) from 1987 to 1991 were used; they were found to be free of missing values, obvious instrument errors and other inconsistencies.

Table 2. Details of rice genotypes in the multilocation trials. Under grain type, LS stands for Long Slender, MS for Medium Slender and SB for Short Bold.

Genotype	Parentage	Duration (days)	Grain type	Mean yield (t ha^{-1})	Year of release
ADT 38	Multiple cross	135	LS	6.2	1987
CO 45 (TNAU 80042)	Rathuheenati/ IR 3403-267-1	135	LS	5.8	1990
CO 43	Dasal/IR 20	130	MS	5.7	1982
IR 20	IR 262/TKM 6	130	MS	5.0	1970
BG 380-2	BG 90/24/OB 677	135-140	SB	6.1	–
BR 153	Multiple cross	135	LS	6.1	–
AD 86749	Ponmani/CO 43	135	SB	5.2	–

Crop data

Three field experiments were conducted at TNRRI in the wet season (WS) of 1990, i.e. from September 1990 to April 1991. A total of seven rice genotypes comprising four released varieties (ADT 38, CO 45 (=TNAU80042), CO 43 and IR 20) and three pre-release rice genotypes (BG 380–2, BR 153 and AD 86749) were planted in a replicated randomized design. The details of the genotypes are given Table 2. The plantings were done on 26 September, 16 October and 15 November 1990. A manuring schedule of 150 kg N: 60 kg P_2O_5: 60 kg K_2O ha^{-1} was followed, in addition to application of 12.5 t ha^{-1} of green leaf manure applied basally. Plant samples (5 m^2 harvested, in four replicates) were collected at time of planting and subsequently at 10 d intervals up to harvest time. The plant organs, i.e., roots, stems and leaves were separated, dried at 70 °C, and weighed. Among the three experiments, the one planted in November showed the highest yield and most consistent trends in organ weights, and was therefore judged to be the best: crop data from this trial were used for calibrations in the present study. The other trials probably experienced unexpected management problems.

From the development of organ dry weights, we derived the carbohydrate partitioning pattern (Penning de Vries et al. 1989). The pattern of carbohydrate partitioning to leaves, stems and storage organs in relation to developmental stages is a key function that distinguishes genotypes and is shown in Figure 1a-g for the different genotypes involved. The irregularity of the curves in these figures is probably an artifact, that could have been removed by deriving these patterns after fitting a line through the original field data, but we preferred to use first the unpolished observations.

Another important parameter is the maximum rate of leaf photosynthesis. Since no direct measurements were taken, its value was calibrated for each of the genotypes; the calibration accounts particularly for any suboptimal treatment the crop might have received. The genotype specific development stage of the seedlings at transplanting was derived from their seedbed history. The genotype specific development rate for vegetative and reproductive phase were derived from observed dates of flowering and maturity (adjusted for effects of temperature); the latter was found not to vary. The fraction of stem reserves at flowering is another important variable to distinguish genotypes, and was measured. The values used are shown in Table 3; all other parameters are as in Penning de Vries et al. (1989).

With the calibrated and measured parameters of the respective genotypes, the grain yields were simulated for the same situation in which the experiment was done. Results (Table 4) show the yields of all genotypes, and the calibrated value of maximum rate of leaf photosynthesis. The high degree of correspondence between simulation and observations is the result of the calibration. The test entries CO 43, AD 86749 and CO 45 are the three best yielding genotypes. In this paper yields are expressed as paddy rice at 14% moisture.

Table 3. The development stage at transplanting (DSI), the genotype specific development rate for the vegetative (DRCV) and reproductive period (DRCR), and the fraction stem reserves (FSTR) of the rice genotypes investigated.

Genotype	DSI	DRCV	DRCR	FSTR
ADT 38	0.354	0.0101	0.0333	0.26
CO 45	0.343	0.0098	0.0333	0.14
CO 43	0.347	0.0099	0.0333	0.13
IR 20	0.343	0.0098	0.0333	0.12
BG 380-2	0.361	0.0100	0.0333	0.20
BR 153	0.354	0.0100	0.0333	0.12
AD 86749	0.361	0.0103	0.0333	0.37

Table 4. Simulated and observed grain yields (t ha^{-1}) in the experiment, Aduthurai, 1990, WS for calibration of the maximum rate of leaf photosynthesis (PLMXP; its standard value is 47 kg ha^{-1} h^{-1}).

Genotype	Observed yield	Simulated yield	Calibrated PLMXP
ADT 38	6.00	6.04	37
CO 45	6.75	6.78	35
CO 43	7.50	7.51	62
IR 20	6.50	6.58	45
BG 380-2	6.00	6.01	64
BR 153	6.40	6.58	42
AD 86749	7.20	7.29	67
Average	6.62	6.68	

The model

The L1D module from the Modules for Annual CROp Simulators (MACROS, Penning de Vries et al. 1989), in the language CSMP (Continuous System Modeling Program) adapted for an IBM/AT personal computer, was selected as the summary model for the present study. It was developed to simulate growth and production of crops in situations with ample water and nutrients, and without pests. Consequently, the simulated yield is the highest that can be obtained: the potential yield.

The core of the model consists of a set of equations which calculate photosynthesis as a function of leaf area and of daily weather conditions. The growth rate depends particularly on current weather, i.e. solar radiation and temperature. Altitude reduces the maximum leaf photosynthetic rate by 12 % for every 1000 m elevation. Respiration processes have first priority for use of assimilated carbohydrates, the remainder is partitioned and converted into plant biomass of leaves, stems and grains. For every day of the growing seasons the rates of all processes are calculated in relation to the current state of the crop and the weather, and then integrated.

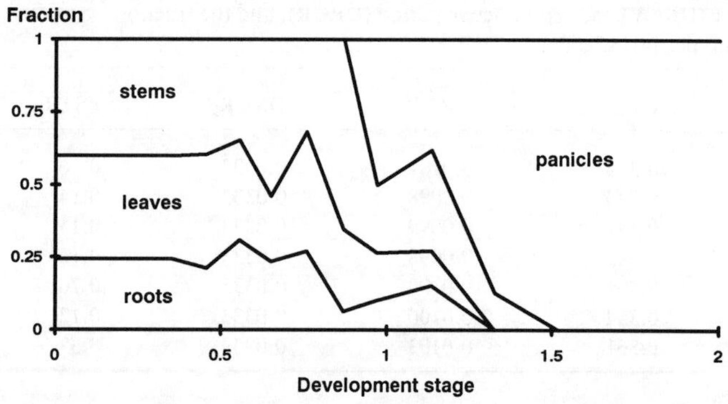

Figure 1a

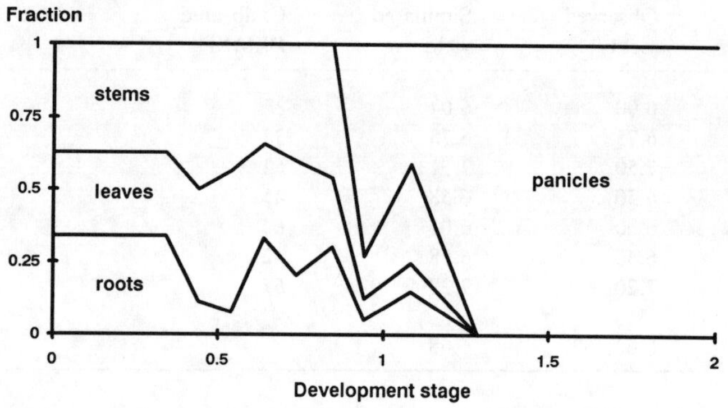

Figure 1b

Figure 1. Biomass partitioning as a function of development stage. The fraction of assimilates for growth that is partitioned to roots, leaves, stems and panicles plus grains as a function of development stage, is presented in a cumulative way for seven genotypes: a. ADT 38, b. CO 45, c. CO 43, d. IR 20, e. BG 380-2, f. BR 153, g. AD 86479. The functions were derived from periodically harvested field crops.

For the present study, it is assumed that the crops are not sensitive to day length, and that the only significant physiological differences between the rice genotypes are the rate of progress towards flowering, the biomass partitioning pattern and the fraction stem reserves; the difference in maximum leaf photosynthesis integrates the unknown differences in all other responses to the environment.

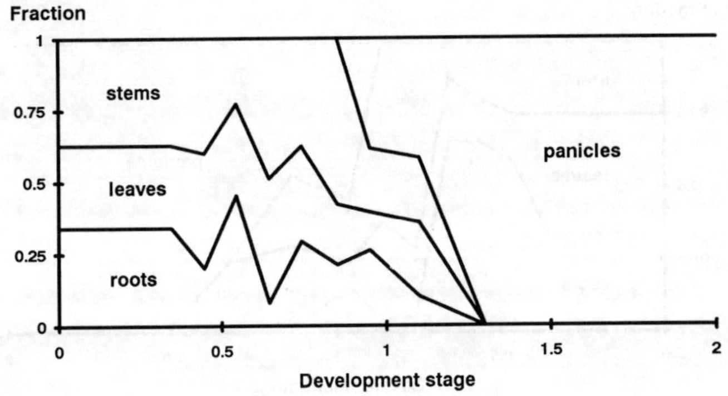

Figure 1c

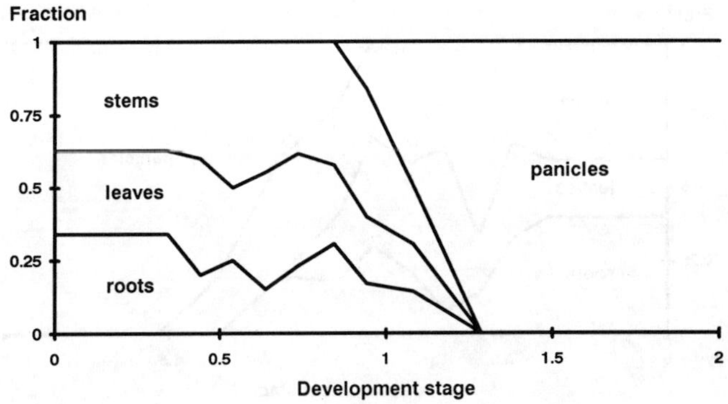

Figure 1d

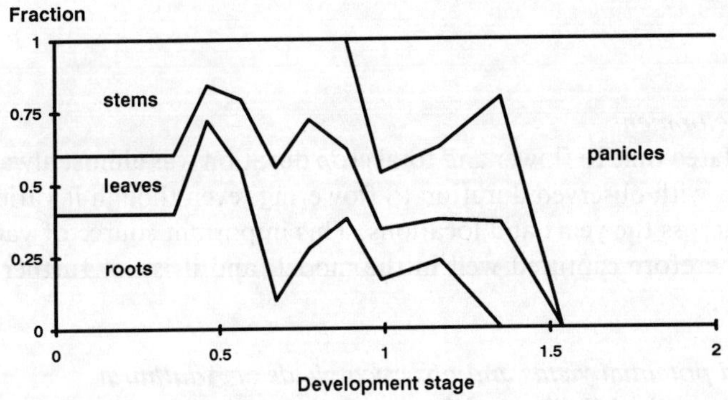

Figure 1e

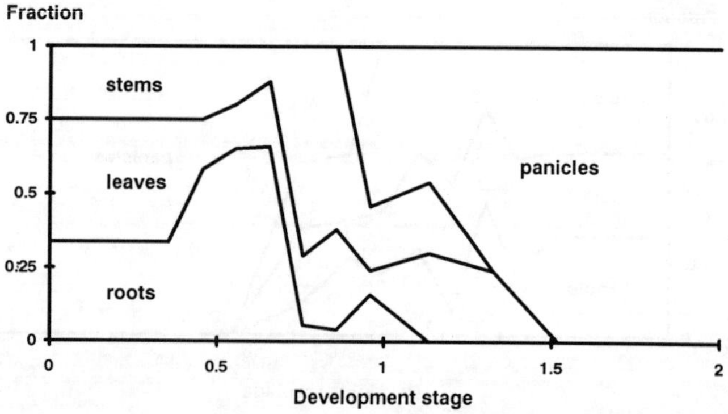

Figure 1f

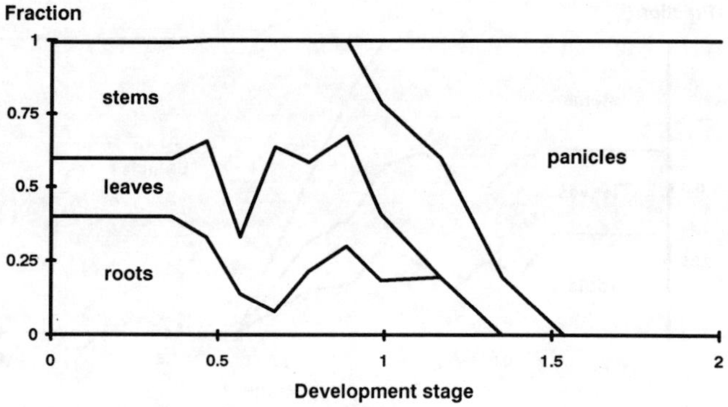

Figure 1g

Results

Crop development

The simulated time to flower and total crop duration was almost always in good agreement with observed duration to flowering, even though it varied by 10% or more across the years and locations. This important source of variability in yield is therefore captured well in the model, and it is not further discussed separately.

Simulated potential yields and observed yields at Aduthurai

The results of the MLT's conducted at Aduthurai during the 1987, 1988, 1989 and 1990 WS are presented in Table 5. With regards to absolute values, the predicted and observed yield averaged across the genotypes, was 6.7 t ha^{-1} in the

Table 5. Observed and simulated grain yields (kg ha^{-1}) in Multi Location Trials for different years. All seasons – apart from Coimbatore 1991 – are wet seasons. (obs – observed yield; sim – simulated yield).

Aduthurai	1987		1988		1989		1990	
	obs	sim	obs	sim	obs	sim	obs	sim
ADT 38	5070	6818	3667	6993	2980	7105	5175	5881
Co 45	6593	7601	4278	7866	–	8160	5702	6621
Co 43	4826	8415	3625	8582	3207	8504	4737	7188
IR 20	3860	7334	2819	7434	3131	7468	4474	6310
BG 380-2	5151	6622	4722	6764	3282	6754	5073	5687
BR 153	5721	7368	4528	7578	3914	7221	6140	6773
AD 86749	–	7945	–	7973	4141	7960	6228	6935

Coimbatore	1987		1988		1989		1991	
	obs	sim	obs	sim	obs	sim	obs	sim
ADT 38	5970	6456	–	6067	–	5929	4125	4016
Co 45	6667	6891	5625	6827	–	6713	5028	4681
Co 43	6074	7850	5685	7658	5745	7435	4514	4894
IR 20	5970	6802	4097	6618	4618	6430	4055	4342
BG 380-2	6222	6083	5257	6245	5662	6464	4361	3887
BR 153	5141	6964	4553	6614	5686	5933	4250	4496
AD 86749	–	7517	–	7112	6173	7104	4333	4868

Madurai	1987		1989		1990	
	obs	sim	obs	sim	obs	sim
ADT 38	–	6162	–	6167	6563	6472
Co 45	5330	6609	–	7208	6328	6713
Co 43	6160	6152	4811	6394	6453	7735
IR 20	5100	5919	4022	6237	5734	6824
BG 380-2	6600	4723	–	5015	6828	6036
BR 153	4530	6319	4922	6193	6828	7134
AD 86749	–	4533	5078	5731	6609	7142

1991 calibration experiment, and 1.11, 1.14, 1.14 and 0.96 × this value in the years 1987–1990, respectively. Variations in weather are the cause of this instability. The observed results are also expected to be related to weather, and to be roughly similar to the simulated ones. However, this was not quite so: the actual yield was only a fraction of 0.71, 0.52, 0.45 and 0.83 of the potential in these years. Because the two years with the highest potential yield had most radiation and the smallest fraction, while that with the lowest potential showed the highest fraction, we expect imperfect irrigation to have played a role.

Table 6. Rating of genotypes according to yield in field observations and by simulation, for three stations and four years.

Year	Rank	Aduthurai		Coimbatore		Madurai	
		Observed	Simulated	Observed	Simulated	Observed	Simulated
1987	I	TNAU 80042	CO 43	TNAU 80042	CO 43	BG 380-2	TNAU 80042
	II	BR 153	TNAU 80042	BG 380-2	BR 153	CO 43	BR 153
	III	BG 380-2	BR 157	CO 43	TNAU 80042	TNAU 80042	CO 43
1988	I	BG 380-2	CO 43	CO 43	CO 43	–	–
	II	BR 153	TNAU 80042	TNAU 80042	TNAU 80042	–	–
	III	TNAU 80042	BR 153	BG 380-2	IR 20	–	–
1989	I	AD 86749	CO 43	AD 86749	CO 43	AD 86749	CO 43
	II	BR 153	AD 86749	CO 43	AD 86749	BR 153	IR 20
	III	BG 380-2	IR 20	BG 380-2	BG 380-2	CO 43	BR 153
1990	I	AD 86749	CO 43	CO 45*	CO 43*	BG 380-2	CO 43
	II	BR 153	AD 86749	CO 43*	AD 86749*	BR 153	AD 86749
	III	CO 45	BR 153	BG 380-2*	CO 45*	AD 86749	BR 153

TNAU 80042 was proposed for release as CO 45 in 1990.
* 1991 DS.

In 1987, the top three ranks were simulated for CO 43, CO 45 and BR 153, while field tests yielded CO 45, BR 153 and BG 380–2 as the best three (Table 5, 6). Of the six genotypes of which both simulated and observed yield were determined, MLT and simulation both showed the same two genotypes among the best three, but disagreed on the third. For 1988, again agreement was on two genotypes to be the best yielders, and the methods disagreed on a third. In 1989, with the very low yields, only one genotype was picked by both simulation and field trial to be among the top three, but in 1990 there was again agreement about two out of three.

Simulated potential yields and observed yields at Coimbatore

The results of MLT trials conducted during 1987,1988, 1989 WS and 1991 Dry Season (DS) at Coimbatore are furnished in Table 5. The WS average yield simulated amounted to 7.7 t ha^{-1}, while the observed was 5.9 t ha^{-1}, or 0.77×; the DS yields were relatively low, but agreed closely. Of the rankings of five to seven genotypes, two, two, three and two were picked by both simulation and trials among the top three in these years, respectively.

Simulated potential yields and observed yields at Madurai

The average simulated yield of MLT in 1987, 1989 and 1990 at the Agricultural College and Research Institute, Madurai, was 6.3 t ha^{-1}; the average observed yield was 0.85× this value (Table 5). Of the rankings, three times two genotypes, were put by simulation and trials both among the top three.

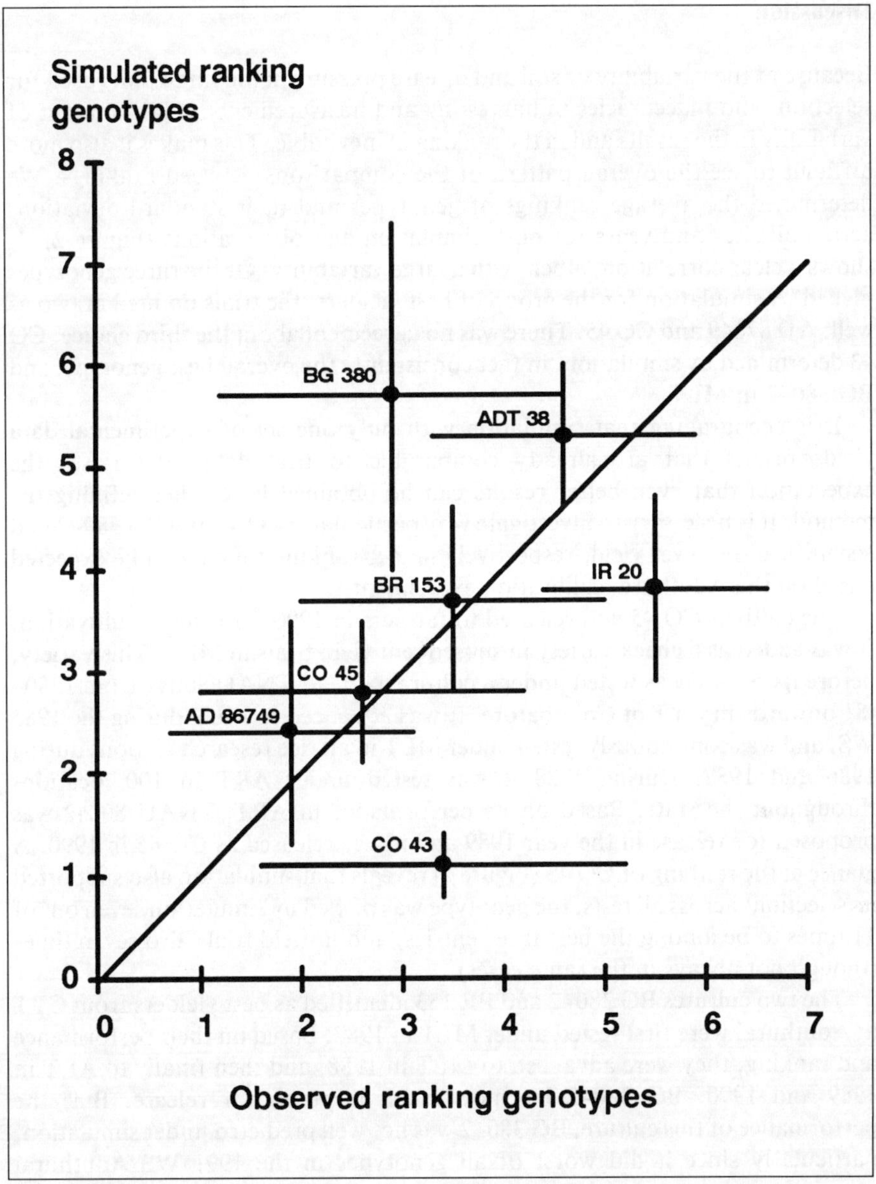

Figure 2. The average ranking of the seven genotypes in the simulation and in the Multi Location Trials, with the standard deviations.

Discussion

Because of the variability of soil and disease pressure in the small plots used for selection, and inaccuracies in harvesting and measurements, a fair amount of variability in the results and in the ranking in inevitable. This makes it also more difficult to see the overall pattern of the comparisons between rankings. We determined the average rankings of genotypes and their standard deviations across all sites and years for both simulation and observations (Figure 2). It shows a clear correlation, albeit with a large variability. Of the three genotypes picked by simulation for the group of best yielders, the trials pointed at two as well: AD 87649 and CO 45. There was no agreement about the third choice: CO 43 determined by simulation, in fact consistently the overall best genotype, and BG 380–2 in MLT.

It is encouraging that simulation with only one set of experimental data yields results that are already comparable to trial data, and raises the expectation that even better results can be obtained by further refining the method. It is necessary to investigate why particularly CO 43 and BG 380–2 had a significantly lower yield, respectively higher ranking than could be expected based on the Aduthurai calibration experiment.

The cultivar CO 45 was released to farmers in 1990 for general cultivation. It was added as a check variety in subsequent yield trials in MLT. This variety, before its release was tested under a culture name of TNAU 80042 from 1980–'81 onwards in CYT at Coimbatore. It was advanced to MLT during the 1985 WS, and was continuously tested under MLT in all rice research stations during 1986 and 1987. During 1988 it was tested under ART in 100 locations throughout the state. Based on its performance in ART, TNAU 80042 was proposed for release in the year 1989 and it was released as CO 45 in 1990. A glance at the ranking of CO 45 (Figure 2) reveals that simulation also supported its selection: across all tests, the genotype was picked by simulation seven out of 11 times to be among the best three entries, and in field trials also seven times (though not always in the same cases).

The two cultures BG 380–2 and BR 153 identified as best yielders from CYT at Aduthurai were first tested under MLT in 1987. Based on their performance and ranking, they were advanced to MLT in 1988, and then finally to ART in 1989 and 1990. BG 380–2 has now been proposed for release. But, the performance of this culture, BG 380–2 was not well predicted under simulation, particularly since it did worst of all genotypes in the 1991 WS Aduthurai experiments that was used for calibration.

Culture AD 86749, identified at Aduthurai, was tested in MLT during 1989 and 1990, and forwarded to ART in 1991. The simulation results also brought its ranking within the top three, in support of its selection (Figure 2).

It is disturbing to note that CO 43 was always the best simulated variety, while it never made the top three in the trials except for the carefully controlled 1991 experiment: in spite of a great potential, it may be more sensitive to management than the other varieties.

Our crop growth model was found to be useful for simulating potential yields (no-input constraints) under existing weather conditions of different rice growing zones. This is in line with the observation that simulation, and systems analysis have become important research tools in the understanding, integration and extrapolation of complex agricultural systems and subsystems such as those associated with productivity and sustainability of rice based cropping systems in Asia (McWilliam et al. 1990). Models represent a valuable means of synthesizing knowledge about different components of system and can help to integrate the complex and highly interactive responses that are involved in crop growth and development. They can replace much of the empirical time consuming trial and error research, and thus improve the efficiency and efficacy of the experimental approach. Modelling does not replace the need for detailed field trials, but it can specify the need for fewer and more critical trials that yield a broader and more valuable perspective on the factors influencing the performance of the crops (McWilliam et al. 1990).

Conclusion

The model MACROS.L1D, with crop data obtained from Aduthurai, could predict the performance of rice genotypes at Coimbatore and Madurai to the extent of selecting the best genotype, based on their yield ranking, with results comparable to the ranking from observed yields. Genotypes TNAU 80042 and AD 86749 came within the first three ranks both under field trials and under simulation. The former has been released as a variety CO 45 and the later is awaiting release shortly.

The present study support our hypothesis that simulation models can be a great help in pre-testing of rice genotypes at different locations, resulting in a substantial shortening of the period of evaluation. However, the experiment to collect the crop data on partitioning patterns for the model has to be refined.

References

Dua A B, Penning de Vries F W T, Seshu D V (1990) Simulation to support evaluation of the production potential of rice varieties in tropical climates. Transactions of the ASAE 33(4): 1185–1194.

Herrera-Reyes C G, Penning de Vries F W T (1989) Evaluation of a model for simulating the potential production of rice. Philipp. J. Crop Sci. 14(1): 21–32.

Penning de Vries F W T, Jansen D M, Ten Berge H F M, Bakema A (1989) Simulation of Ecophysiological processes of growth of several annual crops. Simulation Monographs, PUDOC, Wageningen, The Netherlands, and the International Rice Research Institute, Los Baños, Philippines. 271 p.

McWilliam J R, Collinson M P, Van Dusseldorp D B W (1990) Report of the evaluation of SARP project. Ministry of International Cooperation, DGIS, The Haque, The Netherlands. 39 p.

Genetic specific data for crop modeling

J.T. RITCHIE
Department of Crop and Soil Sciences, Michigan State University, East Lansing, MI 48824, U.S.A.

Key words: crop duration, genotypic traits, juvenile stage, leaf number, maize, phenology, photoperiodism, photothermal time, rice, vernalization, wheat

Abstract

A major objective in crop production is to fit the crop growth cycle into the constraints of the environment. Environments hostile to plant growth such as low soil water, excessively high or low temperature must be avoided to ensure plant survival and productivity. A major objective of crop yield simulation is to predict the duration of growth, the average growth rates, and the amount of assimilate partitioned to the economic yield components of the plant. With such a simulation system, it is possible to optimize the utilization of resources and quantify risk related to weather variation.

Most annual crops develop at rates proportional to the temperature of the growing parts of the plant. The time from seedling emergence to flowering, is also influenced by the length of a juvenile phase, by photoperiod, and in some cases, by vernalization. When these characteristics are known, genetic specific crop coefficients can be derived to enable the simulation of the timing of major events in plants.

The crop coefficients needed to predict crop development rate can be determined most accurately in controlled environment chambers. Less precise coefficients can be determined in field experiments when the temperature is measured and various photoperiods are obtained by selection of appropriate dates of sowing.

Introduction

The simulation of crop growth, development and yield is accomplished through evaluating the stage of crop development, the growth rate and the partitioning of biomass into growing organs. All of these processes are dynamic and are affected by environmental and genetic specific factors. The environmental factors can be separated into the aerial environment and the soil environment. The aerial environment information comes from a record of the weather conditions such as temperature and solar radiation. The soil environment information must come from synthesized approximations of the soil water status, the availability of soil nitrogen and other nutrients, the aeration status of the soil and the spatial distribution of the root system.

The potential biomass yield of a crop can be thought of as the product of the rate of biomass accumulation times the duration of growth. The rate of biomass accumulation is principally influenced by the amount of light intercepted by plants over a fairly wide optimum temperature range. The duration of growth for a particular cultivar, however, is highly dependent on its thermal environment and to some extent the photoperiod during floral induction. Highest potential yields of a particular annual crop are obtained in regions

F.W.T. Penning de Vries et al. (eds.), Systems Approaches for Agricultural Development, 77–93.

where the crop duration is maximized because of relatively low temperatures unless the radiation levels are also low. In regions where the temperature is relatively high, potential yield levels can reach those of cooler temperature regions only by combining yields from two or more crops grown in sequence so that the duration of the total growth periods is about the same in the high and low temperature zones. Thus the accurate modeling of crop duration is as critical as the accurate modeling of crop growth rate to predict crop productivity.

Genetic variation between modern annual cultivars within a species is usually most evident in the duration of growth and least evident in the rate of growth. Older cultivars with approximately the same duration of growth as modern ones differ from the modern ones principally in partitioning less assimilate to the reproductive organs and more to the vegetative organs, causing a lower harvest index. Many modern cereal cultivars have shorter, stronger stems, helping to prevent lodging when high soil fertility levels are used. Little genetic variation exists within a species for the processes of photosynthesis and respiration.

Although the temperature of the plant affects the duration of growth, other factors such as photoperiod and vernalization can strongly modify the duration of growth period. These environmental modifiers usually have their greatest influence on the duration of the vegetative period and the least influence during the growth of the reproductive organs. This genotype and environmentally induced variation in the proportion of time spent in the vegetative and reproductive stages can also influence the partitioning patterns of plants and cause some variation in the harvest index.

The challenge to plant breeders is to develop phenotypes in a particular environment with a growth period that will maximize yields within the constraints of a soil water supply or a favourable thermal environment. The ability to predict the stage of crop development is important for management decisions related to such things as timing of pesticide application, scheduling of orderly crop harvest, or synchronizing the flowering of cross-pollinated crops for hybrid seed production.

This paper emphasizes some of the genotypic variations that occur in crops and how they can be quantified for use in crop simulation to determine how genotype and the environment influence the plant development rate. Although other genotypic variations such as pest resistance are known to exist, less has been done to quantify these factors for simulation purposes.

Temperature influence on plant development

The time scale of plants is closely linked to the temperature of the growing parts of the plant. Thus, before discussing how the genetic characteristics and the environment interact, it is important to understand how some development characteristics are affected by temperature, independent of photoperiod or other environmental constraints.

Réaumur first suggested in 1735 that the duration of particular stages of growth was directly related to temperature and that this duration for a particular species could be predicted using the sum of mean daily air temperature (Wang 1960). This procedure for normalizing time with temperature to predict plant development rates has been widely used in the 20th century. Investigators who have studied the use of the system on the same cultivar in different environments have found several inaccuracies in the system. Most attention given to improving the system has focused on determining the low temperature where development is zero (the base temperature) and the high temperature where development stops increasing or begins to decline.

Several synonymous terms have been used to describe the process of summation of temperatures to predict plant growth duration (Nuttonson 1955). These include the terms degree-days, day-degrees, heat units, heat sums, thermal units and growing degree days. Other terms that include more than a temperature summation but also account for photoperiod or give different weighting to night and day temperatures have been used for specific crops such as Soybean Development Units (Brown and Chapman 1961), Ontario Corn Heat Units (Brown 1975), Biometeorological Time Scale (Robertson 1968), and Photothermal Units (Nuttonson 1948) and development stage (de Wit et al. 1970). A term most appropriate to describe the duration of plant development is "thermal time" as suggested by Gallagher (1979). Because the time scale of plants is closely coupled with its thermal environment, it is appropriate to think of thermal time as a plant's view of time. Likewise, if the photoperiod is used to modify thermal time, the corresponding term is photothermal time (Summerfield et al. 1991).

Thermal time has the convenient units of °C• day. The simplest and most useful definition of thermal time t_d

$$t_d = \sum_{i=1}^{n} (\overline{T}_a - T_b)$$

where $\overline{T}_a$ is daily mean air temperature, T_b is the base temperature at which development stops and n is the number of days of temperature observations used in the summation. The calculation of T_a is usually done by averaging the daily maximum and minimum temperatures. This calculation of thermal time is appropriate for predicting plant development if several conditions are met:

1. The temperature response of the development rate is linear over the range of temperatures experienced.
2. The daily temperature does not fall below T_b for a significant part of the day.
3. The daily temperature does not exceed an upper threshold temperature for a significant part of the day.
4. The growing region of the plant has the same mean temperature as T_a.

When one or more of the above assumptions are not correct, alternative calculations of thermal time are required. Usually for such conditions, the diurnal temperature values must be measured or approximated because the system becomes non-linear or another function of temperature when outside the range of the normal linear function. If the temperature is below T_b, the thermal time is assumed to be zero. If it is above an upper temperature threshold, the thermal time is assumed to be equal to the upper threshold value or some value lower than the threshold (Covell et al. 1986; Ritchie and NeSmith 1991).

When using air temperatures measured at weather stations to infer plant temperature, it is important to understand that there are several possible sources of error that can be introduced into the thermal time calculations. For example, using separate liquid-in-glass maximum and minimum thermometers read and reset once a day, bias in these temperature records can be caused by the time of observation (Schaal and Dale 1977). Minimum temperature for the next observational day cannot be higher than the temperature at the time of observation. Similarly it is possible to have an upward bias in maximum temperature readings when the readings are taken in the afternoon if it is possible for the maximum temperature to occur after the reading is recorded. Biases of these types were reported to average about 0.5°C to 1.0°C per day for several months of the year in Indiana, USA (Epperson and Dale 1983).

Another possible error can also be introduced when averaging the maximum and minimum temperature to represent the daily mean temperature (Arnold 1960). The major discrepancy occurs when a cold or warm air mass moves into the area near the beginning or ending of the daily temperature record. This source of error is usually random and may not be significant under most conditions. Errors of this type in a study reported by Ritchie and NeSmith (1991), averaged about 0.01°C, −0.09°C and 0.25°C in each of three seasons for 100 days during the maize growing season.

Major errors in thermal time calculation are possible when temperatures at screen height do not represent the temperature of the growing part of the plant. There are two sources of this type of error. First if the temperature measurements are made some distance from the field where plants are located, it is possible to introduce an unknown bias due to spatial variations in air temperatures. This source of error can be quite large, especially in regions where elevation is different between the sensor and the field. When temperature sensors are in the field where the crop is growing, the air temperature at screen height may be different from the temperature sensed by the plant. The specific sites on a plant where temperature influences development are the zones where cell division and expansion are occurring (Watts 1972). Several investigators have demonstrated that the near-surface soil temperature where the growing point of a plant was located when it is young was a more accurate indicator of plant development rates than when air temperature was used (Law and Cooper 1976; Cooper and Law 1978; Swan et al. 1987). For most monocot crop plants in the seedling stage, the apical meristem growth zone is often about 1 to 2 cm below the soil surface. Thus, during times when the plant's meristem is at or near the soil surface, the near surface soil temperature is more appropriate for predicting plant development than air temperature.

Differences in air and plant temperature when the growing parts of the plant are in the soil are often biased toward a warmer average temperature in the soil. For a maize crop grown in the early spring in Michigan, USA, the 2.5 cm depth soil temperature averaged 2.2°C warmer than the nearby screen air temperature (Ritchie and NeSmith 1991). Errors of this type can cause a bias of several days delay in the prediction of developmental events such as flowering. Factors such as tillage and residue management can also affect the soil-air temperature difference (Wierenga et al. 1982; Swan et al. 1987). Thus, for the greatest accuracy, the soil temperature at the level of the apical meristem should be used for the prediction of plant development until the internodes elongate enough for the meristem to move above the soil surface. When the meristem of crops are submerged in water, the water temperature will influence the plant apex temperature and thus be the controlling temperature.

Determining temperature response functions

Time between developmental events can be measured and expressed as an equivalent rate by taking the inverse of time as the rate. Developmental events such as time between the appearance of leaf tips or time between emergence and flowering can provide the information needed to develop accurate temperature response functions. Often when these developmental rates are measured at various temperatures, the resulting relationship is linear over a relatively large range of temperatures. Figure 1 provides information for rice of the type needed to develop temperature response functions for vegetative development. Leaf tip numbers were measured on four main stems of plants that were growing at four contrasting growth cabinet temperatures (Unpublished data of the author). The plants were grown at 25°C until the third or fourth leaf tip had appeared, after which they were kept in growth cabinets at various constant temperature. The data show that the rate of leaf appearance is higher for the first six or seven leaves than it is for the remaining leaves and that temperature has a major influence on rate of leaf appearance. The slope of the leaf appearance curves provides the needed rates for an evaluation of the temperature dependent development information. The leaf appearance rates determined from data similar to that in Figure 1, are shown in Figure 2 for two individual varieties that had variations in leaf appearance rates of 12 diverse varieties along with the mean of the 12 varieties. This information demonstrates that the base temperature for vegetative development is approximately 9°C, that a plateau, termed supra-optimal by Summerfield et al. (1992), begins at about 26°C, and that there is little difference between varieties in this relationship although the 12 varieties in the study consisted of Indica, Japonica and mixed types.

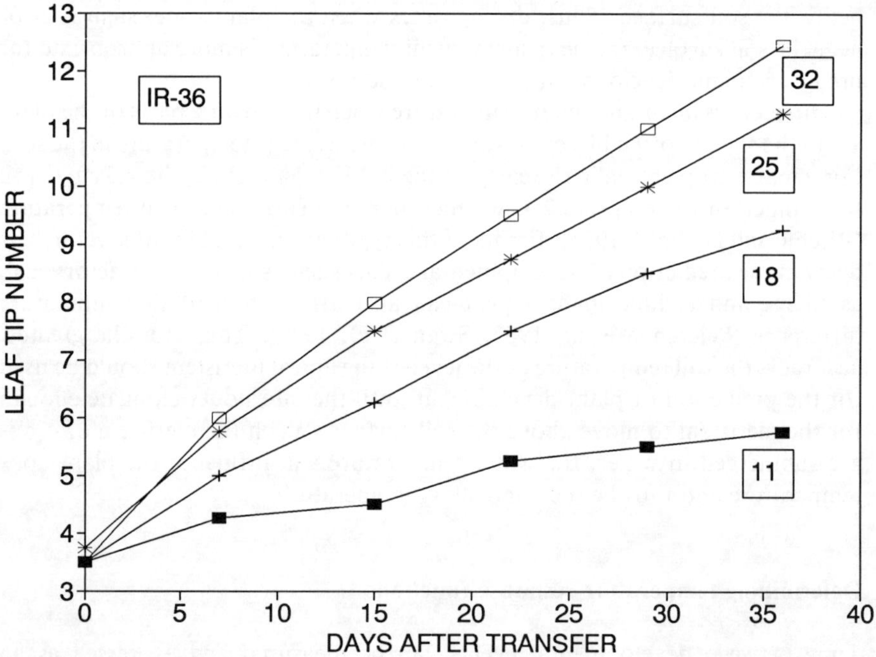

Figure 1. The influence of temperature on leaf tip appearance for IR36 rice plants. The plants were grown at 25°C until the transfer to the various temperatures shown by each curve (unpublished data of J.T. Ritchie, Michigan State University).

Temperature response functions for rice development rates determined from time to flowering when plants were grown in constant (11.5 h) photoperiod reveal a similar temperature response function to that for leaf appearance. Data from Summerfield et al. (1992) converted to a rate of development by taking the inverse of the duration to panicle emergence for 16 contrasting rice varieties grown at several temperatures are presented in Figure 3. The three curves are averages for several varieties that had similar development patterns but contrasting differences in time to panicle emergence principally because of differences in juvenile stage duration. The development data show a quite similar trend to the leaf appearance data of Figure 3 with the slight exception that the base temperature extrapolates to be about 10°C instead of 9°C from the leaf data. The temperature at which development toward ear emergence stops increasing, about 26°C, also agrees with the leaf appearance data.

Several other development events can be used to evaluate temperature response functions. Time of seed germination, seedling emergence, leaf growth, grain filling etc. are useful means of determining development information although precise information is usually somewhat more difficult to extract (Ritchie and NeSmith 1991). The base temperature and supra-optimal temperature may vary somewhat between these processes. The base temperature for grain filling is usually a few degrees higher than that for

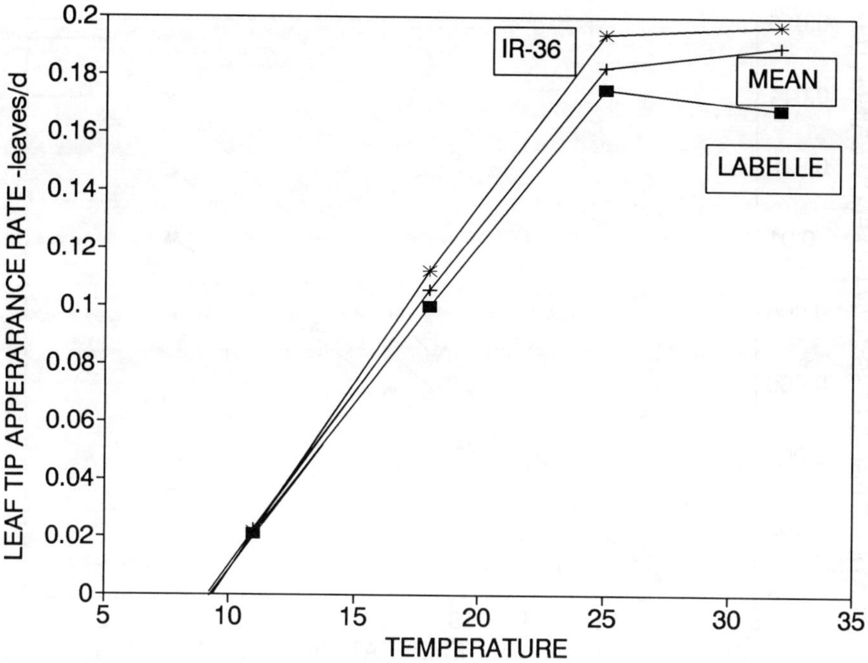

Figure 2. The influence of temperature on the rate of leaf appearance of several rice cultivars. The curve labelled mean is the average for 12 cultivars. The varieties IR36 and Labelle represent the fastest and slowest development rates of the 12 cultivars studies (unpublished data of J.T. Ritchie, Michigan State University).

vegetative development and may not have a measurable supra-optimal value.

Grain filling duration is more difficult to quantify than visual developmental events such as leaf appearance or time to flowering. For many crops, there is a lag period after anthesis before rapid filling begins. The grain filling rate is almost constant if average temperatures are relatively constant until the grain is almost filled unless there is a shortage of assimilate, nitrogen, or stored carbohydrate available for grain filling. The most accurate determination of the beginning and ending of grain filling requires destructive sampling, thus creating a higher level of variability in the data than for non-destructive type observations. Determining the timing of grain filling from visual observations such as flowering to physiological maturity often leads to uncertainty because of the lack of a clearly observable plant feature at the beginning and end of grain filling. Black layer formation has been used in recent years for defining maize maturity. It has not proven to be a highly accurate indication of end of grain filling, although it does provide qualitative evidence of differences between genotypes that would be needed in modeling (Daynard 1972).

The temperature response functions for plants within a species are rather stable because investigators are able to reproduce information about plant development processes and the information is transferable from growth

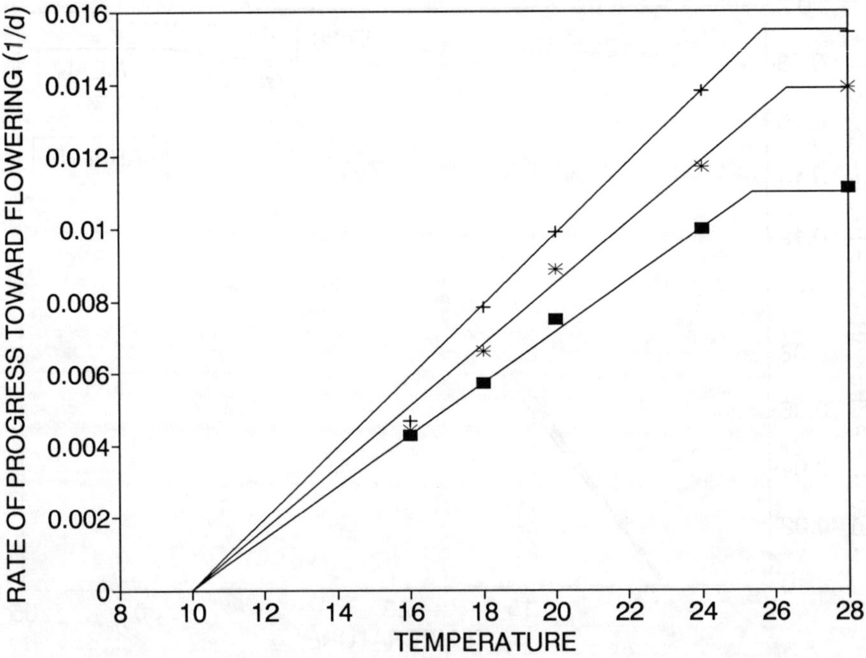

Figure 3. The influence of temperature on the rate of progress towards panicle emergence for 16 rice cultivars. The + symbols represent the average for varieties Eiko, Jkau, Suweon and Stejaree; the * for varieties BPI 76, Barkat, Carreon, and Pinulot; and for varieties Azucena, Intan, IR36, IR5, IR8, IR42, Peta, and TN1. Only data from the constant day-night temperature treatments with 11.5 hours daylength were used in the analysis except for the 18C data which came from the 20/16C day-night treatment. (Data from Summerfield et al. 1981b).

cabinets to the field when the proper precautions are taken to make sure the plant temperature is correct. Thus there is little to be gained by continued research into these developmental processes once they are established. However, more research is needed to understand and predict how extremely low growth rates caused by low levels of light, soil water or nutrition will alter development rates. For tillering plants, there is also a lack of information on factors affecting the development rate of individual tillers.

Genotypic variation in response to photoperiod

The primary difference in development between genotypes within a species occurs in the length of their vegetative phase. This variation can result from genotypic differences in photoperiod sensitivity, maturity type, or vernalization sensitivity. Maturity type variations are exhibited in the length of a juvenile stage of some crop species such as rice, maize, sorghum and millet in which young plants do not begin photoinduction to flowering until they reach a

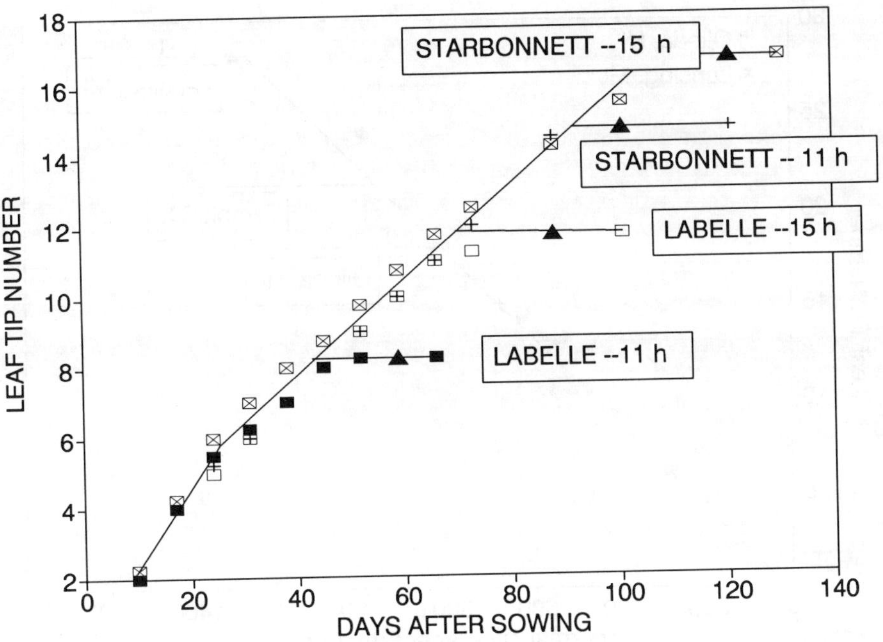

Figure 4. The influence of photoperiod on leaf appearance, final leaf number, and ear emergence for 2 contrasting rice varieties. The solid triangle is the date of ear emergence.

certain leaf number. During this juvenile stage the plants are not sensitive to photoperiod.

The influence of photoperiod on plant characteristics can be demonstrated by comparing leaf appearance, final leaf number and time to flowering for two contrasting genotypes of rice grown at constant temperature and at two daylengths (unpublished data of the author, shown in Figure 4). The rate of leaf appearance had little variation due to photoperiod, but the number of leaves and the time to flowering was strongly influenced by photoperiod and genotype. The short season variety Labelle averaged 8.25 and 11.75 total leaves at 11 and 15 hour photoperiods, whereas the long season variety Starbonnett averaged 14.75 and 16.75 leaves when grown under the same environmental conditions as Labelle. Labelle is very photoperiod sensitive and has a short juvenile period whereas Starbonnett is less photoperiod sensitive but has a long juvenile phase.

The photoperiod influence on leaf appearance and leaf number within or between genotypes can also be demonstrated using field derived information taken from locations where a great contrast exists between photoperiods and temperatures. A photoperiod sensitive tropical maize cultivar, X304C, grown at three locations with large contrasts in daylength during development had almost the same rate of leaf development when expressed in terms of thermal

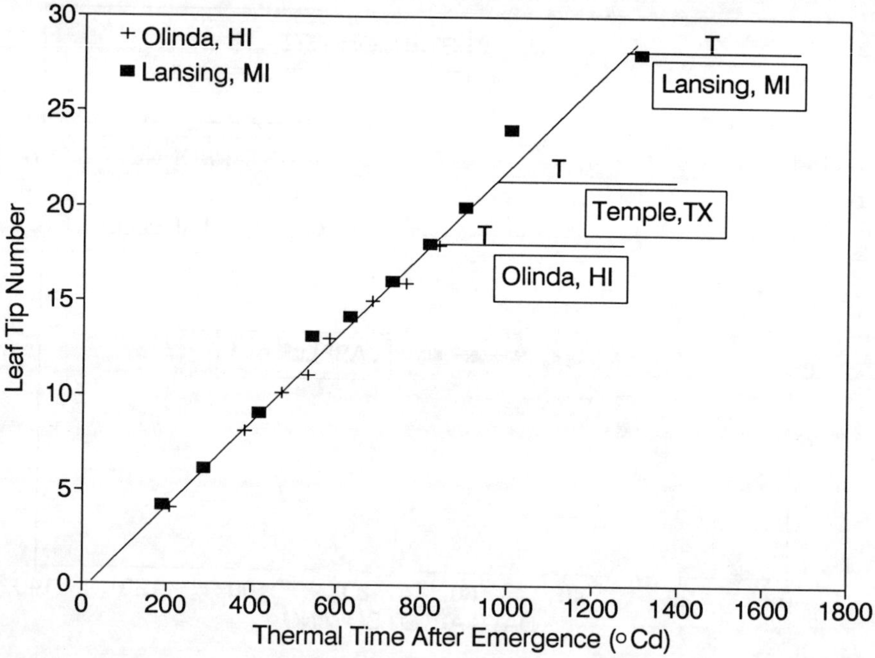

Figure 5. Maize leaf tip number as a function of thermal time after emergence at Olinda, HI; Temple, TX; and East Lansing, MI. Final leaf number is indicated by horizontal lines and time of tasseling is indicated by T. From Ritchie and NeSmith (1991).

time, but quite different leaf numbers as can be seen in Figure 5 (Ritchie and NeSmith 1991).

At the Hawaii site, the photoperiod (daylength and civil twilight) at tassel initiation was about 13.5 hours, at Texas about 14.2 hours and at Michigan about 16.3 hours. These data clearly demonstrate that for photoperiod sensitive genotypes, thermal time to flowering is not a constant. Thermal time would only be constant for plants grown with common photoperiods during floral induction (Kiniry et al. 1983a).

Winter cereal plants respond to photoperiod in the opposite direction to most of the warm season crops. Their development rates are faster in long days. They have no true juvenile phase, thus photoperiod sensitivity to floral induction begins at emergence (Ellis et al. 1989).

Genotypic variation in juvenile stage

The time required for a plant to advance from sowing to maturity is usually controlled by temperature, photoperiod and in some species, the length of a juvenile stage. The juvenile stage can be defined as the pre-induction stage when the plant is not sensitive to variations in photoperiod. The variation in the length of a juvenile stage has often given rise to the classification of plants in terms of maturity types. The maturity type concept is only qualitative and may be only regionally transferable if the maturity type classification is the result of both the juvenile stage duration and the responsiveness to photoperiod. Thus a genotype that is long season in one region can be short season in another region if the photoperiod is quite different at the time of floral initiation for genotypes sensitive to photoperiod.

To simulate the juvenile stage it is necessary to know the variation among genotypes in this characteristic. In short day plants, this period can be observed by growing the plants in short day conditions (usually 12 hours or less) and observing, by plant dissection, the time to floral initiation. The end of the juvenile stage is usually about 5 days before floral induction in maize, a phenomenon that can be determined with special experiments in which plants are interchanged between long days and short days during the late juvenile and floral induction phase (Kiniry et al. 1983b). Photoperiod exchange experiments with rice have demonstrated a similar pattern to maize with a similar minimum induction period (unpublished data of the author). The duration of the juvenile stage is almost totally controlled by temperature. Thus a single genetic coefficient can be determined to describe the duration of the juvenile period with units of degree-days. During the juvenile and floral induction phase, leaf primordia are being developed. When floral induction ends, leaf initiation no longer occurs and the final leaf number has been determined. Because the rate of appearance of the final leaves that have differentiated, but not yet appeared, is controlled by temperature, the thermal time required for the remainder of the vegetative stage is a consequence of what has happened in the juvenile and floral induction phase. Likewise, the duration of the remainder of the life cycle of the plant is principally controlled by temperature, so that thermal time can also be used to describe the plant growth cycle until the plant reaches physiological maturity.

The influence of photoperiod on time from sowing to panicle emergence or similar phenological events can be obtained with rather simple experiments to determine the genotypic variation in photoperiod sensitivity. The experiments should be done at a controlled or known temperature regime. When the time to panicle emergence is known, the thermal time can be calculated and it then becomes possible to approximate the time at which floral induction ended. The thermal time to panicle emergence was determined for rice variety IR36 in two separate and independent experiments where the photoperiod was varied within ranges expected where rice plants are grown. One experiment was done by the author at the Duke University Phytotron (USA) and the other done at the Plant Environment Laboratory in Reading (UK) by Summerfield et al. (1992). The UK experiments were done in varying temperature environments. The thermal time to panicle emergence for several of the temperature and photoperiod treatments is presented in Figure 6. A common feature of the data is that there is a plateau or minimum photoperiod below which plants cease to respond to photoperiod. In the IR36, the value appears to be at about 12.5 hours. This

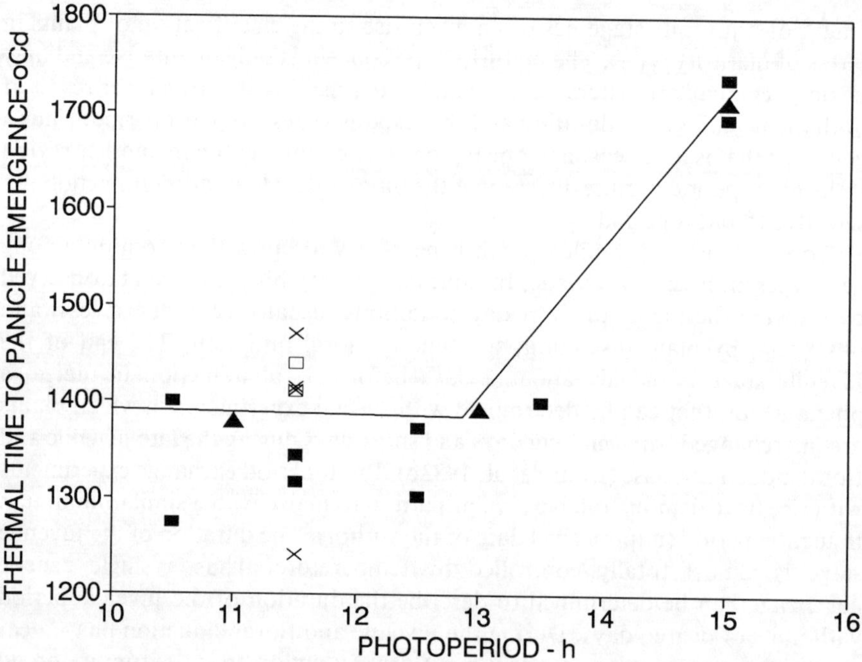

Figure 6. The influence of photoperiod on the thermal time required for panicle emergence in rice variety IR36. All symbols except the triangles are from studies of Summerfield et al. (1992) conducted at various temperatures. The triangles are data from J.T. Ritchie (unpublished data, Michigan State University).

photoperiod threshold varies between genotypes, the range is about 11.5 hours to 14.0 hours for most rice cultivars (Vergara and Chang 1985). The relative duration of the juvenile stages in short day plants can be assessed from the value of thermal time at the minimum photoperiod threshold value. For example, from Figure 6 it can be determined that variety IR36 will always require at least about 1380 degree days(base temperature 9°C). The slope of the response to photoperiod above the threshold photoperiod also varies considerably between genotypes. If the slope is expressed in degree days per hour, this information can become a genotype specific coefficient to use in simulation of the duration of plant vegetative development (Alocilja and Ritchie 1990).

Vernalization influence on winter cereal development

Winter cereals have practically no juvenile phase and they are also long day plants, that is, they develop most rapidly in long days. The true winter type cereals can have delayed floral induction in a similar manner to a juvenile phase in the warm-season, short-day plants as discussed earlier in the paper. This delay is caused by the requirement for the winter cereals to go through a period of relatively cold temperatures before floral induction. This process is called vernalization, and there are large differences in genotypic sensitivity to this process. The differentiation between 'spring' and 'winter' types of winter cereals is principally the result of differences in sensitivity to vernalization. There are intermediate types which have varying degrees of sensitivity to vernalization. These are often referred to as 'semi-winter' types (Kirby and Perry 1987).

For wheat, temperatures above 0°C to about 8°C seem to be the most effective for vernalization (Ahrens and Loomis 1963; Chujo 1966). Vernalization also can occur with lower effectiveness in temperatures up to about 15°C (Chujo 1966). The delay in flowering and increase in leaf number before flowering was demonstrated for winter wheat plants by Chujo (1966). Plants of variety Norin 27 grown for 60 days at 1°C or 4°C and then transferred to a glasshouse at 18°C developed only 7 leaves whereas plants grown at 18°C for 60 days developed 16 leaves. In the same experiment, plants exposed to durations of vernalizing temperatures between 20 and 60 days caused variation in leaf numbers between 7 and 16.

Genetic coefficients for use in crop simulation that describe the way in which vernalization affects the duration of vegetative development in winter wheats have been developed by Ritchie (1991) and Weir et al. (1984).

Simulating the combined influence of environmental factors on plant development

An approximate template to describe the influence of the juvenile stage, the photoperiod, and temperature on rice development is presented in Figure 7. The template shows the development of the crop from sowing to maturity. The principal feature of the system is that crop development duration is primarily controlled by temperature as expressed in thermal time. The deviation from a constant thermal time for all phasic developmental events is caused by the variation in photoperiod during the floral induction stage. It is only during that time that the plant is sensitive to photoperiod and the rate of photoinduction depends on day length. Future phenological events are consequences of the duration of the induction phase. During the emergence phase, the juvenile phase, and the induction phase, leaf primordia are being produced at a rate proportional to the temperature. When floral induction ends, the leaf number for the plant has been determined as well as the thermal time required for the remaining unexposed leaves to appear. For determinate plants, total leaf number will fix the thermal time to flag leaf appearance, panicle emergence, anthesis, grain filling and physiological maturity. Except for the length of the juvenile stage, which varies between genotypes in thermal time, the photoperiod during induction is the primary factor causing differences in thermal time to the major phenological events within a crop species. In winter

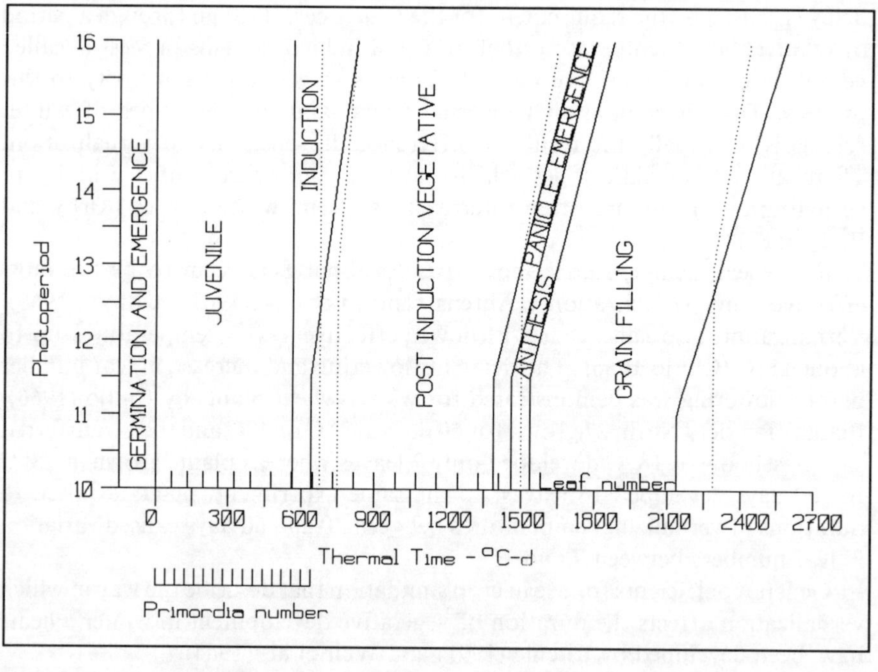

Figure 7. A template demonstrating the duration of different stages of rice development as a function of thermal time and photoperiod. Shown with the thermal time scale is a scale for leaf tip number and leaf tip primordia number. Note that the final leaf primordia number at the end of floral induction is approximately equal to the number of leaves at panicle emergence. The solid lines represent one variety and the dotted line another variety. The variety represented by the solid line has a shorter juvenile phase but greater sensitivity to photoperiod.

cereals, the template is similar to that shown in Figure 7, except that the plants are sensitive to photoperiods in short days. Also the requirement for vernalization can be conceptualized as a factor causing variations in an apparent juvenile stage.

Conclusions

Plant breeders are aware of the potential for selection of phenological components of economic yield because of their prime importance in crop adaptation and productivity. A primary contribution that systems science can make to crop improvement is the ability to predict the environmental modulation of phenological events with the view of optimizing the use of resources for most production and profitability. The ability to predict phenology, using genetic specific coefficients, in different environmental circumstances as they express themselves over space and time would provide a more rapid method for determining where genotypes would be adapted and the climate related risks associated with their production. Such phenological predictions allows the possibility of minimizing adverse environmental consequences on crop constraints such as low temperature and aridity. The ability to predict phenology will also assist in the facilitation of more efficient screening of genotypes in order to select those best adapted to specific target environments.

Although there will surely be advances made in our ability to predict the development patterns of crops, our present knowledge should dictate that for most field trials where genotypes are being evaluated, it is important that the times of major phenological events be recorded, that a weather station with accurate air temperature measurements be nearby, and that ideally the soil temperature at about 2 cm be recorded during early vegetative growth. The prediction of plant response to phasic developmental events that are not observable without plant dissection is greatly aided if the number of leaves produced on the main stem of the plants under study is determined.

This paper has emphasized the genetic specific information that is best understood for predicting plant growth duration. Other genetic specific information that is more difficult to obtain is needed for a more totally integrative systems approach for assessing crop productivity. Genetic specific variations within a species that are known to exist and could be quantified include partitioning related phenomenon such as leaf size, stem dimensions, grain numbers and grain growth rates. Quantitative genotypic variations in response to pest damage, excesses or deficiencies of soil water, and nutrient deficiencies are also needed cases when those stress related phenomenon exist. For prediction of the crop response to these stresses, however, good simulation models of the pest occurrence and soil related stresses are needed.

Acknowledgments

Special thanks is extended to R.J. Summerfield, for providing a preprint of the paper 'Photothermal responses of flowering in rice (Oryza sativa)' from which I extracted some useful information for this paper. I also thank the reviewers of an earlier version of the manuscript for helpful criticism.

References

Ahrens J F, Loomis W E (1963) Floral induction and development in winter wheat. Crop. Sci. 3(6):463–466.

Alocilja E C, Ritchie J T (1990) The application of SIMOPT2:RICE to evaluate profit and yield-risk in upland rice production. Agric. Systems 33:315–326.

Arnold C Y (1960) Maximum-minimum temperatures as a basis for computing heat units. J. Am. Soc. Hort. Sci. 76:682–692.

Brown D M (1975) Heat units for corn in southern Ontario. Ministry of Agric. and Food Factsheet No. 75–077. Ontario Agricultural College, University of Guelph, Canada.

Brown D M, Chapman L J (1961) Soybean ecology. III Soybean development units for zones and varieties in the Great Lakes Region. Agron. J. 53:306–308.

Chujo H (1966) Difference in vernalization effect in wheat under various temperatures. Proc. of the Crop Sci. Soc. of Japan 35:177–186.

Cooper P J M, Law L R (1978) Enhanced soil temperature during very early growth and its association with maize development and yield in the Highlands of Kenya. J. Agric. Sci. (Cambridge) 89:569–577.

Covell S, Ellis R H, Roberts E H, Summerfield R J (1986) The influence of temperature on seed germination rate in grain legumes. J. Exp. Bot. 37(178):705–715.

Daynard T B (1972) Relationships among black layer formation, grain moisture percentage, and heat unit accumulation in corn. Agron. J. 64:716–719.

De Wit C T, Brouwer R, Penning de Vries F W T (1970) The simulation of photosynthetic systems. Pages 47–70 in Setlik I (Ed.) Prediction and measurement of photosynthetic productivity. PUDOC, Wageningen, The Netherlands.

Ellis R H, Summerfield R J, Roberts E H, Cooper J P (1989) Environmental control of flowering in Barley (*Hordeum vulgare* L.). III. Analysis of potential vernalization responses, and methods of screening germplasm for sensitivity to photoperiod and temperature. Ann. Bot. 64:687–704.

Epperson D L, Dale R F (1983) Comparison of climatologies for existing heterogeneous temperature records with those for adjusted records for West Lafayette and Whitestown, Indiana. Indiana Acad. Sci. Proc. 92:395–404.

Gallagher J N (1979) Field studies of cereal leaf growth. J. Exp. Bot. 30(117):625–636.

Kiniry J R, Ritchie J T, Musser R L, Flint E P, Iwig W C (1983a) The photoperiod sensitive interval in maize. Agron. J. 75:687–690.

Kiniry J R, Ritchie J T, Musser R L (1983b) Dynamic nature of the photoperiod response in maize. Agron. J. 75:700–703.

Kirby E J M, Perry M W (1987) Leaf emergence rates of wheat in a Mediterranean environment. Aust. J. Agric. Res. 38:455–464.

Law R, Cooper P J M (1976) The effect and importance of soil temperature in determining the early growth rate and final grain yields of maize in Western Kenya. East Afr. Agri. For. J. 41(3):189–200.

Nuttonson M Y (1948) Some preliminary observations of phenological data as a tool in the study of photoperiodic and thermal requirements of various plant material. Pages 129–143 in Vernalization and photoperiodism. A symposium. Waltham, Mass., Chronica Botanica.

Nuttonson M Y (1955) Wheat-climate relationships and the use of phenology in ascertaining the thermal and photo-thermal requirements of wheat. Amer. Inst. Crop Ecol. Washington, D.C., USA. 388 p.

Ritchie J T (1991) Wheat phasic development. In Hanks and Ritchie J T (Eds.) Modeling plant and soil systems. Agronomy 31:31–54.

Ritchie J T, NeSmith D S (1991) Temperature and crop development. In Hanks and Ritchie J T (Eds.) Modeling plant and soil systems. Agronomy 31:5–29.

Robertson G W (1968) A biometeorological time scale for a cereal crop involving day and night temperatures and photoperiod. Int. J. Biometeorol. 12(3):191–223.

Schaal L A, Dale R F (1977) Time of observation temperature bias and 'climate change'. J. of Applied Meteor. 16:215–222.

Summerfield R J, Roberts E H, Ellis R H, Lawn R J (1991) Towards the reliable prediction of time to flowering in six annual crops. 1. The development of simple models for fluctuating field environments. Expl. Agric. Great Britain. (submitted)

Summerfield R J, Collinson S T, Ellis R H, Roberts E H, Penning de Vries F W T (1992) Photothermal responses of flowering in rice (*Oryza sativa*). Ann. Bot. 69:101-112.

Swan J B, Schneider E C, Moncrief J F, Paulson W H, Peterson A E (1987) Estimating corn growth, yield, and grain moisture from air growing degree days and residue cover. Agron. J. 79:53–60.

Wang J Y (1960) A critique of the heat unit approach to plant response studies. Ecology 41:785–790.

Watts W R (1972) Leaf extension in *Zea mays*. II. Leaf extension in response to independent variation of the temperature of the apical meristem, of the air around the leaves, and of the root-zone. J. Exp. Bot. 23(76):713–721.

Weir A H, Bragg P L, Porter J R, Rayner J H (1984) A winter wheat crop simulation model without water or nutrient limitations. J. Agric. Sci. 102:371–382.

Wierenga P J, Nielsen D R, Horton R, Kies B (1982) Tillage effects on soil temperature and thermal conductivity. Pages 69–90 in Unger and Van Doren (eds.) Predicting tillage effects on soil physical properties and processes. Amer. Soc. Agron. Spec. Publ. No. 44.

Vergara B S, Chang T T (1985) The flowering response of the rice plant to photoperiod: A review of the Literature. 4th ed., Int. Rice Res. Inst., Los Baños, Philippines.

SESSION 2

Crop production: weather constraints

Agro-ecological zoning using crop growth simulation models: characterization of wheat environments of India

P.K. AGGARWAL
Water Technology Center, Indian Agricultural Research Institute, New Delhi 110 012, India

Key words: agro-ecological zones, climate change, crop duration, Geographic Information System, India, land evaluation, potential yield, research priorities, simulation, water-limited yield, wheat, WTGROWS

Abstract
The main objectives of agro-ecological zoning are data inventory of environmental resources, identification of homologous environments, determination of agricultural potential of a region, planning for regional development and identification of research priorities. Conventional methods employed are overlaying of maps and various statistical techniques. The major quantitative output of these approaches is growing period calculated from rainfall and potential evapotranspiration. In today's agriculture, when vast areas are irrigated and multiple cropping is practiced, growing period does not necessarily relate with productivity. Crop growth simulation models provide an alternative method for agro-ecological characterization of a region. These models integrate the effect of dynamic climatic, edaphic and crop management factors in determining grain yield. Coupled with a suitable Geographical Information System (GIS), crop simulation models can be used for determining agricultural potential of a region and to identify major constraints limiting productivity. A case study is described where a well validated model, WTGROWS, has been used to determine productivity of wheat at different locations in India as determined by climate and limited by water availability. Regions with similar productivity have been demarcated with a GIS. Effect of climatic variability on wheat yields is described. The model has also been used to identify shift in wheat productivity zones in a climate change scenario.

Introduction

Carving regions within regions with varying purposes has always attracted human attention. In relation to agriculture, two major categories of natural regions have been identified: areas which are naturally homogeneous and beyond human interfere such as climatic zones, and areas which are homogeneous due to human activity such as agro-climatic zones. The latter are the divisions of a region according to features of climatic and edaphic conditions in relation to agricultural objectives.

A review of literature shows that agro-ecological zoning is exercised with the following objectives:

1. Data inventory of environmental resources, spatial and temporal data analysis for demarcation of regions.
2. Technology transfer within a region of great diversity: Since most experimental results are location- and season-specific, identifying regions with homologous environments where these results could be of use is a primary objective.

F.W.T. Penning de Vries et al. (eds.), Systems Approaches for Agricultural Development, 97–109.

3. Planning for regional development: To identify priorities in resource allocation and optimizing resource use efficiency.
4. Identification of research priorities: Determination of optimal duration of a crop/variety. Also to guide the choice of locations for research.
5. Impact of climatic variability on agricultural production.

Overlaying maps of different climatic and edaphic factors has been the most conventional technique of agro-ecological zoning. Classification of India's agricultural environments into 21 zones is a recent example where this method has been applied (Sehgal et al. 1990). Statistical techniques such as correlations, regressions, and multivariate analysis are also routinely employed. Well established programs of international and national nurseries and yield trials at several key locations are at present the best means of identifying the performance of crops and varieties in different environments. These trials are, however, expensive since they need to be conducted for several seasons. Growing season length, estimated from rainfall and potential evapotranspiration, is also used in several classifications (FAO 1983; Oldeman and Frere 1982; Sehgal et al. 1990).

In India, as in most other countries, several attempts have been made to classify its diverse climates for making effective land use plans. Planning Commission of Government of India carved 15 agro-climatic zones based on physiography and climate (Government of India 1987). Subsequently, these zones have been subdivided into 127 agro-ecological zones by the National Agricultural Research Program (NARP) based on physiography, rainfall, soils, cropping pattern and administrative boundaries (Ghosh 1990). National Bureau of Soil Survey and Land Use Planning superimposed soil texture map, physiography, climate (aridity-humidity) and computed growing period (the time when rainfall exceeds 0.5 × potential evapotranspiration) to demarcate 21 zones (Sehgal et al. 1990). Figure 1 shows the major agroclimatic variables used in this classification. Each zone is a unique combination of four factors, although more than one zone can have one or more criteria in common.

Most agro-ecological zoning efforts have been principally based on considerations of rainfall and to some extent on soil texture. Major quantitative output of these analysis is growing period. From a purely ecological view point, this may be adequate but when the interest is in the knowledge of agricultural potential of a region and the objectives listed earlier, one needs to know much more. For instance, what is the productivity potential of major crops in different regions? What are the major constraints limiting yield? What will be the duration of different crops or cropping systems? Moreover, growing period does not necessarily have a relation with crop productivity. Growing period, for example, is more than 210 days in Assam and West Bengal, but the yield of principal rainy season crop – rice– is very low due to limited radiation and input use. By comparison, growing period is 90–150 days only in Punjab, yet rice yields are high due to significantly higher radiation and greater input use. It is all the more difficult to relate this period with performance of postrainy season crops. Perhaps the growing period based agro-ecological zoning is appropriate

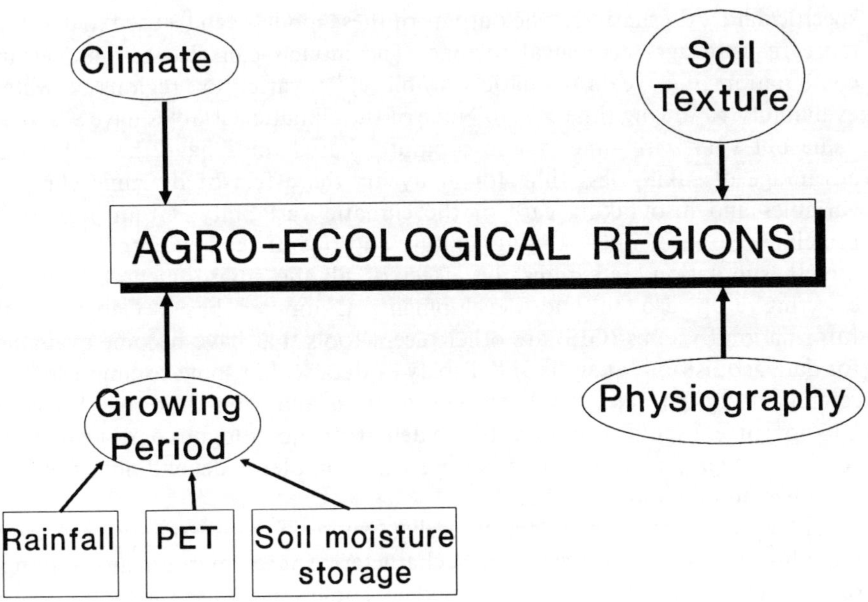

Figure 1. Major parameters used for delineating agro-ecological regions in conventional approaches (adapted from Sehgal et al. 1990).

when agriculture is monoculture and rainfall is the principal variable determining crop yield such as in large tracts of semi-arid tropics. With the increasing availability of irrigation resources in vast areas and increased area under multiple cropping during last few decades, there is now a need to reconsider the main variables effecting crop growth and yield and hence agro-ecological zoning.

Crop growth, development and water use are dependent upon several climatic, edaphic, hydrological and physiological and management factors. The major factors affecting crop growth and development are radiation and temperature (yield determining), water and nutrition (yield limiting), and pests and diseases (yield reducing). In addition, productivity is also determined by many other factors such as variety, its physiology, crop management, which interact with weather and soils to influence yield level. In irrigated and well managed crops, productivity is primarily determined by radiation and temperature whereas in rainfed areas, precipitation and soil moisture storage are also important.

There is a need to couple more closely environment data with effect on crop growth and yield. Systems analysis and crop growth simulation are relatively recent techniques that offer a means of quantitative understanding of the effects of dynamic climatic and edaphic factors, and agronomic management factors on crop growth and productivity in different land evaluation units. They can be applied to characterize agro-climates of any region using its site specific data. Alternatively, the outputs of these models can be used to develop more realistic agro-ecological regions. The models can also test numerous combinations e.g. years (climatic variability) × variety × regions × water availability × sowing time × ...). None of the climatic variables have a 'fixed' value but vary with time. These computer based tools have the additional advantage of taking less time for analyzing the effect of dynamic climatic variables and also taking care of the climatic variability. Grain yield, the principal reason of interest in agriculture and one of the key outputs of crop growth simulation, integrates the effect of all the environmental variables effecting crop growth and development. Computer based Geographical Information Systems (GIS) are other recent tools that have become available for data acquisition, analysis and display of data within some common spatial referencing system (Nix 1987). Systems analysis and simulation together with GIS now offer exciting opportunities to demarcate homologous agro-ecological zones at mega-, macro-, meso- as well as microlevel depending upon the objective and data availability.

There have been few attempts earlier where systems analysis and crop growth simulation have been used to characterize agro-environments in terms of agricultural productivity. Jones and O'Toole (1987) used a crop growth simulation model ALMANAC to illustrate possible applications of crop simulation models for meeting the various objectives of agro-ecological characterization at a micro-level, namely definition of research priorities, matching technology with resources and to describe the impact of climatic variability on crop yields. Angus (1989) analyzed the long-term mean agro-environment of Philippines to estimate opportunities for multiple cropping of rainfed rice using simulation model POLYCROP. Simulated yields in north and central Philippines varied little (3.5 to 4.0 t ha^{-1}) indicating rather uniform availability of water stress free period on a long-term basis. However, the percentage of years in which two wet seeded rice crops could be simulated varied from less than 10 to 85% among the simulated locations indicating considerable climatic variability.

Aggarwal and Penning de Vries (1989) and Aggarwal (1991) used simulation model to characterize agro-climates of south-east Asia and other warm tropical regions, in terms of production possibilities of wheat, a non-traditional crop for the region. Regions which could be more productive in irrigated and/or rainfed conditions were identified.

Crop growth simulation model outputs, coupled with GIS, have been used recently to characterize agro-environments. Van Lanen et al. (1991) used ARC-INFO, a vector based GIS package along with WOFOST, a general purpose simulation model to estimate potential and rainfed yields of winter wheat and of other crops in the European Community.

A case study of characterization of agro-environments of India for wheat productivity

Wheat is an important staple food crop of India and is grown in a diversity of agro-climatic conditions from 15° N to 32° N and from 72° E to 92° E. It is grown after monsoon season with very little seasonal rainfall. Temperatures during crop season are relatively cool. Figure 2 shows weather pattern of New Delhi, a typical wheat producing area. Wheat is produced in very light to very heavy textured soils with a variety of irrigation managements. At present wheat is cultivated in 17.4 Mha irrigated areas and 6.3 Mha rainfed areas. The average productivity ranges from 650 kg ha^{-1} to 4500 kg ha^{-1} depending upon the region. The total production of wheat in 1990 was 54 Mt. Since the demand is sharply increasing, it is important to know the productivity potential and scope of improvement in yield of wheat in different regions of the country.

This case study was undertaken with following specific objectives:

- To characterize environment in terms of potential productivity for different land evaluation units and to demarcate regions with similar potential yield;
- to determine the productivity of rainfed wheat in different districts;
- to propose a classification of wheat growing regions based on productivity;
- to describe the effect of climatic variability on wheat yields;
- to determine areas which will be more effected by climatic change (increased carbon dioxide and temperature).

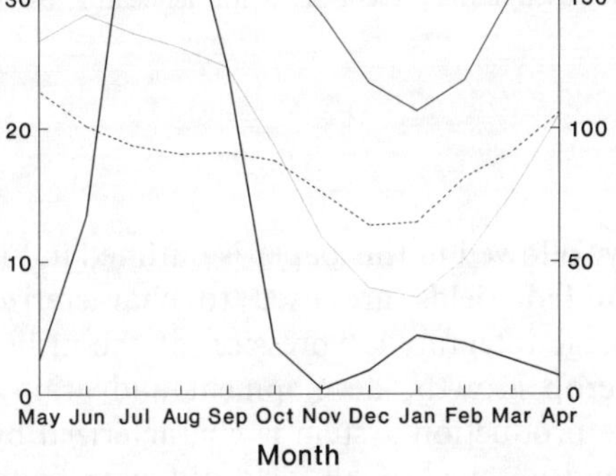

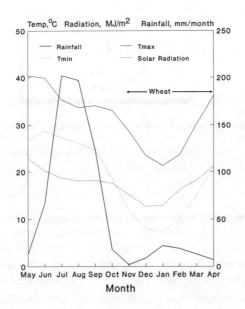

Figure 2. Long-term mean weather data for New Delhi, a typical wheat producing area.

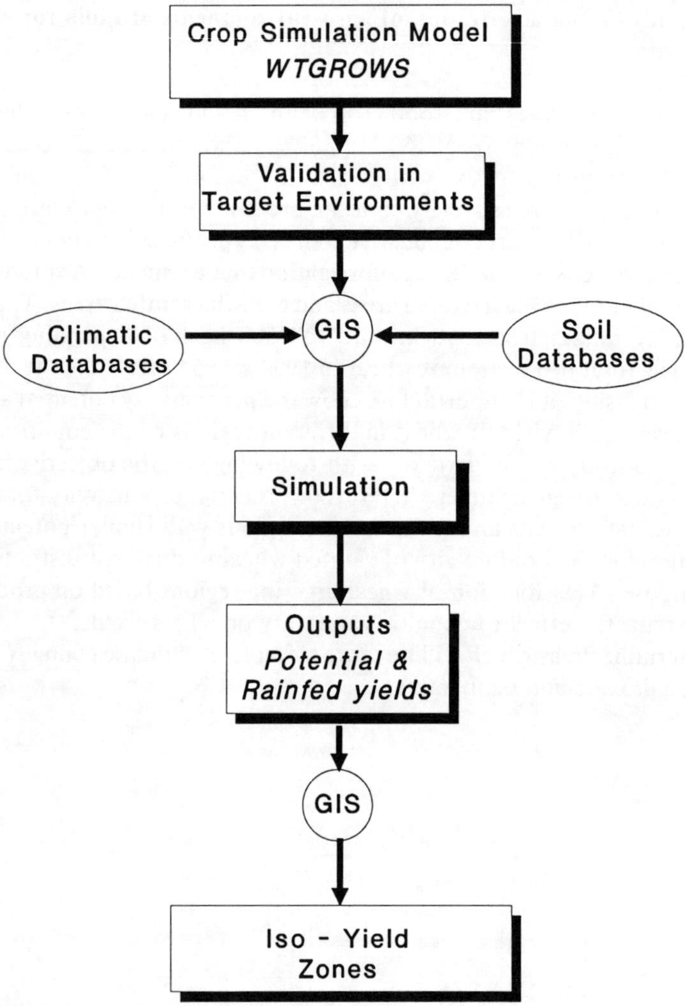

Figure 3. Methodology used in the present paper for delineating Iso-Yield Zones.

Methodology

The methodology followed in this paper is outlined in Figure 3. Two variables potential and rainfed yields are used to characterize agro-environments. Potential yield is the integrated expression of the influence of radiation and temperature on crop growth, development and grain yield of a particular crop/variety. The production system is characterized by adequate water and nutrient supply and absence of all yield reducing factors such as pests and diseases. Potential productivity can be interpreted as the upper limit of yield.

Rainfed productivity characterizes the effect of water deficit on potential productivity. For calculating water deficits, water balance is computed which involves rainfall, soil texture, depth and other soil physical properties.

Primary land evaluation unit

India is divided into 434 administrative regions called districts. This is the smallest unit for which dependable primary data on cropped area, productivity, etc. is available. Government of India has recently set up database centers in each of these districts to collect primary data on soils, weather, land use and crop productivity besides other activities such as urban development, education, water resources and health. These centers are connected via a satellite network, NICNET, based in New Delhi. For the present study, we have, therefore, chosen district as the primary land evaluation unit.

Agro-environment

Weather. Long-term average monthly weather data of 219 locations spread all over the country were used in the model. These locations are spread from 8° N to 31° N and from 73° E to 95° E and from sea level to 2000 m elevation. The daily weather data was estimated by linear interpolation. In addition, daily weather data for a number of key locations was also used for estimating effect of climatic variability on grain yield.

Daily rainfall data is needed for estimating rainfed productivity. The total amount of rainfall during wheat season is very small for most of the major wheat producing areas (Figure 2). Since this data was not available for all locations, representative rainfall data was obtained for at least one location in each of the 21 agro-ecological zones demarcated by Sehgal et al. (1990). This rainfall was assumed to be same for all locations within that zone.

Soil. The geographical distribution of soils was obtained from the agro-ecological zones map of India (Sehgal et al. 1990). By overlaying a map of location of weather stations over this map, soil texture for each location was identified. The dominant textures of wheat growing regions are loam sand, sandy loam, black and red loam. For each of these textures, a representative water retention profile was used (A K Singh, IARI, 1991 pers. comm.). The soils of wheat growing regions are deep (Sehgal et al. 1990). In the present analysis, soil was assumed to be 150 cm deep and at 80% of the field capacity at the time of sowing.

The simulation model

In this case study, we have used WTGROWS, a crop growth simulation model developed to describe the effect of various climatic factors and their variability, soil characteristics, agronomic management and physiological factors on wheat growth, development, water and nitrogen use. The primary structure of the model is based on MACROS (Penning de Vries et al. 1989) as adapted for determining wheat production potential in South-East Asia by Aggarwal and Penning de Vries (1989). The model written at present in CSMP is largely explanatory and is run on an IBM PC/AT or XT compatible computer. It simulates daily dry matter production as determined by radiation intensity and maximum and minimum temperatures and limited by water and nitrogen availability. The model consists of eight submodels: crop development, photosynthesis, respiration, partitioning, crop growth, photosynthetic area, senescence, and grain yield. A soil water and nitrogen balance model is attached to determine the availability of water and nitrogen for crop use. Information on available soil water, soil physical properties, rainfall, vapor pressure and irrigation is required to compute water deficits. The model results are sensitive to agronomic management factors and genotype. The model has been satisfactorily evaluated in key wheat growing regions of India (unpublished results by the author).

Simulation

The potential and water-limited growth, development and grain yield of wheat were simulated for each of the 219 locations. Nine combinations of three sowing dates and three varieties varying in their duration were simulated. The maximum yield obtained among these treatments was considered potential/rainfed yield of that location. Yield for each district was derived by interpolation.

Geographical Information System (GIS)

Digitized maps comprising of district boundaries (polygons), soils, weather stations and agro-ecological region were stored in a GIS package ARC-INFO (ESRI 1987) together with polygon attribute data (potential and rainfed productivity, actual district yield, area). Results are presented as tables and maps using this package.

Results

Potential and water-limited yields

The potential yields varied from 2.0 t ha^{-1} to 8.0 t ha^{-1} depending upon the district (Figure 4). In 100 districts spread over the states of Punjab, Uttar Pradesh, Bihar, Assam, Rajasthan, and Madhya Pradesh, potential yields were 7 t ha^{-1} or more. Another 103 districts have a yield potential between 6 t ha^{-1} to 7 t ha^{-1}. The yield was between 5 t ha^{-1} and 6 t ha^{-1} in 41 districts mostly in middle latitudes and states of West Bengal and Madhya Pradesh. In most districts of Gujrat, Maharashtra and Orissa, yield potential was only 4 t ha^{-1} to 5 t ha^{-1}. The agroclimate of southern states of Andhra Pradesh, Karnataka, southern and coastal Maharashtra did not allow yield potential to exceed 4 t ha^{-1}. It can be summarized that potential yields increased with latitude and elevation. In almost 50% districts of India, agroclimate is favorable for wheat growth and development. Yield potential in these districts is at least 6 t ha^{-1}.

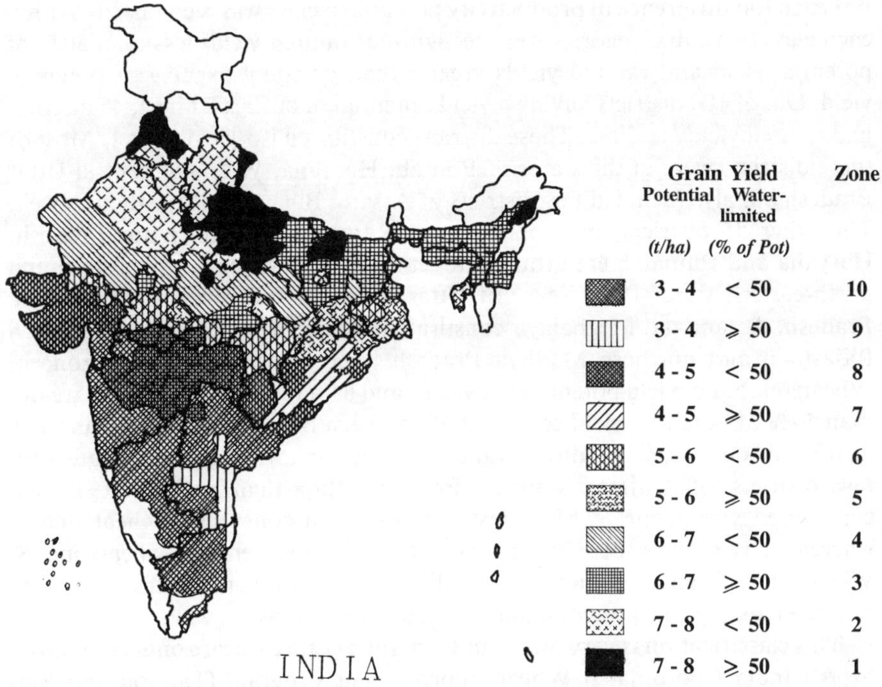

Figure 4. Iso-yield wheat zones of India based on potential and water-limited productivity. Blank areas were not simulated.

The above trend in potential yield associated with latitude were not necessarily maintained in rainfed productivity. Rainfed yield was always much lower than potential yield in all districts. It was highest (4 t ha^{-1}) in 33 districts of Punjab, Bihar and Assam. Most districts of northern India (164) had rainfed yield between 3 t ha^{-1} and 4 t ha^{-1}. For the same latitude, compared to western districts, yield was slightly higher in eastern districts. The rainfed yields in several districts of West Bengal, Rajasthan, Madhya Pradesh and Orissa were between 2 and 3 t ha^{-1} only. In all southern states, rainfed yields were lower than 1.5 t ha^{-1}.

Practically in all districts, potential as well as rainfed yields were higher than actual yields indicating considerable yield gap. There were 64 districts where the gap between potential and actual yield was at least 5 t ha^{-1}. There is significant irrigated wheat cultivation (total of 1.18 Mha) in these districts. In addition, another 8 Mha of irrigated wheat area in 59 districts have a yield gap between 4 and 5 t ha^{-1}. Corrective measures in these districts shall be more rewarding.

Iso-yield wheat zones

Based on potential and rainfed productivity, districts were classified into iso-yield zones (Figure 4). Each zone represents a homologous agro-environment.

For each ton difference in productivity potential a class was recognized. Within each class two sub categories were recognized: rainfed yields less than 50% of potential yield and rainfed yields greater than or equal to 50% of potential yield. Out of 100 districts having a yield potential of at least 7 t ha^{-1}, 45 districts had rainfed yields $\geq$ 50%. These districts constituted iso-yield zone 1. Most of these districts were in the states of Punjab, Haryana, West and central Uttar Pradesh but also included few districts of Assam, Bihar and Rajasthan as well. The other 55 districts, part of states of Rajasthan, Bihar, Uttar Pradesh, Haryana and Punjab were grouped together in wheatzone 2. All eastern and north-eastern states Bihar, eastern Uttar Pradesh, few districts of Madhya Pradesh, Assam and Meghalaya constituted zone 3. The districts of southern Rajasthan and northern Madhya Pradesh largely constituted wheatzone 4. Wheatzone 5, i.e. yield potential between 5 and 6 t ha^{-1} and rainfed yield greater than 50% of potential yield consisted of southern Bihar, West Bengal and few districts of Orissa and Madhya Pradesh. In the same yield potential category, few districts of Gujrat, Madhya Pradesh, Rajasthan and West Bengal constituted wheatzone 6. Most districts of Orissa constituted wheatzone 7, whereas large parts of Gujrat and Maharashtra were in wheatzone 8. Wheatzones 9 and 10 had mostly districts of southern states and were characterized by low potential and rainfed wheat yields.

This classification is very different from the existing wheat zones recognized by All India Coordinated Wheat Improvement Program (Tandon and Rao 1986). For example, northern plains zone of AICWIP can be further subdivided into two productivity zones based on rainfed productivity. Similarly, central zone has areas with potential wheat productivity varying from 5 t ha^{-1} to 7 t ha^{-1}. These zones although supposedly based on agro-ecological considerations do not reflect the true state of environment in terms of crop productivity. There is thus a need to critically examine if the proposed productivity based classification will provide any advantage in terms of wheat research organization in the country.

Production potential

Production potential of each wheat zone can be estimated by multiplying the potential yield of that zone with the area under wheat cultivation. For estimating this we have considered only those districts where there is a significant ($\geq$ 10000 ha) wheat cultivation and which are largely irrigated ($\geq$ 80%). Eighty three districts covering an area of 12.3 Mha under wheat cultivation qualify this criteria. Most of these districts are in the states of Punjab, Haryana, Uttar Pradesh and Rajasthan but a few districts are in the states of Gujrat, Maharashtra and Madhya Pradesh. At present, wheat production in these districts is 35 Mt. It was calculated that these districts have a production potential of 88 Mt. Districts (62) of wheat zones 1 and 2 alone accounted for 79 Mt of potential production. Thus, it is clear that agro-climate of these regions is favorable for production of at least two times the current wheat production. It is important, however, to emphasize that these estimates are based on the assumption that the wheat crop does not suffer from water, nutrition, disease stress or any other management factors. Such a situation is rare even in experimental fields. Grain yield and production of any region is likely to be lower once the effect of soil type and water availability in the individual zone are also considered. Nevertheless, these values do provide a clue to the magnitude of possible increase in wheat production.

Effect of climatic variability

An important aspect in evaluating the suitability of a crop is the stability of performance over the years. Wherever historic weather data for several years is available, year-to-year variation in grain yields can be simulated. Frequency distribution analysis of this data set provides the probability of obtaining a given grain yield as well as the effect of climatic variability on productivity of a variety. The potential and water limited growth of wheat was simulated for 10 to 20 years for New Delhi, Hyderabad, and Coimbatore (Aggarwal 1991). Frequency distribution analysis indicated that in 50% of the years, potential grain yield exceeded 7.3, 5.0, and 3.2 t ha^{-1} at New Delhi, Hyderabad and Coimbatore, respectively. For these locations the differences in grain yields between 75% and 25% probability levels were 0.95, 0.74 and 0.44 t ha^{-1}, respectively. The differences expressed as a proportion of yield at 50% probability level (an index of stability) were only between 13% and 17%, indicating small effect of climatic variation on potential grain yields in different parts of India. This stability of performance in grain yield is apparently due to clear weather during post-rainy growing season.

Unlike potential yields, year-to-year variation in rainfed yields was much greater. At New Delhi, Hyderabad and Coimbatore, there was a 50% probability of exceeding yield levels of 4.2, 2.4 and 2.9 t ha^{-1}, respectively. These yields were 42, 52 and 9% lower than the yields of irrigated wheat at the same probability level.

Effect of increase in carbon dioxide and temperature

The effect of 450 ppm CO_2 and 2°C increase in temperature on potential yield was simulated for all locations. The grain yields were severely reduced in all districts by this scenario of climate change. The results indicated several districts currently producing wheat may become unsuitable for wheat production even in irrigated conditions. The most productive areas such as Punjab, Haryana and West Uttar Pradesh may suffer a yield penalty of almost 35%.

Conclusions

It is concluded that crop growth simulation models together with GIS can significantly improve agro-ecological characterization. These methods can analyze the effect of climatic variability and climatic change as well. The methods presented in this paper quantitatively analyse the agro-ecological properties of different land evaluation units, in terms of soils, climate and crops to determine potential and water-limited productivities. From these results, one can estimate production potential, yield gap and requirement of different inputs to achieve potential yields in different agro-climatic regions. Regions with greater potential for yield increase with a given amount of input can be identified. Once databases are stored in a GIS system, this methodology can rapidly generate new iso-yield zones, or modify old ones as new concepts or technology such as new varieties become available.

The case study on wheat has shown that using simulation models and GIS, different land evaluation units can be grouped together in iso-yield zones. These crop productivity zones are more realistic expression of homologous agro-climatic regions. Potential yield, yield gap and potential production for each district were quantitatively determined which indicated regions with greater scope of improvement. Such information is not normally available through a conventional agroecological characterization approach. However, determination of optimal wheat yield for each district based on input/output efficiency will be more realistic than the potential yield as expression of the scope of improvement.

Acknowledgments

Drs. S.K. Sinha, A. K. Singh and Naveen Kalra of the Indian Agricultural Research Institute, New Delhi, India are also involved in developing the simulation model and building up of databases. Drs R. K. Midha and P.S. Acharya of DST, Govt of India provided GIS facilities.

References

Aggarwal P K (1991) Simulating growth, development and yield of wheat in warmer areas. Pages 429–446 in Wheat for the nontraditional warm areas. Saunders D A, (Ed.), CIMMYT, Mexico D.F., Mexico. 549 p.

Aggarwal P K, Penning de Vries F W T (1989) Potential and water-limited wheat yields in Southeast Asia. Agric. Sys. 30:49–69.

Angus J F (1989) Simulation models of water balance and growth of rainfed rice crops grown in sequence. ACIAR Monograph, Australia.

ESRI (1987) ARC/INFO: The geographical information system software (users guide): Environmental systems research Institute, California, USA.

FAO (1983) Guidelines: Land Evaluation for Rainfed Agriculture. Soils Bull No 52. FAO, Rome, Italy. 237 p.

Ghosh S P (Ed.) (1990) Agro-climatic zone specific research. Indian perspective under NARP. Indian Council of Agricultural Research. New Delhi, India.

Government of India (1987) Agro-climatic regions planning: An overview. Planning Commission, New Delhi, India.

Jones C A, O'Toole J C (1987) Application of crop production models in agro-ecological characterization: Simulation models for specific crops. Pages 199–209 in Bunting A H, (Ed.), Agricultural environments- characterization, classification and mapping. CAB International, UK.

Nix H (1987) The role of crop modeling, minimum data sets and geographical information systems in the transfer of agricultural technology. Pages 113–117 in Bunting A H, Ed., Agricultural environments- characterization, classification and mapping. CAB International, UK.

Oldeman L R, Frere, M (1982) A study of the agroclimatology of the humid tropics for south-east Asia. Tech Rep FAO/UNESCO/WMO Interagency Project on Agroclimatology, FAO, Rome, Italy.

Penning de Vries F W T, Jansen D M, Ten Berge H F M, Bakema A H (1989) Simulation of ecophysiological processes in several annual crops. Simulation Monographs No. 29. PUDOC, Wageningen, The Netherlands and IRRI, Los Baños, Philippines. 271 p.

Sehgal J L, Mandal D K, Mandal C, Vadivelu S (1990) Agro-ecological regions of India. Tech. Bull. NBSS Publ. 24, 73p.

Tandon J P, Rao M V (1986) Organisation of wheat research in India and its impact. Pages 1–33 in Tandon J P, Sethi A P, (Eds.), Twenty-five years of co-ordinated wheat research. Wheat Project Directorate, AICWIP. New Delhi, India. 287 p.

Van Lanen H A J, Van Diepen C A, Reinds G J, De Koning G H J, Bulens J D, Bregt A K (1992) Physical land evaluation methods and GIS to explore the crop growth potential and its effects within the European Communities. Agric. Syst. 39:307–328.

An agroclimatic approach to agricultural development in India

R.K. SINGH[1] and D.N. SINGH[2]
[1] *C.S.A. University of Agriculture and Technology, Kanpur 208002, India*
[2] *Directorate of Pulses Research (ICAR), Kanpur 208024, India*

Key words: agricultural development, agro-climatic characterization, India, land evaluation, planning, research priorities, socio-economic characterization, subtainability, yield potential

Abstract
After independence, successive planning efforts for nearly four decades have resulted in many successes as well as shortcomings in Indian agriculture. More than three-fold increase in production has been achieved by commodity-oriented programmes involving major technological breakthrough. This resulted in heavy exploitation of natural resource stock without looking at alternative perspectives. Thus major associated problems like linear flow of natural resources (depletion or degradation) and issues related to equity, in terms of inter-regional developmental disparities and inter-generational injustice, have recently received attention and recognition by Indian policy makers and planners. While realising these systemic problems, the application of systems approaches offered suitable option to the improvement of problem situations in agriculture. Accordingly, the Planning Commission of India, framed an agroclimatic approach that assesses potential and problems in agroclimatically uniform resource regions. This paper discusses in depth the agroclimatic approach covering aspects of (i) agroclimatic information system and its application in agricultural development; (ii) characterisation of agroclimatic zones; and (iii) agroclimatic agricultural regional planning in India. Also, emphasis has been placed on understanding of broader environment of systems approaches dealing with not only farming systems research that forms an integral part of a long-term regional wide development effort, but also covers the agricultural technology management system in agroclimatic zone specific research efforts in India.

Introduction

During the last three and a half decades, efforts on planning had led to notable successes as well as few shortcomings in the development of agriculture sector in India. Spectacular increase in food grains production, from 55 to 170 million tonnes, particularly rice and wheat, has taken place. In the Sixties, technological breakthrough in breeding programmes helped to create the new high yielding cultivars. These new types are highly responsive to chemical fertilizers and water under favoured agroclimatic regions and realize high yields. Adopting a change in strategy from an area approach to commodity-oriented mandates with some special production drives for rice and pulses, the Technology Mission for oilseeds and the National Project for Rainfed Agriculture among others have been launched during 1980's. However, the shift to exploitative agriculture from a traditional one, without proper understanding of the various consequences for each of the changes introduced

F.W.T. Penning de Vries et al. (eds.), Systems Approaches for Agricultural Development, 111–125.

makes that country is confronted with an array of environmental and social problems. Major problems are the linear flow of natural resources (depletion or degradation) and issues related to intergenerational equity. The first involves conserving the ecological foundations of sustainable advances in biological productivity. Problems arising from exploitative agriculture, thus necessitated an effective approach to agricultural development in the country. Adopting a holistic view point, the agroclimatic system approach has been accepted in India for regional planning and agricultural development. Maintaining the momentum of agricultural development through an optimal blending of agricultural and allied activities is a major aspect of the Eighth Five Year Plan (1992–1997). In this paper, it is attempted to elaborate the concepts and use of the agroclimatic approach in agricultural planning along with its relevance for sustainable agriculture in India. The objectives are: i) to present a conceptual framework of agroclimatic information system and its applicability in agriculture; ii) to outline certain processes used by the Planning Commission; iii) to relate agroclimatic information with the strategic planning adopted for the Eighth Five Year Plan; and iv) to pin-point specific agricultural research under the National Agricultural Research System (NARS) for seeking solution to problems of sustainability in India.

The agroclimatic information system and its application in agricultural development

The agroclimatic information system is a system that incorporates the physical properties of atmosphere-land surface-soil and hydrology-vegetation interactions into a tool for planning and management of agricultural products (Unninayar 1989). Farmers, commodities dealers, water managers, and others around world have sought to plan agricultural production through the use of models, requiring agroclimatic information. The agroclimatic information system is intended to help NARS to improve the production efficiency and also to target their research efforts according to their specific agroclimatic situation. Further, it will assist in modifying the impact of new technology in each country. The goal of such a system is to achieve an optimum sustainable yield through better use of weather and climatic information and to permit for skillful crop production. This knowledge is also important to quantify the degree of acceptable risk in sustained development of production systems including forestry, fisheries and aquaculture. At the country level, use of an agroclimatic information system in the analysis of sustainability has already achieved an unique importance (Table 1). Figure 1 is a scheme indicating the functional units of an agroclimatic approach. It outlines procedures and interactions involved in implementing an agroclimatic system for short or long-term economic benefits of a particular zone or country. While describing the functions of this approach, Unninayar (1989) indicated the need for the following:

Table 1. Analysis of sustainability.

Level of analysis	Typical characteristics of sustainability (cumulative)	Typical determinants
Field production unit	Productive crops and animals; conservation of soil and water; low levels of crop pests and animal diseases.	Soil and water management; biological control of pests; use of organic manure, fertilizers, pesticides, crop varieties and animal breeds.
Farm	Awareness by farmers; economic, social needs satisfied; viable production systems.	Access to knowledge, external inputs and, markets.
Country	Public awareness; sound development of agro-ecological potential;conservation of resources.	Policies for agricultural development; population pressure; agricultural education, research and extension.
Regional/ continent/world	Quality of natural environment; human welfare and equity mechanisms; international agricultural, research and development.	Control of pollution; climatic stability; terms of trade; distribution.

Source: FAO 1989.

- observatories to measure various climatic elements (like minimum and maximum temperature, precipitation, and solar radiation);
- a biological and geographical monitoring system to determine the original character of land, soil and vegetation;
- an assessment system that helps to determine the optimum land use and agricultural practices;
- a data processing and information dissemination system to guide operational as well as planning decisions, and,
- research on climatic variables and on establish the relation of weather and climate to soil and hydrology for various crop varieties.

Under more diverse environmental conditions a wide meteorological network is required to generate data on various elements of climate and weather. India has achieved success in establishing well equipped weather stations under the Indian National Satellite System for remote sensing. Weather reports are prepared, after rigorous scientific scrutiny with the newly installed Cray supercomputer of Indian Meteorological Department and the Department of Science and Technology (DST), and are providing an essential service to farmers and other people. Over the 30 months of its operation, the Agroclimatic Regional Planning Unit (ACRPU) has opened several links with the Ministry of Agriculture, and several of its specialized organisations for secondary data as well as some channels for primary data. Agrobase is one of the major system

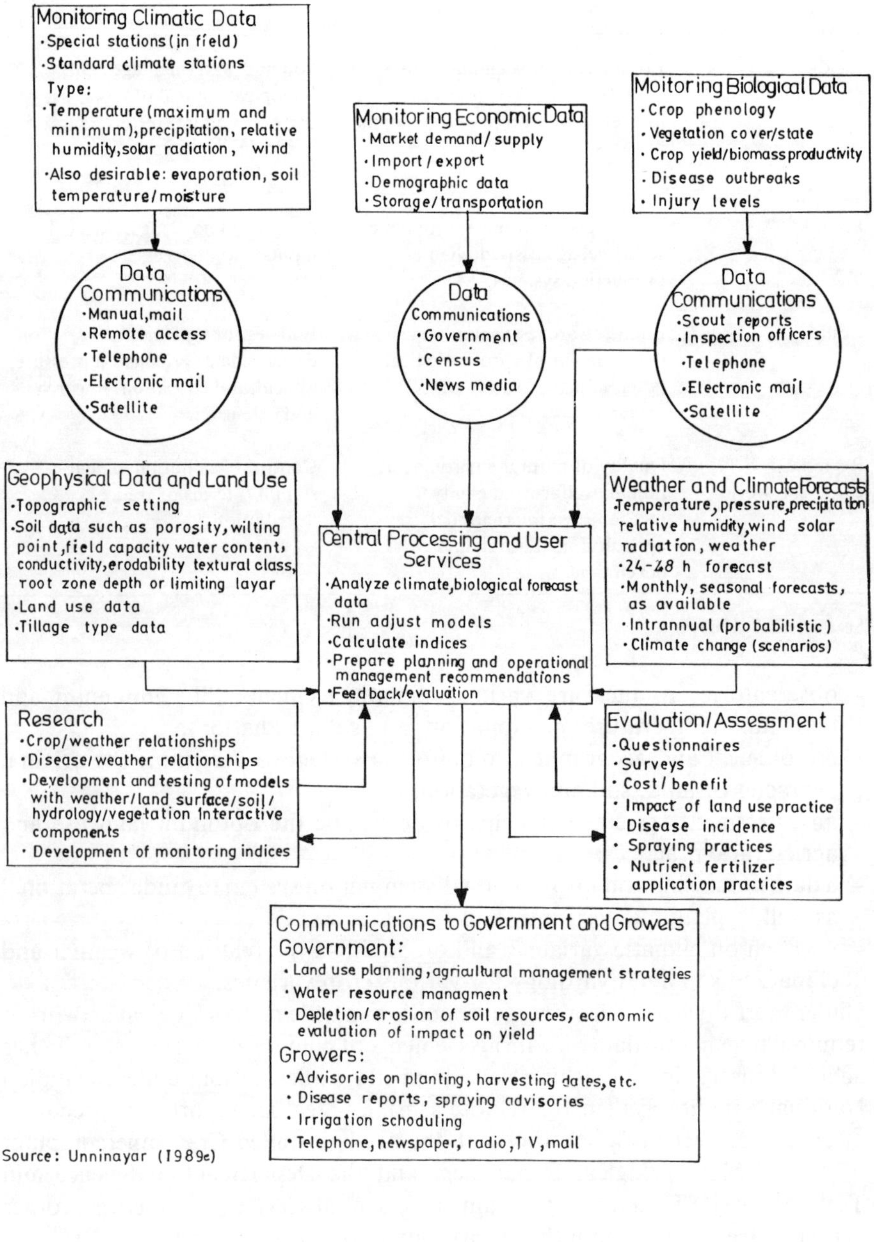

Figure 1. The elements of an agroclimatic approach.

developed by ACRPU. One part of this system presents the profile of the agroclimatic (sub-) zones providing information, interactively, on subjects relevant to agricultural planning like crop, land use, natural resources, inputs and socio-economic parameters. The other part derives results from the analytical sub-system and provides information on various resource and performance indicators, as well as targets and projections. The system is built around a comprehensive database and a query system permitting database updating.

Agroclimatic characterisation

Earlier approaches

Many attempts have been made and criteria suggested in the literature to classify the land area into agroclimatic regions suitable for a certain range of crops and cultivars (FAO 1983). Carter (1954) demarcated broadly six climatic regions in India ranging from arid to per-humid, using the criteria of Thornthwaite system of climatic classification. The Indian Council of Agricultural Research (ICAR) (1979) has grouped the country into 120 agroclimatic zones based on rainfall distribution, existing cropping pattern and administratively functional units for conducting agroclimatic zone specific research at various agricultural universities in different parts of the country. Krishnan (1988) delineated the area into 40 soil-climate zones based on major soil types and moisture index. The Planning Commission of India (1989) divided the country into 15 agroclimatic zones based on physical conditions, topography, soil, geological formation, rainfall pattern, cropping system, development of irrigation and mineral resources at a district level. Using a classification system based on physiography, climate, soil, and effective growing days, Sehgal et al. (1990) identified 21 homogeneous agroecological zones and 54 sub-zones. These, classifications differed widely as each of the studies applied different criteria. Since agroclimatic classification completed in 15 zones by the Planning Commission is too broad to serve the purpose of agricultural planning, the present discussion mainly follows a new approach of characterization of zones and provides information about the planning exercise adopted during Eighth Five Year Plan.

Genesis of current approach for regionalisation

After achieving encouraging results in mid-term appraisal of the Seventh Plan (1985–'90), the Planning Commission and the Government of India reoriented the planning strategy to be used to a systems approach for agricultural development. It was developed on the basis of an analysis of soil conditions, water availability, technological package and the existing level of development and rate of growth. Hundred sixty nine districts in 14 states, covering five major crops were identified and selected for exploring the production potential. At the same time agricultural planners also realised the need for strategies which are dependable and relying on alternative perspectives.

The approach involved in regionalisation

Encouraged by the results of earlier studies (FAO 1985; Sengupta 1968), the Planning Commission, in coordination with officials of the Census of India, Central Water Commission, ICAR, and other professional bodies and scholars, has delineated agroclimatic regions. They form the basis for agricultural planning. Important attributes like soil types, climate, rainfall and water condition were identified and used. The primary regional zoning occurs on the basis of topography and five major ecozones have been identified:

- The Himalayas and associated Hills;
- The Peninsular Plateau and Hills;
- The Northern Plains consisting of Indo-gangetic Plains;
- The East-coast Plains, and
- The West-coast Plains.

Criteria of homogenity in characteristics important for agriculture such as soils, geological formation, rainfall, cropping pattern, development of irrigation and farming systems, have been used to demarcate 15 agroclimatic zones: two in the Himalaya region, five in the Northern Plain, four in the peninsular Plateau and Hill, one in the East-coast Plain and two in the West-coast Plains; the 15th region covers Islands (Andaman, Nicobar and Lakshadweep islands).

Though these 15 zones broadly represent a particular agroclimatic situation and are indicative of agricultural potential, they will not stand the test of homogeneity when micro-variables are considered for the purpose of detailed operational planning. Keeping that in view, the 15 zones were divided into sub-zones on the basis of relevant agrometeorological characteristics namely soil type, climate (particularly temperature with its variation and rainfall with its distribution), and water demand and supply characteristics including quality of water and aquifer conditions. The 120 zones of ICAR's National Agricultural Research Project (NARP) were also taken into consideration, and thus 15 zones were finally sub-divided into 73 sub-zones to overcome the danger of over-generalisation.

In ICAR's (NARP) scheme, states are the indivisible units and they tend to cut across district boundaries. But in agroclimatic regional planning (ACRP), a district is the smallest unit for demarcation of zonal boundaries and tends to cut state boundaries. Fifteen agroclimatic zones along with salient characteristics are described in Figure 2 and Table 2 respectively. Among them, constraints and thrust areas differ. The valleys of the hill regions of Western and Eastern Himalayas constituting zones 1 and 2, receive adequate rainfall and have good soils. Temperate parts of Western Himalayas along with Sikkim and Darjeeling in Eastern Himalaya have inadequate moisture and low temperatures that restrict the crop growing period. Gangetic Plain areas,

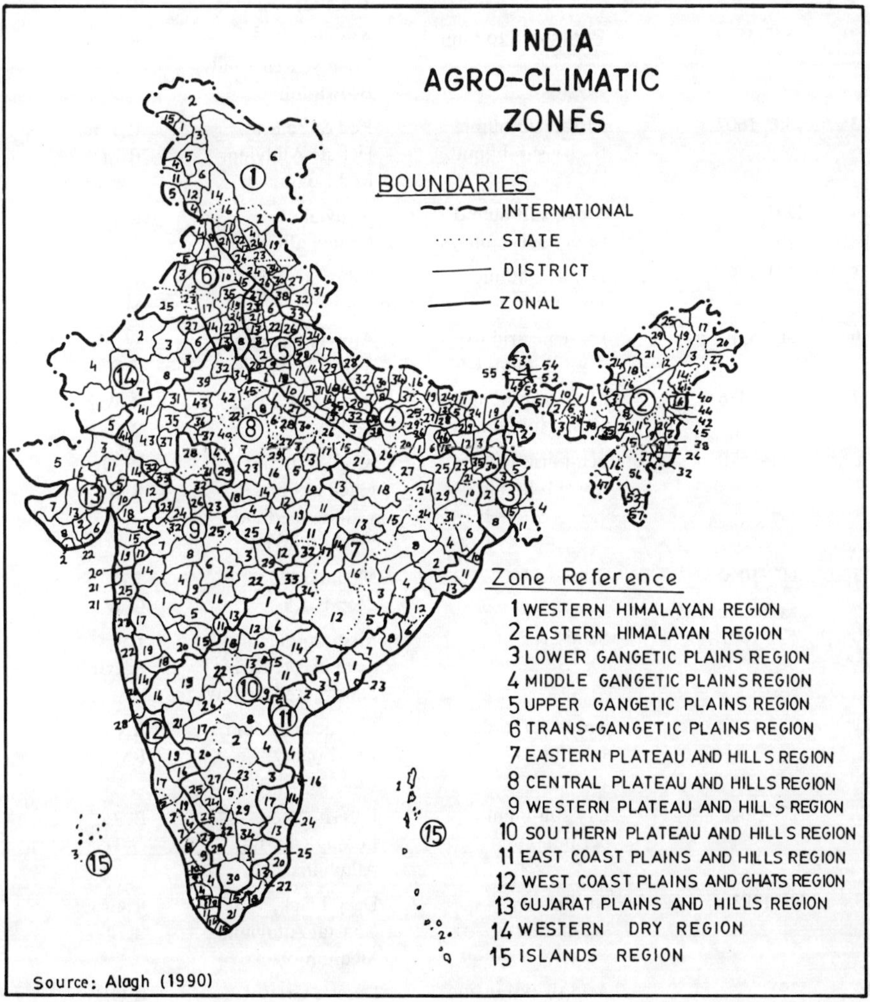

Figure 2. The 15 main agroclimatic zones and their subdivisions.

Table 2. Agroclimatic features of 15 zones in India.

Zone no.	Annual rainfall (mm)	Climate	Representative soil	Major crops
1	165-2000	Cold arid to humid	Brown hill alluvial Meadow, Skeletal,	W,M, R,P
2	1840-3528	Per humid to humid	Alluvial, Red Loamy, Red Sandy, Brown hill	R,M,J R,M
3	1302-1607	Moist sub-humid to dry sub-humid	Red & Yellow Deltaic Alluvium Red Loamy	R,J.W, R,M
4	1211-1470	Moist sub-humid to dry sub-humid	Alluvial, Loamy alluvial	R,W,M
5	721-979	Dry sub-humid to semi-arid	Alluvial	W,R, M,T
6	360-890	Extreme arid to dry sub-humid	Alluvial(recent), Calcareous, desert	W,M, R,S
7	1271-1436	Moist sub-humid to dry sub-humid	Red Sandy, Red, Yellow	R,W,M Ra
8	490-1570	Semi-arid to dry sub-humid	Mixed red and black red and yellow, medium black, alluvial	W,G,J, R,B
9	602-1040	Semi-arid	Medium black, deep black	J,B, C,W
10	677-1001	Semi-arid	Medium black, deep black, Red Sandy, Red Loamy	J,R, Ra,Gr
11	780-1287	Semi-arid to dry sub-humid	Deltaic Alluvium, Red Loamy Coastal Alluvium	R,Gr, Ra,J,B
12	2226-3640	Dry sub-humid per humid	Laterite, Red Loamy Coastal Alluvium,	R,Ra, G,To
13	340-1793	Arid to dry sub-humid	Deep black, Coastal Alluvium Medium black	R,Gr,C B,W
14	395	Arid to extremely arid	Desert, Grey, Brown	B,G, W,Ra
15	1500-3086	Humid	Desert, Grey, Brown	Co

B = Bajra; C = Cotton; G = Gram; Gr = Groundnut; J = Jowar; M = Maize; P = Potato; R = Rice; Ra = Ragi; R&M = Rapeseed and Mustard; S = Sugarcane; T = Tur; To = Tapioca; W = Wheat and Co = Coconut.
Data from Vaswani (1990).

demarcating zones 3 and 4, are characterized by high rainfall, alluvial and deltaic soils with good ground water. The Upper and Trans Gangetic Plains (zones 5 and 6) have rich alluvial soil with medium to low rainfall and a very good availability of water. The Plateau and Hill regions (zones 7, 8, 9 and 10) have relatively poor soils (steep slopes with high run-off) and medium rainfall. The Plain and Hill regions (zones 11, 12 and 13) harbour large variations in soil and rainfall. The Western Dry Region (zone 14) receives highly erratic and uncertain rain, and the deep ground water is scarce and saline. The Islands region (zone 15) has a typical equatorial type of climate with annual rainfall of over 3000 mm.

Interestingly enough, agroclimatic (sub-) zoning has helped the Planning Commission to illustrate the basic characteristics of agriculture in the region, its achievement, recent changes, and assessing about the needs to be done in the region within which agriculture operates. In essence, the present approach is being used in positive and normative way to serve as indicators of agricultural potential and its performance in a particular agroclimatic zone.

Agroclimatic planning

The Planning Commission has taken steps for effective operationalization of the agroclimatic regional planning in India. Zonal Planning Teams (ZPT's) were set up for each zone consisting of experts, voluntary organizations, a representative of financial institutions and the concerned officials headed by the Vice-Chancellor of a State Agricultural University in the zone. An ACRPU has been constituted at Ahmedabad to guide and coordinate activities of the ZPT's. In turn, ZPT's are also working in close coordination with the state agencies with necessary support from Union Government. Such integration gives useful results and provides information related to the Eighth's Plan. Thus with the involvement of scientists, development personnel and administrators, the procedure involves preparation of a zonal profile, development and operationalization of strategies at the (sub-) zone level.

Regional planning concerns primarily the initial stages. Agricultural activities are to be integrated later with non-agricultural programmes.

Concept and objectives

The agroclimatic regional planning approach aims at scientific management of regional resources to meet the needs of growing human population for food, fibre, fodder, and fuelwood without affecting the status of natural resources and environment. The Agroclimatic approach is thus treated as a means for effective natural resource management. The agricultural development will have to be achieved through an appropriate mix of crop production and allied activities including horticulture, forestry, animal husbandry, agroprocessing and a host of other combinations. Major objectives of this approach are:

- to deal with supply-demand balances which are useful to maintain the productive potential of natural resources on long term basis;
- to examine and compare the crop and non-crop based developmental possibilities for individual zones;
- to provide a sound basis for an investment proposal.

The planning approach

A number of factors influence the technical performance of farm production systems like rainfall pattern, soil fertility, temperature and solar radiation. These vary widely with latitude, and are broadly classified as hydrological, pedological and climatic constraints. At disaggregated levels (districts) resources become relatively more specific and their optimal use needs action oriented programmes for balanced development in agriculture. The performance of agriculture in an individual region is based on its resources, its capabilities and use-pattern along with other complementary inputs. Thus after analysing the potential limits of a region, planning decisions were suggested.

Balanced development in an agricultural economy requires the analysis of processes involved in resource attributes and their performance on inter-regional basis. It takes into account a set of selected indicators which relate to resource, economic, and demographic variables. A zone/sub-zone is characterized on the basis of twenty indicators. They include ten resource related variables such as sown area, forest area, land available for cultivation (all expressed on per capita terms); soil index, mean and coefficient of variation (CV) of rainfall; share of problem soil in a sown area and percent of irrigation in sown area; ground water balance per unit sown area and its available soil moisture. Some socio-economic variables like literacy, population density, share of male agricultural workers in total workers, share of unemployment and marginal workers in total work force are considered. Besides these, three performance variables are considered: yield of major crops and of food grain crops, and land productivity, and two input variables: bank credit and fertilizer consumption rate per unit land. The degrees of association between land productivity and other indicators such as sown area per capita, fraction irrigated, ground water balance, literacy, population density, male labour in agriculture, and fertilizer and bank credit used per unit area were statistically determined by correlation analysis.

In a typological study relevant variables were selected to develop a base for grouping of regions. This has shown different patterns of resource endowment and development (Basu and Rajagopalan 1990).

During evolvement of strategies for a region's resource management areas of intervention were identified for which relevant projects are formulated and diagnosed externalities (unpriced cost) have been considered. Thus, agroclimatic regional planning exercises have made substantial progress and have reached the stage of programme formulation in the country. The emerged strategies pertaining to agriculture and allied activities are described in brief below.

Outlines of strategies

Region 1: Western Himalayan region

- Intensification of soil and water conservation programme through watershed management with an integrated approach to forestry, horticulture, commercial crops and fodder for animals.
- Land-use planning based on the topographical concept that land with a slope upto 30% is suitable for agriculture on terraces, with 30–35% for horticulture and silvipastoral programmes, and above 50% for forestry.
- There should be high and low value crops for each sub-zone. High value crops include fruit trees, oil crops, flowers, medicinal and aromatic plants and plantation crops such as tea.
- Providing efficient market services, including diversified utilization of fruits and vegetables.

Region 2: Eastern Himalayan region

- Adoption of soil and water conservation measures for each watershed to avoid heavy run-off, massive soil erosion and floods in the lower reaches and basins.
- Prevention of shifting cultivation by evolving suitable cropping systems.
- Supplying complete package of inputs (e.g. quality seeds, saplings, fertilizers) coupled with marketing and processing, for each sub-zone.

Region 3: Lower Gangetic plains

- Increasing of rice crop productivity as it accounts for about 12% of the country's rice production but at a very low yield level (i.e. 1.1 t ha^{-1}).
- Major thrust on crop improvement, soil and water management besides institutional support and input delivery systems.

Region 4: Middle Gangetic plains

- Improvement of crop productivity along with supply of seeds of new varieties and transfer of technology.
- Establishing poultry, dairy and inland riverine fishery production.

Region 5: Upper Gangetic plains

- Increasing fertilizer use efficiency at par with the Trans Gangetic Plains.
- Reclamation of problem soils.
- Special programmes for vegetable and fruit production.
- Increasing milk production through cross breeding, balanced feeding and efficient management systems.

Region 6: Trans-Gangetic plains

- Major thrust involves improvement in water management, reclamation of salt-affected soils, efficient use of ground water, intensive crop diversification, increasing fodder production, expansion of area under fruits and

vegetables and better live-stock management.
- Intensification of advanced research.

Region 7: Eastern plateau and hills
- Improving water harvesting techniques to avoid run-off losses.
- Minor irrigation programmes.
- Intensification of horticultural development programmes along with strengthening of marketing and processing facilities.
- Improving acidic soil areas.
- Streamlining rehabilitation of degraded peripheral forests on a large scale.

Region 8: Central plateau and hills
- Developing integrated watershed management programmes in rainfed area.
- Substituting low value cereals crops with high value ones.
- Introduction of fruits and vegetables crop.
- Reclamation of ravine land for resettlement.
- Optimising the livestock development in integrated manner.

Region 9: Western plateau and hills
- Improving crop productivity particularly for sorghum and cotton.
- Expanding the hectarage of high value crops like oranges, grapes and bananas.
- Increasing ground water availability through dug and shallow tubewells in areas without other alternative for irrigation.
- Raising the milk production of cattle through cross breeding and balanced feeding.

Region 10: Southern plateau and hills
- Increasing water use efficiency through adoption of dry land technology.
- Intensifying fertilizer use in the rainfed agriculture through minikit demonstrations at a large scale.
- Fixing annually the target area for cultivation of mango, citrus, jujube, pomegranate, guava along with suitable incentives for plantation.

Region 11: East coast plains and hills
- Reclaming salt affected soils.
- Putting major thrust on fisheries with brood stock of tiger prawn, which proved successful in brackish water indoor tanks.
- Adopting multidimensional programmes such as desilting of tanks, renovation of irrigation systems by strenghtening bunds and structures, and improvement of field channels.
- Developing integrated irrigation-cum-drainage systems to save the rice crop from recurrent floods.
- Extending fishery programmes for inland waterways; increasing marine fish production, brackish water fisheries and aquaculture.

Region 12: West coast plain ghats

- Conserving rain water through construction of tanks and micro reservoirs.
- Increasing productivity of rice and millets and diversifying horticulture to crops such as mango, banana and coconut.
- Providing mechanized fishing boats for deep sea fishing.
- Raising brackish water fisheries for prawn culture and providing incentives for breeding and stock production of the prawn *Penaeus monodon*.
- Reclaming of saline lands after allotment to farmers with ownership.

Region 13: Gujarat plains and hills

- Developing integrated watershed programme and managing rain water harvesting.
- Developing suitable technology for dryland farming to maximise production per unit of land and water.
- Developing agroforestry and arid horticulture.

Region 14: Western dry region

- Increasing yield potential of major crops like millet, cluster bean, and moth bean in kharif (rainy season) and wheat and gram during rabi (post-rainy season).
- Increasing the use of scarce resource (like water) in correspondence with the possibilities in a fragile ecosystem and a harsh climate.
- Increasing tree cover to check desertification; provide fodder to livestock; meet the fuel need, and provide timber for implements.

Region 15: Islands region

- Crop improvement, efficient water management and fisheries.

These zone specific strategies are of great importance to establish both short-term and long-term perspectives for sustained agricultural development in the country.

Agroclimatic zone specific agricultural research in India

ICAR (NARP) and State Agricultural Universities are joining efforts on agroclimatic characterization of key research sites to develop an ecosystem-based framework for studies on farming systems. This aims at identifying problems dealing primarily with adaptation of existing agricultural research to provide technology relevant to low resource levels and low levels of external inputs (Shaner et al. 1982). Encouraged by achievements of phase I (1979–1985), phase II of NARP (1986–1993) has extended research mandates, covering horticulture, agroforestry, animal-drawn farm implements, animal nutrition and irrigated farming. It is clearly an action-research approach, with the researcher acting as a facilitator of the learning processes of the farmer participants, and focusses on agriculture as a human activity system (Bawden et al 1985). This calls for adoption of a Farming Systems Research approach in which nature of the problem can be systematically investigated so that the approach becomes an integral part of long-term, region-wide development efforts. Here emphasis lies on developing the potential of a (sub-) region in which technology provides a starting point. Thus indigenous knowledge systems, be they traditional or advanced, have developed over time and reflect the agroecological and agroeconomic condition of that zone, as well as the technical managerial skill of the farm population (UNDP 1991). But the entire research work in support of agricultural planning and development must operate within the framework of the agroclimatic approach, delineated by the Planning Commission in India.

However, technology alone can not induce the full scope of zonal development. A range of institutional involvement, supporting policies, and infrastructure investment must occur, if it has to develop and the benefits are to be spread widely among the rural population (Cummings Jr 1990). Agricultural Technology Management System (ATMS) a broader environment of systems approaches, can provide guidelines (Elliott et al. 1985). It brings together all institutions, individuals and their interdependent relationships aimed at generation, assessment, and diffusion of improved agricultural technologies.

Conclusion

Integration of research strategies and public policies evolved by Planning Commission for specific agroecological zone can play major role in agricultural development. The provision of a fair deal to regions that are economically and ecologically disadvantaged permits sustained development. However, the most important contribution of the agroclimatic systems approach is that it can make through optimizing inputs at a national level, absorbing risks evolved in effective management of resources, and conservation of resources for achieving intergenerational equity. Though the efforts of Planning Commission have helped already in elucidating several aspects basic to agricultural performance of a region for normative use, many vital but still not quantifiable indicators demand attention: reversibility of degradation, the critical threshold of decline in natural resource stock, and the level of genetic diversity. These challenging tasks need better understanding for increased and sustainable production and land use.

References

Alagh Y K (1990) Agroclimatic zonal planning and regional development. Indian J. Agric. Econ. 45(3):248.

Basu D N, Rajgopalan V (1990) Agroclimatic regional planning: regional indicator and topologies. Indian J. Agric. Econ. 45(3):271–272.

Bawden R J, Raymond L I, Macadan R D, Pancham R G, and Valintine I (1985) A research paradigm for systems agriculture. Page 39 in Remenyi J V (Ed.) Agricultural Systems Research for Developing Countries. ACIAR Proceedings No.11. Australian Centre for International Agricultural Research, Canberra, Australia.

Cummings R W (Jr.) (1990) Agricultural technology management: Draft guidelines. Page 16 in Echeverria R G (Ed.) Methods for Diagnosing Research System Constraints and Assessing the Impact of Agricultural Research Vol. I. International Service for National Agricultural Research, The Haque, The Netherlands.

Carter D B (1954) Climate of Africa and India according to Thornthwaite's classification. The John Hopkins University, Publication in Climatology 7(4).

Elliott H (1990) Applying ATMS approaches in widely different systems: Lessons from ISNAR's experience. Page 33 in Echeverria R G (Ed.) Methods for Diagnosing Research System Constraints and Assessing the Impact of Agricultural Research Vol. I. International Service for National Agricultural Research, The Haque, Netherlands.

FAO (1983) Guidelines: Land evaluation for rainfed agriculture. Soil Bull. No.52. FAO, Rome, Italy. 273 p.

FAO (1985) Towards improved multi-level planning for agricultural and rural development. FAO economic Social Development paper 52. Rome, Italy.

FAO (1989) Sustainable agricultural production: Implications for international agricultural research, FAO Research and Technology Paper No.4. Rome, Italy. 5 p.

Ghosh S P (1988) Philosophy, mandate and experiences of NARP. Pages 3–9 in Venkateswarlu K (Ed.) Reading Material cases. Agricultural Research Project Management; NAARM, Hyderabad, India.

Government of India (1989) Agroclimatic regional planning. An overview, Planning Commission, New Delhi. Government Printing Press, New Delhi, India.

Krishnan A (1988) Delineation of soil climatic zones of India and its practical application in agriculture. Fertilizer News 33(4):11–19.

Sehgal J L, Mandal D K, Mandel C and Vadivelu S (1990) Agroecological regions in India, NBSS Bull. Publ. 24:3–5.

Sengupta P (1968) Agricultural regionalization of India: Agricultural regions and their use efficiency. Pages 101–138 in Sengupta P, Sdasyuk G (Eds.) Economic Regionalization of India: Problems and Approaches, Vol. I, published by Registrar General of India, New Delhi, India.

Shaner W W, Philip P F and Sehmehl W R (1982) Farming Systems Research and Development: Guidelines for developing countries. Boulder Westview Press Colorado, USA. 16 p.

Unninayar S (1989) Basic data requirements of an agroclimatic system. in Climate and Food Security. International Rice Research Institute, Manila, Philippines.

United Nations Development Programme (1991) Programme advisory notes on agricultural extension: Technical advisory division; Bureau for programme policy and evaluation, New York. 2 p.

Vaswani L K (1990) Agroclimatic regional planning: Concept and approach. Fertilizer News, 35(6):12–13.

Optimising harvest operations against weather risk

G.Y. ABAWI
Agricultural Engineering Section,
Queensland Department of Primary Industries,
P.O. Box 102, Toowoomba 4350, Australia

Key words: Australia, climate variability, crop duration, GRAIN HARVESTING MODEL, grain loss, grain quality, harvest management, harvesting, simulation, weather forecast, wheat

Abstract

Most wheat in Australia is harvested at a maximum moisture content of 12% wet basis (w.b.). Delaying harvest until the grain is naturally dried to 12% results in yield and quality losses. Losses due to rain on the mature crop cost the grain industry around \$A 30 million annually (\$A 1 = \$US 0.75). These losses are significantly higher in the northern wheat belt of Australia due to a dominant summer rainfall which coincides with the harvest period. If a wet season can be predicted, growers could reduce grain losses by implementing strategies such as early harvesting, contract harvesting, additional grain drying and harvesting at a faster rate.

Weather prediction is possible several months ahead by the analysis of sea surface temperature and air pressure differences between Tahiti and Darwin, as identified by the Southern Oscillation Index. Long range weather forecasts could enable farmers to maximise profits by altering their management practices according to seasonal variations in rainfall.

This paper examines the interactions between weather, crop, machinery selection and management practices during wheat harvesting in northern Australia. A simulation model was used to investigate the effect of harvesting at high grain moisture content, then drying artificially to reduce field and quality losses. The model predicts that additional returns of up to \$A 140 ha^{-1} are possible by adopting long term strategic harvesting decisions in the northern wheat belt. Returns may be further increased by \$A 20 to \$A 40 ha^{-1} through optimising harvest operations during each season. Optimum machinery capacities and harvest strategies for a range of crop areas are outlined.

Introduction

Wheat is Australia's most important crop in terms of production and export. Annual wheat production exceeds 12 Mt and is worth about \$A 2.2 billion dollars in export earnings. Australia exports 70 to 80% of its wheat.

Wheat production in Australia has steadily declined over the past seven years. This is attributed to depressed world commodity prices, lower overseas demand, higher production costs and a move by growers away from traditional wheat farming to alternative crops. While demand for average quality wheat has declined, demand for premium grades of wheat remains high and exceeds production.

The diversity of wheat varieties, soil fertility and seasonal conditions encountered throughout the Australian wheat belt enable a wide spectrum of recognised wheat types to be produced. Wheat is classified into four main

F.W.T. Penning de Vries et al. (eds.), Systems Approaches for Agricultural Development, 127–143.

categories: Prime Hard (PH); Australian Hard (AH); Australian Standard White (ASW); and General Purpose (GP) which includes weather damaged wheat. Each classification reflects grain quality in terms of grain hardness, protein content, milling performance and physical dough properties. Typical grain prices are: PH \$A 170 t^{-1}, AH \$A 160 t^{-1}, ASW \$A 150 t^{-1} and GP \$A 120 t^{-1}.

One of Australia's strengths in the market has been the range of wheat grades it has been able to offer. However, in the competitive world market, the marketability of lower grade wheat is conditional on being able to satisfy the demand for premium grades. Only 6% of Australian wheat is prime hard and it is estimated that the production of PH wheat can double without affecting current pricing levels. Based on the present pricing structure for different categories of wheat, the survival of some wheat farming enterprises in the northern region of Australia will depend on their ability to produce PH wheat.

Pre-harvest sprouting of grain is a major cause of wheat being downgraded to GP level. Greater than 2% sprouting in grain delivered to wheat receiving points results in a GP classification and a financial loss of up to \$A 60 t^{-1}. In 1983–'84, a record 20 Mt of wheat was produced in Australia but over 20% was downgraded to GP due to sprouting and weather damage.

The northern wheat belt (Queensland and northern New South Wales), which produces over 80% of the Australian premium grades (PH and AH) of wheat, also has the highest percentage of GP wheat (Table 1). This is due to summer dominant rain on the mature crop. Table 1 shows that in the southern and western wheat growing states where rainfall is winter dominant the percentage of GP wheat is considerably lower. Figure 1 shows the intake of GP wheat in Queensland plotted against October–November rainfall for the past 20 years. The data shows a significant correlation between summer rain and grain quality in this region. With increasing demand for premium quality wheat on international markets, it is clear that this level of weather damage is significant and has a major impact on the economics of Australia's wheat industry.

Table 1. Average wheat intake by class for each state (1970-1990), expressed as a percentage of total production.

	PH	AH	ASW	GP
New South Wales	15.7	24.2	48.9	11.1
Victoria	–	4.4	88.4	7.1
South Australia	–	18.9	75.3	5.7
Western Australia	–	4.1	89.6	6.3
Queensland	27.7	33.3	23.4	15.5
AUSTRALIA	6.3	14.1	70.6	9.0

Source: Australian Wheat Board.
PH - Prime Hard; AH - Australian Hard; ASW - Australian Standard White; GP - General Purpose.

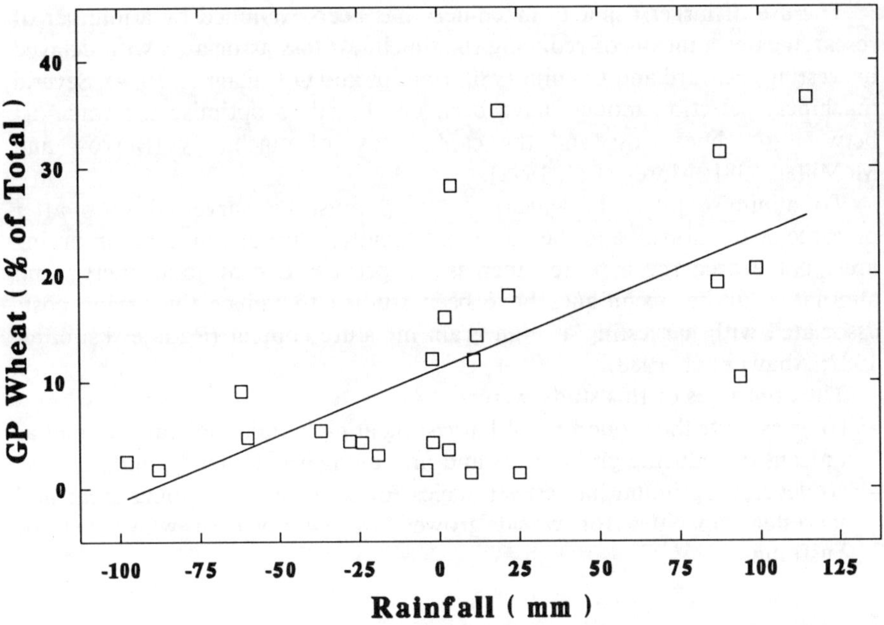

Figure 1. Correlation between intake of GP wheat in Queensland and summer rainfall (October–November). Above and below average rain are shown as positive and negative respectively.

In addition to losses in grain quality, delayed harvesting also results in yield losses. Growers without grain drying facilities must wait until the grain moisture content is suitable for storage before they harvest. The safe storage life of wheat is dependant on the ambient air temperature and grain moisture content. As temperatures during harvesting in Australia are significantly higher than those experienced in many other countries, the recommended moisture content for storage is comparatively lower. Historically, growers have been required to deliver wheat at 12% moisture content (w.b.). At this moisture content the grain is less susceptible to damage due to mould and insects, hence the 12% moisture limit has been adopted as an industry standard throughout Australia. Boxall and Calverley (1986) have, however, shown that the susceptibility of stored grain to damage is correlated to the equilibrium relative humidity and not to the grain moisture content. In some parts of Australia it is possible to store grain at moisture contents of up to 14% (w.b.) without risk of grain deterioration.

The limitation of 12% moisture content results in considerable delays to harvesting. Grain moisture content regularly exceeds 12% (w.b.) during harvest. The formation of overnight dew and high humidity early in the day can often delay harvest until midday. Moist cool sea breezes frequently restrict harvesting to only a few hours per day in coastal wheat growing areas. Yield and quality losses are time dependent and therefore any delay in harvesting increases the level of these losses.

The use of larger capacity machinery has been examined by a number of researchers as a means of reducing the timeliness loss associated with delayed harvesting (Edward and Boehlje 1980; Phillips and O'Callaghan 1974). Several machinery selection models have been developed to optimise the trade-off between timeliness loss and the capital cost of machinery (Burrow and McMillan 1981; Morey et al. 1972).

To minimise losses in wheat, the crop must be harvested soon after physiological maturity and then dried artificially. The costs incurred in drying must not exceed the expected increase in profits. Use of solar energy and alternative drying techniques have been studied to reduce the drying costs associated with harvesting at high grain moisture content (Radajewksi et al. 1987; Abawi et al. 1988).

The objectives of this study were:

- To investigate the economics of harvesting at high grain moisture content as a means of reducing grain losses and improving grain quality; and
- To develop optimum harvest strategies for a range of cropping areas and harvester capacities for wheat growers in the northern wheat belt of Australia.

Agronomic background

The planting and harvesting windows for major summer and winter crops (sorghum and wheat respectively) in the northern grain belt of Australia are shown in Figure 2. Also shown are the climatic conditions under which these crops are grown.

Wheat is planted in winter from May to July and grows mainly during spring from August to October. Harvest commences in Queensland in October and gradually progresses southwards, finishing in Victoria and the southern parts of Western Australia in January.

The planting date is influenced by the incidence of rain and the need to avoid frost at flowering. Within these constraints, wheat is sown as early as possible to allow head filling during the cool part of the season. The date of anthesis and the prevailing environmental condition are major factors that influence grain yield. Kohn and Storrier (1970) and Ridge and Mock (1975) reported yield decreases of 4–7% per week of delay in planting after a certain date. Woodruff and Tonk (1983) showed that in the absence of frost the crop yield decreased linearly, averaging 1.2% d^{-1}, when the anthesis date departs from the optimal mid-winter period. Rainfall during planting is usually low and unreliable. Planting opportunities arise only after significant rain is received. Some flexibility in planting can be achieved by selecting crop varieties that best suit the time of sowing.

From a harvesting and management perspective, early anthesis date is desirable. Early maturing crops avoid the onset of summer storms at harvest, thus reducing the risk of yield and quality losses. Under a continuous cropping

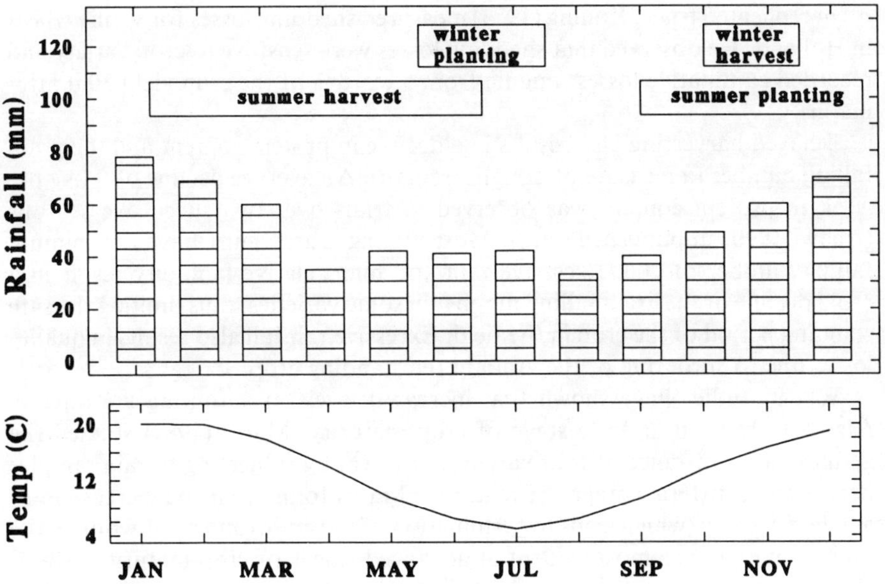

Figure 2. Planting and harvesting windows for major winter and summer crops in northern Australia. Rainfalls are monthly averages (1879–1986) and temperature are minimum monthly averages (1960–1986) at Goondiwindi.

system, advancing the harvesting window maximises summer planting opportunities and thereby reduces the insect risk and yield losses associated with delayed planting. However, in most wheat growing regions of Australia the need to avoid frost at flowering delays grain maturity until early October and the maturity date cannot be brought forward by early planting. Therefore, changes to existing management strategies are needed to reduce losses during harvest. A discussion of these strategies is preceded by a description of the magnitude and causes of grain losses.

Grain losses

Grain crops in the field are exposed to a wide range of environmental influences. From physiological maturity until harvest, dry matter yield diminishes. Among the factors that contribute to crop losses are changing crop characteristics with advancing maturity, climatic conditions (such as rain, hail, wind and temperature), crop variety and damage from insect, birds and wildlife.

Most losses occur in the field before harvest. Bolland (1984) reported yield losses of up to 0.5% d^{-1} of the potential crop yield due to delayed harvest. Abawi (1990, unpublished data) measured losses ranging from 0.3% to 0.9%

d^{-1} in replicated trials. Koning (1973) measured shedding losses for spring wheat in Holland. He observed that shedding losses were sensitive to crop variety and recorded cumulative losses ranging from 5 to 18% of the crop yield 30 d after maturity.

Delayed harvesting also results in a decline in protein content and the grain falling number (a measure of dough strength). An average decline of 0.2% per week in protein content was observed in trials over two successive seasons (Abawi 1990, unpublished data). Most milling wheat must have a minimum falling number of 200. Excessive rain and high relative humidity result in a rapid decline of the grain falling number in some varieties, thus limiting the safe standing period of the crop in the field. Excessive rainfall also results in quality losses due to sprouting of the grain in the standing crop.

Wheat studies have shown that increased levels of sprouting are directly related to rainfall and the stage of crop maturity. Mares (1987) studied the germination resistance of four varieties of wheat by subjecting to rain samples harvested at different stages of maturity. Mares found that the increase in α-amylase activity (which leads to germination of the grain) and a reduction in the falling number was most evident at advanced stages of crop maturity. Mares also determined the time required for grain moisture content to reach a threshold level of 45% dry basis (d.b.). This level is considered necessary for germination to proceed (Mares 1987; Lush et al. 1981). The time ranged from 30h for less resistant varieties to 60h for more tolerant varieties.

In trials conducted both in field and under controlled environments, Gale et al. (1987) observed similar drying rates and an increase in the level of α-amylase activity as a function of time after anthesis. Gordon et al. (1979) studied the germination potential (risk of sprouting) and base α-amylase activity of four varieties of white grained wheat in the northern wheat-belt of New South Wales. Germination potential, α-amylase activity and grain moisture content were determined from the time of anthesis and then at weekly intervals. They found that germination potential and α-amylase activity increased exponentially with time after anthesis. This increase was directly correlated to a decline in the grain moisture content. The average germination potential increased from 50% at 20% moisture content (w.b.) to 73% at 12.5% moisture content (w.b.). These results show that most harvesting is carried out when moisture levels are at a point where the grain is most susceptible to sprouting and quality damage.

The results of investigations by Gordon et al. (1979) and Mares (1987) are useful indicators of grain quality damage, but the majority of results are based on studies in controlled environments where samples are subjected to constant temperatures and moisture levels. A considerable amount of research is needed to quantify grain quality damage under variable environmental conditions.

Model development

To achieve the objectives of this study a wheat harvesting model was developed to investigate the effect of different harvesting strategies and machinery mixes on the overall cost of wheat production in northern Australia.

The model examines a typical wheat farm with one or more harvesters, a wet storage facility and a grain drying facility. The crop is assumed to have reached maturity (30 % moisture content w.b.) and to have attained its maximum potential in yield and quality on a known date. The mathematical equations describing the economics, machinery costs and performances and crop loss characteristics are described by Abawi (1992).

Many farm decisions are affected by short term constraints such as the trafficability of land, grain moisture content, matching harvesting capacity with storage and drying facilities, and the hours of operation available on a given day. For the model to reflect the farm operation adequately, it was necessary to simulate the harvesting and drying system on an hourly basis.

The wet storage facility is included to provide a buffer between harvesting and drying. It is therefore possible for daily harvest to exceed the drying capacity. The balance of harvest is stored in wet storage and dried at night or when harvesting is not possible due to field constraints or high grain moisture content. Aeration of grain in the wet store is not considered in this model, so the period over which the grain can be stored without deterioration depends on the initial grain moisture content, ambient air temperature and relative humidity. The safe storage period is determined from functions described by Ziauddin and Liang (1986). Any grain stored for a longer period than calculated is assumed spoiled and attracts feed quality value. This situation could arise if the drying capacity is inadequate to match the harvesting capacity or the grain is harvested at a very high grain moisture content. Aeration of grain can extend the storage time before drying is required. If aeration is used, a reduction in the drying capacity may be possible.

The rate of artificial drying is determined from the initial grain moisture content, the drying air temperature and the drier characteristics, as described by Radajewski et al. (1987). Grain that does not require drying is delivered to wheat receiving points directly. On-farm storage of grain is not considered in this model and it is assumed that grain can be transported readily to wheat receiving points. The cost of transporting grain from the farm gate to public receiving points is not included as this operation is common to all harvesting systems.

Grain moisture content

Determination of grain moisture content is important to simulating the harvesting system. It not only affects the harvesting time available, but also influences the drying cost of harvested grain. Several models have been developed that relate grain moisture content in a standing crop to ambient weather data such as rainfall, temperature, relative humidity, wind velocity and radiation. (Van Elderen and Van Hoven 1973; Brück and Van Elderen 1969;

Crampin and Dalton 1971). Van Elderen and Van Hoven (1973) reviewed these models and suggested that the Crampin and Dalton model best predicted the relationship between ambient weather conditions and grain moisture content. In this study, the Crampin and Dalton model was chosen because it required few historical weather parameters (temperature, relative humidity, rainfall) and daily values of these variables were readily available for wheat growing areas in northern Australia. Hourly values of temperature and relative humidity were generated from daily data using the procedures described by Kimball and Bellamy (1986).

The Crampin and Dalton model was validated for the northern wheat-belt of Australia over three successive seasons. Hourly samples of grain from the standing crop were collected for the duration of harvest and their moisture contents plotted against those predicted by the model. The correlations in each season were 0.86, 0.84, and 0.83.

Grain losses

Grain losses include losses in yield from shedding, quality losses due to rain and machine losses resulting from gathering and separation of the grain. Gathering and separation losses are described by Abawi (1992).

Shedding losses are dependent on crop variety and seasonal weather conditions and generally accelerate as the crop matures. In this study yield loss due to delayed harvesting is defined by the following function;

$$L_s = \begin{cases} 0.0004 \cdot n \cdot Y & \text{if } n < 10 \\ 0.004 \cdot n \cdot Y & \text{otherwise} \end{cases}$$

Where L_s is the cumulative grain loss (t ha^{-1}), Y is the crop yield (t ha^{-1}) and n is the number of days past maturity (maturity is assumed at 30% moisture content w.b.). In the simulation, the crop yield at maturity is assumed to be normally distributed across seasons with a mean of 3 t ha^{-1} and standard deviation of 0.25 t ha^{-1}. The correlation between winter rain and crop yield was not considered in these simulations.

Due to a lack of published information relating grain quality to rainfall under field conditions, the criteria shown in Table 2 were used. At maturity the grain is assumed to be of PH quality and is downgraded in five discrete steps from PH to Feed, according to crop age and amount of rainfall. Rainfall amounts in Table 2 are accumulated rain over two consecutive days. Any grain not harvested 60 d from maturity is assumed to be of an ASW classification or lower.

The simulation results using the criteria in Table 2 were compared with the historical intake of wheat in various classes and found to reflect quality losses due to rain with reasonable accuracy.

Table 2. Criteria used to model grain quality as a function of maturity and rainfall. Quality is downgraded in four discrete steps (PH, AH, ASW, GP, FEED).

	Days past maturity (n)		
Rainfall (mm)	Less than 7	Between 7 and 30	Greater than 30
30	no effect	no effect	1 step
40	no effect	1 step	1 step
50	no effect	1 step	2 steps
60	no effect	1 step	2 steps
70	1 step	2 steps	3 steps
80	1 step	2 steps	3 steps
90	1 step	3 steps	3 steps

Simulation

The harvesting and drying model was simulated over 30 years of historical weather data for the Darling Downs. The simulation is carried out for a specified grain moisture content during all years. Harvesting commences whenever grain moisture content in the standing crop falls below this level. All grain losses and costs are accumulated whenever harvesting takes place. A range of grain moisture contents from 12% to 22% w.b. was used in the simulation. At 12% w.b. the capital cost of the grain drying system is excluded from the analysis.

The harvesting operation is subject to the following constraints:

- Harvesting of grain only proceeds if grain moisture content falls below the pre-specified level for at least two hours within a given working day;
- the length of a working day is limited to the period between sunrise and two hours after sunset;
- harvest is delayed if ground conditions are not suitable for harvesting machinery. This delay is related to antecedent rainfall as described by Abawi (1992);
- harvesting is delayed if the wet storage and drying facilities are fully utilised; and
- the maximum available period for harvest is set at 80 d from the time the crop reaches maturity. If harvesting is not completed within this time the remaining crop is assumed lost.

All constraints must be satisfied simultaneously for harvesting to progress. Within these constraints, the return for a farm was optimised using different harvesting strategies and machinery capacities. The return, R_v (\$A ha^{-1}), is defined as:

$$R_v = (C_v - T_s - L_p)/A$$

where C_v is the value of crop sold (\$A), T_s is the annual fixed and variables costs (\$A), L_p is the value of grain losses (\$A) and A is crop area in ha.

Results and discussions

The moisture content regime over the harvesting period was simulated using 30 years of historical weather data as shown in Figure 3. On average the grain moisture content fell below 12% w.b. for about 150 h each season, whereas the grain moisture content fell below 18% w.b. for some 600 h in each season. This is a four-fold increase in operating hours if grain is harvested at a maximum of 18% w.b. rather than 12% w.b. This increase in operating time can be viewed as an equivalent increase in harvesting capacity, thus reducing timeliness losses and improving grain yield and quality.

The effect of harvest moisture content on return was investigated for crop areas ranging from 200 to 800 ha. A single harvester with a rated capacity of 4.6 ha h^{-1} was used in the analysis. The forward speed of the harvester in all simulations was assumed to be 6 km h^{-1}. The results in Figure 4 show that significant improvements in return are possible by harvesting at moisture contents above 12% w.b. and then artificially drying the grain. The optimum moisture content increases with an increase in crop area, but is limited to 19% w.b. for the range of areas considered in this simulation. Harvesting at moisture contents exceeding 19% w.b. does not result in higher returns because of higher drying cost and machine losses.

The lower return at 12% w.b., particularly for large cropping areas, is due to higher losses and an inadequate harvesting capacity to complete the harvest

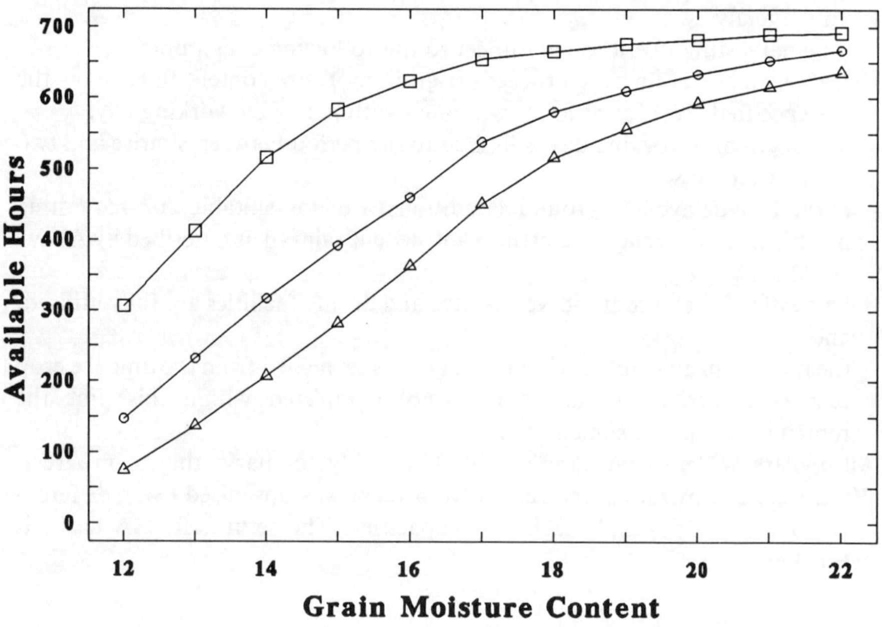

Figure 3. Available operating hours at a given grain moisture content. Δ Wettest year; ○ Average of all seasons; □ Driest Year

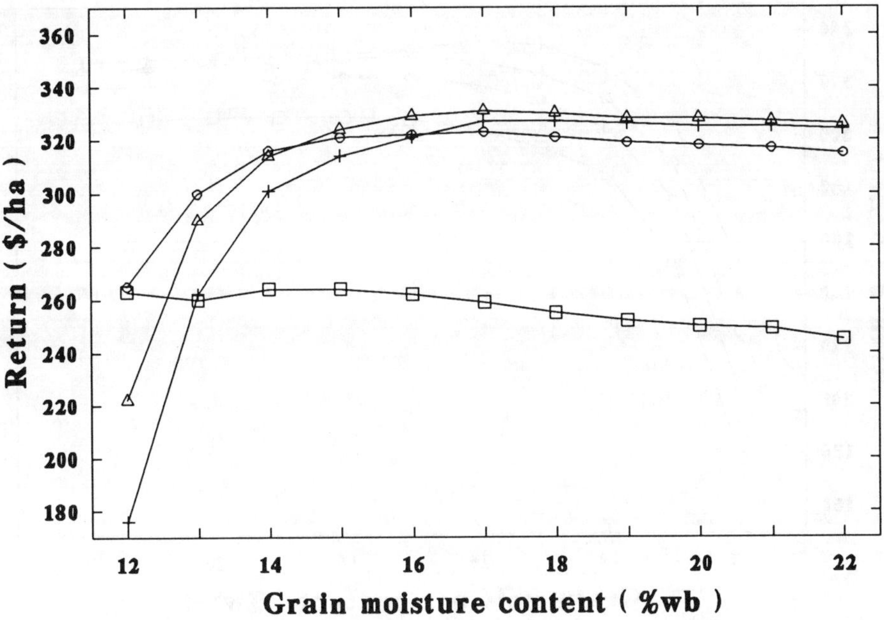

Figure 4. The effect of cropping area and harvest moisture content on return. □ 200 ha farm; ○ 400 ha farm; Δ 600 ha farm; + 800 ha farm; Harvester capacity 4.6 ha h^{-1}

within the nominated time. The results also show that for 200 ha the maximum return at 14% moisture content is only marginally higher than the return at 12% w.b. and implies that a minimum cropping area of 200 ha is necessary for grain drying to result in any significant improvement in return.

The results, while highlighting the benefits of harvesting at high grain moisture content, also accentuate the importance of matching harvesting capacity with cropping area.

The effect of harvesting capacity on return was examined for a range of grain moisture contents. Simulation runs were carried out for four harvesters with capacities of 3.2, 3.9, 4.7 and 7.0 ha h^{-1} and a crop area of 600 ha. The results in Figure 5 show that for the given cropping area the optimum harvest moisture content is affected significantly by the available harvesting capacity. With low harvesting capacity, harvesting must start at a high grain moisture content to minimise losses. With high harvesting capacity, harvesting may be delayed to lower grain moisture contents to minimise drying costs. The figures also show that for the same cropping area the maximum return results from the use of a large harvester and suggests that large harvesters are economically profitable even for small areas.

The economics of owning more than one harvester was examined for a range of cropping areas between 200 and 1000 ha. Two harvesters each with a capacity of 4.7 ha h^{-1} were used in the simulation.

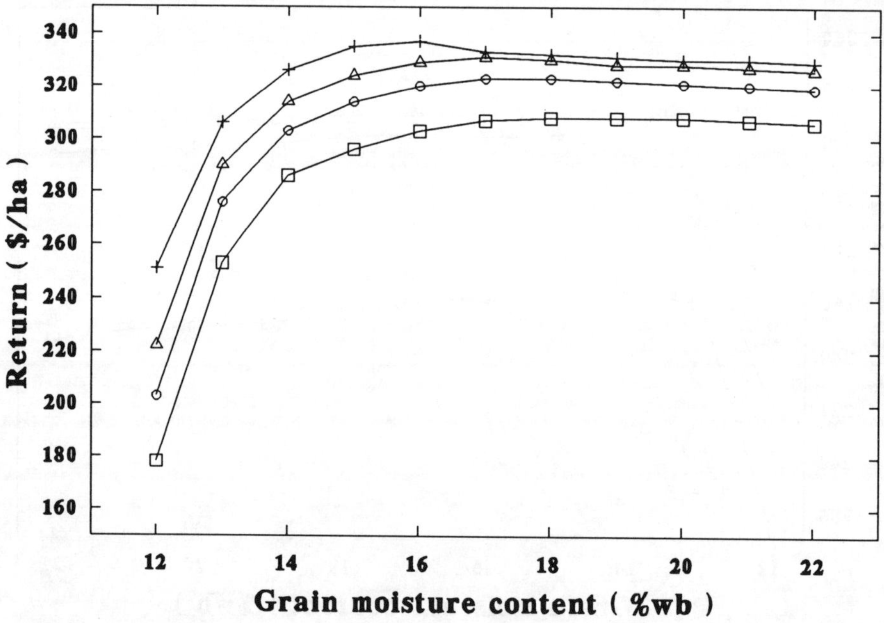

Figure 5. The effect of harvester capacity and grain moisture content on return for a crop area of 600 ha. Lines represent the following harvesting capacities: □ 3.2 ha h^{-1}; ○ 3.9 ha h^{-1}; Δ 4.7 ha h^{-1}; + 7.0 ha h^{-1}

Figure 6 shows the maximum return for a given area for one and two harvesters. For cropping areas less than 400 ha it is clear that one harvester is more economical. Conversely, two harvesters result in higher profits for cropping areas exceeding 400 ha. In both cases the maximum return resulted at moisture contents exceeding 12% w.b. The results show that the effect of machine capacity, cropping area and harvesting strategies are interrelated, so all variables must be optimised simultaneously to gain the optimum results.

Use of long range weather forecasts (LRF) in harvest management

The harvesting strategies outlined in the preceding discussions are based on long-term historical weather data and are independent of seasonal variations in rainfall. Under these strategies, identical machinery and management decisions are implemented in each season. To maximise profits the seasonal variation in rainfall must also be considered. If the probability of extreme rain during harvest is high, alternative strategies such as the use of additional machinery or advancing the commencement of harvest can be taken to minimise losses. Under favourable weather conditions the harvest can be delayed to minimise the drying costs.

Prediction of rain is possible several months ahead by the analysis of sea

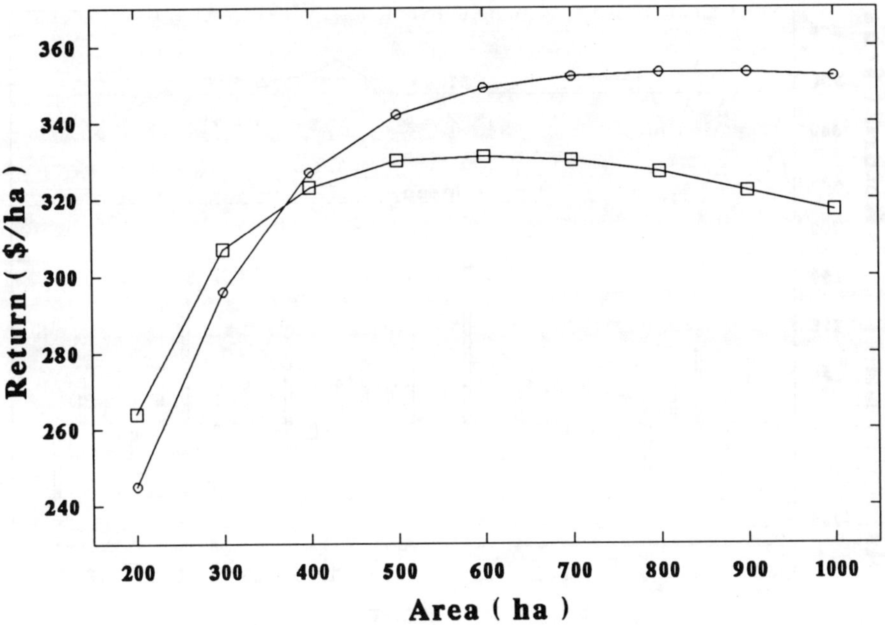

Figure 6. Comparison of return for one and two harvesters over a range of cropping areas. □ One harvester, capacity of 4.7 ha h^{-1}; ○ Two harvesters, capacity of 4.7 ha h^{-1} each

surface temperature (SST) and air pressure differences between Tahiti and Darwin, as identified by the Southern Oscillation Index (SOI). Detailed discussion of the Southern Oscillation phenomena and its relationship with Australian rainfall are given by McBride and Nicholls (1983), Nicholls (1988), and Pittock (1975). Associated with the Southern Oscillation phenomenon are large seasonal variations in sea surface temperature, rainfall and wind strength over much of the Pacific. Streten (1981) reported that variability of the SST in the southern hemisphere influences inter-annual changes in Australian rainfall. Streten also indicated that the extent of above-normal SST over the oceans surrounding the Australian continent is closely associated with extreme rainfall years.

McBride and Nichols (1983) constructed maps for simultaneous and lag correlations between seasonal rainfall in Australia and the Troup SOI (mean sea-level pressure difference between Tahiti and Darwin, normalised for each calendar month). They found a positive correlation between the SOI and rainfall, but reported a substantial seasonal variation in the magnitude and the areal extent of the correlation. The strongest correlation occurred in spring (using average spring SOI as a predictor of summer rainfall) over the north-east of Australia, encompassing the northern wheat belt. Clewet et al. (1990) found that an increase in the spring value of SOI from strongly negative (less than -5) to strongly positive (greater than $+5$) was associated with an increase in mean

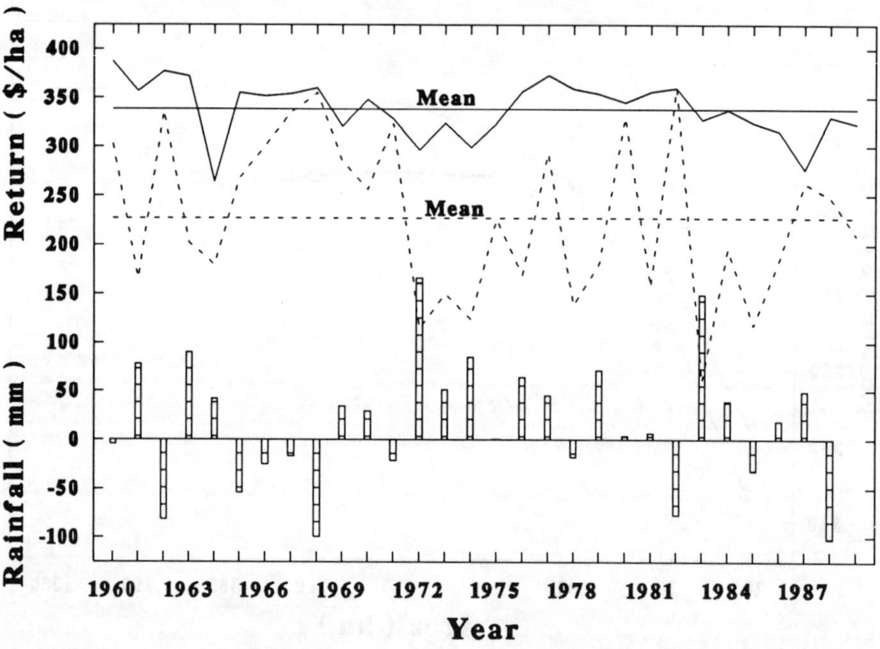

Figure 7. Predicted seasonal return under two harvesting strategies. Solid line: Harvesting at 17% moisture content (w.b.); Dashed line: Harvesting at 12% moisture content (w.b.); Positive and negative bars represents above and below average rainfall (October–November) respectively.

summer rainfall of about 30% for most locations in Queensland.

Given that most quality damage in wheat occurs at above average rainfall, the predictive relationship between spring SOI and summer rainfall is a useful aid to wheat harvest management in northern Australia.

Figure 7 shows the predicted seasonal returns for 30 years of simulation (1960–1990) using historical rainfall data. The simulations are shown for harvest moisture contents of 12 and 17% (w.b.). The results, while showing a significant improvement in return at 17% moisture content, suggest that returns could have further increased by $A 40 to $A 60 ha^{-1} in seasons with high rainfall (for example 1964, 1969, 1972, 1974, 1983) if measures had been implemented to reduce the deleterious impact of extreme rain. Growers could respond to LRF by implementing one or more of the following short-term tactical strategies:

(i) Harvest at a moisture content that optimises the trade-off between drying cost and grain losses in each season;
(ii) Harvest at a faster rate to minimise timeliness loss, though at the expense of additional machine losses and drying costs;
(iii) Increase the harvesting and drying capacity (contract work) to minimise grain losses.

Implementation of each decision would incur a cost (C) and result in an additional return (R). The decisions are ranked in order of increasing costs and would be implemented according to increasing levels of rainfall intensity.

Simulations using historical rainfall data ('perfect knowledge') showed that higher returns result from implementing decisions of higher order. For example, optimising the harvest moisture content in each season (strategy 'i') increased the average return by \$A 5 ha^{-1} above the long-term mean of \$A 338 ha^{-1} (obtained from harvesting at 17% moisture content), whereas increasing the harvesting and drying capacities, increased the net return by \$A 20 to \$A 40 ha^{-1} in some seasons.

In practice, perfect forecasts are never achieved so higher returns are associated with higher risks. For example if additional machinery is acquired in response to a wet season forecast but the season remains dry, then additional costs have been incurred at a lower than expected return ($C > R$). Currently, the best correlation between SOI and rainfall is about 0.5. This is equivalent to predicting rainfall correctly, in two out of three seasons. For LRF to be considered an effective aid to decision making at harvest the mean return over a time period must be much greater than the costs incurred from the use of LRF ($R_{mean} > C_{mean}$).

The impact of long range forecasting as a decision aid in harvest management is currently under investigation. Simulation results using historical rainfall data are being compared with 'hind-cast' rainfall probabilities based on classes of SOI, to determine the usefulness of LRF in harvest management. Heuristic strategies are being developed to aid farmers in their management during wheat harvesting in northern Australia.

Conclusions

Harvest is the highest risk period in wheat production, particularly in northern Australia. Losses in yield and quality have a significant impact on the profitability of wheat farming in this region. This study focused on the effect of harvesting at high grain moisture content to reduce losses. The results showed that harvesting at 14 to 17% (w.b.) then drying artificially results in improved grain quality and higher returns to growers. Use of large capacity harvesters was shown to be economically profitable. The simulation also indicated that a minimum cropping area of 200 ha is needed to gain a significant improvement from harvesting at high grain moisture content.

The use of long range weather forecasts in wheat harvest management has considerable potential. Simulation results have shown that returns could be further increased by \$A 20 to \$A 40 ha^{-1} if a wet season can be predicted and measures taken to reduce its impact. For long range forecasts to be seen as an effective aid in harvest management the long-term benefits must be proven to outweigh the costs that are associated with imperfect forecasts. This aspect of long range forecasting is currently being studied.

Acknowledgments

The results presented in this paper is part of a project funded by the Australian Wheat Research Council (AWRC). The financial assistance of the council is greatly appreciated. Travel costs for attendance at the SAAD conference were also provided by the AWRC. I wish to thank my colleagues Alan Andrews and Brian Robotham and my assistant Tom Bott for their helpful comments and assistance.

References

Abawi G Y, Radajewski W, Jolly P (1988) Artificial drying as a means of minimising crop losses. Pages 413–416 in National conference on agricultural engineering. The Institution of Engineers, Australia Publication 88/12. Australia.

Abawi G Y (1992) A simulation model of wheat harvesting and drying in northern Australia. J. Agric. Eng. Res. (In press).

Bolland M D A (1984) Grain losses due to delayed harvesting of barley and wheat. Aust. J. Exp. Agric. Anim. Husb. 24:391–395.

Boxall R A, Calverly D J B (1986) Grain quality considerations in relation to aeration and in-store drying. Pages 17–23 in Proceedings of the international seminar on preserving grain quality by aeration and in-store drying. Kuala Lumpur, Malaysia.

Brück I G M, Van Elderen E (1969) Field drying of hay and wheat. J. Agric. Eng. Res. 14(2):105–116.

Burrow J L, MacMillan R H (1981) A computer simulation for the selection and management of wheat harvesting machinery. University of Melbourne, Agric. Eng. Rep. 55/81. Australia.

Clewet J F, Howden S M, McKeon G M, Rose C W (1990) Optimising farm dam irrigation in response to climatic risk. Pages 307–328 in Muchow R C, Bellamy J A (Eds.) Climatic risk in crop production: Models and management for the Semi-arid Tropics and Subtropics. CAB International, Wallingford-UK.

Crampin D J, Dalton G E (1971) The determination of moisture content of standing grain from weather records. J. Agric. Eng. Res. 16(1):88–91.

Edwards W, Boehlje M (1980) Farm machinery selection in Iowa under variable weather conditions. Iowa Agric. Home Econ. Exp. St. Rep. 85. USA.

Gale M D, Salter A M, Leuton J R (1987) The induction of germination alpha-amylase during wheat grain development in unfavourable weather conditions. Pages 273–283 in Fourth international symposium on pre-harvest sprouting in cereals. Westview Press, Colorado, USA.

Gordon I L, Balaam L N, Derera N F (1979) Selection against sprouting damage in wheat. II Harvest ripeness, grain maturity and germinability. Aust. J. of Agric. Res. 30:1–17.

Kimball B A, Bellamy L A (1986) Generation of diurnal solar radiation, temperature and humidity pattern. Energy Agric. 5:185–197.

Kohn G D, Storrier R R (1970) Time of sowing and wheat production in Southern New South Wales. Aust. J. Exp. Agric. Anim. Husb. 10:604–609.

Koning K (1973) Measurement of some parameters of different spring wheat varieties affecting combine harvesting losses. J. Agric. Eng. Res. 18:107–115.

Lush M W, Groves R H, Kay P E (1981) Pre-sowing hydration-dehydration treatments in relation to seed germination in early seedling growth of wheat and ryegrass. Aust. J. Plant Physiol. 8:409–425.

Mares D J (1987) Pre-harvest sprouting tolerance in white grained wheat. Pages 64–75 in Fourth international symposium on pre-harvest sprouting in cereals. Westview Press, Boulder, Colorado, USA.

McBride J L, Nicholls N (1983) Seasonal relationship between Australian rainfall and the southern oscillation. Monthly Weather Review 111:1998–2004.

Morey R V, Peart R M, Zachariah G L (1972) Optimal harvest policies for corn and soy beans. J. Agric. Eng. Res. 17:139–148.

Nicholls, N (1988) El Nino–southern oscillation and rainfall variability. Journal of Climate 1:418–421.

Phillips P R, O'Callaghan J R (1974) Cereal harvesting – a mathematical model. J. Agri. Eng. Res. 19:415–433.

Pittock A B (1975) Climatic change and the patterns of variation in Australian rainfall. Search 6:498–504.

Radajewski W, Jolly P, Abawi G Y (1987) Optimisation of solar grain drying in a continuous flow drier. J. Agri. Eng. Res 38(2):127–144.

Ridge P E, Mock I T (1975) Time of sowing for barley in the Victorian Mallee. Aust. J. Exp. Agric. Anim. Husb. 15:830–832.

Streten N A (1981) Southern hemisphere sea surface temperature variability and apparent associations with Australian rainfall. J. Geophys. Res. 86-C1:485–497.

Van Elderen E, Van Hoven S P J H (1973) Moisture content of wheat in the harvesting period. J. Agric. Eng. Res. 18:71–93.

Woodruff D R, Tonks J (1983) Relationship between time of anthesis and grain yield of wheat genotypes with differing development patterns. Aust. J. Agric. Res. 34:1–11.

Ziauddin A M, Liang T (1986) A weather risk model for grain drying system design and management in a developing region. Am. Soc. Agric. Eng. 29(5):1434–1440.

The impacts of climate change on rice yield: evaluation of the efficacy of different modeling approaches

D. BACHELET[1], J. VAN SICKLE[1] and C. A. GAY[2]

[1] *ManTech Environmental Technology Inc., US EPA Environmental Research Laboratory, 200 SW 35th Street, Corvallis OR 97333, U.S.A.*

[2] *Sveriges Lantbruksuniversitet, Institutionen for Ekologi och Miljovard, Box 7072. S-750 07 Uppsala, Sweden*

Key words: Asia, CERES-Rice, climat change, cultivated area, General Circulation Model, MACROS, model validation, pests, rice, rice ecosystem, RICESYS, sensitivity analysis, simulation, weed, yield levels, yield trend

Abstract

Increasing concentrations of carbon dioxide (CO_2) and other greenhouse gases are expected to modify the climate of the earth in the next 50–100 years. Mechanisms of plant response to these changes need to be incorporated in models that predict crop yield to obtain an understanding of the potential consequences of such changes. The objectives of this paper are (i) to review climate change predictions and their reliability, (ii) to review the major hypotheses and/or experimental results regarding rice sensitivity to climate change and (iii) to evaluate the suitability of existing rice models for assessing the impact of global climate change on rice production in the rice-growing areas of Asia. A review of physiologically-based rice models (CERES–Rice, MACROS, RICESYS) illustrates their potential to predict possible rice responses to elevated CO_2 and increased temperature. Both MACROS and CERES (wetland rice) responses to temperature and CO_2 agree with recent experimental data from Baker et al. (1990c). RICESYS is an ecosystem model which predicts herbivory and inter-species competition between rice and weeds but does not include CO_2 effects. Its response to increasing temperature also agrees with experimental findings. Models using empirical relationships between climate and yield have been used to predict country-scale changes following climate change. Their simplicity is an asset for continental-scale assessments but the climatic effects are often overshadowed by stronger technological or political effects. In conclusion, each modeling approach has its value. Researchers should choose or build the most appropriate model for their project's objectives.

Introduction

Rice is a staple food crop for over half of the world's population. Demographic forecasts indicate that rice supplies world-wide will need a 1.6% annual increase until year 2000 (IRRI 1989) to match population growth estimates. The major rice-consuming countries of south and southeast Asia will need a 2.5% annual increase in rice yield. Much of that increase is expected to come from irrigated areas which are responsible for 73% of the world's rice production (IRRI 1989). Continued increases in rice production depend on favourable climatic and environmental conditions.

CO_2 concentration in the atmosphere has increased considerably since pre-industrial times with current mean annual increments of about 1.5 ppm

F.W.T. Penning de Vries et al. (eds.), Systems Approaches for Agricultural Development, 145–174.

(Keeling et al. 1984). According to current model predictions, the increase in CO_2 and other greenhouse gases is likely to induce surface air temperature rises over the next 50 years and concurrent changes in precipitation patterns. While an increase in CO_2 is usually associated with an increase in photosynthesis for most crops (Cure 1985), an increase in mean daily temperatures may lower rice yield potential in many areas. The combined effects of elevated CO_2 and increased temperatures are for the most part unknown and are currently under investigation. Models are the best tools to synthesize current knowledge and hypotheses to estimate directional trends for potential changes in rice yield following climate change.

Several publications have described how rice yield may be affected by climate change. Yoshino et al. (1988), using Horie's (1987) rice yield model, predicted that lowland rice yields could increase in Japan by about 9% following a doubling of CO_2 and subsequent climatic changes as predicted by the Goddard Institute of Space Studies (GISS) general circulation model. An increase of 0.75°C in mean temperature (scenario derived from 'expert opinion'), with a 2% increase in precipitation is expected to bring a 0.3% decrease in rice yield in India and a 0.6% decrease in China (National Defense University 1980). Yield estimates for a range of sites varying in latitude from 6°N to 31°N were reduced by 10 to 20% by a rise in harvest (November–December) temperature of 1 to 2°C (Seshu and Cady 1984). From these estimates, it is clear that a wide range of crop yield predictions have been made using climate change scenarios.

Our objective is (i) to review climate change predictions and their reliability, (ii) to review the major hypotheses and/or experimental results regarding rice sensitivity to climate change, and (iii) to evaluate the suitability of existing rice models for assessing the impact of global climate change on rice production in the rice-growing areas of Asia.

Climate change scenarios

General circulation models (GCMs) have been used extensively to provide potential climate change scenarios (Cohen 1989; Grotch et al. 1988; Gutowski et al. 1988; Parry et al. 1988; Smith and Tirpak 1989). According to current model predictions, the increase in CO_2 and other greenhouse gases is likely to induce surface air temperature rises over the next 50 years and concurrent changes in precipitation patterns (Table 1). In this study, four GCM models (Table 2) – United Kingdom Meteorological Office model (UKMO), Goddard Institute for Space Studies model (GISS), Geophysical Fluid Dynamics Laboratory model (GFDL), Oregon State University model (OSU)– were compared with regard to their predictions of current climate for Asia. Predictions of current precipitation and temperature regimes by the GCMs were also compared to the means of these variables from climatological records for all the meteorological stations (US airfield summaries, Hatch 1986) present

Table 1. Estimates of changes in means of surface air temperature and precipitation over South East Asia (5–30°N x 70–105°E), from preindustrial times to 2030, assuming the Intergovernmental Panel on Climate Change "business as usual scenario" (Houghton et al. 1990). (DJF = December-January-February; JJA = June-July-August; T = temperature; PPT = precipitation). General circulation models cited below are from the United Kingdom Meteorological Office (UKMO), the Geophysical Fluid Dynamics Laboratory (GFDL) and the Canadian Climate Centre (CCC).

Model	Reference	Change in T (°C)		Change in PPT (%)	
		DJF	JJA	DJF	JJA
CCC	Boer et al. 1989	+1	+1	−5	+5
GFDL	Wetherald, Manabe 1989	+2	+1	0	+10
UKMO	Mitchell et al. 1989	+1	+2	+15	+15

Table 2. Major features of the four general circulation models (GCMs) used in this study.

GCM Acronym	GFDL[1]	GISS[2]	OSU[3]	UKMO[4]
Horizontal resolution (lat. x long.)	4.44°x7.5°	7.83°x10.0°	4.0°x5.0°	5.0°x7.5°
Vertical resolution (# of layers)	9	9	2	11
Diurnal cycle	no	yes	no	yes
Base 1 x CO_2 (ppm)	300	315	326	323

Source Laboratory:
1. Geophysical Fluid Dynamics Laboratory
2. Goddard Institute for Space Studies
3. Oregon State University
4. United Kingdom Meteorological Office

References:
1. Wetherald and Manabe (1989)
2. Hansen et al. (1988)
3. Schlesinger and Zhao (1989)
4. Wilson and Mitchell (1987)

in each GCM grid cell (Table 3). The average number of meteorological stations per GCM cell was 4.89, 10.06, 5.77 and 2.85 for the GFDL, GISS, UKMO and OSU models respectively. The grid cell size varied from 4° latitude by 5° longitude (OSU) to 8° latitude by 10° longitude (GISS) (Figure 1). Spatial homogeneity was assumed inside each GCM grid cell. Baseline runs assumed current atmospheric CO_2 levels (1xCO_2). The models were run until climatic equilibrium was reached following an instantaneous equivalent doubling of CO_2 (2xCO_2). Equivalent doubling refers to CO_2 and other greenhouse gases having a radiative effect equivalent to the doubling of the pre-industrial value of CO_2. Model monthly statistics correspond to an average of 10 years of predictions (15 for UKMO).

While predictions of current temperature are relatively accurate (r^2 between 0.8 and 0.9), predictions of monthly rainfall are obviously incorrect and vary widely between models (Table 3). A number of problems arise when general circulation model output and data are compared over a region. First, models

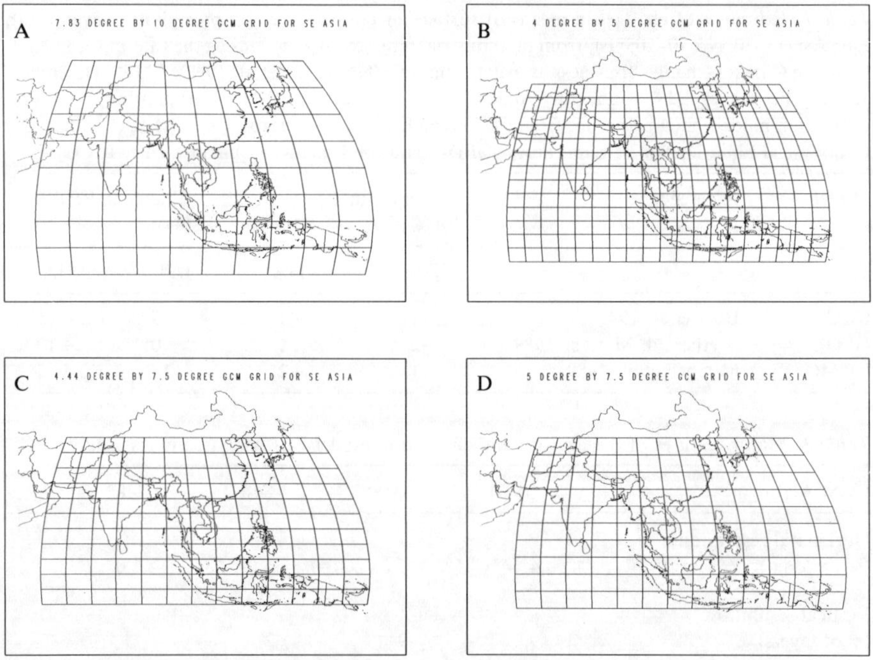

Figure 1. Map of the study area with general circulation models (GCM) grids overlay: *A.* GISS (Goddard Institute of Space Studies); *B.* OSU (Oregon State University); *C.* GFDL (Geophysical Fluid Dynamics Laboratory); *D.* UKMO (United Kingdom Meteorological Office).

Table 3. Results from a regression analysis comparing general circulation model (GCM) predictions of monthly temperatures and rainfall to long-term averaged climatic records for all the meteorological stations (airfield summaries) included in each of the GCM grid cells over Thailand. (OSU: Oregon State University, UKMO: United Kingdom Meteorological Office, GFDL: Geophysical Fluid Dynamics Laboratory, GISS: Goddard Institute for Space Studies). The average number of meteorological stations included in each GCM cell is also included.

1. Daily average temperature (°C)				
GCM	# of Stations	Slope	Y-Intercept	r^2
GFDL	4.89	1.29	−10.71	0.86
GISS	10.06	0.97	− 0.33	0.90
OSU	2.85	1.21	− 5.51	0.86
UKMO	5.77	1.24	− 7.02	0.83
2. Monthly rainfall (mm)				
GCM	# of Stations	Slope	Y-Intercept	r^2
GFDL	4.89	0.22	98.08	0.15
GISS	10.06	0.17	122.74	0.06
OSU	2.85	0.15	57.29	0.20
UKMO	5.77	0.42	82.00	0.12

generally produce results at grid points that, due to computational costs, are typically separated by hundreds of kilometers. In such cases it is not immediately obvious what the temperature or rainfall at a gridpoint really represents. Because of this relatively crude areal resolution, only a few gridpoints (80 to 247 for Asia) are used in estimating averages for subcontinental intercomparisons (Joyce et al. 1990) and are compared to data from 3 to 10 times as many meteorological stations.

The models also represent orography in a smoothed way that may affect comparison with data and with other models. Alterations in climate predicted within a GCM cell would likely apply fairly uniformly across regions that have relatively uniform land surface characteristics. On the other hand, steep topography or lakes smaller than GCM grids can mediate climate. Therefore, even if GCM predictions were accurate at grid scale, they would not necessarily be appropriate to local conditions (Schneider et al. 1989). Since Asia's relief is varied and includes the Himalayan region, predictions of climate change are likely to be inaccurate because models average conditions ranging from deep valleys to high mountain slopes.

Another problem arises because models differ from one another. First, model results are available on regular grids but of different sizes (Figure 1). This precludes a direct point by point intercomparison of GCM results. However, various interpolation schemes (Grotch 1988) are available to alleviate this problem. Secondly, the resolution used to represent oceans and the cloud parameterization vary between models and make comparisons between models more difficult. Finally, the effect of averaging (last 10 years of 35 GFDL, GISS, and 20 OSU year runs) model results must be carefully considered in model intercomparisons (Grotch et al. 1988).

The predictions used in this study represent average changes that might be expected in the future following climate change. Changes in the frequency and intensity of hurricanes, the frequency of floods, and the intensity of the monsoons, are much more important to the rice farmer than the average increase in monthly precipitation or temperature. However, given state-of-the-art climate models, it is still uncertain how climate variability will vary as a consequence of the increase in greenhouse gases. To date, only one analysis of the importance of climate variability to agriculture (Mearns 1991) has been published. Its preliminary results illustrate how CERES–wheat responds to changes in the interannual variability of temperature in Goodland, Kansas. Further work is currently under way to complete the study and no firm conclusions have been drawn yet. It is not the purpose of this paper to discuss this issue but only to suggest its importance when examining the impacts of climate change.

More importantly, there are systematic errors in simulating current climate because of the models' weaknesses in representing physical processes in the atmosphere relating to clouds (Cess and Potter 1987; Mearns 1990) and the lack of biofeedback (Dickinson 1989). Even if more sophisticated models have now been designed, their predictions are not readily available outside the climatological community. Another cause of inaccuracy is the fact that we used equilibrium runs assuming an instantaneous doubling of CO_2 and a subsequent equilibrium state which, in reality, might not occur in the next fifty years if atmospheric CO_2 were to double. Given these limitations, one may wonder why GCM predictions are used to estimate future changes in regional climate. The GCMs provide the best and only estimates of how global climate may be affected by increased CO_2. They can help managers and policy makers to contemplate the future with some awareness of the trends future climate might bring, and thus prepare for it. Moreover, advances in the reliability of regional values produced by GCMs can be anticipated within the next 5–10 years. These advances will stem from better parameterization of the climatic processes, better incorporation of the effects of the surface boundaries, especially oceanic processes, and better specification of cloud-radiation feedbacks. Temporal and spatial scales of analysis should also be refined in the near future. To improve the spatial resolution by a factor of two requires an increase by a factor of eight in computer demand (Hulme, U.K. Meteorological Office, pers. comm., 1989). However, a rapidly advancing methodology to solve this problem is the use of mesoscale models nested within GCMs to simulate the detail of a selected region (Giorgi 1990; Giorgi and Mearns 1991). Since reliable and accurate predictions at the regional and local scales may be available only after atmospheric CO_2 has already doubled, we should make the best of what is currently available and ascertain the weaknesses of the predictions.

Review of effects of climate change on rice and paddy ecosystem

CO_2 effects

The possibility that a rise in atmospheric CO_2 may significantly increase agricultural yield sparked a wave of activity from the agricultural research community (Adams et al. 1988; Bolin et al. 1986; Cooter 1990; Crosson 1989; Goudriaan and Unsworth 1990; Hammer et al. 1987; Parry et al. 1988). A general survey of crop response to a doubling of CO_2 (Cure 1985) reviewed studies on rice (Table 4) and showed increases in rice yield and biomass due to an increase in the net assimilation rate. The same studies showed a decrease in leaf conductance and transpiration due to CO_2 enhancement. More recent research by Baker et al. (1990a) showed an increase in grain yield and biomass as CO_2 concentration increases (Figure 2), together with an increase in canopy net photosynthesis and water use efficiency. Baker et al. (1990b) found that elevated CO_2 accelerated rice development and shortened total growth duration (Figure 2). Strain (1985) also stated that the net effect of CO_2 on plant reproduction is to accelerate all events from the dates of anthesis to the date of seed maturation.

At the rice ecosystem level, the indirect impacts of CO_2 are important. In general, when photosynthesis is enhanced by increased CO_2, C/N ratio also increases (Strain 1987) which reduces the nutritional quality of leaves (Overdiek

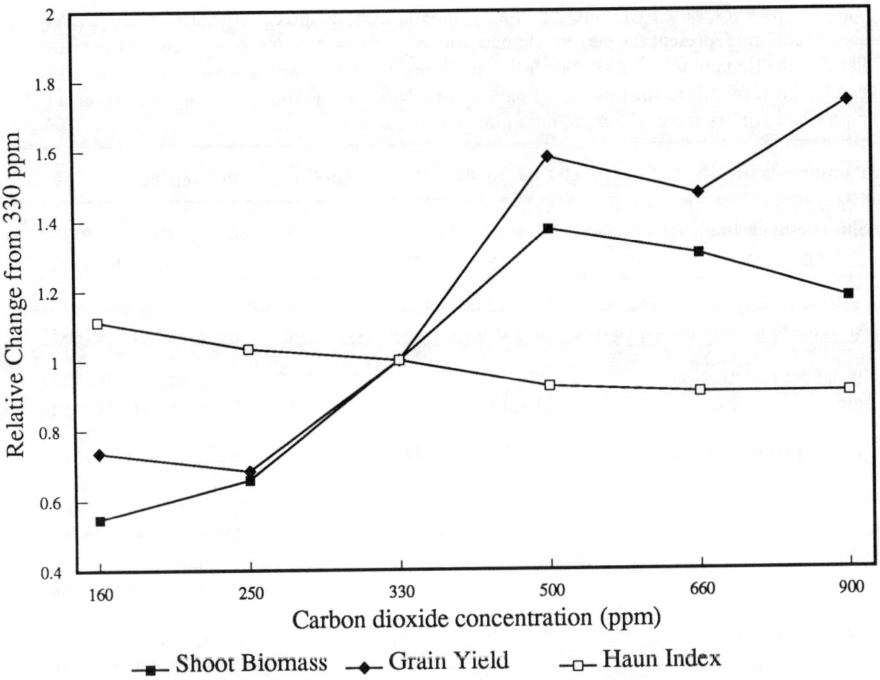

Figure 2. Relative change in shoot biomass, grain yield and Haun index at various carbon dioxide concentrations (Baker et al. 1990 a,b,c) relative to those at 330 ppm. Haun index is calculated by dividing laminar length of the last expanded leaf by the visible laminar length of the growing leaf and adding this ratio to the leaf number of the last expanded leaf.

et al. 1988) and therefore increases feeding by herbivorous insects (Lincoln et al. 1984, 1986; Osbrink et al. 1987; Fajer 1989; Johnson and Lincoln 1990) before reducing their fitness. A high C/N ratio also slows down plant tissue decomposition and could reduce methane emission. Bazzaz and colleagues have published a number of studies examining the effects of CO_2 concentration on competing assemblages of different species (as reported by Patterson and Flint 1990). The competitive ability of the C_3 species improved in assemblages grown in enriched CO_2 as compared with ambient level atmospheres. In the case of rice cultivation, this means that rice, which is a C_3 plant, would compete advantageously with common weeds, such as *Echinochloa* spp., which have the C_4 pathway. Finally, both symbiotic and nonsymbiotic nitrogen fixation have been shown to increase with increased CO_2 (Strain 1985). Spontaneous biological nitrogen fixation currently represents from 1–7 kg nitrogen ha^{-1} $crop^{-1}$ (Roger and Ladha 1990), free living blue green algae can fix as much as 30 kg ha^{-1} $crop^{-1}$, and legumes and azolla up to 100 kg ha^{-1} $crop^{-1}$, while the average fertilizer use in tropical Asia was 30–55 kg NPK (nitrogen-phosphorus-potassium) ha^{-1} arable land in 1978 (Roger and Watanabe 1986). Microbiological activity in the rhizosphere may be enhanced at higher CO_2

Table 4. Rice response to a doubling of atmospheric CO_2, adapted from Cure (1985). Data in the second column represent the percent change from 300–350 ppm to 680 ppm in each of the variables listed in the first column. The 95% confidence limits on those changes are also included. The third column corresponds to the number of data points (N) used, the fourth column corresponds to the number of studies from which the data points were extracted.

Parameters	Change–%	N	Studies	References
Short-term carbon exchange rate	43 ±19	8	3	Akita, Tanaka 1973 Morison, Gifford 1983 Imai, Murata 1978b
Acclimated carbon exchange rate	46 ±0	2	1	Imai, Murata 1978b
Initial net assimilation rate	26 ±11	8	3	Imai, Murata 1976 Imai, Murata 1979a,b
Biomass accumulation	27 ±7	22	11	Akita, Tanaka 1973 Imai, Murata 1976 Imai, Murata 1978a,b Cock, Yoshida 1973 Yoshida 1973 Morison, Gifford 1984
Root-shoot ratio	-4.0 ±0	2	1	Imai, Murata 1976 Morison, Gifford 1984
Harvest index	1.9 ±0.6	6	3	Yoshida 1973 Cock, Yoshida 1973
Stomatal conductance	-33 ±7	5	3	Imai, Murata 1978b Morison, Gifford 1983
Transpiration	-16 ±9	7	3	Akita, Tanaka 1973 Imai, Murata 1976 Imai, Murata 1978b
Yield	15 ±3	6	3	Yoshida 1973 Cock, Yoshida 1973

concentrations due to an increased exudation from plants (Lamborg et al. 1983).

Temperature effects

Critically low and/or high temperatures define the environment where the life cycle of the rice plant can be completed. Critical temperature thresholds are (i) low temperatures around 15°C from seedling stage to panicle initiation, (ii) low temperatures around 20°C or high temperatures around 35°C at flowering which could induce sterility during pollinization (about 80 d after planting). Within the critical low and high temperature range, temperature influences the rate of development of leaves and panicles and the rate of ripening, thereby affecting the growing season length of a rice variety and determining the

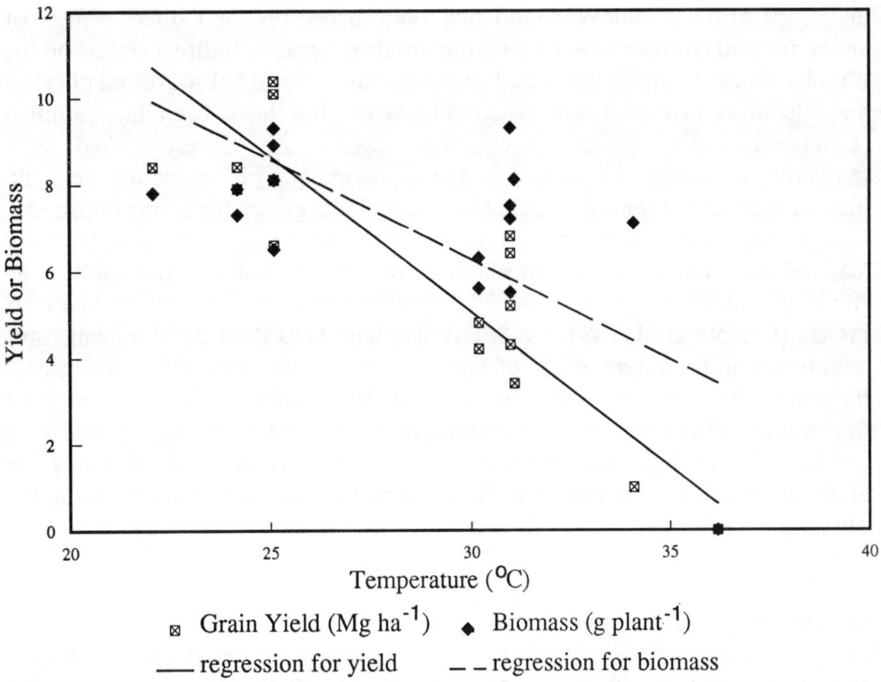

Figure 3. Sensitivity of rice grain yield and biomass to temperature (Baker et al. 1990c). Regressions were calculated for yield ($y = -0.7T + 28.8$, $r^2 = .89$) and for biomass ($b = -0.1T + 10.8$, $r^2 = 0.10$).

suitability of that variety to the environment (Yoshida 1976). Increased temperature speeds up plant development but decreases the length of the grain filling period. According to Stansel and Fries (1980) for a 120–day rice variety, the average daily temperature during the 55 day–long vegetative period should be 22°C, the temperature during the reproductive period (about 21 d) should be 24°C. Harvest occurs following a 35–day period of grain filling and maturation with an average daily temperature of 24°C.

In principle, global warming may contribute to a northward expansion of rice growing areas and a lengthening of rice growing seasons which are now constrained by low temperatures (Seshu et al. 1989). This is particularly important for northern countries such as Japan where rice cultivation is constrained by low temperatures. In some areas, it may even be possible to grow two annual rice crops instead of only one. However, a shortening of the development process can also result in incomplete grain filling and reductions in yield. Baker et al. (1990c) found that increased temperature reduced maturation time, resulting in seed sterility and a linear decrease in yield ($r^2 = 0.89$) while biomass remained mostly unaffected (Figure 3). In this figure, the regression curve between biomass and temperature deceptively shows a strong relationship while the actual correlation was weak ($r^2 = 0.10$).

Temperature effects on rice yield are not simply restricted to direct effects of air temperature on aboveground metabolic processes, and direct effects of water and soil temperatures on root metabolic processes. Indirect effects on the population dynamics of plant pathogens and insects and belowground physical and chemical processes are potentially large, but have been little studied (Campbell 1989). Insect development cycles are closely linked with temperature, and a shortening of the development phases could mean an increased potential for outbreaks. However, plants grown for a long time under increased temperature also have a reduced nutrient content and a lower specific leaf weight, when compared to plants grown under cooler temperatures. The resulting reduction in nutritional quality increases feeding by herbivorous insects (Lincoln et al. 1984, 1986) but also lengthens their development, thus counteracting the direct effect of temperature on insect growth. Rice disease incidence may also be severely affected by modifying the frequency of occurrence of the optimal infection temperature. Finally, the rhizosphere flora are sensitive to high temperatures. Because free-living nitrogen fixers are abundant in the rice root zone, a change in temperature could affect the paddy nitrogen cycle.

General review of rice models

We compared three modeling approaches. The first one uses mechanistic models simulating processes of plant growth and development (MACROS, CERES). The second uses ecosystem models integrating the direct and indirect effects of crop-weed competition and herbivory on yield (RICESYS). The third approach uses correlation and regression models and identifies relationships between yield and summary statistics of climatic variables during the growing season.

Physiological models

We chose two physiologically-based models (Table 5) simulating paddy rice: CERES–Rice (Alocilja and Ritchie 1988; Godwin et al. 1990) and MACROS (Penning de Vries et al. 1989). They are both used by scientists at the US Environmental Protection Agency (EPA–Corvallis, EPA–Washington D.C.) to predict future rice yields.

CERES–Rice

CERES–Rice (Alocilja and Ritchie 1988; Godwin et al. 1990) emphasizes the effects of management and the influence of soil properties on crop performance. The model was designed to assess yield as constrained by crop variety, soil water and nitrogen, for alternative technology and for new growing sites. It was designed to reduce time and cost of agrotechnology transfer of new varieties and management practices. Weather inputs include daily solar radiation, maximum and minimum temperatures and precipitation. It ignores potential effect of typhoons and assumes no pests. However, some work has

Table 5. Characteristics of three rice yield simulation models (d = day; PAR = photosynthetically active radiation; Tmin/Tmax = minimum/maximum temperature; N = nitrogen; LAI = leaf area index; TR = transpiration ratio; T = temperature; RH = relative humidity; CHO = carbohydrates; Pmax = maximum photosynthetic rate; PPT = precipitation; d/4 = 1/4 of a day or 6 hours).

Feature	CERES–Rice	MACROS	RICESYS
Reference	Alocilja, Ritchie 1988, Godwin et al. 1990	Penning de Vries et al. 1989	Graf et al. 1990a,b, 1991
Time step	d	d; d/4	d
Meteorological Inputs	PAR, Tmin, Tmax, PPT	PAR, Tmin and Tmax, PPT, vapor pressure, daylength, wind speed	PAR, Tmin and Tmax
PHOTOSYNTHESIS			
climatic controls	PAR, T, TR	PAR, T, RH, TR	PAR, T
plant controls	N, plant density, LAI	N, LAI, leaf thickness, CHO accumulation, phenology, stomatal resistance	N, CHO demand
canopy structure	LAI	5 layers with Pmax and light use efficiency	
RESPIRATION			
maintenance	not explicit	T	T
TRANSLOCATION	phenology, source size, sink demand, N, soil water, priority rule.	phenology, priority rule, source size	priority rule, sink size, N

(Table 5 is continued on the next page.)

been done to simulate the impact of the disease on yield (J. Ritchie, Michigan State University, East Lansing, pers. comm., 1991). The influence of CO_2 on photosynthesis (enhancement) and transpiration (decrease) has been recently added (scalar modifier) to assess the impact of increased CO_2 on rice yield in a US EPA sponsored international project (C. Rosenzweig, NASA Goddard, New York, pers. comm., 1991). Subroutines simulating the impact of major diseases on rice yield are being developed (J. Ritchie, Michigan State University, East Lansing, pers. comm., 1991).

In CERES–Rice, photosynthesis is expressed as daily biomass production and calculated as a function of leaf area index (LAI), solar radiation and a radiation use efficiency constant. Biomass can be reduced by high temperatures and nitrogen deficit. LAI is a function of leaf appearance and expansion which can also be reduced by nitrogen deficit. Allocation depends on the phenological stage of the plant which is primarily a function of thermal time (or day-degrees,

Table 5. (continued). Characteristics of three rice yield simulation models. (LAI = leaf area index; T = temperature; photoP = photoperiod; # = number; PAR = photosynthetically active radiation).

Feature	CERES–Rice	MACROS	RICESYS
ECOSYSTEM LEVEL			
Soil	input file with 5–10 layers, initial nutrient content, texture, water holding capacity	soil water balance	soil nitrogen balance
Management practices	N fertilization, land preparation, irrigation	irrigation, fertilization	fertilization
Herbivores	no	no	yes
Weeds	no	no	yes
PLANT DEVELOPMENT	genetic file, T	Veg. stage: T, photoP Reprod. stage: T	PAR, T, bookkeeping

GROWTH	sink demand, T	T; sink # and reserve
		height = f(phenology)
			supply/demand driven

ASSUMPTIONS	no typhoons, no pests potential yield = panicle weight at anthesis	no typhoons, no pests biochemical constraints	no typhoons, no pests except leafhoppers.

accumulated temperatures above a temperature threshold). Translocation between shoots and roots can be altered by water and nutrient deficiencies. Phenological development and crop duration are related to temperature, photoperiod and genotype. Phenological stages include sowing, germination, emergence, juvenile period, panicle initiation, heading, beginning of grain fill, end of grain fill and physiological maturity. Photoperiod sensitivity influences thermal time requirement during floral induction. Specific genetic information needed to run the model includes the thermal time (degree-days) between emergence and the end of the juvenile period, the photoinduction rate, the optimum photoperiod, the thermal time required for grain fill, the efficiency of sunlight conversion to assimilates, the tillering rate and the grain size.

A modified Priestly and Taylor's (1972) potential evapotranspiration method is used to calculate the water balance separating soil evaporation and transpiration. The soil is characterized by its initial nitrogen content, water-holding properties, texture, profile and field topographic position. The user determines the number of soil layers to be included in the model.

No calibration is required when cultivar characteristics are known. When they are unknown, genetic parameters can be approximated from field results through a calibration exercise.

MACROS

MACROS (Modules of an Annual Crop Simulator) emphasizes the biochemical aspect of plant physiology based on the concept of growth-limiting factors (de Wit and Penning de Vries 1982; Penning de Vries and Van Laar 1982) and is intended for educational purposes, especially for developing countries. Weather inputs include solar radiation, maximum and minimum temperatures, precipitation, daylength, vapour pressure and wind speed. The model includes the effect of CO_2 on photosynthesis. MACROS is set up to run in one of three distinct modes: (i) water and nutrient levels are assumed optimal; (ii) water stress due to limited water availability is assumed to occur; (iii) nitrogen or water are limiting to plant production during part of the year. In all cases, the carbon fraction in dry matter and the biochemical composition are fixed, and pest damage is assumed negligible.

The canopy is divided into five layers with the radiation at the bottom of the upper layer used as input to the lower layer. Each layer has a maximum leaf photosynthesis rate and initial light use efficiency. The distribution of leaf angles for each layer is specified. Photosynthesis includes photorespiration. The rate is expressed per unit leaf area, considering only the area of the upper surface of each leaf. Changes in ambient CO_2 concentration are simulated by adjusting the maximum photosynthetic rate relative to the ratio of the new CO_2 concentration over 340 ppm. Maximum photosynthetic rate is also reduced by glucose accumulation in leaves, leaf thickness, leaf nitrogen content, temperature and water stress. Maintenance respiration is a direct function of temperature.

The phenological stage determines biomass partitioning to leaves, roots, stems, and storage organs using locally calibrated constants. During the vegetative phase, development rate is a non-linear function of temperature and photoperiod. After flowering, development rate is only dependent on temperature. Carbohydrate requirements to the various plant parts are determined by the typical biochemical composition of each component. Growth stops when carbohydrate reserves represent less than 5% of leaf dry weight, while maintenance respiration continues. With an ample supply of water and nutrients, crop growth rate can be reduced by low temperatures or by an insufficient number of carbohydrate sinks. Excess carbohydrate is stored in a reserve pool.

Photosynthetic rate is largely determined by transpiration rate when water supply is low; when it is high, the inverse is true. Transpiration is a function of radiation and evaporative demand (Penman 1956). Reduction in photosynthesis due to sink size limitation, aging, or low relative humidity all result in increased stomatal resistance and subsequent decrease in potential transpiration rate. Water vapor movement is reduced by the leaf, the canopy, and the leaf boundary layer resistances which are calculated for the upper canopy layers which contribute most.

The model includes two different water balance modules. The first one is for free-draining soils, and the second for soils with impeded drainage. In the case of unsaturated soils, the approach is similar to that used in SAHEL (Semi-Arid Habitats that Easily Leach) (Van Keulen 1975; Stroosnijder 1982; Jansen and Gosseye 1986). For soils with impeded drainage, which frequently occur where lowland rice is grown (a hard pan forms after puddling), Simulation Algorithm for Water Flow in Aquatic Habitats (SAWAH) is used (Penning de Vries et al. 1989).

MACROS was calibrated for the rice cultivar IR–36 grown at several locations across the Philippines.

Sensitivity analysis

We ran both models with the same climatic dataset (Los Baños, Philippines, year 1984, mean annual temperature 26.7°C) using changes in temperature of the same order of magnitude as in climate change scenarios predicted to accompany a doubling of atmospheric CO_2 (Houghton et al. 1990). Maximum and minimum temperatures were simultaneously increased by 0.5°C increments up to a total of 5°C. Atmospheric CO_2 was varied from 330 ppm (ambient) to 660 ppm (double CO_2) using 30 ppm increments.

In CERES–Rice, some interactions exist between CO_2 levels and temperature regimes (Figure 4A). An increase in temperature generally corresponds to a decrease in grain yield, while an increase in CO_2 corresponds to an increase in yield. In MACROS there is a smooth relationship between potential grain yield, CO_2 and temperature with maximum potential yield at ambient temperatures and doubled-CO_2 levels (Figure 4B). Ultimately, in both models, the direction of the relationships between yield, temperature and CO_2 is the same.

We compared observed rice yield (Baker et al. 1990a, c) with yield predicted by both MACROS and CERES under variable temperature regimes as represented by the regression curves obtained for model output and temperature range. Without calibrating the models to fit the observed data, we observed a closer fit of MACROS' results (within the 95% confidence interval for the slope of Baker et al.'s (1990a, c) data) than CERES's (Figure 5). At high temperatures, MACROS simulates a sharper decrease in potential yield than CERES and this correlates better with Baker et al.'s (1990a, c) experimental data. CERES–Rice predicts a lower impact of temperature on potential rice yield (18% from 25°C to 30°C) than MACROS does (62% from 25°C to 30°C) which corresponds better with Baker et al.'s (1990c) experimental data (50% decrease from 25°C to 30°C). However, while CERES–Rice predicts a 15% increase in potential yield due to a doubling of CO_2, MACROS only predicts a 9% increase, well below the 47% increase observed by Baker et al. (1990c). Both CERES–Rice and MACROS have already been used to estimate possible changes in potential rice yield following a doubling in atmospheric CO_2 concentration. Panturat and Eddy (1990), using CERES–Rice (upland version), predicted a 20% decrease in upland rice yield (due to a decrease in irrigation water availability) in Thailand following a doubling of CO_2 and subsequent climatic changes as predicted by the GISS general circulation

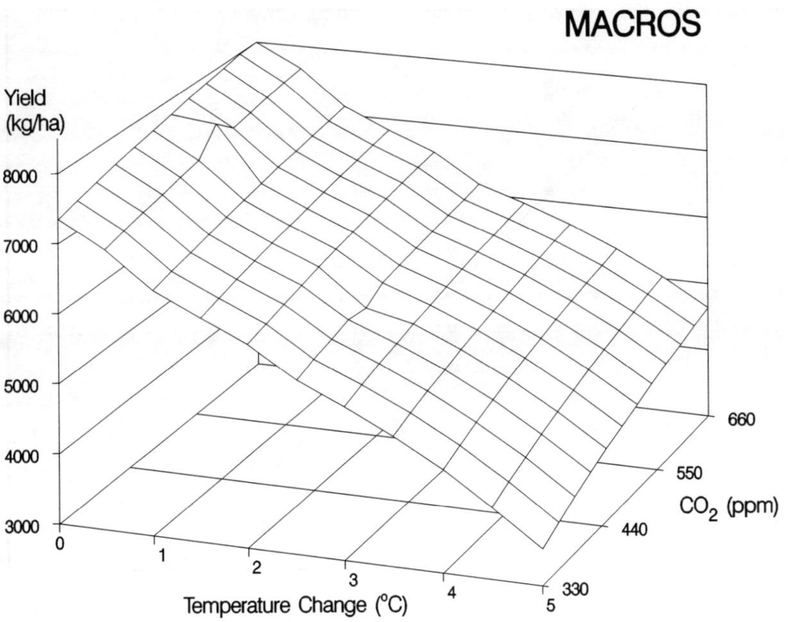

Figure 4a

Figure 4b

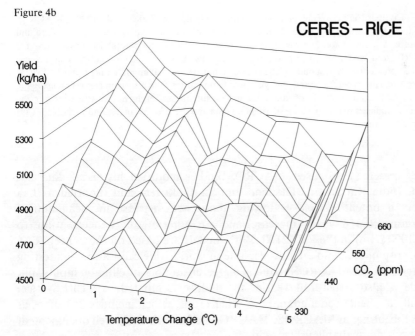

Figure 4. Sensitivity of MACROS (top) and CERES-Rice to temperature (same scenario as described in Figure 3) and CO_2 changes. CO_2 is increased from current levels (330 ppm) to a doubled-concentration scenario (660 ppm) by increments of 20 ppm.

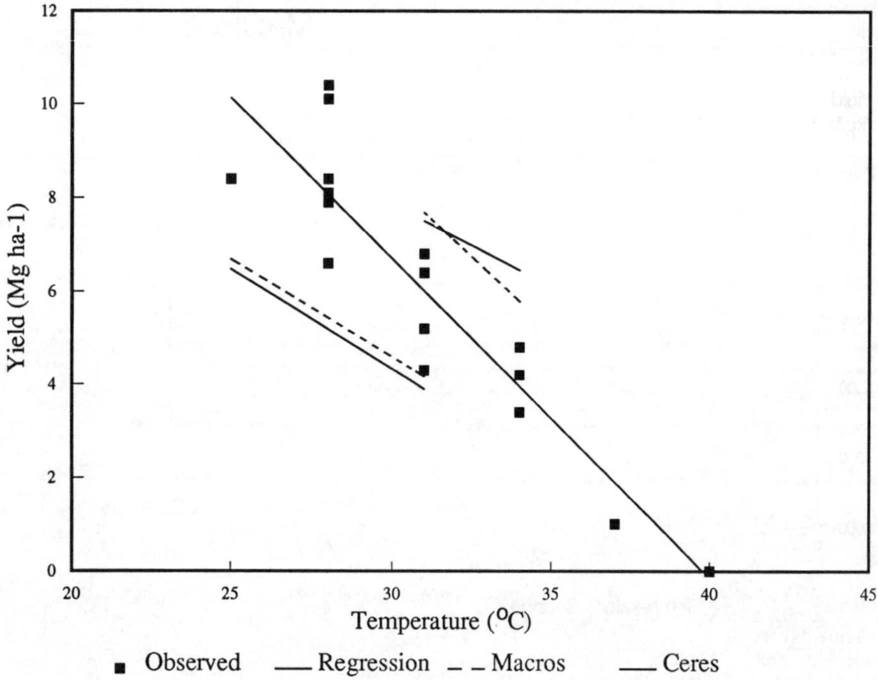

Figure 5. Comparison between experimental data (symbols) from Baker et al. (1990c), MACROS (dashed line) and CERES–Rice (solid line) predictions of the relationship between rice yield and temperature. The regression line was drawn using the observed data points. The lower curve for both CERES and MACROS corresponds to the relationship between minimum temperatures and yield. The upper curve corresponds to the relationship between maximum temperatures and yield. The regression curves between predicted yield and temperature were drawn on this graph using the minimum and maximum values of minimum and maximum temperature records used to run the models as boundaries to the curves. Experimental data and predictions correspond to ambient CO_2 conditions (330 ppm).

model. Jansen (1990) used MACROS and a modified historical database (China, India, Indonesia, Thailand and South Korea) and reported a 10% increase in potential yield following about 4°C increase in mean annual temperature (0.3°C yearly increase) and a tripling of CO_2 concentration (to about 1000 ppm) in the year 2100. Both models are based on the concept of a generic crop model, and are thus capable of world-wide calibration and testing for several crops. MACROS requires a greater number of climatic inputs, but its code is also structurally easier to modify, compared to CERES–Rice, originally a wheat model adapted to other crops, which includes 2–3 times as many subroutines as MACROS. MACROS'code is relatively well documented, but there is no available user's guide.

Ecosystem model
RICESYS (Graf et al. 1990a, b) is a demographic model for rice growth and development as affected by temperature and solar radiation. It was written to study crop-pest interactions and provide a useful tool for the development of sound integrated pest management strategies. It was designed to capture the dynamics of the crop-weed competition and to predict dry matter allocation with respect to different weeding strategies (Graf et al. 1990b). It does not include the effect of ambient CO_2 concentrations on plant physiology. A distributed delay model is used to describe the dynamics of tiller production and culm, leaf, root and grain mass growth (Graf et al. 1990a). It also includes several species of leaffolders feeding on leaves and weeds competing with rice for limiting solar irradiance or nitrogen.

All the nutrients are assumed to be non-limiting, except for nitrogen, and water is assumed to be abundant in the irrigated paddies. The model includes a simple soil nitrogen balance. RICESYS also assumes a negligible effect of typhoons and no pests other than herbivorous leafhoppers.

The Frazer-Gilbert functional response model is used to predict photosynthesis. This approach allows the simulation of energy acquisition at different trophic levels. Photosynthesis is a function of solar radiation, temperature, leaf area index and of the plant's demand for carbohydrate. Nitrogen stress (ratio between actual and potential nitrogen content) also modifies photosynthesis. Maintenance respiration is a direct function of temperature. Photosynthate accumulates in a metabolic pool and is used to satisfy respiration first, reproduction, growth and reserves with decreasing priorities. Carbohydrate and nitrogen are partly recycled from dying plant material. The demographic component of the model is a bookkeeping device for births, deaths, growth and aging of mass and numbers of plant subunits.

RICESYS was originally calibrated to represent the variety Makalioka 34 from Madagascar in irrigated paddies that were fertilized before transplanting. It was secondarily validated for high-yielding varieties (IR54 and IR68) in the Philippines (Graf et al. 1991c).

Since RICESYS does not include CO_2 effects, only its sensitivity to temperature is presented (Figure 6). It predicts a 43% decrease in yield following a 5°C temperature increase, assuming no pest damage, which compares well with Baker et al.'s (1990c) 50%. RICESYS offers the opportunity to look at the effect of climatic change at the system level. It is a well-structured, well-documented program that lends itself easily to inclusions and modifications. It also constitutes a radically different approach to partitioning carbohydrates from the three other models. Demand drives allocation. In the context of climate change, crop development calendars will change and the modeling technique of associating a different allocation coefficient for each phenological stage may not be adequate, since the duration of these changes may be unknown. RICESYS, which includes competition between rice and its natural weeds and herbivory, simulates actual yield while both CERES–Rice and MACROS estimate potential yield, which is rarely

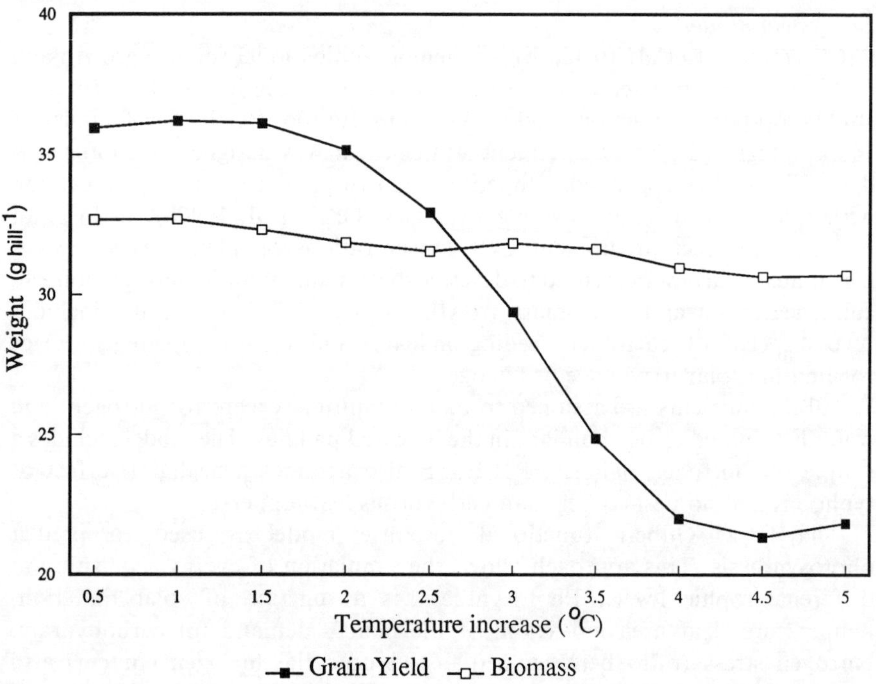

Figure 6. Sensitivity of RICESYS to temperature using climatic inputs from Los Baños (1984) with a mean average annual temperature of 26.7°C, arbitrarily increased by 0.5°C increments.

attained in field conditions in the tropics. Using a climate scenario with RICESYS can actually give the rice farmer a better idea of how yields may change over the next 50 years.

Empirical models

The complexity of the interactions between climatic variables and soil and plant variables makes an accurate analysis of relationships quite difficult. Several attempts have been made to use correlation and regression methods to study the effect of climate on rice yield (Table 6). These agrometeorological models may be either short-term, to forecast yield of current crops, or longer-term, based on multiple-year weather and yield time series. In this paper, we report results for Thailand from exploratory regressions of observed annual yield on climate, since our goal was to assess the strengths of yield-climate relationships rather than to develop predictive models. Thailand was chosen as a test country since it is a major rice exporter, ranking 5th in total production and accounting for about 31% of the total world export.

Definition of candidate regressors is a major issue in weather-based regression models. Specifically, the analyst must define critical periods within the crop cycle and then decide how best to summarize temperature, radiation, and rainfall during those periods; the resulting set of derived weather variables

Table 6. List of a few empirical models used in Asia to predict rice yield and of the climatic variables (PPT = precipitation, T = air temperature, I = solar radiation, other = others such as price of rice) used as driving variables.

REFERENCES	COUNTRY	PPT	T	I	Other
Khan and Chatterjee, 1987	India	+			+
Gao et al., 1987	China		+	×	+
Agrawal and Jain, 1982	India	+	+		+
Jain, Agrawal and Jha, 1980	India	+	+	+	+
Da Mota and da Silva, 1980	Brazil		+	+	
Yao and Leduc, 1980	Japan	+	+	+	
	Taiwan	+	+	+	
	China	+	+	+	+
Tsujii, 1978	Thailand	+			+
Huda et al., 1975	India	+	+		+
Kudo, 1975	Japan		×	+	
Stansel, 1975	Texas		×	+	
Yoshida and Parao, 1976	Philippines		+	+	
Munakata, 1976 and et al. 1967	Japan		+	+	
Murakami et al., 1973	Japan		+	+	+
Ueno, 1971	Japan		+	+	
Das, 1971	India	+	+		+
Murata and Kudo, 1968	Japan		+	+	
Moomaw et al., 1967	Japan			+	
Hanyu et al., 1966	Japan		+	+	+
Murata, 1964	Japan		+	+	

should be highly correlated with crop yield (Murata 1975; Seshu and Cady 1984; Huda et al. 1975). Taking this idea to an extreme, one can systematically search a large set of time windows and corresponding aggregation methods to find that combination whose weather statistics give maximum correlation with yield (Coakley et al. 1988).

Most agrometeorological models include a technology trend regressor when modeling multi-year yield time series (e.g. Yao and LeDuc 1980; da Mota and da Silva 1980; Agrawal and Jain 1982). The time trend expresses the increasing technology variable frequently seen in crop yields, particularly for emerging nations, and it acts as a surrogate for those nations' increasing usage of technologies such as high-yield rice varieties, fertilizers, pesticides, and irrigation. An increasing time trend is apparent for Thailand rice yields, and was included in our models by using time (Year) as a candidate regressor.

We fit the following model separately to wet and dry season yields:

$$Y_{ij} = \left(\sum_{i=1}^{18} \beta_i Z_i \right) + \alpha_1 X_{1ij} + \alpha_2 X_{2ij} + \alpha_3 X_{3ij} + \alpha_4 X_{4ij} + \alpha_{5j} \qquad (1)$$

where:

i = Station index,		j = Year index,	
Y = Annual yield,		X_1 = Mean T_{max},	
X_2 = Mean T_{min},		X_3 = Mean Daily Radiation,	
X_4 = Total Rainfall,		α, β = regression coefficients,	
Z_i = 1 for Station i, 0 for other stations.			

This equation specifies a different average yield level for each station, but all stations have a common slope with respect to each of the other regressors (Carter and Zhong 1988).

For both wet and dry seasons, none of the climate variables made a significant contribution to overall yield variation; only the time trend was significant (Model 1, Table 7). Climate variables themselves did not show noticeable time trends (the correlation between each variable and Year was less than < 0.25 in all cases), so the time trend was due to non-climate factors such as technology. The negligible contribution of climate variables is emphasized by comparing r^2 for Model 1 with that of Model 2, which contains the time trend alone (Table 7). Finally, the results for Model 3, which contains no temporal explanatory variables, show that most of the yield variation, especially in the wet season, may be due to station-to-station differences. Stansel and Fries (1980) point out that solar radiation is not an important yield limitation for regions like Thailand where mean yield does not exceed 4 t ha^{-1} and, because of Thailand's extensive irrigation system, it was expected that annual rainfall would not be strongly related to yield. The lack of a significant temperature effect (Table 7) however, is in sharp contrast to the temperature responses of the simulation models.

These are crude models. It is quite possible, for example, that temperature, radiation, or rainfall might show a measurable impact on yield if critical averaging periods were chosen more carefully. However, we wish simply to illustrate that Thailand yield statistics show an ambiguous climate signal, against a backdrop of other sources of variation. Total yield variation is dominated by station-to-station differences across the country, due in part to spatial variation in long-term average climatic conditions. However, the spatial component of variation also might be due to economic and political factors, such as landholding patterns and availability of agricultural inputs. Regardless of its source, spatial variation can mask temporal trends in crop production.

Most important, attempts to model long-term trends in yields as a function of climate change will be seriously inhibited by the strong technology trends which are already seen in rice production time series (Seshu and Cady 1984). For example, the time trend regression coefficients of Model 1 above correspond to a yield increase of about 5% and 10% per decade, for wet and dry seasons respectively. In contrast, Jansen's (1990) simulations predict a maximum increase in rice yield of 10% over the next four decades.

It could be argued that Thailand rice yields would not be expected to show a distinct connection with temporal weather patterns, due to the region's stable weather history at levels of rainfall and temperature which are well suited to rice

Table 7. Regressions of observed annual yield on climatic variables, for 18 Thailand stations during 1975–1987 (wet season crop) and 1980–1988 (dry season crop). All models contain indicator variables for individual stations, and table entries are P-values for additional regressors; X's denote excluded regressors.

	P-value for additional regressors					
	Mean Tmax	Mean Tmin	Mean rad.	Total PPT	Multiple year	r^2
Wet season (n = 218)						
Model 1	0.15	0.87	0.57	0.16	0.0004	0.878
Model 2	x	x	x	x	0.0001	0.876
Model 3	x	x	x	x	x	0.865
Dry season (n = 121)						
Model 1	0.84	0.30	0.54	0.72	0.008	0.599
Model 2	x	x	x	x	0.002	0.589
Model 3	x	x	x	x	x	0.550

growing. In contrast, the rice-growing systems of other countries, such as China, have long been known to be strongly affected by weather (Kueh 1984). China's annual crop production has a boom and bust history, with relatively frequent crop failures attributed to extreme weather events such as droughts and floods (Xiang and Griffiths 1988; Feng and Fu 1989). The reality of highly-variable crop yields being driven by unpredictable, extreme events is ill-matched by the simulation and empirical modeling approaches described above, both of which focus on marginal changes in yield as a function of small changes in seasonally integrated climate variables. Moreover, time series of observed rice yield and cultivation area (IRRI 1991) also show strong technology (see Japan increasing yields and decreasing cultivation areas on Figure 7) and political (see Cambodia and China rice yield fluctuations against political events on Figure 7) trends which will mask potential climate-induced trends.

Conclusions

Our study suggests that reliable, long-term forecasts of climate-induced changes in Asian rice crop yields on regional and national scales may be difficult to produce. Although growth simulation models may perform well for site-specific field trials, they do not reflect farmers' reported yield, either in magnitude or in temporal pattern. For the time being, their utility is limited to suggesting how climate may alter potential rice yield. The overall impact of CO_2 on plant growth and yield remains uncertain because most extrapolations are still limited to model-based estimates that use data from greenhouse

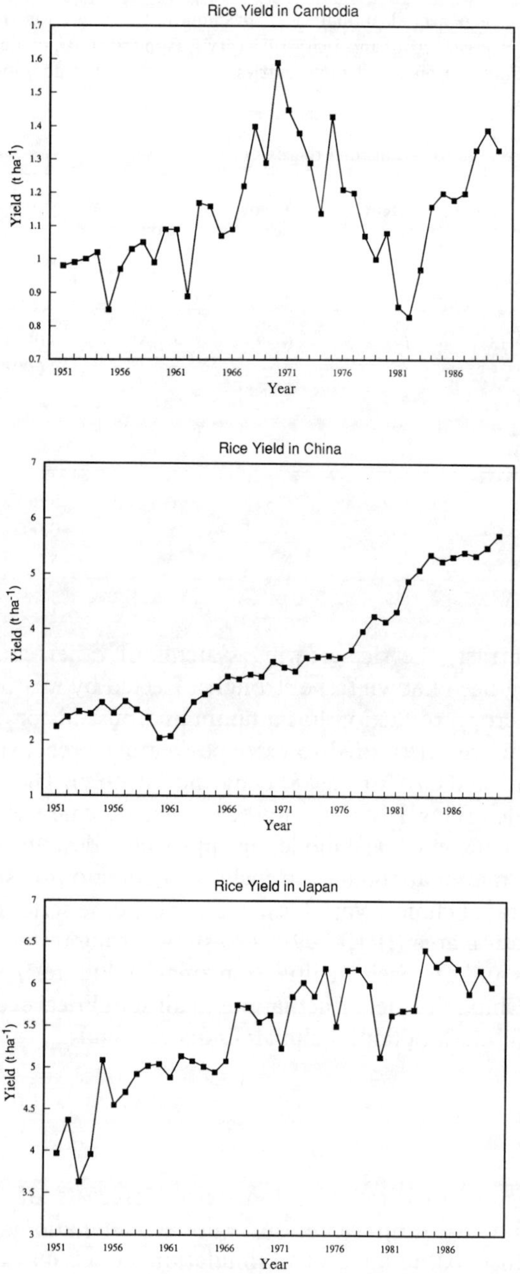

Figure 7. Rice yield (left hand side graphs) and rice cultivation area (right hand side graphs) for three countries in Asia: Cambodia, China and Japan. Note with regard to Cambodia: 1970 corresponds to the abolishment of the monarchy and the start of a 5 year-long civil war that corresponds to a precipitous drop in the size of the area cultivated and in yield. 1978 corresponds to the invasion by Vietnam that overthrows the Khmer Rouge, again with a corresponding drop in

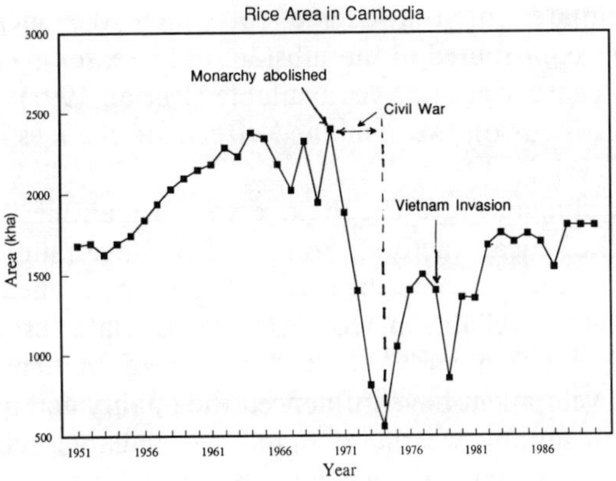

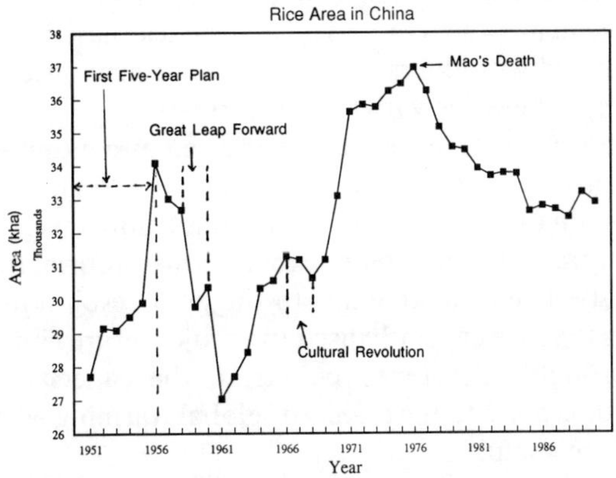

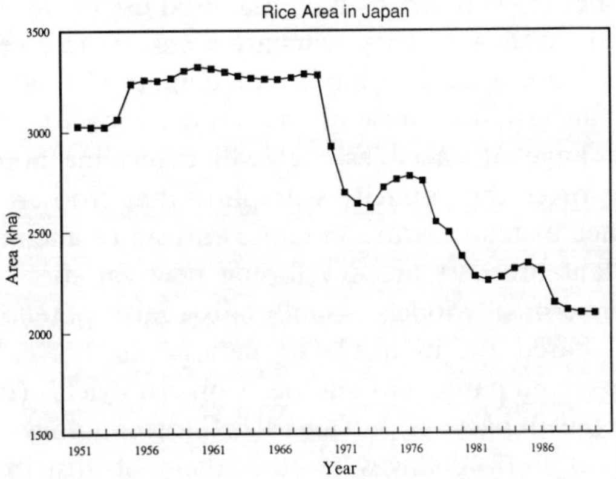

area cultivated with rice. Note with regard to China: 1953–1957 is the first 5-year plan; 1958–1960 the "Great Leap forward" with the abandonment of the soviet model; 1966–1968: the cultural revolution which starts the beginning of an exponential increase in the rice cultivation area; 1976: With the death of Mao Tse Tung, rice area decreases rapidly.

experiments. Estimates of the degree to which the increases in CO_2 concentration have contributed to the substantial increases in crop yields since the middle of the century are not yet available (Ruttan 1991). Moreover, few greenhouse studies focus on the combined effects of changes in temperature, rainfall and CO_2 levels.

The effects of climate are difficult to detect, let al.one quantify, in the presence of technological factors. Some other uncertainties inherent to predicting climate change impacts on yield need to be mentioned here. First, the essential requirement of reliable meteorological and related crop yield data for a series of years is not frequently met in Asia, where political instability and rapid economic development have influenced the quality and quantity of data collected. Although suitable weather data may be available, reliable long-term regional yield estimates may be unavailable because production is difficult to estimate when most grain is consumed on the farm. Furthermore, recent changes in management practices and rice varieties may mean that weather effects are confounded with the rate of adoption of new systems and with the different responses of new rice varieties to weather.

Effects of climate change are generally expressed by continuous functions of the CO_2 increase or of temperature (Okamoto et al. 1991). In reality, even if climatic means remain the same, more frequent and more severe fluctuations of temperature and precipitation could occur. This information, critical to agronomic forecasts, is unavailable due to the poor resolution of the general circulation models that are currently used to predict future climate. Moreover, even in the most highly developed countries, the capacity of agricultural research systems to respond to the effects of global warming will depend on the rate at which climate change occurs.

Finally, we believe that rather than trying to improve our estimate of future yield production, better estimates of potential land use changes and economic development in Asia in the next fifty years are necessary to give any credibility to the future yield forecasts. The impact of climate change on agricultural potential remains uncertain because of the limited capacity of the models to consider the full range of variables that will determine potential shifts in cropland. There is often the implicit assumption that crop area and yield are narrowly determined by temperature, soil moisture and sometimes, soil quality (Ruttan 1991). Plant breeders are developing new varieties which push the climatic limits of crops. Models usually disregard potential impacts of biotechnology – -based on molecular genetics– and of advances in the protection of crops from pathogens and pests on crop yields (Ruttan 1991). It seems likely that the impacts of advances in biotechnology on agricultural potential, including both changes in the regional distribution of crop production and increases in yields, will dwarf the impact of climate change currently predicted from a doubling in CO_2 (Ruttan 1991).

To explore at regional and greater scales, how the boundaries of rice cultivation areas may shift with changes in climate, seems an uncertain and arduous enterprise. However, the foundations for this approach are already in place, in the form of regional-scale agroclimatic classification schemes (Oldeman 1980; Gao et al. 1987). We envision expanding the classification approach, through the use of remote sensing technology to identify the present spatial distribution of Asian rice on a regional and continental scale, perhaps to the level of differentiating among various cropping systems (Malingreau 1986). We would then recommend studying the correlations between existing rice distribution maps and maps of long-term climatic means and variances, with the goal of constructing a classification model. Eventually, it should be possible to predict how agroclimatic regions suitable for rice might expand, contract or migrate in response to alternative climates caused by the increase in greenhouse gases.

Acknowledgments

We greatly appreciate the helpful comments from Linda Mearns, Klaas Metselaar, Kate Dwire, Ann Hairston, and Steve Wondzell.

The research described in this article has been funded by the U.S. Environmental Protection Agency (EPA). This document has been prepared at the EPA Environmental Research Laboratory in Corvallis, Oregon, through contract #68–C8–0006 to ManTech Environmental Technology, Inc. It has been subjected to Agency review and approved for publication.

References

Adams R M, McCarl B A, Dudek D J, Glyer J D (1988) Implications of global climate change. West. J. Agric. Econ. 13:348–356.

Agrawal R, Jain R C (1982) Composite model for forecasting rice yield. Indian J. Agric. Sci. 52:177–181.

Akita S, Tanaka I (1973) Studies on the mechanism of differences in photosynthesis among species. IV. The differential response in dry matter production between C_3 and C_4 species to atmospheric carbon dioxide enrichment. Proc. Crop Sci. Soc. Jap. 42:288–295.

Alocilja E C, Ritchie J T (1988) Upland rice simulation and its use in multicriteria optimization. IBSNAT Research Report Series 01. 96 p.

Baker J T, Allen L H, Boote, K J (1990a) Growth and yield responses of rice to carbon dioxide concentration. J. Agric. Sci. 115:313–320.

Baker J T, Allen L H, Boote K J, Jones P, Jones, J W (1990b) Developmental responses of rice to photoperiod and carbon dioxide concentration. Agric. For. Meteorol. 50:201–210.

Baker J T, Allen L H, Boote K J, Rowland-Bamford A J, Waschmann R S, Jones J W, Jones P H, Bowes (1990c) Response of vegetation to carbon dioxide: Temperature effects on rice at elevated CO_2 concentration. DOE Progress report No. 60. 71 p.

Boer G et al. (1989) Equilibrium response studies with the CCC climate model. Pers. comm. as cited in Cubash U, Cess, R D, Processes and modelling. Pages 69–91 in Houghton J T, Jenkins G J, Ephraums J J (Eds.) Climate change, the IPCC scientific assessment. University Press, Cambridge, UK.

Bolin W, Doos B R, Jager J, Warrick R A (1986) The greenhouse effect, climatic change, and ecosystems. Scope 29. John Wiley and Sons. 541 p.

Campbell I M (1989) Does climate affect host-plant quality? Annual variation in the quality of balsam fir as food for the spruce budworm. Oecologia 81:341–344.

Carter C A, Zhong F (1988) China's grain production and trade: An economic analysis. Westview Press, Boulder, CO. 124 p.

Cess R D, Potter G L (1987) Exploratory studies of cloud radiative forcing with a general circulation model. Tellus 39a:460–473.

Coakley S M, McDaniel L R, Line R F (1988) Quantifying how climatic factors affect variation in plant disease severity: A general method using a new way to analyze meteorological data. Clim. Change 12:57–75.

Cock J H, Yoshida S (1973) Changing sink and source relations in rice (*Oryza sativa* L.) using carbon dioxide enrichment in the field. Soil Sci. Plant Nutrit. 19:229–234.

Cohen S J (1990) Bringing the global warming issue closer to home: The challenge of regional impact studies. Bull. Am. Met. Soc. 71:520–526.

Cooter E J (1990) The impact of climate change on continuous corn production in the southern USA. Clim. Change 16:53–82.

Crosson P (1989) Climate change and mid-latitudes agriculture: Perspectives on consequences and policy responses. Clim. Change 15:51–73.

Cure J D (1985) Carbon dioxide doubling responses: A crop survey. Pages 99–116 in Strain B R, Cure J D (Eds.) Direct effects of increasing carbon dioxide on vegetation. U.S. Dept. of Energy, DOE/ER-0238.

Da Mota F S, da Silva J B (1980) A weather-technology model for rice in southern Brazil. Pages 235–238 in Agrometeorology of the rice crop. World Meteorological Organization and the International Rice Research Institute, Los Baños, Philippines.

Das J C, Mehra A K, Madnani M L (1971) Forecasting the yield of principal crops in India on the basis of weather–paddy/rice. Indian J. Met. Geophys. 22:47–58.

De Wit C T, Penning de Vries F W T (1982) La synthèse et la simulation des systèmes de production primaire. Pages 23–27 in Penning de Vries F W T, Djiteye M A, (Eds.). La productivité des pâturages sahéliens. Agricultural research reports 918. PUDOC Wageningen, The Netherlands.

Dickinson R E (1989) Uncertainties of estimates of climatic change: A review. Clim. Change 15:5–13.

Fajer E D (1989) The effects of enriched CO_2 atmospheres on plant-insect herbivore interactions: Growth responses of larvae of the specialist butterfly, *Junonia coenia* (Lepidoptera: Nymphalidae). Oecologia 81:514–520.

Feng D, Fu L (1989) Main meteorological problems of rice production and protective measures in China. Int. J. Biometeorol. 33:1–6.

Gao L Z, Jin Z Q, Li L (1987) Photo-thermal models of rice growth duration for various varietal types in China. Agric. For. Meteorol. 39:205–213.

Giorgi F (1990) On the simulation of regional climate using a limited area model nested in a general circulation model. J. of Climate 3:941–963.

Giorgi F, Mearns L O (1991) Approaches to the simulation of regional climate change: A review. Reviews of Geophysics 29:191–216.

Godwin D C, Singh U, Buresh R J, De Datta S K (1990) Modeling of nitrogen dynamics in relation to rice growth and yield. Pages 320–325 in Proceedings 14th International Congress of Soil Science Transactions, Kyoto, Japan, Aug. 1990, Vol IV, Int. Soc. of Soil Science.

Goudriaan J, Unsworth M H (1990) Implications of increasing carbon dioxide and climate change for agricultural productivity and water resources. Pages 111–129 in Impact of carbon dioxide, trace gases and climate change on global agriculture, ASA Special Publication no. 53.

Graf B, Rakotobe O, Zahner P, Delucchi V, Gutierrez A P (1990a) A simulation model for the dynamics of rice growth and development. I. The carbon balance. Agric. Syst. 32:341–365.

Graf B, Gutierrez A P, Rakotobe O, Zahner P, Delucchi V, (1990b) A simulation model for the dynamics of rice growth and development. II. The competition with weeds for nitrogen and light. Agric. Syst. 32:367–392.

Graf B, Dingkuhn M, Schnier F, Coronel V, Akita S (1991) A simulation model for the dynamics of rice growth and development. III. Validation of the model with high yielding varieties. Agric. Syst. 36:329–349.

Grotch S L (1988) Regional intercomparisons of general circulation model predictions and historical climate data. US Dept. of Energy, Carbon Dioxide Research Division, Washington, DC. DOE/NBB–0084, USA.

Gutowski W J, Gutzler D S, Portmand D, Wang W C (1988) Surface energy balance of three general circulation models: Current climate and response to increasing atmospheric CO_2. US Dept. of Energy, Carbon Dioxide Research Division, Washington, DC. DOE/ER/60422–H1, USA. 119 p.

Hammer G L, Woodruff D R, Robinson J B (1987) Effects of climatic change on reliability of wheat cropping–a modelling approach. Agric. For. Meteorol. 41:123–142.

Hansen J, Fung I, Lacis A, Rind D, Lebedeff S, Ruedy R, and Russell G (1988) Global climate changes as forecast by Goddard Institute for Space Studies three-dimensional model. J.Geophys. Res. 93:9341–9364.

Hanyu J, Uchijima T, Sugawara S (1966) Studies on the agro-climatological method for expressing the paddy rice products. I. An agro-climatic index for expressing the quantity of ripening of the paddy rice. Bull. Tohoku Nat. Agric. Exp. Sta. 34:27–36.

Hatch, W (1986) Selective guide to climatic sources. Key to meteorological records documentation. National Climatic Data Centre, Asheville, NC, USA.

Horie, T (1987) A model for evaluating climatic productivity and water balance of irrigated rice and its application to southeast Asia. Southeast Asian Studies 25:62–74.

Houghton J T, Jenkins, G J, Ephraums J J, (Eds.) (1990) Climate change, the IPCC scientific assessment. University Press, Cambridge, UK. 365 p.

Huda A K S, Ghildyal B P, Tomar V S, Jain R C (1975) Contribution of climatic variables in predicting rice yield. Agric. Meteorol. 15:71–86.

Imai K, Murata Y (1976) Effect of carbon dioxide on growth and dry matter production of crop plants. I. Effects on leaf area, dry matter, tillering, dry matter distribution ratio, and transpiration. Proc. Crop Sci. Soc. Jap. 45:598–606.

Imai K, Murata Y (1978a) Effect of carbon dioxide on growth and dry matter production of crop plants. III. Relationship between CO_2 concentration and nitrogen nutrition in some C_3 and C_4 species. Japanese J. Crop Sci. 47:118–123.

Imai K, Murata Y (1978b) Effect of carbon dioxide on growth and dry matter production of crop plants. V. Analysis of after-effect of carbon-dioxide-treatment on apparent photosynthesis. Japanese J. Crop Sci. 47:587–595.

Imai K, Murata Y. (1979a) Effect of carbon dioxide on growth and dry matter production of crop plants. VI. Effect of oxygen concentration on the carbon-dioxide-dry matter production relationship in some C_3 and C_4 species. Japanese J. Crop Sci. 48:58–65.

Imai K, Murata Y. (1979b) Effect of carbon dioxide on growth and dry matter production of crop plants. VII. Influence of light intensity and temperature on the effect of carbon dioxide-enrichment on some C_3 and C_4 species. Japanese J. Crop Sci. 48:409–417.

International Rice Research Institute (1989) Implementing the strategy, Work Plan for 1990–1994, International Rice Research Institute, Los Baños, Philippines.

International Rice Research Institute (1991) World Rice Statistics. International Rice Research Institute, Los Baños, Philippines.

Jain R C, Agrawal R, Jha M P (1980) Effect of climatic variables on rice yield and its forecast. Mausam 31:591–596.

Jansen D M (1990) Potential rice yields in future weather conditions in different parts of Asia. Netherlands J. Agric. Sci. 38:661–680.

Jansen D M, Gosseye, P (1986) Simulation of growth of millet (*Pennisetum americanum*) as influenced by water stress. Simulation Reports CABO-TT No. 10. Centre for Agrobiological Research, Wageningen, The Netherlands.

Johnson R H, Lincoln D E (1990) Sagebrush and grasshopper responses to atmospheric carbon dioxide concentration. Oecologia 84:103–110.

Joyce L A, Fosberg M A, Comanor J M (1990) Climate change and America's forests. USDA Forest Service. General Technical report RM–187. 12 p.

Keeling C D, Carter A F, Mook W G (1984) Seasonal, latitudinal, and secular variations in the abundance and isotopic ratios of atmospheric CO_2. J. Geophys. Res. 89:4615–4628.

Khan S A, Chatterjee B N (1987) Crop weather analysis for forecasting the yield of autumn rice in West Bengal. Indian J. Agric. Sci. 57:791–794.

Kickert R N (1984) Names of published computer models in the environmental biological sciences: A partial list and new potential risks. Simulation 22–39.

Kudo K (1975) Economic yield and climate. Pages 199–220 in Y. Murata (Ed.) Crop productivity and solar energy utilization in various climates in Japan. Tokyo University Press, Tokyo, Japan.

Kueh Y Y (1984) A weather index for analyzing grain yield instability in China, 1952–81. China Quarterly 97:68–83.

Lamborg M R, Hardy R W F, Paul E A (1983) Microbial effects. Pages 131–176 in Lemon E R, (Ed.) CO_2 and plants: The response of plants to rising levels of atmospheric carbon dioxide. AAAS selected Symp. Ser. 84. Westview Press, Boulder, CO.

Lincoln D E, Couvet D, Sionit N (1984) Response of an insect herbivore to host plants grown in carbon dioxide enriched atmospheres. Oecologia 69:556–560.

Lincoln D E, Sionit N, Strain B R (1986) Growth and feeding response of *Pseudoplusia includens* (Lepidoptera: Noctuidae) to host plants grown in controlled carbon dioxide atmospheres. Environ. Ent. 13:1527–1530.

Malingreau, J P (1986) Global vegetation dynamics: Satellite observations over Asia. Int. J. Remote Sensing 7:1121–1146.

McMennamy J A, O'Toole J C (1983) RICEMOD: A physiologically based rice growth and yield model. IRRI Research Paper Series No. 87. 33 p.

Mearns L O (1990) Future directions in climate modeling: A climate impacts perspective. Pages 51–58 in Wall G, Sanderson M (Eds.) Climate change: Implications for water and ecological resources. Proceedings of an International Symposium, Dept. of Geography, Univ. of Waterloo, Canada, Publishers.

Mearns L O (1991) Climate variability: Possible changes with climate change and impacts on crop yields. Pages 147–157 in Wall, G, (Ed.) Symposium on the impacts of climatic change and variability on the Great Plains. Dept. of Geography, Univ. of Waterloo, Canada, Publishers.

Mitchell J F B, Senior C A, Ingram W J (1989) CO_2 and climate: A missing feedback? Nature 341:132–134.

Moomaw J C, Baldazo P G, Lucas, L (1967) Effects of ripening period environment on yields of tropical rice. IRC Newsletter, Spec. Issue 18–25.

Morison J I L, Gifford R M (1983) Stomatal sensitivity to carbon dioxide and humidity. A comparison of two C_3 and two C_4 grass species. Plant Physiol. 71:789–796.

Morison J I L, Gifford R M (1984) Plant growth and water use with limited water supply in high CO_2 concentrations. II. Plant dry weight, partitioning and water use efficiency. Aust. J. Plant Physiol. 11:375–384.

Munakata K (1976) Effects of temperature and light on the reproductive growth and ripening of rice. Pages 187–207 in Climate and rice, IRRI, Los Baños, Philippines.

Munakata K, Kawasaki T, Kariya K (1967) Quantitative studies on the effects of the climatic factors in the productivity of rice. Bull. Chugoku Agric. Exp. Sta. A. 14:59–96.

Murakami T, Wada M, Yoshida Z (1973) Quantitative study in paddy rice plant growth versus climate in cool weather districts. Bull. Tohoku Agric. Exp. Stat. 45:33–100.

Murata Y (1964) On the influence of solar radiation and air temperature upon the local differences in the productivity of paddy rice in Japan. Proc. Crop Sci. Soc. Jap. 33:59–63.

Murata Y (1975) Estimation and simulation of rice yield from climatic factors. Agric. Meteorol. 15:117–131.

Murata Y, Kudo S (1968) On the influence of solar radiation and air temperature upon the local differences in the productivity of paddy rice in Japan. Proc. Crop Sci. Soc. Jap. 38:179.

National Defense University (1980) Crop yields and climate change to the year 2000. Volume I.

Report on the second phase of a climate impact assessment. USDA, NOAA, Institute for the Future, Defense Advanced Research Projects Agency. Fort Lesley J. McNair, Washington D.C. 20319, USA. 128 p.

Okamoto K, Ogiwara T, Yoshizumi T, Watanabe, Y (1991) Influence of the greenhouse effect on yields of wheat, soybean and corn in the United States for different energy scenarios. Clim. Change 18:397–424.

Oldeman, L R (1980) The agroclimatic classification of rice-growing environments in Indonesia. Pages 47–56 in Agrometeorology of the rice crop. World Meteorological Organization and the International Rice Research Institute, Los Baños, Philippines.

Osbrink W L A, Trumble J T, Wagner R E (1987) Host suitability of *Phaseolus lunata* for *Trichoplusia ni* (Lepidoptera: Noctuidae) in controlled carbon dioxide atmospheres. Environ. Entomol. 16:639–644.

Overdiek D, Reid C, Strain B R (1988) The effects of preindustrial and future CO_2 concentrations on growth, dry matter production and the C/N relationship in plants at low nutrient supply: *Vigna unguiculata* (cowpea), *Abelmoschus eschulentus* (okra) and *Raphanus sativus* (radish). Angew. Bot. 62:119–134.

Panturat S, Eddy A (1990) Some impacts on rice yield from changes in the variance of precipitation. Clim. Bull. 24:16–27.

Parry M L, Carter T R, Konijn N T, (Eds.) (1988) The impact of climatic variations on agriculture. Kluwer Academic Publishers, Dordrecht, The Netherlands. 2 Volumes. 876 p. and 764 p.

Patterson D T, Flint, E P (1990) Implications of increasing carbon dioxide and climate change for plant communities and competition in natural and managed ecosystems. Pages 83–110 in Kimball B A, Rosenberg N, Allen L H, Heichel G H, Stuber C W, Kissel D E, Ernst S, (Eds.) Impact of carbon dioxide, trace gases, and climate change on global agriculture. ASA Special Publication No. 53, ASA, CSSA, SSSA.

Penman H L (1956) Evaporation: An introductory survey. Netherlands J. Agric. Sci. 4:9–29.

Penning de Vries F W T, Van Laar, H H, (Eds.) (1982) Simulation of plant growth and crop production. Simulation Monographs. PUDOC, Wageningen, The Netherlands.

Penning de Vries F W T, Jansen D M, Ten Berge H F M, Bakema A, (Eds.) (1989) Simulation of ecophysiological processes of growth in several annual crops. IRRI, Los Baños. Simulation Monographs 29, PUDOC, Wageningen, The Netherlands. 272 p.

Priestly C H B, Taylor R J (1972) On the assessment of surface flux and evaporation using large scale parameters. Monthly Weather Rev. 100:81–92.

Roger P A, Watanabe I (1986) Technologies for using biological nitrogen fixation in wetland rice: Potentialities, current usage, and limiting factors. Fert. Res. 9:39–77.

Roger P A, Ladha J K (1990) Estimation of biological N_2 fixation and its contribution to nitrogen balance in wetland rice fields. In Proceedings 14th International Congress of Soil Science Transactions, Kyoto, Japan, Aug. 1990, Int. Soc. of Soil Science.

Ruttan V W (1991) Review of Climate Change and World Agriculture. Environment 33:25–29.

Schlesinger M E, Zhao Z C (1989) Seasonal climatic change introduced by doubled CO_2 as simulated by the OSU atmospheric GCM/mixed-layer ocean model. J. Climate 2:429–495.

Schneider S H, Gleick, P H, Mearns, L O (1989) Prospects for climate change. Pages 41–74 in Waggoner P E (Ed.) Climate change and US water resources, Wiley and Sons, New York, USA.

Seshu, D V, Cady F B (1984) Response of rice to solar radiation and temperature estimated from international yield trials. Crop Sci. 24:649–654.

Seshu D V, Woodhead T, Garrity D P, Oldeman L R (1989) Effect of weather and climate on production and vulnerability of rice. Pages 93–121 in Climate and food security. International Rice Research Institute, Los Baños, Philippines.

Smith J B, Tirpak D A, (Eds.) (1989) The potential effects of global climate change on the United States. U.S. Environmental Protection Agency, Chapter 4, Methodology, Vol. 1.

Stansel J W (1975) Effective utilization of sunlight. in Six decades of rice research in Texas. Texas A&M. Univ. Bull., USA. 136 p.

Stansel J W, Fries R E (1980) A conceptual agromet rice yield model. Pages 201–212 in

Agrometeorology of the rice crop. World Meteorological Organization and the International Rice Research Institute, Los Baños, Philippines.

Strain B R (1985) Physiological and ecological controls on carbon sequestering in terrestrial ecosystems. Biogeochemistry 1:219–232.

Strain B R (1987) Direct effects of increasing atmospheric CO_2 on plants and ecosystems. Trend Ecol. Evol. 2:18–21.

Stroosnijder L (1982) Simulation of soil water balance. Pages 175–193 in Penning de Vries F W T and van Laar H H (Eds.) Simulation of plant growth and crop production. Simulation Monographs. PUDOC, Wageningen, The Netherlands.

Tsujii H (1978) Effect of climatic fluctuation on rice production in continental Thailand. Pages 167–179 in Takahashi K, Yoshino M M (Eds.) Climatic change and food production, Proceedings of International Symposium on Recent Climate Change and Food Production. October 4–8, 1976, Tsukuba and Tokyo, Japan.

Ueno Y (1971) A study on the method for estimating rice yield from climatic factors. PhD Thesis, Kyoto Univ., Japan. 70 p.

Van Keulen H (1975) Simulation of water use and herbage growth in arid regions. Simulation Monographs, PUDOC, Wageningen, The Netherlands.

Wetherald R T, Manabe S (1989) Pers. comm, as cited in Cusbash U, Cess R D, Processes and Modelling. Pages 69–91 in Houghton J T, Jenkins, G J, Ephraums, J J, (Eds.) Climate change, the IPCC scientific assessment. University Press, Cambridge, UK.

Wilson C A, Mitchell J F B (1987) A doubled CO_2 climate sensitivity experiment–with a global climate model including a simple ocean. J. Geophys. Res. 92:13315–13343.

Xiang S, Griffiths, J F (1988) A survey of agrometeorological disasters in South China. Agric. Forest Meteorol. 43:261–276.

Yao A Y M, LeDuc S K (1980) An analogue approach for estimating rice yield in China. Pages 239–248 in Agrometeorology of the Rice Crop. World Meteorological Organization and the International Rice Research Institute, Los Baños, Philippines.

Yoshida S (1976) Rice. Pages 57–87 In: Ecophysiology of tropical crops. Alvim P de T, Koslowski T T (eds.) Academic Press, New York, USA.

Yoshida S (1973) Effects of CO_2 enrichment at different stages of panicle development on yield components and yield of rice (*Oryza sativa* L.). Soil Sci. Plant Nutrit. 19:311–316.

Yoshida S, Parao F T (1976) Climatic influence on yield and yield components of lowland rice in the tropics. Pages 471–494 in Climate and rice. International Rice Research Institute, Los Baños, Philippines.

Yoshino M, Uchijima Z, Tsujii H (1988) The implications for agricultural policies and planning. Pages 853–868 in Parry M L, Carter T R, Konijn N T (Eds.) The Impact of Climatic Variations on Agriculture. Volume 1: Assessments in cool temperate and cold regions. Kluwer Academic Publishers, Dordrecht, The Netherlands.

Rice production and climate change

F.W.T. PENNING DE VRIES
Center for Agrobiological Research (ABO-DLO), P.O. Box 14,
6700 AA Wageningen, The Netherlands

Key words: climate change, crop duration, cropping system, cultivated area, food demand, MACROS, plant breeding, rice, yield ceiling, yield potential, yield trend

Abstract
An eventual change of the climate in the next decades will have an impact on productivity of crops. Predicting effects of climate change on crop yields and area harvested therefore receives much attention. Yet, yield levels and area harvested will be affected much more by our management than by climate change.

A crude estimate of how much rice yields will increase in the next decades shows that in very large areas, particularly in East Asia, farmer's crops will produce at levels that now correspond with the ceiling yield of current rice varieties. This underlines the need for varieties with a significantly higher yield potential.

The effect of climate change on the ceiling yield was investigated. For 2020, with a rise in ambient CO_2 concentration of 50 ppm and in temperature of 1.0 °C, potential rice yields will increase on average by a few percent. The increase in the warmest regions is smallest. Lack of accurate predictions of climate change contribute to a fair amount of uncertainty of the conclusions. Yet, the small response implies that climate change in itself is not an issue in relation to the challenge of developing new varieties of irrigated tropical rice with a significantly higher yield potential. Optimal use of irrigation water will become an important issue for rice production.

Research should identify the regulation of the processes that make rice production sensitive to high temperature, and create varieties that are less sensitive.

Introduction

Increasing demand, increasing yield

By the year 2020, 65% more rice should be produced in the world to cope with expanding populations and with a raise in per capita income and food consumption (IRRI 1989). There are major regional differences. In Asia, where 90% of the world's rice is produced and the average food energy intake from rice ranges from 30–60%, four regions are distinguished:
- East Asia, represented by China,
- South Asia, represented by India,
- South East Asia, represented by Indonesia, the Philippines and Vietnam, and
- North East Asia, represented by Japan, North and South Korea.

These regions produce 40, 21, 15 and 6% of all rice in Asia, respectively, and differ significantly in population growth and changing of diets (Table 1). The regional average yields (Table 2) are related to the fraction of the rice land irrigated, because irrigated rice land generally yields much more than rainfed rice; irrigated areas in all four regions contribute most to national rice production (IRRI 1989).

F.W.T. Penning de Vries et al. (eds.), Systems Approaches for Agricultural Development, 175–189.

Table 1. Key numbers on the increasing demand for rice in the period 1989–2020. East Asia: data for China; South Asia: data for India; South East Asia: data Indonesia, Philippines, and Vietnam (weighted 3,1,1); North East Asia: data for Japan, North and South Korea (weighted 13, 3 and 5). Fraction irrigated refers to area. (Data from IRRI 1990).

Region	Population growth (10^6)	Population growth (% y^{-1})	Consumption years (% $capita^{-1}$)	After 30 irrigated (y)	Fraction (-)
E Asia	1104	1.0	0.15	1.41×	0.93
S Asia	835	1.6	0.6	1.92×	0.35
SE Asia	317	1.8	0.6	2.04×	0.54
NE Asia	189	0.6	−0.5	1.03×	0.92

Table 2. Actual yield level (1990) as compared to 2020 yield levels projected in two ways: by extrapolation with an annual yield increase of 80 kg ha^{-1} (column 2), and by dividing the required regional production by the expected area harvested (column 3). The simulated potential yield of crops without (column 4) and with the impact of climate change (column 5) are also shown. All values in t ha^{-1} rough rice (14% moisture).

Region	Current yield	Trend yield	Demand yield	Potential yield	Potential plus climate change
E Asia	5.3	7.7	8.9	10	10.5
S Asia	2.3	4.7	4.4	8	8.3
SE Asia	3.5	5.9	5.3	8	8.3
NE Asia	6.3	8.7	8.9	12	13

Expansion of the area under rice in Asia is generally not possible. The harvested area still increases in S and SE Asia (IRRI 1991) due to an increase in double cropping (often made possible by irrigation). But the rice area in E and NE Asia declines, as land is lost to non-agricultural purposes or used for other crops (IRRI 1991a; Chomchalow 1992), and most of the area is already under double cropping. This is probably an irreversible trend. Assuming that no large scale import will replace national production, the projected annual increase in demand for rice (Table 1) in E Asia (1.15%), in S Asia (2.2%), and in SE Asia (2.4%), should therefore be realized at a constant or slightly declining area for rice crops. The required annual production increase in NE Asia is only 0.1%. Yield increase can probably compensate for area reduction, and this topic is not further pursued. The combined increase in demand for rice in the four regions at these growth rates for 30 years adds up to the 65% mentioned earlier.

The required annual increase in rice production of 1.15–2.4% does not appear excessively high in view of the average annual growth over the past two decades in Asia of 2.3% (IRRI 1991b). Indeed, one generally finds growth rates of national average crop yields of 80 kg ha^{-1} y^{-1} or more (1.5–3% y^{-1} at current yield levels) once the mark of 2 t ha^{-1} is exceeded (Park et al. 1990; De Wit 1992). However, we should ask the question: can yields continue to increase at such a rate?

There are two ways to estimate future yield levels in the regions. One: extrapolate current yield levels for 30 years at the typical rate of progress in agriculture summarized in the value of 80 kg ha^{-1} y^{-1}; the result may be called a 'trend yield'. Two: divide the required rice production per region by the area harvested. This may be called a 'demand yield'. The change in area harvested was estimated by extrapolating the trend of the past 20 years (IRRI 1991). This leads to a change in area of rice harvested by 2020 of 0.84 × in E Asia, no change in S Asia, 1.35 × in SE Asia and 0.75 × in NE Asia. It is reassuring that both independent projections, shown in Table 2, are not very different. No attempt was made to improve these rough projections, as only the order of magnitude of future rice production levels is sought here.

Yield ceiling

We must realize, however, that such projections of future crop yields ignore limits. Yet, there is an ecophysiological maximum yield level, related to properties of rice crops and local weather. This ceiling, or potential yield, is reached when management is optimum and the crop utilizes radiation as efficiently as possible. The potential yield of current rice varieties is about 10 t ha^{-1} in the tropics and over 13 t ha^{-1} in temperate regions (Yoshida 1981), exact values depending on the weather during the season and the specific varieties used (Dua et al. 1990; Dingkuhn et al. 1992). The potential yield level of the new varieties bred for the tropics over the past 15 years has remained constant (IRRI 1991b), indicating that 'ceiling' is a proper term.

Measured maximum rice yields for a number of countries are presented in Table 3. Although these values are not very precise, they support Yoshida's values. They are from experiments done under the best known management conditions. Top yields reported in Japan exceed 12 t ha^{-1} (Horie 1992), confirming that the current yield plateau in Japan (6 t ha^{-1}) is not a biological, but an economic ceiling.

Table 3. Potential rice yields (rough rice, t ha^{-1}, 14% moisture), simulated and measured. The wide range in values within countries is due to differences in locations, seasons and varieties.

Country	Simulated	Measured	Reference
China	7–11	7.5	Pan Deyun et al. 1991
		8	Defeng et al. 1989
India	6–10	6–10	Mohandass, Murthy Ramasamy, TNAU, pers.comm.
Bangladesh	6–9	6–8	Dua et al. 1990
Indonesia	6–12	5–8	Daradjat and Fagi 1991
			Sutoro and Makarim 1991
Malaysia	7	4–7	Singh et al. 1991
Philippines	7–9	10	Herrera-Reyes and Penning de Vries 1989
Thailand	5–8	4–7	Dua et al. 1990

The level of potential yield can also be calculated, based on the key processes of growth and their relation with the environment. The potential yield is the resultant of production, translocation and conversion processes, and can be approximated by the equation:

$$PY = ((CP \cdot PE + TR) \cdot CF) \cdot 1.16$$

where:

PY	= potential grain production, in t ha^{-1}
CP	= net rate of canopy photosynthesis (around 0.3 t CH_2O ha^{-1} d^{-1})
PE	= grain filling period (25–60 d)
TR	= translocated stem reserves (up to 2.5 t ha^{-1})
CF	= growth conversion factor (about 0.75 t biomass t^{-1} CH_2O)
1.16	= conversion factor to express yield at 14% moisture.

Of these variables, the conversion efficiency is probably the most constant, and stem translocation and grain filling period the most variable (Penning de Vries 1991). However, in the tropics this filling period does not yet really exceed 30 days and is often considerably shorter due to early loss of leaf nitrogen (Penning de Vries et al. 1990a). At low temperatures in temperate regions, the filling period can reach 60 d (Yoshida 1981). It is the combination of a restricted filling period, maximum rate of net photosynthesis and limited translocation that set the ceiling yield at around 10 t ha^{-1} in the tropics. More elaborate calculations with a simulation model and local weather data provided the potential yields for different regions, shown in Table 3.

The maximum yield attained by farmers is below the ecophysiological ceiling. In actual situations, grain yields do not exceed 80–90% of the potential, because realizing potential yields requires an uneconomic intensity of management so that soil conditions are not always optimal, nutrients not always in ample supply and often pests are present. Some stress and imperfect crop protection are accepted in farmer's practice.

When the projected trend and demand yields for 2020 are compared with the ceiling yields, it appears that the two are uncomfortably close in E Asia, and in NE Asia to a lesser extent. The projected yields in S and SE Asia are still below the levels that could be attained when all rice fields can be fully irrigated, but the quantity of water available for agriculture may be insufficient for such an extensive irrigation (Wang 1992). But it is probable that irrigated rice crops in SE and S Asia also approach the levels of current potential yields in the next decades. Hence, the issue of a yield ceiling for rice is of major importance.

The smaller the yield gap (the difference between actual and potential production), the more difficult it is to further improve the yield level. Gryseels et al. (1992) argue that the International Agricultural Research Centres have a special responsibility for research to raise yield ceilings. Indeed, IRRI, reasoning that most of the progress in world rice production will come from the productive, irrigated areas, has set itself the major task of creating new types of rice crops for irrigated conditions with a potential yield of 15 t ha^{-1} in the tropics (IRRI 1989).

Relation to climate change

Designing rice varieties with a higher yield potential has already started (IRRI 1989, 1991b; Penning de Vries 1991; Dingkuhn et al. 1992). In this paper, it is investigated whether the expected climate change will make the task of raising the potential production even more difficult, or will facilitate it. With that in mind, the impact of climate change on potential production of current rice varieties is discussed. Neither water or nutrient limited yield levels or impacts of pest will be considered, nor the water and nutrient requirements to achieve such yields. The focus will be on E Asia, S Asia and SE Asia for 2020 as the target year. Hence, we discuss changes over a period similar to the planning horizon of most persons currently active in rice production and research. Moreover, the climate change will still be limited by 2020, so that uncertainties in prediction of climate change do not erode the conclusions completely.

Model and data

Data on climate change

Of the features of climate that may alter in the next decades, the ambient CO_2 concentration, temperature and radiation are particularly important for potential crop production. The eventual role of increased levels of UV_b is still uncertain, but it is assumed that its impact will not be important for crops over the next decades. By 2020, the ambient CO_2 concentration will have reached about 400 ppm. The average annual temperature is projected to have gone up by 0.9 °C in the middle temperature increase scenario, or at least by 0.2 °C, or by 2.4 °C at the most (Jaeger 1988). There are recent indications that 80% or more of the increase in temperature is due to an increase of the minimum temperature, with little or no increase in the daytime maximum temperature (Beardsley 1992). Future radiation levels have not yet been predicted by General Circulation Models, and might either decrease or increase. Whether a warmer climate will alter the sea level is still disputed, changes in snowfall in the Antarctic being a major uncertainty. Though this may ultimately have a significant effect locally on the area available to grow rice, the sea level changes slowly and its consequences are not considered here.

Models for potential production of rice crops

The ecophysiological potential production of crops is defined as the yield in conditions where water and nutrients are continuously in ample supply, and where no insect pests, diseases or weeds interfere with growth. The growth rate and final yield level are determined by the response of the crop's physiological processes to radiation, temperature and CO_2. Potential production situations can be created for intensively managed crops.

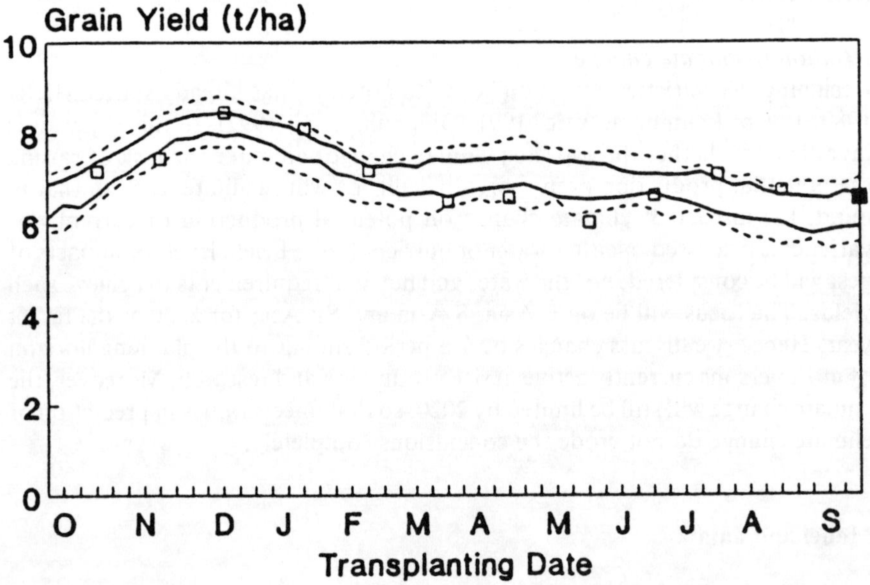

Figure 1. The potential yield of IR36 in Los Baños, the Philippines, in response to the weather, as a function of month of transplanting. The points indicate yields predicted for historic weather patterns of the past months; the lines, from top to bottom, represent the levels of 90, 75, 25 and 10% probability of lower yields, and were obtained by simulation for 29 historical years. (Data from Climate Unit IRRI, pers. comm.).

Simulation models for potential production have been developed for many years, and are well documented and tested. They are the only tools with which we can predict impacts of climate change (Bachelet et al. 1992). In this study, a model was used similar to that of Jansen (1990), which in turn was based on a set of crop modules with crop data for rice (Penning de Vries et al. 1989).

The model consists of a set of equations that describe the relations between the main physiological processes, i.e. photosynthesis, respiration, biomass accumulation and partitioning, phenological development and leaf area development, the structure of the canopy, and the environment, represented by radiation, temperature and CO_2 concentration. Radiation and CO_2 affect photosynthesis, temperature affects photosynthesis, respiration and the rate of phenological development. Parameter values were derived from the *indica* rice variety IR36. The performance of the model was evaluated in the Philippines (Herrera-Reyes and Penning de Vries 1989; Dingkuhn et al. 1990) and other locations in Asia (Dua et al. 1990). Limitations of the model are particularly due to limited knowledge of crop performance (fertility, net photosynthesis) at temperatures exceeding 30–35 °C. This model is run every month by IRRI's

Climate Unit to determine the absolute and relative level of the potential yield of the past growing period (Figure 1); the overall pattern and the yield variability at specific transplanting dates are due to trends and variations in radiation and temperature only.

The model with IR36 parameters was used in this study to simulate the impact of climate change on yield of an early, high yielding crop in Nanjing (China), a wet and a dry season crop in Muara (Indonesia), an early and late crop in Coimbatore (India), and a late season crop in Joydebpur (Bangladesh). Original weather data, daily observations, from these locations were applied. These locations are fair approximations for rice production in E, S and SE Asia (Jansen 1990). Responses to climate change at these locations are not identical due to different distributions of radiation and temperatures during the season. Weather conditions during the cropping season permit use of an *indica* variety at all stations.

For the simulations in a future climate, 1.0 °C was added to all of the historic records (Jansen 1990). The ambient CO_2 concentration was set at a constant 400 ppm. Radiation was unchanged. To establish the sensitivity of the results for these variables in the 2020 climate, all three variables were also modified around this point. To simulate the pronounced increase in night temperature (Beardsley,1992), the daily minimum temperature was increased by 1.6 × the average temperature increase, and the daily maximum temperature by 0.4 × this value.

Impact of climate change on potential production

Impact on rice crops

Jansen (1990) simulated potential yield, water use efficiency and nitrogen limited yield of IR36 for three temperature and CO_2-scenario's for 2020 and 2100 for Asia. For 2020, he concluded that, at an unchanged level of radiation, the change in potential production of rice due to climate change in Asia is negligible for the middle temperature scenario, about +5 to −10% for the high temperature, and about +5 to +10% for the coolest scenario. The increase in the latter case is brought about by the higher CO_2 concentration, the reduction in the first case by the reduction in length of the growing season. Yield variability remains the same in the low increase scenario, but increases significantly in all other cases. There is little difference among the regions in this respect. Penning de Vries et al. (1990b) showed much larger responses to climate change, but for situations that may arrive much later then 2020, i.e. when CO_2 in ambient air has doubled.

Our first simulations for the 2020 scenario yield a picture similar to that of Jansen (1990). Results from all situations are merged into a single response curve of yield to temperature (Figure 2). To better examine the effects, the range of simulated temperature changes for each station was widened to −5 to +5 °C. The overall result shows a continuous decline of yield as temperature

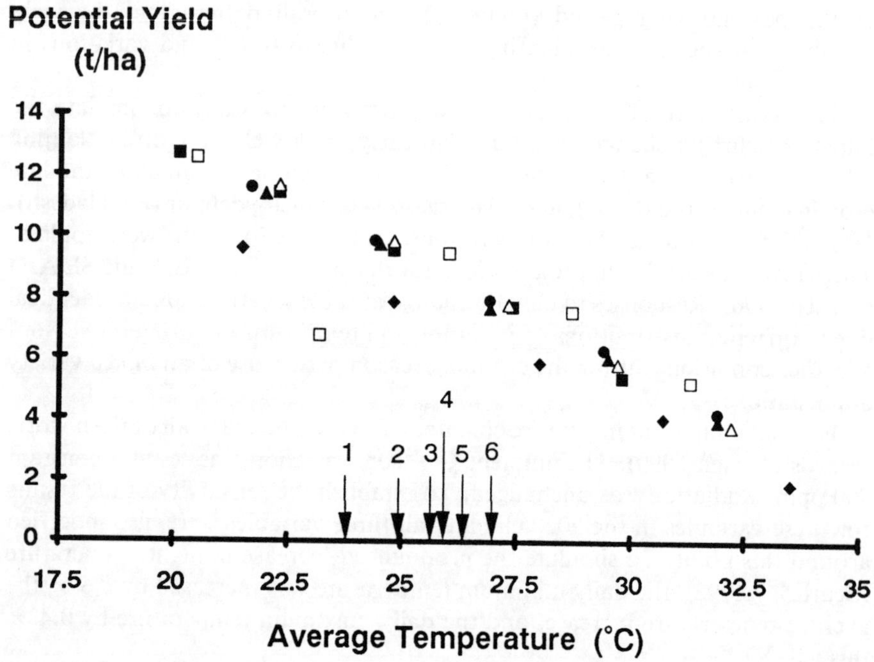

Figure 2. A response curve of rice grain yield to average temperature during the growing season, assembled from simulations for different locations under 400 ppm CO_2 and a temperature equal to a historic record plus or minus a certain value. Radiation level as in 1990. The marks indicate the average temperature of the 1990 growing seasons. Symbols: ● Nanjing (3), ■ Coimbatore late (1), □ Coimbatore early (2), ♦ Joydebpur (6), ▲ Muara late (5), △ Muara early (4).

increases. Baker et al. (1990) established experimentally almost the same response in the 24–35 °C temperature range.

Impact on rice cropping systems

Because land use for rice production is generally very intensive, it is argued that future farmers will choose their varieties such that these occupy the same positions in the cropping calendar as current varieties, and neither for longer periods (as this would reduce the possibility for profitable multiple cropping) nor for shorter periods (as this would result in lower rice yields, Akita 1989). For a given variety, the growing period shortens when temperature increases (Yoshida 1981; Summerfield et al. 1992). But farmers can replace them with varieties of longer duration, compensating the direct effect of temperature on development rate, so that crop duration becomes independent of temperature (Figure 3); this process of variety selection was mimicked by eliminating the direct effect of temperature on phenological development. The simulation results with this assumption (Figure 4) show a modified response curve: there is

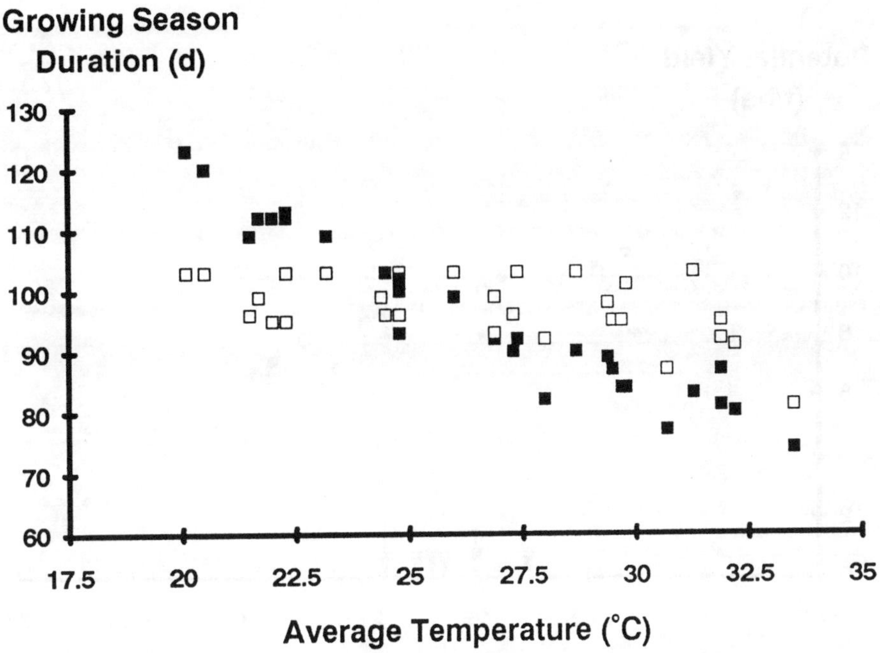

Figure 3. The simulated growing season duration at different temperatures with (filled symbols) and without (open symbols) a direct effect of temperature on the rate of morphological development.

less benefit of an extended cool season, but also the yield reduction at higher temperatures is smaller, so that a broad plateau presents itself. But rice is mostly grown at the upper end of this plateau (Figure 4, arrows), i.e. in climates where increasing temperatures will lead to yield reduction.

At the northern margins of rice cultivation, where only one crop per year can be grown, temperature rise will allow a longer growing season. Farmers may want to select varieties of longer duration that have higher yields (Akita 1989). It is difficult to estimate how much longer the growing season will be, and no attempt was made to quantify this potential benefit. Since rice production in northern zones is only a small fraction of the world production, this aspect is neglected. It may be important locally.

For constant durations of the growth cycle, simulated potential rice yields for 2020 were found to go up by 3%, in average, over the 1990 levels, because the positive CO_2-effect dominates the negative temperature effect. Simulation further showed that the average yield increase grows to about 4% when daytime temperature will rise much less than night time temperature. Such yield increases are very small indeed.

Taking into account the uncertainty about the temperature scenario leads to a range of values for the expected increase in potential yields from – 9% for the high temperature scenario to + 11% for the low temperature scenario (Table 4).

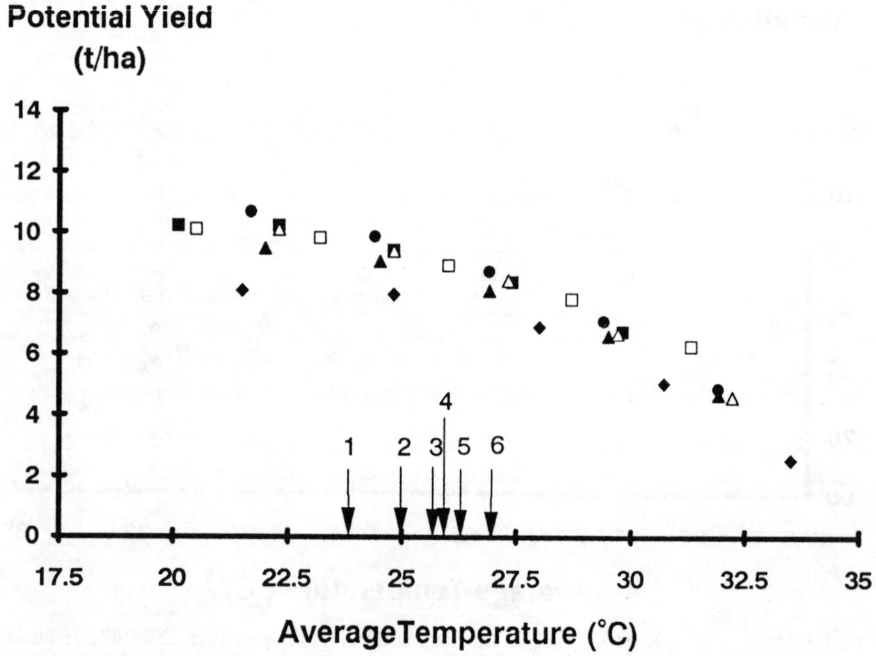

Figure 4. The response curve of rice grain yield to average temperature during the growing season, assembled from simulations for different locations under 400 ppm CO_2 and a temperature equal to a historic record plus or minus a certain value (as Figure 2) but with a constant length of the growing season. Radiation level as in 1990. The marks indicate the average temperature of the 1990 growing seasons. Symbols: ● Nanjing (3), ■ Coimbatore late (1), □ Coimbatore early (2), ♦ Joydebpur (6), ▲ Muara late (5), △ Muara early (4).

Still, these are small values compared to the major yield increases that will have to be realized in most regions (Table 1). These simulation results are only very global, and cannot replace a thorough study in which many more local conditions and varieties are considered in detail.

Components of the response to climate change

To take a simple view on the impact of climate change on rice crops, the three main effects can be considered separately. The CO_2 concentration effect is generally positive (Figure 5). The average response amounts to an increase of potential yield of about 10 kg $ha^{-1}ppm^{-1}$ CO_2, or about 15 kg ha^{-1} y^{-1}. This value is almost independent of location and cropping season. The ambient CO_2 concentration was only around 300 ppm at the end of the last century. The increase since then has already helped rice breeders to raise the yield ceiling by about 0.5 t ha^{-1}.

The effect of increasing temperature on rice potential production is generally

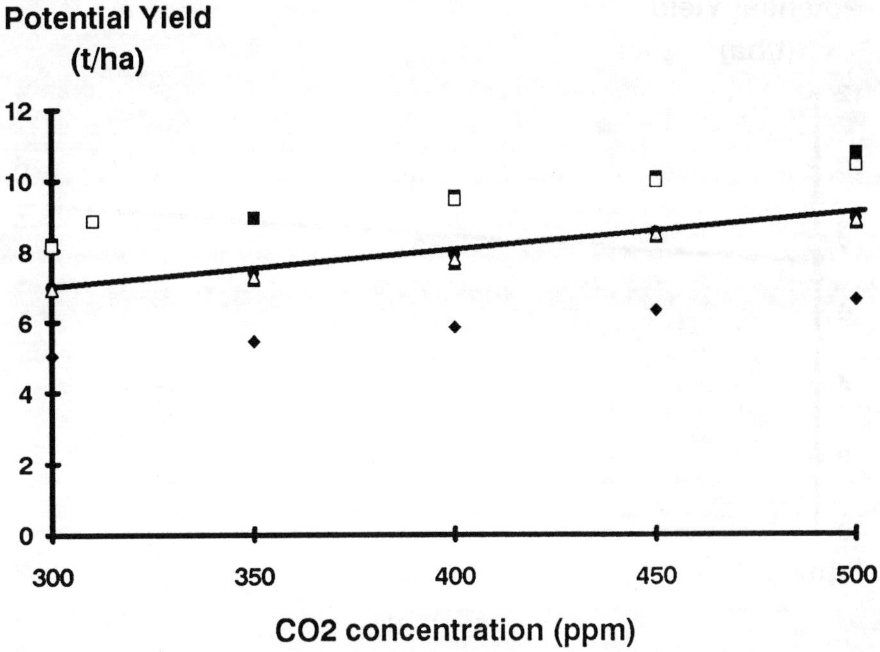

Figure 5. Potential rice yield as a function of ambient CO_2-concentration for different locations in 2020 weather. Symbols: ● Nanjing, ■ Coimbatore late, □ Coimbatore early, ♦ Joydebpur, ▲ Muara late, △ Muara early.

Table 4. Relative effects (%) of three temperature scenarios on the potential yield of rice crops in 2020 relative to the 1990 situation.

Region	Low	Middle	High
East Asia	11	5	−9
South Asia	6	4	−6
SE Asia	8	4	−9

negative. It is brought about in three ways: it decreases photosynthesis, increases respiration, and shortens the vegetative and grain filling period. Only the latter can be mitigated by choosing varieties with a longer grain filling period. The temperature effect amounts to about 400 kg ha^{-1} $°C^{-1}$ if temperature increases in day and nighttime (Figure 4), and is slightly less (300 kg ha^{-1} $°C^{-1}$) when mainly nighttime temperature increase. If current varieties would not be replaced by varieties with the 'original' growing period, the temperature effect would be almost twice as large (Figure 2). Differences among regions are small compared to the overall trend.

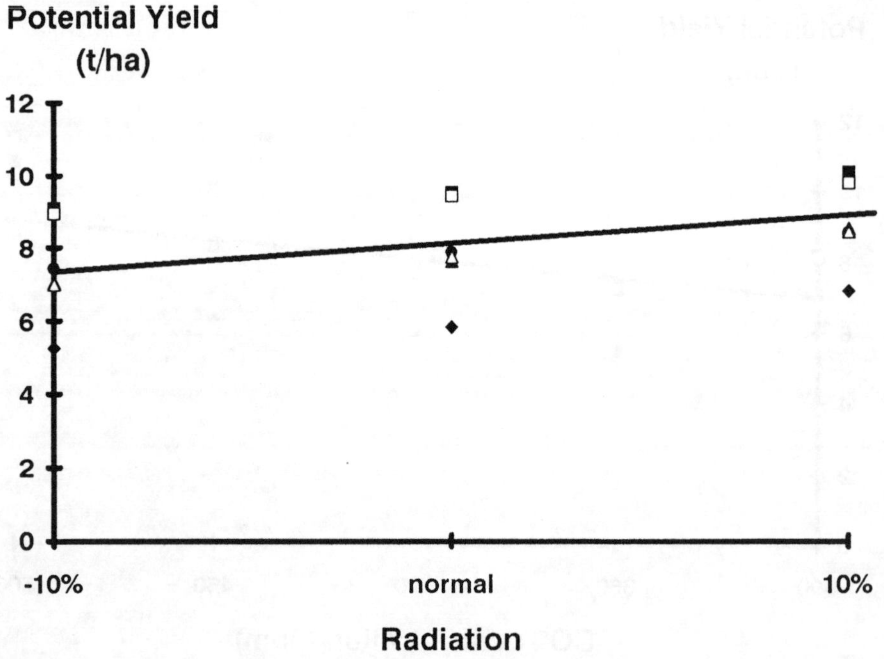

Figure 6. Figure 6. Potential rice yield as a function of radiation, expressed as a relative change, for different locations in 2020 weather. Symbols: ● Nanjing, ■ Coimbatore late, □ Coimbatore early, ♦ Joydebpur, ▲ Muara late, △ Muara early.

As a small sidestep of the analysis, the impact of the eruption of the volcano Pinatubo in the Philippines in May 1991 can be approximated. It probably causes a temperature drop around the world of around 0.5 °C in 1992 (Kerr 1992). Assuming that radiation levels are not significantly affected, the rule of thumb of the temperature effect suggests that the cooling corresponds with an increase in potential yield of 400 kg ha^{-1}. As the current average rice yield of 3.3 t ha^{-1} (IRRI 1991a) is below the potential due to medium levels of inputs and crop protection, the actual benefit of the eruption will only be a fraction of this, and is estimated at 130 kg ha^{-1} (cf. Jansen 1990). This translates into an additional 19 Mt of rice, worldwide.

Response of yield to radiation levels is shown in Figure 6. One percent change in radiation level corresponds with about 0.5% change in potential yield, or 40 kg ha^{-1}. This effect is probably the smallest of the three, but note that changes in radiation level are also the most uncertain.

To offset the effect of very high temperatures in parts of the humid tropics and in the hot season of warm temperate regions, e.g. China, plant types should be identified that stand these temperatures better, particularly with respect to fertility, photosynthetic efficiency and respiration, and leaf area duration. Higher yielding plant types will be characterized by a longer grain filling period

Higher yielding plant types will be characterized by a longer grain filling period and more translocation of pre-flowering storage to grains. Any temperature increase will make it somewhat more difficult to increase grain filling duration.

Climate change might include alterations in level and distribution of precipitation, and affect the extent of the area at which rice crops can be irrigated. This very important aspect of rice production needs careful study, taking ground water resources and alternative use of water into account.

Conclusions

For 2020, due to a rise of CO_2-level by 50 ppm and of temperature by 1.0 °C, potential rice yields increase on average by a few percent. The increase is smallest in the warmest regions. Lack of accurate predictions of climate change add some uncertainty to this conclusions. Yet, the finding implies that climate change in itself is not an issue in relation to the major challenge of increasing the yield potential of irrigated tropical rice.

The yield extrapolations and area reduction estimates employed in this paper are obviously crude. Also the level of the potential yield needs further refinement. However, these uncertainties do not obscure the conclusion that yields of rice crops in large areas will approach within decades the ceiling of our best current crops. At a small scale the ceiling is already reached by the most rice productive farmers under optimum conditions (Ramasamy 1990, TNAU, pers. comm.).

In 2020, a large fraction of the rice fields should produce grains at levels that approach the potential of current rice crops. This statement applies particularly in E Asia, but also to S and SE Asia (depending on how much irrigation can be extended further) and to NE Asia. This underlines the importance of breeding new crop types with a significantly higher potential. Establishing a strategy for rice production under limited availability of irrigation water will become very important in E and SE Asia.

For research, important topics are to identify the regulation of the processes that make rice plants sensitive to high temperature, and to breed or select varieties that are less sensitive.

In summary, developing new varieties of irrigated tropical rice with a significantly higher yield potential is urgently required. Expected changes in temperature, radiation and CO_2 concentration by 2020 will generally affect potential rice yields very little. Optimal use of irrigation water will become a very important issue for rice production.

Acknowledgements

Critical comments by Drs. H. van Keulen and S.C. van de Geijn are kindly acknowledged.

References

Akita S (1989) Improving yield potential in tropical rice. Pages 41–73 in Progress in Irrigated Rice Research, International Rice Research Institute, Los Baños, Philippines.

Bachelet D, Van Sickle J, Gay C A (1992) The impacts of climate change on rice yield: evaluation of the efficacy of different modeling approaches. Pages 145–175 in Penning de Vries P W T, Teng P S, Metselaar K (Eds.) Systems Approaches for Agricultural Development, Proceedings of the International Symposium on Systems Approaches for Agricultural Development, 2–6 December 1991, Bangkok, Thailand (this volume).

Baker J T, Allen L H, Boote K J, Rowland-Bamford A J, Waschmann R R S, Jones J W, Jones P H, Bower G (1990) Response of vegetations to CO_2 – 0600. Temperature effects on rice at elevated CO_2 concentration. Report no. 89. Joint program of the Crop Research Division, U.S. Department of Energy and Agriculture Research Service, USDA, with University of Florida, Gainesville, Florida, USA. 60 p.

Beardsley T (1992) Night heat. Sci. Am. 226(2):10–11.

Chomchalow N (1992) Agricultural development in Thailand. Pages 429–445 in Penning de Vries F W T, Teng P S, Metselaar K (Eds.) Systems Approaches for Agricultural Development, Proceedings of the International Symposium on Systems Approaches for Agricultural Development, 2–6 December 1991, Bangkok, Thailand (this volume).

Daradjat A A, Fagi A M (1991) Effect of seasonal variation in weather on crop potential of rice cultivar IR36 in the north coastal region of West Java. Pages 79–84 in Simulation and systems analysis for rice production (SARP), Penning de Vries F W T, Van Laar H H, Kropff M J (Eds.) PUDOC, Wageningen, Netherlands. 369 p.

Defeng Zhu, Cheng Chihua, Zhang Xiufu, Pan Jun (1989) Simulation of potential production of rice using L1D. Pages 35–52 in Case studies in crop growth simulation, Van Laar H H et al. (Eds.) Internal report Multiple Cropping Department, International Rice Research Institute, Los Baños, Philippines.

De Wit C T (1992) Resource use efficiency in agriculture. Agric. Syst. (in press)

Dingkuhn M, Penning de Vries F W T, De Datta S K, Van Laar H H (1990) Concepts for a new plant type for direct seeded flooded tropical rice. Pages 17–38 in Direct seeded flooded rice in the tropics. International Rice Research Institute, Los Baños, Philippines.

Dingkuhn M, Penning de Vries F W T, Miezan K M (1992) Improvement of plant type concepts: systems research enables interaction of plant physiology and breeding. Pages 19–35 in Penning de Vries F W T, Teng P S, Metselaar K (Eds.) Systems Approaches for Agricultural Development, Proceedings of the International Symposium on Systems Approaches for Agricultural Development, 2–6 December 1991, Bangkok, Thailand (this volume).

Dua A B, Penning de Vries F W T, Seshu D V (1990). Simulation to support evaluation of the potential production of rice varieties in tropical climates. Transactions of the ASAE 33(4):1185–1194.

Gryseels G, De Wit C T, McCalla A, Moyno J, Kassam A, Craswell E, Collinson M (1992) Setting agricultural research priorities for the CGIAR. Agric. Syst. (in press).

Herrera-Reyes C G, Penning de Vries F W T (1989) Evaluation of a model for simulating the potential production of rice. Philipp. J. Crop Sci. 14(1):21–32.

Horie T, Yajima M, Nakagawa H (1992) Yield forecasting. Agric. Syst. (in press).

International Rice Research Institute (1989). IRRI towards 2000 and beyond. International Rice Research Institute, Los Baños, Philippines. 80 p.

International Rice Research Institute (1990) Rice Facts, Feb. 1990. International Rice Research Institute, Los Baños, Philippines. 6 p.

International Rice Research Institute (1991a) World Rice Statistics 1990. International Rice Research Institute, Los Baños, Philippines. 320 p.

International Rice Research Institute (1991b) A continuing adventure in rice science. International Rice Research Institute, Los Baños, Philippines. 88 p.

Jaeger J (1988) Developing policies for responding to climate change. WMO/TD 225, Geneva, Switzerland. 53 p.

Jansen D M (1990) Potential rice yields in future weather conditions in different parts of Asia. Neth. J. Agric. Sci. 38:661–680.

Kerr R A (1992) 1991, Warmth, chill may follow. Science, 17 January 1992, 281.

Pan Deyun, Wang Zhaoqian, Yan Lijiao (1991) Simulation of tillering and potential production of indica rice. Pages 94–101 in Penning de Vries F W T, Van Laar H H, Kropff M J (Eds.) Simulation and systems analysis for rice production (SARP), PUDOC, Wageningen, Netherlands. 369 p.

Park R K, Cho S Y, Moon H P, Choi H C, Park N K, Choi Y K (1990) Rice varietal improvement in Korea. Crop Experiment Station, Rural Development Administration, Suweon, Korea. 109 p.

Penning de Vries F W T (1991) Improving yields: designing and testing VHYV's. Pages 13–20 in Penning de Vries F W T, Kropff M J, Teng P S, Kirk G J D (Eds.) Systems Simulation at IRRI. IRRI Research Paper Series no 151, IRRI, Los Baños, Philippines.

Penning de Vries F W T, Van Keulen H, Alagos J C (1990a) Nitrogen redistribution and potential production in rice. Pages 513–520 in Sinha S K, Lane P V, Bhargava S C, Agrawal R K (Eds.) International Congress Plant Physiology. Society for Plant Physiology and Biochemistry, Water Technology Centre, Indian Agricultural Research Institute, New Delhi 110 012 India.

Penning de Vries F W T, Van Keulen H, Van Diepen C A, Noy I G A M, Goudriaan J (1990b) Simulated yields of wheat and rice in current weather and in future weather when ambient CO_2 has doubled. Pages 347–358 in Climate and Food Security, International Rice Research Institute, Los Baños, Philippines.

Penning de Vries F W T, Jansen D M, Ten Berge H, Bakema A H (1989) Simulation of Ecophysiological Processes in several annual crops. PUDOC, Wageningen, Netherlands, and International Rice Research institute, Los Baños, Philippines. 271 p.

Singh S, Ibrahim Y, Rajan A (1991) Simulation of potential yield of three rice cultivars with different patterns in carbohydrate partitioning at three locations in Peninsular Malaysia. Pages 124–131 in Penning de Vries F W T, Van Laar H H, Kropff M J (Eds.) Simulation and systems analysis for rice production (SARP), PUDOC, Wageningen, Netherlands. 369 p.

Summerfield R J, Collinson S T, Ellis R H, Roberts E H, Penning de Vries F W T (1992) Photothermal responses of flowering in rice (*Oryza sativa*). Ann. Bot. (Lond) 69:101–112.

Sutoro S T, Makarim A K (1991) Potential production of six rice varieties at two different locations in West Java. Pages 13–24 in Van Laar H H et al. (Eds.) Case studies in crop growth simulation. Internal report International Rice Research Institute, Los Baños, Philippines. 224 p.

Yoshida S (1981) Fundamentals of rice crop science. International Rice Research Institute, Los Baños, Philippines. 270 p.

Wang Tian-Duo (1992) A systems approach to the assessment and improvement of water use efficiency in the North China Plain. Pages 195–207 in Penning de Vries F W T, Teng P S, Metselaar K (Eds.) Systems Approaches for Agricultural Development, Proceedings of the International Symposium on Systems Approaches for Agricultural Development, 2–6 December 1991, Bangkok, Thailand (this volume).

SESSION 3

Crop production: soil constraints

A systems approach to the assessment and improvement of water use efficiency in the North China plain

TIAN-DUO WANG
Shanghai Institute of Plant Physiology, Chinese Academy of Sciences, 300 Fenglin Road, Shanghai 200032, China

Key words: cotton, ground water, irrigation, North China Plain, peanut salinity, socio-economic constraints, water balance, water resources, water use efficiency, wheat, Yellow River

Abstract
The ways in which the total yield of crops in the North China Plain can be improved by optimizing the use of water are explored. The approach differs from many current ones in that, firstly, only the water entering the system as precipitation and inflow is considered water income, and only the water leaving the system through evapotranspiration and into the sea is considered lost; secondly, the dependence of yield on water supply is analyzed on the basis of physical and physiological mechanisms involved in water transport and yield formation, instead of on empirical relations. The whole Plain is divided into three main categories, and different strategies are suggested for them, i.e., (1) Areas with poor water supply, in which the limited amount of water recharge is to be optimized; (2) Areas with considerable surface water supply, where allocation of incoming water to fields with different distances from the water sources is to be optimized; (3) Areas with abundant water supply, means are to be found to excavate the underground free space to accommodate the excess water during the rain season.

Introduction

The North China Plain is the largest plain of agricultural importance in China. It covers about 18.30 million hectares of farm land, or 18.6% of the national total, and it produces about 18.3% of the total grain yield and more than 55% of the total cotton yield of the country (Xi et al. 1985). It is also one of the major production areas of soybean and peanut.

It is an almost ideal subject for a systems approach study, because: (1) the boundaries are well defined; (2) the interactions between many factors influencing crop yield are well known; and (3) the treatment of the area as a whole with water as the input and yield as the output differs from the consideration of water balance of individual farms or fields in several important ways. Based on the analysis of the economy of water use in three categories of areas with widely different water supply levels, strategies for improving the water use efficiency (WUE) in these categories can be made using systems approach.

F.W.T. Penning de Vries et al. (eds.), Systems Approaches for Agricultural Development, 193–205.

General description of the area

Geography

The Plain is surrounded in the north and west by the Yanshan and Taihangshan mountains, and is bordered in the east by the seas (Bohai and Huanghai) and Taishan mountains. The southern border is somewhat arbitrarily taken as the Huaihe River, a tributary of the Yangtze River (Changjiang). The plain is very flat, with a slope ranging from 0.01% to 0.05%.

The soils

The soils of the Plain are formed from the sediments carried by the several rivers entering it, with the Yellow River (Huanghe) contributing more than all the others combined. The soils are mostly calcareous, hence alkaline, and rich in phosphorus and potassium, but the availability of phosphorus is low due to the high pH. Because of the meandering of the rivers and frequent breaking of the levees, the variability of soil texture is large both horizontally and vertically, forming alternating areas or layers of soils with widely different textures and permeabilities.

The layer of sediments is very thick, usually tens of meters, which provides a large space for ground water storage.

The climate

There is a marked seasonal variation in temperature, with a warm summer even at the northern border, and in rainfall. From south to north, the rainfall decreases from 900 mm y^{-1} near Huaihe to a low 480 mm y^{-1} in the southern part of Hebei Province, but rises to 650–700 mm y^{-1} near the Yanshan mountains. It is highly concentrated in the summer months, and spring drought is frequent.

The crops

Wheat is the main winter crop, with some rapeseed. There are many summer crops, i.e. maize, sorghum, soybean, millet, peanuts, and others. Cotton is the most important cash crop, which, with its long growth period, can rotate with wheat only when it is intercropped with wheat during its early growth.

The water supply

The total annual local rainfall on the Plain amounts to 214.3 10^9 m^3. There is also some water entering the Plain through the rivers. The total amount of inflow is 81.7 10^9 m^3 y^{-1}, of which 45.6 10^9 m^3 y^{-1} is contributed by the Yellow River alone, the remaining part comes from the rivers entering the Plain from the mountains surrounding the Plain. Nearly all the water in rivers other than the Yellow River is used in the Plain for irrigation and other purposes; while the better part (about 30 10^9 m^3 y^{-1}) of the water in the Yellow River flows into the sea (Liu and Wei 1989).

The supply of river water differs from local rainfall in that it is dependent not only on the climatic conditions (in the mountainous catchments upstream) but also on the amount dammed and used in the upstream areas. The amount of water diverted from the Yellow River is under the control of the Yellow River Authority, and depends on the demand and supply, both varying from year to year.

There is an additional water source, i.e. the lateral flow of water from the mountainous catchments and from the Yellow River through aquifers. The extent of this replenishment depends on the nature of aquifers and the local demand on water resources.

According to the status of water supply, the areas in the Plain can be roughly divided into three categories: (1) Areas with a low water supply, mainly the southern part of Hebei province at some distance from the Taihangshan mountains, where the local rainfall, about 500 mm y^{-1}, the lowest in the Plain, is almost the sole source of water; (2) areas with a considerable surface water supply, which include the areas along both sides of the Yellow River, and those along the Yanshan and Taihangshan mountains, most of which have also a slightly higher rainfall (650 mm y^{-1}); and (3) areas with an abundant water supply, which lie some way from the Yellow River to Huaihe River, with a rainfall of 700–900 mm y^{-1}.

Systems approach to the assessment of water resources

There are two main approaches which prevailed for a long time. The first is the approach adopted by some who regard underground water resource as a kind of mine, and terms used in mining are adopted. Thus people often talk about the richness of underground water resources in some places. But in a region with similar rainfall, the richness often turns out to be one of the following cases: (a) in places near mountain ranges, ground water is fed by water collected in the mountainous catchments; (b) in places near the seashores, underground water is often saline or brackish, and therefore cannot be pumped for irrigation; and (c) in lowlands where there were frequent floods, and deep pumping has not yet been practiced. Whenever pumping is started, and an irrigation norm higher than the average annual rate of recharge is adopted, the water table falls.

The second approach is adopted by people who manage irrigation schemes or install or design irrigation canals, and who consider all water lost through percolation, either along the canals or in the fields under luxury irrigation, to be permanently lost. But, in fact, a considerable part of it eventually finds its way to underground aquifers, and can be – and is – used by pumping. The systems approach distinguishes itself from such approaches with regard to the water balance in that (1) inflow and outflow (loss), as flow variables, are treated separately from water stored in any form, which is a stock variable (state variable); and (2) only rainfall and surface and underground water coming from outside the system are considered inflow. Only the water lost to the atmosphere through evapotranspiration and to the sea by surface flow is really lost; all movements of water within the system are taken to be redistribution. These differences will affect the ways in which water use is optimized.

The principles of WUE improvement

The various factors influencing the WUE and the ways to improve WUE by manipulating these factors are well documented (see, for example, Taylor et al. 1983). Here our attention will be focused on those aspects in which the systems approach of a large area differs significantly from the many approaches which rely more on empiricism and where the water balance is made for too small a unit, like a plot or a farm.

An analysis of WUE

The WUE for a field can be analyzed at three levels. On the leaf level, the ratio P/T (where P is photosynthesis and T is transpiration) is the amount of photosynthates formed per unit amount of water transpired. On the biomass level, respiration loss (R) must be subtracted from the biomass formed by photosynthesis, and soil evaporation (E) will add to water loss, and the ratio of biomass/water loss will then be $(P-R)/(T+E)$, where $T+E$ is evapotranspiration (ET). On the yield level, since only a part of biomass is in the harvestable parts, typically in the grain, the overall WUE will be $Ke \cdot (P-R)/(T+E)$, where Ke is the harvest index. Rough figures for the various ratios for wheat and maize are given in Table 1 (Wang 1992).

The rates of P, T, and R are all affected by climatic factors and genetic characteristics of crop varieties, but E is very much dependent on irrigation and

Table 1. Water use efficiency of wheat and maize at different levels (g kg^{-1}).

Crop	Wheat	Maize
Leaf level	8	15
E = 0		
Canopy level	4.8	9
Yield level	2.16	2.88
E = 0.5T		
Canopy level	3.2	6
Yield level	1.44	1.92
E = T		
Canopy level	2.4	4.5
Yield level	1.08	1.44

Notes: Values of E expressed as fractions of T indicate the assumption of different soil evaporation rates. The harvest index is assumed to be 0.45 for wheat, 0.32 for maize, and 0.45 for rice. Respiratory loss is assumed to be 40% of total photosynthesis.

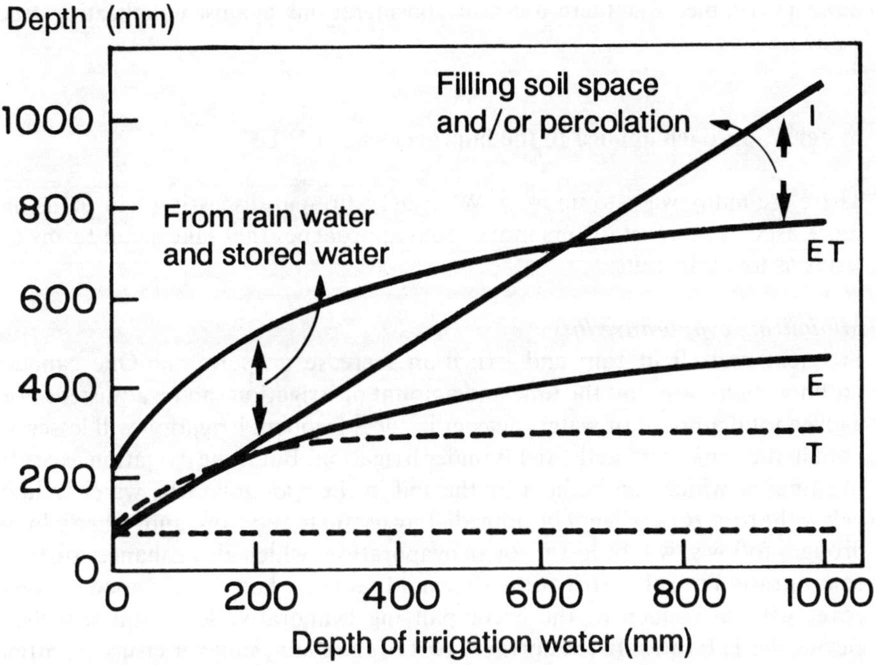

Figure 1. The relation of evaporation (E), transpiration (T), and evapotranspiration (ET) to irrigation (I) and the waterbalance.

canopy cover, both of which are under human control or influences. The harvest index is influenced by many factors, among which the water status during the time of floral differentiation and anthesis has the greatest effect. This sensitivity to water status is especially important for wheat, which develops its inflorescence and comes to flower in late spring when the evaporative demand is becoming high, and water in the soil has almost been exhausted.

The biomass and yield are usually linear functions of T (de Wit 1958), and almost linear ones of ET (Hanks 1983). Their dependence on irrigation depth is far from linear (Vaux and Pruitt 1983) and shows marked diminishing return because irrigation (I) must exceed a threshold to overcome the serious water deficit during the critical stage while an excessive amount of irrigation water will cause evaporative loss and percolation loss. But the latter is a loss to the field only, and not a loss to the agroecosystem. Figure 1 illustrates schematically the dependence of E, T, and ET on I, and the water balance under different I values.

But in a systems approach to the improvement of WUE in the whole Plain, there is another level, i.e. the regional level. The average WUE for a region will be improved by regulating the water flow within the region and allocating water to various fields in a way that the marginal revenue is the same everywhere. In case water is used in a luxurious way close to the water source, equalizing the rate of irrigation across all fields may make the water now giving a low return more profitable. But there are also considerations against equalization (see below).

Systems approach applied to the improvement of WUE

There are many ways to improve WUE. The present discussion will focus on those aspects where it differs much from approaches that take small farms or fields as its basic units.

Minimizing evaporative loss
Frequent and slight rain and irrigation increase evaporation. One can do nothing about rain, but the time and amount of irrigation can be adjusted. For a given total amount of water, heavier but less frequent irrigation will lessen E both in the canals and in the fields under irrigation. But if the irrigation exceeds the amount which can be held by the soil in the root zone, the water falling below the root zone will not be immediately useful to the crop, and, when a long drought follows, will be lost through evaporation, which also enhances surface concentration of salts. Irrigation after mild rain or when there is good canopy cover will save much of the accompanying evaporative loss, and will thus elevate the T/E ratio. In the northern part of the Plain, summer crops are often sown into the dry soil after a protracted spring drought. At this time, one has often to make a compromise between not irrigating, which delays the emergence and growth of the crop, resulting in poor canopy cover during the rainy season, and irrigating too much to cause excessive evaporation from the then bare soil surface.

Alfalfa and other perennial crops can form a better crop cover before summer rains come at the expense of deep soil moisture, and therefore have higher T/E ratios during early summer rains.

Evaporative loss along the canals during irrigation is also to be minimized. This is the motive to use less permeable or impermeable materials to make, build or line the conduits and canals. But here one must take into account of the cycling of a part of the water seeping out through deep percolation. Otherwise, if the water balance is based on taking all the water saved from seepage as water gained, as is often done, too high a cost for investment may be accepted.

Optimizing irrigation time
Beside minimizing evaporative loss, the effect of irrigation on Ke is of prime concern, especially for wheat during the critical period. The optimization of irrigation time is a complex problem. Because it is closely related with the problem of optimization of the amount of irrigation, and the number of meaningful combinations of different time and quantities is enormous. Because of the variability of rainfall from year to year, the results of irrigation experiments have low reproducibility. For optimizing irrigation time in years with different amounts and patterns of rainfall , some guiding principles are

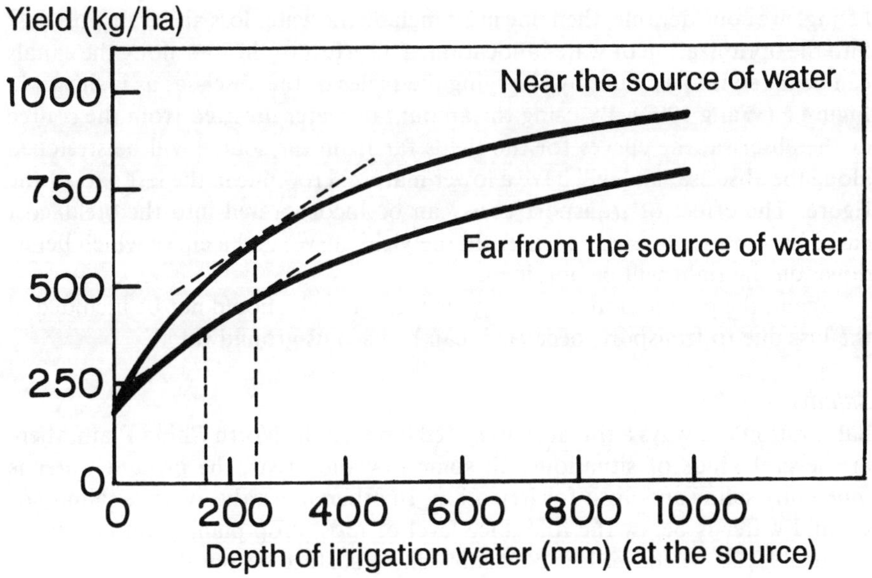

Figure 2. The difference in irrigation norm for equal marginal output of fields near or far from the source of water

needed so that simulations are to be made only for a limited number of meaningful options.

Optimal allocation of water to different fields or crops.
There are three levels of optimization of irrigation. First, with a given acreage of crops that need irrigation, the problem is optimal allocation of a given total amount of water to the different fields. In that case, the solution will be one in which the marginal revenues for all the response curves are the same. Second, since the acreage of crops which need irrigation and higher input of labor and fertilizer can also be changed, the possibility arises that when the total acreage of such crops is curtailed, and higher irrigation rates are used on a smaller area, the saving of fixed cost may exceed the loss due to a slight lowering of marginal output at higher irrigation rates. Third, since some other inputs, notably fertilizers, have strong positive interactions with water supply, there is the possibility that the ridge of the response surface of yield in a three-dimensional space may be concave upward, so that joint use of higher amounts of water and fertilizer to part of a given area may produce more total yield than the more equalized allocation to all the area as commonly recommended. This is a case of the advantage of intensification.

Optimal allocation to fields at different distances
When the distance of water transport is long, the loss of water along the canals, and the cost of building and maintaining the canals, and, sometimes, of water

lifting, are considerable, then one must include the water loss and transport cost into the optimization of water allocation. The effect of the loss along the canals can be taken into account by changing the scales of the abscissa, as is shown in Figure 2 (Wang 1992). By using the amount of water diverted from the source as the abscissa, the curves for the fields far from the source will be stretched along the abscissa, and will have a lower marginal revenue in the left part of the figure. The effect of transport costs can be incorporated into the yield/cost curve by subtracting these costs from the yield curve, and a curve which bends down on the right will be obtained.

Again, the percolation loss along the water canals should not be included in the loss due to transport, because it can be used as groundwater.

Salinity problems

Salinization is always a threat to irrigated land. In the North China Plain, there are several kinds of situations. In some lowland areas, the ground water is inherently saline and unfit for irrigation. In other areas, the average salinity of ground water is below the tolerance level of most crop plants, but can reach harmful levels whenever it is concentrated by surface evaporation.

In the former case, ground water must be pumped out, and the upper layers shall be washed with fresh water to ensure low salt concentration down to a certain depth.

In the latter case, it is imperative that there is enough downward flow of water to counteract the effect of upward flow and surface concentration. Furthermore, the water table must not be shallower than a critical depth, usually taken as 2 m, depending on soil texture and the season. So some pumping is always necessary, even in areas with ample water supply.

In the areas with a considerable water supply, the incoming water has a low salt content, normally below 0.02%. Yet if salt is allowed to accumulate, eventually there will be a day when the salt concentration in soil water is too high for crop growth. Therefore, the removal of a certain amount of salinized water is indispensable. Since concentration does not occur in deep layers, the salt content in drainage water can at best be the average one. Theoretically, one need only to spare a fraction of water which is equal to the ratio between the concentration in the inflowing water and the tolerance level of the crops.

Since the use of separate canals will use more land, the same canals are often used for both irrigation and drainage. It is important that the water drained from some lands should not be used again for irrigation. It is possible to separate irrigation and drainage in time with the same canals, but that needs good planning and management.

Strategies designed for the three categories of areas

The strategy for each category is a choice of one or a combination of the above principles. But each category has its peculiarities. The main points of the strategy will be delineated for each category.

Areas with a low water supply
With a rainfall of 480–500 mm y^{-1}, and facing an evaporative demand of 1100 mm y^{-1}, water is in short supply. But in the summer months, there is often quite intensive rainfall which causes some surface runoff. Non-agricultural areas, especially the roofs, are much less permeable and therefore produce much more surface runoff. As the land is very flat, almost all of the runoff is collected in the ditches and pools, and, apart from the evaporative loss, eventually becomes ground water. The total amount may correspond to 50–100 mm y^{-1}, which may produce a marked increase of yield if applied in the critical period. When the irrigation water is preferentially applied to only part of the farmland, the irrigation depth can be higher. The non-irrigated fields can be left to perennial leguminous forage crops.

Strategies for areas with a considerable water supply
Until very recently, river water was supplied at very low prices, which encouraged excessive use by those who got access to it. One of the difficulties in allocating water rationally lies in that usually irrigation depth is taken to be the amount of water loss, but the true loss is ET, which is difficult to measure and to be separated from deep percolation.

Recycling will cause a waste of energy in pumping. But as water is in short supply at the source during spring drought when water is most needed, the use of underground water provides the farmers with a higher degree of freedom than with surface water.

In using groundwater, which can be recharged by excessive irrigation water or by sideways percolation, or both, it is important that the use shall not exceed the amount of recharge. Since the critical depth can be taken as the safe upper limit of the water table, the depth of the lower boundary can be calculated from the amount of recharge and the water storing capacity of the soil. A deeper lower boundary allows more water to be stored, which may be needed for equalizing the water supply over years, but too deep a lower boundary may lead to excessive energy use for water lifting. Since the value of water should be taken to be the marginal output (or, more precisely, the marginal revenue), and not just the cost of water lifting and transport, the price of water should be set at such a level as to guide water use to maximize total yield. Though it is the surface water use that is regulated directly by water pricing, a higher water price will encourage the use of underground water, and also decreases recharge from excessive irrigation. This may be important in regulating the water table to prevent salinization.

Strategies for areas with an abundant water supply
These areas get their water chiefly from rainfall, but the northern parts also get water from the Yellow River. In principle, a rainfall of 850 mm y^{-1} is about enough to meet an evaporative demand of 1000 mm y^{-1}. But there is spring drought, though not as serious as in areas of the other categories. Because of the low price of river water, the short duration and moderate severity of the

drought, and also because of the insufficient supply and the high price of electricity, farmers are reluctant to drill and use wells. As the ground water is little used and fully recharged, heavy rains in the summer often cause much surface runoff, which causes serious flooding downstream.

Accordingly, the strategies for these areas should be: (a) To raise the price of water from the Yellow River, so that it be saved for use in northern areas with a steeper marginal output, and thereby to encourage the use of groundwater; (b) to plant more water-consuming crops, chiefly rice, which has also a higher price, to enhance water use.

About half of these areas have soils with a high montmorillonite content and have concretions of different sizes and abundance in various soil layers. Both of them have adverse effect on the permeability of the soil. There are several ways of facilitating infiltration: apply large amount of organic matter, drill holes and fill them with porous material, dig ditches, and remove the concretions. All these measures are labor-consuming.

Diversion of water over long distances

In principle, the discussions about the areas with a considerable water supply are also valid here. But when the distance is exceptionally long, and the canals pass through many administrative units (provinces, for instance), there may be more serious problems. Some of them are:

(a) Silt and mud contained in river water, especially in that of the Yellow River (with an average of 29 kg m^{-3}), will cause heavy sedimentation in the canals. This has been partly solved for short-distance transport, which is within single administrative units.

(b) The season when water is needed most in the Plain, i.e. spring and early summer, happens to be also the time of low supply in both the Yellow River and the Yangtze River. There is the problem of the available quantity for the former, and one of backwash of sea water for the latter.

(c) If water is to be transported directly to the fields that need it because of the absence of any reservoir, there will be a long time lag due to the long distance of transport. Reliable mid-term weather forecast is needed.

To solve the last two problems, it is desirable that water be stored locally where it is needed in ditches and pools, and, after downward percolation, as ground water.

(d) Since there are positive interactions between irrigation and other agricultural inputs, notably chemical fertilizers, the allocation of these inputs, where they are not locally produced, should be considered together with the allocation of water. One way often recommended in areas with a low water supply to increase crop yield is to use more fertilizer to increase the WUE. But the marginal return of fertilizers is greater when there is more water. So it is more profitable to allocate more fertilizers to areas with ampler water supply. In solving this problem from the point of view of planners of the national economy, the transport of water, grain, and fertilizer, or the location of new fertilizer plants, may supplement each other, or serve as alternatives to each other, so the relative advantages or disadvantages of them should be weighed.

Simulation and optimization problems

Systems approach in general requires simulation and optimization as tools to deal with large systems and the many interactions between the factors, components and processes in the systems. Here are a few problems relevant to the Plain.

The complexity of the situation
The yield–ET relation is the basic relation in any optimization program for water use. There are many variables which may change from field to field and from year to year. The uniformity of the Plain on the surface is deceptive. Underground, there are layers differing widely in thickness and texture, which makes the water-holding capacity and conductivity widely different and the pattern of underground water flow highly complicated. The computer time needed to simulate all fields in a given area will be formidable even with high speed computers. The collection of basic data is also a major problem.

The problems with meteorological variables are slightly different. Most of the variables vary little over space in the Plain. But rainfall is an exception in that it often varies very much over short distances, and also over the years.

Though one of the principles of simulation is not to make drastic simplifications by using sums or averages, one cannot totally avoid making simplifications, because of the infinite numbers of combinations of the meteorological variables. There must be some interactions or fluctuations which are not important and can be neglected without invalidating the conclusions. It is interesting to know to what extent the productivity of a year can be correlated with any statistics, such as total annual rainfall or rainfall in the growing season of any crop.

Boundary of a system
This is an important problem for the systems approach to any problem. There is not a fixed size of a system which is good for all problems. It has been said that a plot is too small a unit. But too large a unit has its problems too. In areas with a low water supply, because of the slowness of the horizontal underground flow, a distance of, say, 200 m may be large enough for all horizontal flow of water to be negligible. But the situation is very different in areas near the mountainous catchments, where horizontal flow can go quite far.

In the allocation of water diverted from large rivers, in contrast, the unit taken as a system should be much larger. Thus, for the diversion of water from the Yangtze River to Hebei province, the system will consist of two areas a thousand kilometers apart. In such a case, though the extent of the system is large, the components considered must be limited to those closely related to water use, and ways are to be found to give the average or overall response of yields of the main crops to irrigation, so that valid conclusions can be obtained about the feasibility of long-distance diversion of water.

The interdependence of different parts of a scheme

Optimization with respect to many variables at once is difficult enough. But it will even be more difficult when the elements of an optimization problem belong to different disciplines. Such is the case with the value of alfalfa, which depends upon the number as well as the species of animals, each with different feeding habit and efficiency in feedstuff conversion. The same animals may give different conversion efficiency when used as draft animals or producers of dairy products, wool or beef and mutton.

Economic aspects

The price factor may not appear explicitly when water is allocated to the same crop in different fields, where the sum of the yields is the variable to be maximized. But when it is allocated to different crops, the ratio between the prices of these crops must be known. This is more complex when some crops, like alfalfa, are used in the household without being sold. Then the values of such crops must be estimated.

As regards the cost of water transport, there are other problems. Take, for example, the cost of removing the silt in the turbid water diverted from the Yellow River. The fine components are fertile, while the coarse ones are not. The latter, however, can be used to make bricks or concrete, but only in limited quantity for any location. Therefore, the cost of dealing with silt is a complex problem and must be determined by comparing many options.

The use of water in agriculture should be considered jointly with that in industry and urban life, which compete with agriculture in a stronger position. But industry can also help irrigation by providing the capital needed for the installation of power plants, since industry, unlike irrigation, uses water more evenly over the months. Agriculture, therefore, is interconnected with the rest of the economy in many ways

Administrative and institutional aspects

For any strategy to be effective, it must first be adopted by a decision maker at an appropriate level. It will therefore be desirable that the boundary of the system coincide with some administrative unit. For the present, most farmers are responsible for their own decisions as to the time and amount of water to use in their fields. But large irrigation systems usually serve regions each covering tens of thousands of hectares of farmland and a corresponding number of households. The water authority, however, can regulate the use of water by constructing irrigation works and water pricing, both can regulate the use of ground water in an indirect way.

Subsidy in the form of funds or cutting of prices of fuels and electricity can encourage the use of underground water. In the areas with a low water supply, it has gone too far so that deep well construction and intensive pumping created many underground 'funnels'. It remains to be seen whether merely discontinuing such subsidies can bring the water table back to the right depth. But in the areas with an abundant water supply, if subsidy for pumping may help to bring the water table down to leave room for accepting the water from summer rains, it can be a good thing.

Concluding remarks

As has been seen above, the use of water is influenced by many factors. And there are other users sharing water with agriculture. There are many problems untouched in this paper. Water pollution is an outstanding one. It is not possible for any individual or single institution to take up these problems all at once. What is important is that in making any systems analysis about the water problem one must be aware of the interconnections of water use with other things, so that the conclusions and the advices derived will be in the right context, and that they are communicable to people in the other fields.

Acknowledgements

This paper is based on research supported by a fund from the Office of Agricultural Projects, Chinese Academy of Sciences. The author is indebted to Prof. H. Y. Luo, the Institute of Geology, National Bureau of Earthquakes, and Prof. Y. Y. Wang, the Institute of Systems Science, Chinese Academy of Sciences, for many of the basic ideas. The figures and part of the table are borrowed from a paper by the author with the kind permission of the publisher, Science Press, Beijing.

References

Hanks R J (1983) Yield and water-use relationships: An overview. Pages 393–412 in Limitations to efficient water use in crop production. Taylor H M et al., (Eds.), American Society of Agronomy, Crop Science Society of America, Soil Science Society of America, Madison, Wisconsin.

Liu C M, Wei Z Y (1989) Agricultural hydrology and water resources of the North China Plain [in Chinese]. Science Press, Beijing. 236p.

Taylor H M et al., (Eds.), (1983) Limitations to efficient water use in crop production. American Society of Agronomy, Crop Science Society of America, Soil Science Society of America, Madison, Wisconsin. 538p.

Vaux H J, Pruitt W O (1983) Crop-water production functions. Adv. Irrigation 2:61–97.

Wang T D (1992) Ways to improve the efficiency of use of water resources in the Huang-Huai-Hai Plain: Analysis and suggestions. In Studies on crop water consumption and water use efficiency on the Huang-Huai-Hai Plain [in Chinese]. Chen Z X, ed., Science Press, Beijing, (in press).

Xi C F, Deng J Z, Huang R H (1985) Problems about the integrated management and agricultural development of the Huang-Huai-Hai Plain [in Chinese]. Science Press, Beijing. 151p.

Soil data for crop-soil models

J. BOUMA[1], M.C.S. WOPEREIS[2,3], J.H.M. WÖSTEN[3] and A. STEIN[1]

[1] *Department Soil Science and Geology, Agricultural University, P.O.Box 37 6700 AA Wageningen, The Netherlands*

[2] *International Rice Research Institute, P.O.Box 933 1099 Manila, Philippines*

[3] *Winand Staring Center for Integrated Land, Soil and Water Research (SC-DLO), P.O.Box 125, 6700 AC Wageningen, The Netherlands*

Key words: Geographic Information System, geostatistics, hydraulic conductivity, land evaluation, moisture retention, pedotransfer function, potato, profile, rice, sampling, simulation, spatial variability, sugarbeet, water balance

Abstract

Systems approaches for agricultural development can be realized at different levels of detail which are associated with different data needs. For soils five levels were defined, ranging from farmer's knowledge and expert systems to use of complex simulation models. Any problem to be studied should be analyzed thoroughly beforehand and the most appropriate level of study should be established for each discipline. Field monitoring of soil physical conditions, which is crucial for model calibration and validation, becomes increasingly important at higher levels of detail. Monitoring should be increased, paying due attention to soil profile characteristics during installation of equipment. Measurements of physical parameters needed for simulation should preferably be made in situ, using methods that are relatively simple, accurate and low-cost, such as the crust-test infiltrometer. Pedotransfer functions, which relate available soil data to parameters needed for simulation, are a potentially important source of basic soil data and should be further developed. One type of pedotransfer function uses pedogenic soil horizons from soil surveys as 'carriers' of data. This approach is illustrated for both a small scale and a large scale soil survey carried out in the Philippines. Field work is needed to better characterize 'non-ideal' soil behaviour due to soil heterogeneity, which is quite common and is as yet not covered by existing models which assume soils to be homogeneous. Soil input is not only relevant for obtaining point data but also to assist in obtaining results that are representative for areas of land. Geostatistics can make contributions towards developing efficient sampling schemes which base the number of observations on spatial heterogeneity and not on the scale of the map to be made. Geostatistics was used successfully to interpolate point data to areas of land, including estimates of accuracies, as is illustrated for a case study at the International Rice Research Institute.

Introduction

Simulation modelling of crop growth is becoming an important part of a systems approach towards the formulation of alternative scenarios for agricultural development. Increasing emphasis on environmental side-effects of agricultural production practices has led to the need to not only consider crop yields but also the associated solute fluxes and their ecological side-effects. Soils, obviously, play an important role in such simulation models as they provide both a medium for root development and for solute flow and transformation.

F.W.T. Penning de Vries et al. (eds.), Systems Approaches for Agricultural Development, 207–220.

This paper focuses on the question which soil data are needed to allow successful operation of crop-soil simulation models. Attention in this paper will be confined to soil physical aspects.

Two types of soil data may be distinguished:

(1) Basic parameters in equations that characterize dynamic physical processes in crop-soil systems. Examples are hydraulic conductivity and moisture retention for process-oriented mechanistic models and field capacity and wilting point for capacity-oriented models.

(2) Data derived from monitoring in situ the dynamics of actual physical conditions, for calibration and validation of models. Such data are also useful for defining physical boundary conditions for model-runs for both actual and potential conditions. Examples are observations of water contents during the growing season in different soil horizons, watertable levels and changing rooting patterns during the growing season.

For both types of data considered, both measurement and estimation procedures have to be evaluated. Measurements will usually be more precise and accurate than estimates. However, depending on the degree of detail of the question being pursued, estimates of soil data may be quite satisfactory. Such a judgement can only be made, however, when the relative importance of the variable is known, and when the accuracy of an estimate of its value is known.

Soil science can also contribute to effective crop-soil modelling by assisting in transforming point data into areal data. Practical questions often deal with areas of land. Soil surveys produce delineated areas of land representing different types of soil which often act significantly different with respect to different land use types. Usually, point simulations or measurements are considered, assuming that the selection of the location and the number of points are based on efficient sampling schemes. Sampling can often be improved using geostatistics which defines soil heterogeneity and allows quantitative estimates of variability obtained as a function of observation densities. Recent advances in geostatistics allow application of quantitative interpolation procedures, such as kriging, to obtain areal estimates of important soil variables from point data. These estimates include a measure for the associated precision.

When considering systems analysis for agricultural development, we should realize that soil is only one part of the overall system. Climate, crops, hydrology and pests and diseases are important components of the complete agricultural system which need to be modelled as well. Any general model, to be used for a particular application, should be internally consistent in the sense that its submodels interact in the most functional and effective way. Depending on the question being pursued, a required degree of detail or generalization will have to be defined for each submodel on the basis of discussions within a multidisciplinary research team. The challenge for each discipline is, therefore, to define parameters for modelling in different categories of detail and with different precisions.

The above discussion indicates that there is no single crop-soil model with an exclusive set of required data. The question being posed by the user of soil crop data is central to all that follows. Sometimes a question is general to the extent that model calculations are not applicable. Then, use of farmer interviews or an expert system may be adequate. Sometimes, questions are quite specific and only modelling is likely to provide the necessary quantitative answers.

On-site monitoring

Monitoring physical conditions often includes measurement of soil water contents by periodic sampling, by neutron probe or, more recently, by TDR (Time Domain Reflectometry). Also, pressure heads can be measured in situ by transducer tensiometry (e.g. Klute 1986). Modern data storage and transmission techniques make in situ monitoring less labor intensive which is relevant for developed countries. Of particular interest for crop-soil studies are observations on rooting patterns during the growing season. Some models simulate root growth, which may be reasonable in homogeneous soils with a low resistance. However, in soils with well developed structures, rooting patterns are often quite irregular as roots may follow larger pores in the soil such as animal burrows or shrinkage cracks, thereby by-passing the soil matrix. Under such conditions, 'accessibility' of water and nutrients may be a problem rather than 'availability' which is usually considered in water uptake studies (e.g. Bouma 1990). Dye studies can be used to identify the soil volume through which fluxes of water occur. Morphological descriptions of roots and soil structures allow estimates of the 'active' volume of soil, which is often a fraction of the total soil volume in well structured soils (e.g. Bouma and van Lanen 1989).

Basic data

When simulating rice growth under wetland conditions, the effect of the puddled layer on soil hydrology is of crucial importance. Rather than trying to simulate the hydraulic resistance of this layer with a soil-mechanics model, it is preferable to measure in-situ conditions with the objective to obtain this resistance. Wopereis et al. (1991a) made measurements at 36 sites within a rice field (for the experimental setup see Figure 1) and reported an average resistance of the 6 cm thick least permeable layer of 212 days with 95% confidence limits between 158 and 266 days. Corresponding hydraulic conductivities were 0.27 mm d^{-1} with confidence limits between 0.22 and 0.37 mm d^{-1}. Such measured values can be input in simulation models for lowland rice.

For simulation of water movement in soils and water uptake by crops, different procedures are being followed. Capacity models (e.g. Wagenet et al. 1991) consider the soil to consist of a rootzone and a subsoil, whereby the rootzone functions as a 'box' which contains water within two predefined

Figure 1. Experimental set-up to measure the in-situ hydraulic resistance of the puddled layer in lowland rice soils. (After Wopereis et al. 1991a).

critical pressure heads: field capacity and wilting point. When water is applied to the soil, it is assumed that vertical water movement is rapid at water contents above field capacity and that water is only retained in the 'box' to field capacity level. The rest flows downwards. Water can be extracted to the wilting point; water held at lower pressure heads is unavailable for plants. The International Benchmark Sites Network for Agrotechnology Transfer (IBSNAT) manual (IBSNAT 1988) describes a procedure to determine 'field capacity' under field conditions by flooding a field and measuring water content after two days while avoiding evaporation. Similarly, the wilting point is estimated by growing a crop and by observing the moisture content at which wilting occurs. This procedure to estimate the wilting point is very laborious and hard to control. Formerly, the water content at 15 bar pressure was taken to represent the permanent wilting point but this value has correctly been criticized because wilting depends not only on the soil water state but also on the type of plant and the evaporative demand, as well as weather conditions in terms of relative humidity and wind velocity. Distinction of arbitrary 'field capacity' and 'wilting point' values poses a scientific problem as we deal in nature with continuous processes of water transport and uptake. Mechanistic models allow dynamic characterization and their use is therefore preferable.

Mechanistic models (Wagenet et al. 1991), need hydraulic conductivity (K–h) and moisture retention data (h–θ) to calculate real fluxes between layers in the soil (where h = pressure head; θ = water content). Many methods have been published (e.g. Klute 1986). Little attention is paid to operational aspects of methods such as costs, complexity, accuracy and specific applicability. Considering such aspects we use the one-step outflow method for measuring both conductivity and retention characteristics (Kool et al. 1987). The crust test suction infiltrometer is suitable to measure K near saturation, particularly when macropores are present (Booltink et al. 1991). Inverse modelling can also be used to obtain 'equivalent' hydraulic characteristics by running a simulation model for crop growth and by varying hydraulic input data, while keeping all other factors constant. Calculations are continued until a set of hydraulic characteristics is found which yields output data that matches independently measured data. Good results with this method were obtained by Wopereis et al. (1991c) although direct measurements yielded better data. Derivation of soil physical data by inverse modelling includes all uncertainties associated with crop modelling. When considering all necessary data for crop-soil models, it would appear to be preferable to measure as many parameters as possible and to reserve inverse modelling for those parameters that are very difficult to measure or that are poorly defined. In this context, measurement or estimates of soil physical data rather than use of inverse methods is preferable. Specific soil data associated with different levels of study detail, are summarized in Table 1 which will be discussed in the last section of this paper.

Table 1. Soil-data needs for the five input levels distinguished in systems analysis. (COLE: Coefficient of linear extension.)

Level 1	Local experience; difficult to describe in reproducible terms and to extrapolate.
Level 2	Soil texture; systemized local experience in terms of descriptive land qualities (water availability, workability, trafficability, rootzone aeration etc.).
Level 3	Field capacity; wilting point; thickness rootzone; bulk density, watertable depth, if any.
Level 4	Hydraulic conductivity, moisture retention, sinkterm for root activity, watertable depth if any, empirical relations for bypass flow; COLE values.
Level 5	As 4, but with deterministic bypass flow and effect of dead end pores; soil-root contact; hysteresis.

Pedotransfer functions

Pedotransfer functions relate available soil data to parameters that are needed for modelling and that are relatively difficult to measure (e.g. Bouma 1989). Much work has been reported to relate soil characteristics, such as texture, bulk density and organic matter contents to soil hydraulic characteristics. Examples were presented by Wagenet et al. (1991) and Vereecken et al. (1990). For example, Cosby et al. (1984) through regression analysis of nearly 1500 soil samples, estimated hydraulic conductivity and moisture retention data for h >

-20 kPa based on the following equations:

$$\text{moisture retention:} \quad h = a\,(\theta/\theta_s)^{-b}$$
$$\text{hydraulic conductivity:} \quad K = K_s\,(\theta/\theta_s)^{2b+3}$$

where: h = pressure head (m); θ = water content ($m^3\ m^{-3}$); θ_s = saturated moisture content ($m^3\ m^{-3}$); K_s = saturated hydraulic conductivity ($m\ s^{-1}$); and a and b are parameters describing the shape of the moisture retention and hydraulic conductivity curves. They found, for example:

$$\log a = -0.0095\ (\%\ \text{sand}) - 0.0063\ (\%\ \text{silt}) + 1.54\ (r^2 = 0.85)$$
$$b = 0.1570\ (\%\ \text{clay}) - 0.0030\ (\%\ \text{sand}) + 3.10\ (r^2 = 0.966)$$

Recently, successful efforts have been made to use pedogenetic soil horizons as 'carriers' of physical information, such as conductivity and retention (Wösten

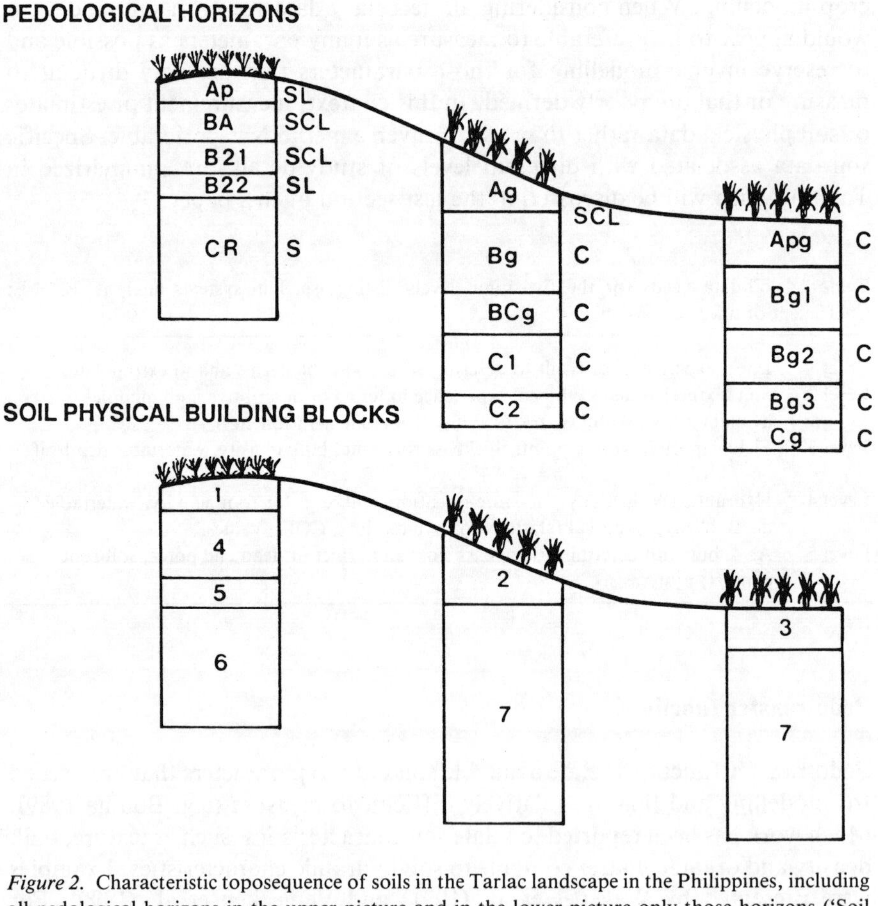

Figure 2. Characteristic toposequence of soils in the Tarlac landscape in the Philippines, including all pedological horizons in the upper picture and in the lower picture only those horizons ('Soil physical building blocks') that had significantly different physical properties. (After Wopereis at al 1991d).

et al. 1985). The procedure implies identification of soil horizons in the context of a soil survey. Next, measurements are made in at least five randomly selected locations within each horizon and average curves are determined for each horizon. Differences of curves between horizons are tested by functional criteria and horizons that do not show significantly different behaviour after statistical testing, are grouped together. Thus, a limited number of horizons are distinguished, each with a certain thickness and depth of occurrence which are ideal regional variables. A case study in the Tarlac Province of the Philippines can be used to illustrate the procedure (Wopereis et al. 1991d). One characteristic toposequence in the Tarlac landscape is shown in Figure 2. The distinguished pedogenic horizons in three characteristic profiles are indicated, as well as the horizons that acted differently by soil physical criteria. The example shows a reduction in the number of different horizons following the functional classification: fifteen pedological horizons corresponded with seven functional horizons. This reduction is attractive from an operational point of view. The procedure to link physical data to major soil horizons has been applied in the Netherlands, where a standard series of 36 conductivity- and retention-curves for surface- and subsoils of some 150 soil profiles are successfully being used in models describing regional soil water regimes (Wösten 1987; Wösten et al. 1990). A comparable series should be developed for rice growing areas, relating well defined horizons in soil taxonomic units to physical and chemical parameters.

Many other pedotransfer functions should be established, because they render existing soil data (which are often not used) more useful. By using regression equations or by expressing data for horizons in statistical terms, the accuracy of the functions can be well expressed. This, in turn, is important for expressing the accuracy of modelling efforts using this type of estimates.

Soil conditions in the field

One might think that realistic crop-soil models based on sound physical principles, are ready and waiting to be fed with parameters to be obtained by measurement or estimation. This impression is correct but only for those soil conditions that are defined by standard flow theory based on Richards' equation, which requires soils to be homogeneous and isotropic. Unfortunately, such soils hardly exist. Many examples of preferential flow of water in soil were recently summarized (Van Genuchten et al. 1990). Water may follow macropores such as animal burrows or cracks; irregular wetting may result from infiltration from a rough soil surface; organic matter may be associated with hydrophobic properties and irregular soil horizons may induce complex wetting and drying patterns. These processes have no marginal effects; in fact they often govern soil water regimes under natural conditions (see examples in Van Genuchten et al. 1990). Much work has yet to be done to characterize flow processes under field conditions. We believe that morpho-

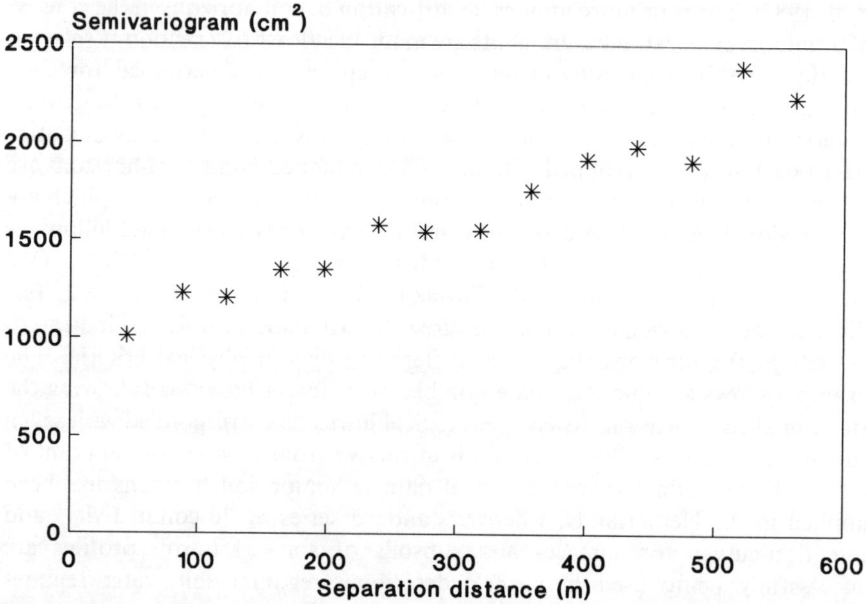

Figure 3. Figure 3. Semi-variogram of the depth to the tuff layer at the IRRI farm, Philippines, showing a clear spatial structure indicating a linear increase of semi- variance as a function of separation distance between observation points. (After Wopereis et al. 1991e).

logical descriptions of soil structure and use of tracers may be helpful to better understand natural flow processes, as was recently illustrated by Bouma (1991). One example from a rice-growing country may serve to further illustrate the importance of macropore flow. Wopereis (unpublished) studied water movement in a previously puddled clay soil in the Philippines that was used for growing mungbean after rice. The soil dried out and cracked to a depth of 100 cm. Prior to transplanting the next rice crop, the field was submerged again, to facilitate puddling of the topsoil by the farmer. High losses of water occurred during submergence due to rapid flow of water through the cracks into the subsoil. Shallow surface tillage after the mungbean harvest could have been helpful to avoid crack-flow to the subsoil by making cracks discontinuous. Modelling of this process, to explore the possible effects of different management scenario's, is impossible with standard flow theory. However, it is feasible when cracking patterns are taken into account (e.g. Bronswijk 1988).

From point to area

The number of soil observations needed in the context of a soil survey has traditionally been determined by the scale of the map to be made in terms of a certain number of observations per cm² map area. It is, however, more efficient to make the number of observations a function of soil spatial variability.

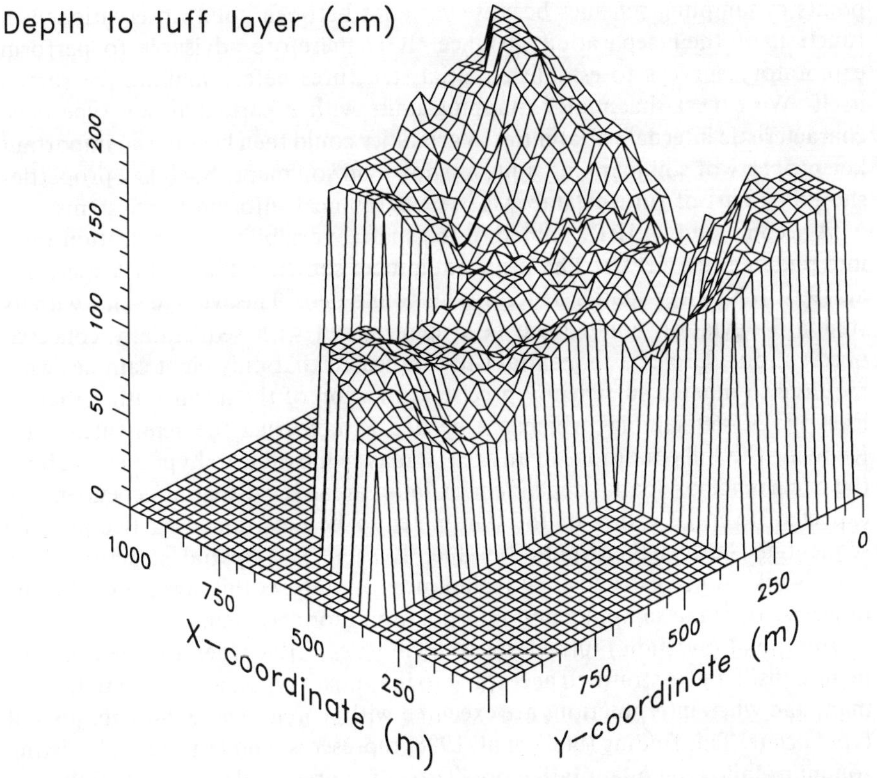

Figure 4. Block diagram showing depth to the tuff layer at the IRRI farm, Philippines. This diagram was obtained by applying the interpolation technique kriging using the semi-variogram of Figure 3.

Modern geostatistical techniques are available to express spatial variability in terms of the semi-variogram. The example in Figure 3 was derived from measurements of depth and thickness of major soil horizons at the farm of the International Rice Research Institute (IRRI) in the Philippines (Wopereis et al. 1991e). As discussed in the section on pedotransfer functions, these characteristics can be interpreted as regional variables. The semi-variogram of Figure 3 was used to interpolate depth values of horizon number 5, using kriging (Figure 4). Predicted values were compared with measured values in test points indicating that predictions could be made with a precision of 45 cm. Kriging allows interpolations of points towards areas of land, yielding the associated accuracy of predictions as well. This is attractive for modern quantitative questions relating to land use alternatives in a given area of land. Predictions with an identical degree of accuracy could have been obtained with 50% less observations. Semi-variograms with a clear spatial structure allow efficient predictions of soil characteristics by kriging with a minimum number of observations. They also allow determination of the density of observation points in sampling schemes because variance between point observations is a function of their separation distance. It is therefore advisable to perform exploratory surveys to establish spatial structures before making the survey itself. We expect that major mapping units with a particular soil type have characteristic internal variabilities. Variability could then become an important key property of soil mapping units as shown on soil maps. Such key properties should be part of future databases in Geographical Information Systems.

A second study a IRRI related to the spatial variability of infiltration rates into puddled rice field of 2 ha using equipment shown in Figure 1 (Wopereis et al. 1991b). No clear spatial structure was found here. The average value with its standard deviation could have been determined with sequentially collected observations. Use of a sequential t-test indicated that only eight samples were required to make a sufficiently precise estimation of the mean, rather than at least the 50–60 samples which are needed to construct the semi-variogram. Knowing the soil conditions in an area, it may be possible to hypothesize about the occurrence of spatial dependencies in an area. If no spatial dependence is expected, average values of the parameter to be determined suffice and the sequential t-test is then quite suitable. If there is a spatial structure, it is worthwhile to use geostatistical techniques to interpret this structure with the objective to make sampling and interpolation more efficient.

Interpolation studies turned out to be most effective when the area of land being considered is stratified according to major soil types, according to the soil map, and when interpolations are executed within areas where those major soil types occur. This finding (Stein et al. 1988) represents innovative use of existing soil maps, allowing quantitative predictions for areas of land. Predictions can relate to infiltration rates and depths of horizons, as discussed, but also to more complex entities such as the moisture supply capacity or crop yield.

Use of geostatistics has become important in modern soil science allowing quantitative expressions of variation in space. Simulation results for long time periods allow quantification of temporal variability, which can also be interpolated from points to areas. This does, of course, not only relate to aspects of soil science but to other aspects of crop production as well.

Soil descriptions in systems analysis

As stated before, the soil forms only part of the total agricultural production system. Comprehensive crop-soil models, which are part of a systems approach to research, need to also consider climate and crop data, as well as data on pests and diseases. Consideration of a socio-economic component is necessary when realistic land use alternatives have to be defined. The main theme of this paper is the role of soil science within a comprehensive systems approach for agricultural development. This approach is not restricted towards modelling. Five levels may be distinguished for soils input into systems approaches for agricultural development (Bouma 1989):

Level 1 Farmer's knowledge
Level 2 Expert knowledge and associated data needs.
Level 3 Simple capacity models and associated data needs.
Level 4 Complex mechanistic models and associated data needs.
Level 5 Very complex models for subprocesses and associated data needs.

Data needs increase with increasing level which is illustrated in Table 1. Selection of the proper level for any particular study will depend on an analysis of the agricultural problem to be studied. Once this has been done, the research approach and the appropriate degree of detail have to be established. The various subprograms being used should have the appropriate degree of detail, and every discipline should therefore define different sets of data for the different levels discussed above. Combinations of different levels are often desirable. For example, when defining land use alternatives for the European Economic Community, using crop-soil simulation models, (Van Lanen et al. 1990) used first an expert system to screen out areas of land that would not be suitable for crop production. Next simulations were made only for those areas where conditions were considered to be potentially suitable. This approach, combining levels 2 and 3, saved 40% of the resources that would have been needed to make simulations for all areas. Wopereis et al. (1991d) used a comparable approach for a study on land use in Tarlac province, Philippines. However, they did not develop an expert system but initiated close and frequent interaction with local experts from the National Bureau of Soils using maps produced with a Geographical Information System as an effective means of communication for each phase of development.

Another example deals with the effects of soil compaction on agricultural production, which was a real problem for farmers in sandy loam polder soils in the Netherlands. The level 2 approach was not successful because the experts did not know the answers. Using capacity models at level 3 was unsatisfactory because all important horizons and rooting patterns could not be adequately represented in the model. Successful use was made of a level 4 model which required some specific physical measurements of hydraulic conductivity and moisture retention data (e.g. Van Lanen et al. 1987). In addition, level 5 work was needed to specifically define the effects of irregular rooting patterns in soil with large peds which made some of the water inside the peds inaccessible. A new sink term was defined (Bouma 1990).

The scheme with the five levels of soils input, discussed above, does not imply that approaches at each level and the associated data demands are specifically defined. Questions remain at each level and continued research is needed to fill gaps in knowledge and understanding. Also, interaction among the various submodels in multidisciplinary studies needs attention in future research, particularly with regard to the required degree of detail of each submodule. When soil problems are being pursued in the context of a systems analysis that covers the entire agricultural production system, applied soil research is bound to be better focused and therefore more efficient as compared with conditions where research is only inward looking and discipline oriented.

Indeed: applied soil research will never be the same again!

Considerations for future activities

1. Systems approaches for agricultural development can be realized at different levels of detail which are associated with different data needs. For soils five levels were defined, ranging from farmer's knowledge to use of complex simulation models for important subprocesses. A similar analysis should be made for other disciplines. Any problem to be studied should be thoroughly analysed beforehand and the most appropriate level of study should be established for each discipline.
2. Field monitoring is crucial to allow model calibration and validation. Placement of equipment should be based on an analysis of soil profile characteristics to avoid unrepresentative measurement. Automatic techniques for measuring e.g. watertable levels and soil water contents should be more widely applied because manual monitoring may become too expensive in some countries.
3. Measurement techniques for basic soil physical parameters for crop-soil models should be summarized in manuals that emphasize not only technical but also operational aspects, such as costs, complexity, accuracy and applicability for different sites and seasons.
4. Pedotransfer functions can be used successfully to apply existing soil information for predicting parameters needed for simulation. Of particular interest is the use of pedogenic soil horizons as 'carriers' of physical (and chemical) data. Thus, existing soil survey information can be utilized more effectively.
5. Most soils in the field show 'non-ideal' behaviour in terms of preferential flow along macropores, irregular wetting due to effects of microrelief or hydrophobicity and irregular flow patterns due to occurrence of contrasting soil horizons. Clayey soils with natural, large aggregates often have rooting patterns implying inaccessibility of water inside aggregates. Standard flow theory and existing models for water extraction by roots do not consider these phenomena. Process oriented field studies are needed to better characterize these soil conditions, rather than 'calibration' of standard models.
6. Geostatistics provides quantitative techniques to transform point data into data representative for areas of land. Predictions include an estimate for the accuracy, which is essential as it allows quantitative integration of data from different sources in systems analysis, assuming that other data are defined in terms of their accuracy as well. Exploratory surveys of study areas are effective in defining efficient sampling schemes with minimized cost.

Acknowledgements

Frequent reference has been made in this paper to results of the SMISRIP project (Soil Management for Increased and Sustainable Rice production) which was financed by the Netherlands' minister for Development Cooperation (Project no. RA 89/950). This project is a collaboration between researchers from the International Rice Research Institute, Los Baños, Philippines, the Winand Staring Centre for Integrated Land, Soil and Water Research, and the Department. of Soil Science, Agricultural University, Wageningen, the Netherlands.

References

Booltink H G W, Bouma J, Gimenez D, (1991) A suction crust infiltrometer for measuring hydraulic conductivity of unsaturated soil near saturation. Soil Sci. Soc. Am. J. 55:566–568.

Bouma J (1989) Using soil survey data for quantitative land evaluation. Pages 177–213 in Advances in Soil Science. Vol. 9, Stewart B A, (Ed.), Springer-Verlag, New-York Inc.

Bouma J (1990) Using morphometric expressions for macropores to improve soil physical analysis of field soils. Geoderma 46:3–13.

Bouma J (1991) Influence of soil macroporosity on environmental quality. Adv. Agron 46:1–37.

Bouma J, van Lanen H A J (1989) Effects of the rotation system on soil physical properties and the relation with potato yield. Pages 77–89 in Vos J, van Loen C D, Bollen G J (Eds.) Developments in plant and soil sciences. Vol. 40. Kluwer Publ. Co., Dordrecht, The Netherlands.

Bronswijk J J B (1988) Effect of swelling and shrinkage on the calculation of water balance and water transport in clay soils. Agric. Wat. Manage. 14:185–193.

Cosby B J, Hornberger G M, Clapp R B, Ginn T R (1984) A statistical exploration of the relationship of soil moisture characteristics to the physical properties of soil. Water Resour. Res. 20:682–690.

International Benchmark Sites Network for Agrotechnology Transfer (IBSNAT) (1988) Experimental design and data collection procedures for IBSNAT. Technical Report no.1 (Third Edition). University of Hawaii. 73 p.

Klute A (1986) Methods of soil analysis, 1. Physical and mineralogical methods. Agronomy 9, 2nd ed., Am. Soc. Agron. Madison, Wisc.

Kool J B, Parker J C, van Genuchten M Th (1987) Determining soil hydraulic properties from one-step outflow experiment by parameter estimation, I. Theory and numerical studies. Soil Sci. Soc. Am. J. 49:1348–1353.

Stein A, Hoogerwerf M, Bouma J (1988) Use of soil-map delineations to improve (co-)kriging of point data on moisture deficits. Geoderma 43:163–177.

Van Genuchten M Th, Ralston D E, Germann P F (Eds.) (1990) Transport of water and solutes in macropores. Geoderma 46(1–3):1–297.

Van Lanen H A J, Bannink M H, Bouma J (1987) Use of simulation to assess the effects of different tillage practices on land qualities of a sandy loam soil. Soil & Tillage Res. 10(4):347–361.

Van Lanen H A J, Van Diepen C A, Reinds G J, De Koning G H J (1991) Comparing qualitative and quantitative physical land evaluations using the assessment of the growing potential for sugarbeet in the European Communities. Soil Use Manage 8:80–89.

Vereecken H, Maes J, Feyen J (1990) Estimating unsaturated hydraulic conductivity from easily measured soil properties. Soil Sci. 149(1):12–32.

Wagenet R J, Bouma J, Grossman R B (1991) Minimum datasets for use of taxonomic information in soil interpretive models. Pages 161–183 in Mausbach M J, Wilding L P (Eds.) Spatial variabilities of soils and landforms. Soil Sci. Soc. Am. Special Publication 28.

Wopereis M C S, Wösten J H M, Bouma J, Woodhead T (1991a) Hydraulic resistance puddled rice soils: Measurement and effects on water movement. Soil & Tillage Res. (in press).

Wopereis M C S, Stein A, Bouma J, Woodhead T (1991b) Use of spatial variability analysis to increase sampling efficiency when measuring infiltration rates into puddled clay soils. Agric. Water Manage. (in press).

Wopereis M C S, Wösten J H M, Ten Berge H F M, De San Agustin E, Woodhead T (1991c) A comparison of three methods to derive soil hydraulic conductivity functions: A case study for upland rice. J. Hydrol. (subm.).

Wopereis M C S, Sanidad W, Alajos M J A, Kropff M J, Wösten J H M, Bouma J (1991d) Case study on combining crop modelling and a geographic information system: Tarlac Province, Philippines (in prep.).

Wopereis M C S, Stein A, Bouma J (1991e) Use of geostatistics to optimize sampling strategies for soil hydraulic characterization of rice growing regions. (in prep.).

Wösten J H M, Bouma J, Stoffelsen G H (1985) Use of soil survey data for regional soil water simulation models. Soil Sci. Soc. Am. J. 49:1238–1244.

Wösten J H M, Bannink M H, Beuving J (1987) Water retention and hydraulic conductivity characteristics of top- and subsoils in the Netherlands: The Staring series. Report No. 1932, Soil Survey Institute, Wageningen, The Netherlands.

Wösten J H M, Schuren C H J E, Bouma J, Stein A (1990) Functional sensitivity analysis of four methods to generate soil hydraulic functions. Soil Sci. Soc. Am. J. 54:827–832.

Root ventilation, rhizosphere modification, and nutrient uptake by rice

G.J.D. KIRK
International Rice Research Institute, P.O. Box 933, 1099 Manila, Philippines

Key words: iron, nutrient uptake, oxidation, pH-change, reduction, rhizosphere, rice, root aeration, root structure, simulation, soil chemistry

Abstract
A model is described to examine how nutrient uptake by rice roots growing in flooded soil is influenced by the ability of the roots to ventilate themselves, the root structure needed for this ventilation and the associated root-induced changes in the rhizosphere. The model predicts the axial diffusion of O_2 from the atmosphere through the growing root, its consumption in respiration and radial loss to the soil where it reacts with mobile ferrous iron, and the generation and removal of root and soil CO_2. It allows for the variations in root parameters with root length and time, and for parameters controlling the transport and reaction of O_2, CO_2 and Fe in the soil. A more detailed but CPU-time intensive model also allows for the pH changes resulting from the soil reactions and their consequences for reactant mobility. Some model predictions are given and their implications for nutrient uptake and the design of nutrient-efficient germplasm and improved nutrient management are discussed.

Introduction

The operation of rice roots growing in flooded soil in relation to nutrient uptake is very poorly understood, and this limits both the development of nutrient-efficient germplasm and the development of soil management practices. Central to the operation of rice roots is their ability to supply O_2 to respiring tissues and to exhaust respired CO_2. This must be done internally because gas transport through flooded soil is very slow. The root structure required for this ventilation influences the root's ability to absorb nutrient ions from the soil at its surfaces, and gas transfer between the root and soil markedly alters conditions in the narrow zone of soil that is root-influenced, with corresponding changes in nutrient availability. The root-influenced zone extends just a millimetre or so into the soil, but existing experimental methods cannot analyse the soil with resolutions much finer than a millimetre. Because of this experimental difficulty and because the system involves the simultaneous operation of a series of complex, linked processes, simulation models are invaluable in studying the system.

This paper describes the characteristics of the rice root system and outlines a model to predict how root and soil characteristics influence the root's operation and how root-modification of the rhizosphere influences nutrient uptake. Much of the paper is speculative; it is intended that the model be used

F.W.T. Penning de Vries et al. (eds.), Systems Approaches for Agricultural Development, 221–232.

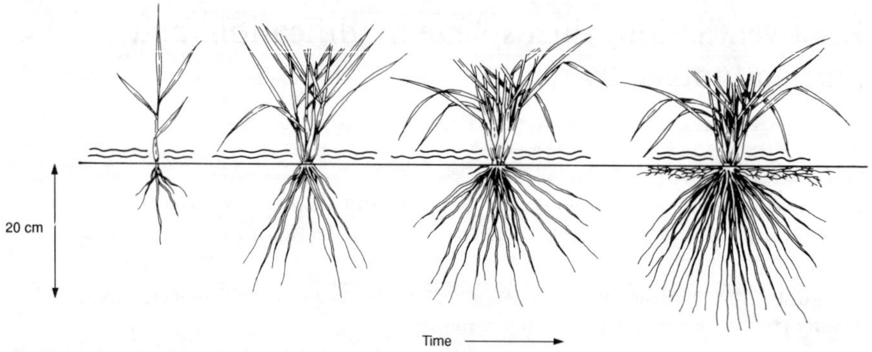

Figure 1. Development of the lowland rice root system.

to generate hypotheses that can be tested experimentally. The model is based particularly on the theory of solute diffusion in soils. This theory has been thoroughly reviewed by Nye (1979); other examples of its application are given by Kirk (1990).

Rice root structure and development

Figure 1 gives the main features of the developing rice root system during the vegetative growth stages. The root system mainly consists of 'nodal' roots originating in the lower nodes of the plant, and these initially grow almost straight down and are unbranched. New nodal roots form throughout tillering.

As a root grows, the cortex in the region closest to the base begins to disintegrate forming continuous gas channels between the base and tip. This both permits gas transport and reduces the amount of respiring tissue per unit root volume. Low root wall gas-permeability reduces O_2 loss to the soil. The structure of the individual rice root is therefore apparently dominated by the need to allow internal gas transport. But, on the face of it, this structure does not make for efficient nutrient uptake. The disintegration of the cortex seems likely to impair the ability of the older parts of the root to take up nutrients and convey them to the stele, and barriers to O_2 loss across the root wall presumably also impede nutrient transport. Furthermore, oxidation of the rhizosphere may impede nutrient transport to the surfaces of the older parts of the root, as discussed below. Therefore, nutrient uptake may be limited to a few cm close to the growing tip. Drew and Saker (1986) explored this point for maize roots grown under low O_2 conditions in nutrient solutions, and found no impairment of nutrient uptake in sections of root in which up to 80% of the cortex was disintegrated. I know of no similar studies on rice or with plants growing in soil, but the disintegration of the rice cortex and development of barriers in the root wall are much greater.

In the later vegetative growth stages, abundant fine secondary roots generally form on the upper parts of the nodal roots (Drenth et al. 1991). The function of these rootlets is unknown but may be related to the development of the internal gas channels. Thus, they may compensate for the reduced mechanical strength of the root resulting from disintegration of the cortex; they may also compensate for impaired nutrient uptake. There is evidence that the rootlets release O_2 into the soil (I. Watanabe, IRRI, pers. comm. 1991), which would have implications for nutrient uptake.

The distribution of primary root angles from the horizontal at heading has a maximum in the range 30–60 degrees with relatively few roots (3–10%) in the range 0–20 degrees (Abe et al. 1990); there are some varietal differences in this characteristic. Broadcast and incorporated N fertilizer will usually be concentrated in the surface 2 to 4 cm of soil. Considering how the roots grow in relation to the position of the fertilizer, and the likelihood that only a fraction of the total root length is active in nutrient uptake, it is not surprising that broadcast and incorporated N fertilizer is not used very effectively. The fertilizer is subject to large losses by volatilization and perhaps nitrification-denitrification, and it is in a position where only small amounts are in the proximity of active roots.

During the reproductive growth stages, the main body of the root system is largely degraded and is thus probably not very active in nutrient uptake. At these stages, there is often a well developed surficial root system at the soil-floodwater boundary (Figure 1). These roots do not contain gas channels and are more like the roots of upland crops. The role of the surficial root system is therefore interesting. If the bulk of later nutrient-uptake is due to these roots, then any remaining deep-placed fertilizer would be unavailable.

Gas transport through the roots and soil

From the above considerations it is clear that understanding gas transport through rice roots and the surrounding soil is central to understanding nutrient uptake. Gas transport is principally by diffusion along concentration gradients in the root gas channels.

Loss of respired CO_2 across the root wall may cause mass flow of air down the root because gaseous O_2 consumed in respiration would not then be replaced by a volumetrically equivalent release of gaseous CO_2. However, it can be shown that the supply of O_2 by this means must be subordinate to the diffusive supply (Beckett et al. 1988). But very high CO_2 concentrations arise in flooded soils because CO_2 produced in anaerobic microbial respiration and fermentation escapes from the soil only very slowly, and thus CO_2 may be passively taken up by roots (Penning de Vries (1991) has evidence that this is an important carbon source for photosynthesis under maximum yield conditions). Absorption of CO_2 and other soil gases could promote a mass flow of air up the root. To avoid undue complication, the model described here does not allow for this, but a further-developed version of the model shows that under particular conditions these effects are important.

The throughflow ventilation systems that operate in waterlilies and certain other wetland plants – driven by gradients in temperature and water vapour between surfaces of the upper plant and the atmosphere – do not arise in rice.

The rate of gas transport depends on the cortical porosity and hence cross-sectional area for diffusion, the rates of respiration in different root tissues, and the rates of radial transfer across the root wall. In rice, tissue respiration rates and root wall permeability decline markedly from the root tip upwards (Luxmoore et al. 1970; Armstrong and Beckett 1987). The cortical respiration rate per unit cortical volume will further decline as the cortical porosity increases.

There have been few satisfactory measurements of O_2 leakage from roots under realistic conditions and it is not known how much O_2 leaks out of the parts of the root that are most active in nutrient uptake. Oxygen leaking into the soil will react with ferrous iron – the main O_2-reacting material in most flooded lowland rice soils – according to the scheme

$$4Fe^{2+} + O_2 + 10H_2O \rightleftharpoons 4Fe(OH)_3 + 8H^+$$

(O_2 consumption in microbial respiration is much slower). As a result, Fe^{2+} will diffuse towards the oxidation zone. Thus, to evaluate the rate of O_2 loss to the soil it is necessary to account for the transport and reaction of both O_2 and Fe^{2+}.

The rate of gas transport through the root will be very rapid compared with the rates of root growth and solute transport in the soil. Thus we may use the steady-state equation

$$\frac{d}{dz}\left[D_G \Theta_G f_G \frac{dC_G}{dz}\right] - A = 0 \tag{1}$$

where C_G = gas concentration in the cortex, mol dm^{-3} (gas space),
D_G = diffusion coefficient of the gas in air, dm^2 s^{-1},
Θ_G = cortical porosity dm^3 (gas space) dm^{-3} (cortex),
f_G = diffusion impedance factor,
A = rate of gas consumption (A positive) or addition (A negative) in respiration in different root tissues and in transfer across the root-wall, mol dm^3 (cortex) s^{-1},
and z = axial distance from the root base, dm.

In evaluating A at a particular axial distance, the flux, F (mol per dm^2 of root surface area), of each gas across the root surface is obtained from an equation of the form

$$F = -\lambda\left[C_{L0} - \frac{C_G}{K_H}\right] - q_s C_{L0} \tag{2}$$

where λ = a coefficient for the effective resistance of the root wall to dissolved gas diffusion, s dm^{-1},
C_{L0} = gas concentration in the soil solution at the root surface, mol dm^{-3} (solution),

K_H = solubility of gas in water (dimensionless), and
q_S = water flux into the root, dm s^{-1};

and the rates of solute movement through the soil by mass flow and diffusion are obtained from equations of the form

$$\frac{\partial C_T}{\partial_\tau} = \frac{1}{r}\frac{\partial}{\partial_r}\left[rD_L\Theta_L f_L \frac{\partial C_L}{\partial_r} + r_O q_S C_L \right] - R \tag{3}$$

where C_T = total solute concentration in the soil, mol dm^{-3} (soil),
C_L = solute concentration in the soil solution, mol dm^{-3} (solution),
D_L = solute diffusion coefficient in free solution, dm^2 s^{-1},
Θ_L = soil volumetric moisture content, dm^3 (solution) dm^{-3} (soil),
f_L = soil solution diffusion impedance factor,
R = rate of consumption (R positive) or production (R negative) in reactions, mol dm^3 (soil) s^{-1}, and
r = radial distance from the root, subscripted 0 for the root surface, dm.

In equation (3), the first term in the square bracket describes diffusion and the second term mass flow. The diffusion term is with respect to the solute concentration in solution because solutes may diffuse through the solution but those sorbed on soil minerals and organic matter are immobile.

For loss of O_2 from the root into the soil, there is an equation of this type for O_2, in which R is divided by 4 because of the Fe^{2+} oxidation stoichiometry, and another equation of this type for Fe^{2+}. Contrary to O_2, which is not adsorbed, Fe^{2+} is adsorbed by soil solids to a degree depending on the acidification resulting from oxidation and on $Fe(OH)_3$ precipitation. However, although the profile of Fe in the soil is thus greatly influenced by the effects of pH and $Fe(OH)_3$ precipitation, the net rate of O_2 consumption is little influenced (Kirk et al. 1990) and, for the purpose of calculating gas transport through the root, as a first approximation the effects can be ignored. In this case, Fe sorption can be described by the Freundlich equation $[Fe^{2+}]_S = a[Fe^{2+}]_L{}^m$ and $[Fe^{2+}]_T = \Theta_L[Fe^{2+}]_L + \varrho[Fe^{2+}]_S$, where the subscript S refers to the soil solid and ϱ is the soil bulk density.

The rate equation for Fe^{2+} oxidation in soil is $R = k_{ox}[O_2]_L[Fe^{2+}]_S$ (Ahmad and Nye 1990), where k_{ox} is a rate constant. When the O_2 supply from the root declines and $[O_2]_L$ falls to a low value, it is also necessary to allow for re-reduction of Fe^{3+}. In the rhizosphere, carbon is unlikely to limit Fe re-reduction. Available data suggest that the process is roughly of first order kinetics in $[Fe^{3+}]_S$: $R_{red} = k_{red}[Fe^{3+}]_S$.

For CO_2 uptake from or loss to the soil, the CO_2 sources (decomposition of root exudates and soil organic matter, dissolution of carbonates, root release) and sinks (reduction to methane, precipitation as carbonates, root uptake) roughly balance so as to maintain a constant mean CO_2 pressure in the soil. Thus it is generally observed that the bulk soil pH reaches a steady state during rice growth. Because CO_2 is not adsorbed by soil solids, the rate of CO_2

transport is high in relation to the distance between the root surface and the mid-point between adjacent roots. Therefore, the steady state is reached rapidly.

The model allows for axial changes in Θ_G, λ, and respiration rates, and for differences in respiration rates between the stele, cortex and root wall.

Figure 2 shows predicted profiles of O_2 and CO_2 in the root cortex and the corresponding reactant concentration profiles in the adjacent soil. For the chosen conditions, which represent the upper range of soil O_2 consumption rates, the maximum length of root that can be aerated is 21 cm (Figure 3b); this is in the range predicted by other models of gas transport through roots (Bouldin 1966; Luxmoore et al. 1970; Armstrong and Beckett 1987). The model shows that substantial amounts of Fe are transferred towards the root surface resulting in a well-defined zone of ferric hydroxide accumulation; the extent of Fe accumulation changes along the root length and with time. A zone of Fe^{2+} depletion arises where oxidation is intense, but is rapidly filled in when the O_2 supply decreases. Re-reduction of Fe(III) is slow compared with oxidation, although the chosen reduction rate is the highest likely. CO_2 is lost from the root to the soil near the root tip, but taken up by the root near the base; the resultant CO_2 concentration profiles in the soil are small.

Rhizosphere conditions

Three processes may markedly modify the pH of the soil near rice root surfaces and hence the supply of nutrient ions. First, concomittant with the reaction of O_2 with Fe^{2+} is a release of acidity: the oxidation of each mole of Fe^{2+} produces 2 moles of H^+.

Secondly, because the roots take up a considerable excess of cations over anions – N being taken up chiefly as NH_4^+ ions under the chemically-reduced conditions of flooded soils – they release H^+ ions into the soil to maintain electrical neutrality across the root-soil interface. Yoshida (1981, Table 3.11) gives the total uptake of cations (NH_4^+, K^+, Ca^{2+} and Mg^{2+}) as 2.4×10^4 eq ha^{-1} and the total uptake of anions ($H_2PO_4^-$, SO_4^{2-} and Cl^-) as 5.2×10^3 eq ha^{-1}; these are data for a 130-day lowland rice variety grown under optimal conditions. Note that Si crosses the root as the uncharged molecule H_4SiO_4. Assuming (a) the total uptake period corresponds to half the growing season, (b) the whole root length is active in uptake (if it is not then the H^+ efflux will be concentrated accordingly) and (c) a mean root radius of 0.1 mm and rooting density of 2×10^3 dm dm^{-3}, the flux of H^+ across the root surface is 3×10^{-10} mol dm^{-2} s^{-1}. This is comparable to the value calculated by Nye (1981) for well nodulated legume roots which take up their N as uncharged N_2 molecules.

Thirdly, as we have seen, the roots may either release or take up CO_2 from the soil. Depletion of CO_2 in the rhizosphere would tend to raise the pH. However, the CO_2 gradients predicted by the model are small.

The magnitude of the pH changes caused by these processes will depend on

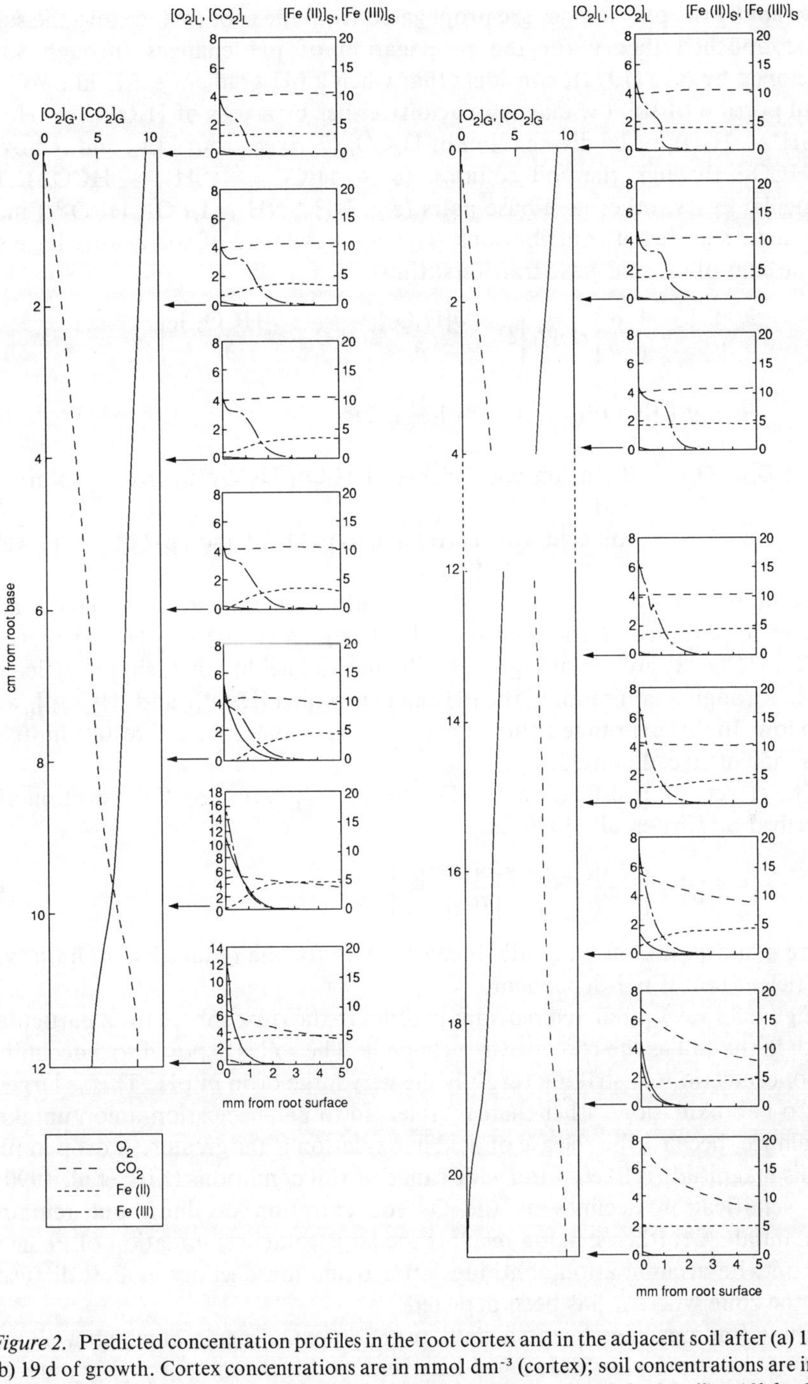

Figure 2. Predicted concentration profiles in the root cortex and in the adjacent soil after (a) 10 and (b) 19 d of growth. Cortex concentrations are in mmol dm^{-3} (cortex); soil concentrations are in mol dm^{-3} (solution) x 10^5 for O_2, mmol dm^{-3} (solution) for CO_2, and mol dm^{-3} (soil) x 10^2 for Fe(II) and Fe(III). The parameter values are given in the Appendix.

how rapidly the pH changes are propagated from the root surface into the soil. An established theory for the propagation of pH changes through soil, developed by Nye (1972), considers that when a pH gradient exists in a soil, a small portion of soil (S) may gain acidity either by access of H_3O^+ ($S + H_3O^+ \rightleftharpoons SH^+ + H_2O$), or by dissociation of H_2CO_3 (derived from CO_2) and removal of HCO_3^- through the soil solution ($S + H_2CO_3 = SH^+ + HCO_3^-$). In particular cases, other acid-base pairs (e.g. NH_4^+-NH_3, $H_2PO_4^-$-HPO_4^{2-}) may also make significant contributions. The net movement of soil acidity is given by the sum of all acid-base transfers; thus

$$\frac{\partial[HS]}{\partial t} = \frac{1}{r}\frac{\partial}{\partial r}\left[r\Theta_L f_L\left\{D_{LH}\frac{\partial[H_3O^+]_L}{\partial_r} - D_{LC}\frac{\partial[HCO_3^-]_L}{\partial_r}\right\}\right.$$

$$\left. + r_O q_S\left\{[H_3O^+]_L - [HCO_3^-]_L\right\}\right] + 2R \qquad (4)$$

where D_{LH}, D_{LC} = diffusion coefficients of H_3O^+, HCO_3^- in free solution, and

HS = soil acid which is related to pH by the soil pH buffer power, b_{HS} ($= -$ d[HS]/dpH).

(R is multiplied by 2 because of the oxidation reaction stoichiometry.) The relative importance of the pair H_3O^+-H_2O is greater at low pH and that of H_2CO_3-HCO_3^- is greater at high pH. The net soil acidity diffusion coefficient passes through a minimum in the pH range in which $[H_3O^+]_L$ and $[HCO_3^-]_L$ are both low. In this pH range a flux of acid or base through the soil results in steep pH gradients (see Figure 3).

The effects of acidification and $Fe(OH)_3$ precipitation on Fe^{2+} sorption are described by Kirk et al. (1990) as

$$[Fe^{2+}]_L = \left[\frac{[Fe]_{S\,total}}{a}\right]^{1/m}\left[\frac{[Fe^{2+}]_S}{[Fe]_{S\,total}}\right]^{1/n} \qquad (5)$$

where a and m are the Freundlich coefficients for the reduced soil and n is a coefficient that is pH-dependent.

Figure 3 shows predicted reactant profiles in the rhizosphere for a particular depth in the soil as the root grows through it. The soil is exposed to a declining O_2 concentration. A striking result is the very large drop in pH. This is largely due to Fe^{2+} oxidation rather than H^+ released to balance cation-anion uptake-imbalance, because the release of acid in oxidation is far greater. A drop in pH of this magnitude is likely for a wide range of soil conditions (Kirk et al. 1990). The acidification declines as the O_2 concentration declines, but remains substantial. A further striking result is the large total accumulation of Fe near the root. Fe accumulation continues after oxidation declines as Fe^{2+} diffuses into the zone where it has been depleted.

These findings have been partially tested in experiments in which a planar layer of rice roots was sandwiched between two blocks of thoroughly reduced soil, so that after a few days the soil blocks could be thinly sectioned parallel to

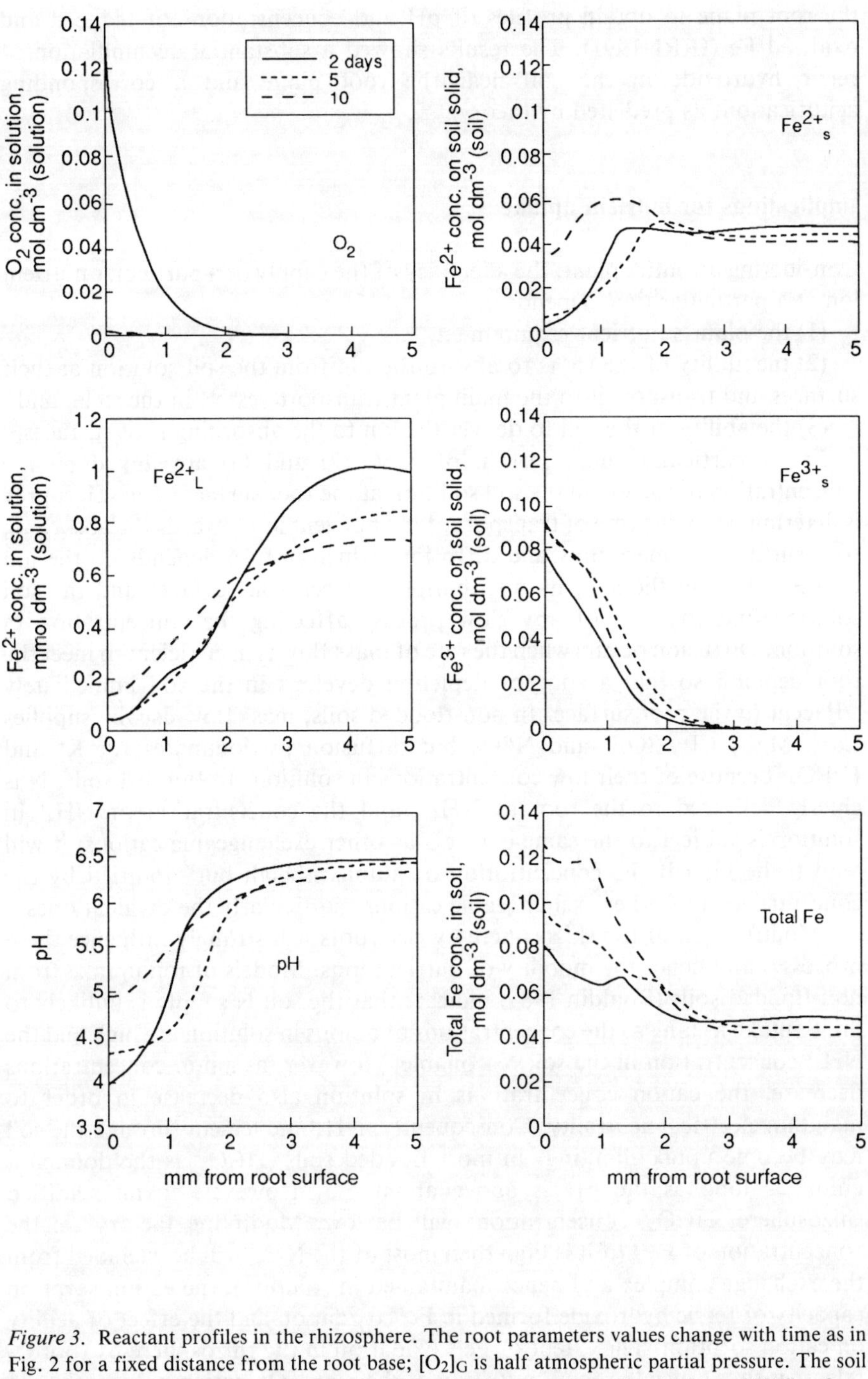

Figure 3. Reactant profiles in the rhizosphere. The root parameters values change with time as in Fig. 2 for a fixed distance from the root base; $[O_2]_G$ is half atmospheric partial pressure. The soil parameters are as in Fig. 2 and Kirk et al. (1990) with b_{HS} = 0.05 mol kg^{-1} pH^{-1} and H^+ flux from the root = 3 x 10^{-9} mol dm^{-2} s^{-1} declining 10^3-fold as λ declines.

the root plane to obtain profiles of pH and concentrations of reduced and oxidized Fe (IRRI 1991). The results showed a substantial accumulation of ferric hydroxide in the soil near the root plane and a corresponding acidification, as predicted by theory.

Implications for nutrient uptake

Considering an entire plant, the adequacy of the supply of a particular nutrient ion at a particular time depends on

(1) the plant's nutrient requirement,

(2) the ability of the roots to absorb the ion from the soil solution at their surfaces and transport it to the main plant transport vessels in the stele, and

(3) the ability of the soil to deliver the ion to the absorbing root surfaces.

For a particular small section of root, (2) and (3) are linked by the concentration of the ion in the soil solution at the root surface. The soil supply is determined by the rate of transport of the nutrient ion towards the absorbing root surfaces by mass flow and diffusion, which in turn depends on the ion concentration in the soil, its distribution between soil surfaces and the soil solution and the rate of any slow process affecting the concentration in solution. Diffusion occurs when the rate of mass flow is insufficient to meet the root demand so that a zone of depletion develops in the soil immediately adjacent to the root surface. In non-flooded soils, mass flow usually supplies Ca^{2+}, Mg^{2+}, Cl^-, SO_4^{2-} and NO_3^-, but diffusion predominates for K^+ and $H_2PO_4^-$ because of their low concentrations in solution. In flooded soils, N is chiefly delivered to the root as NH_4^+, and the concentration of NH_4^+ in solution is subject to the same controls as other exchangeable cations: it will tend to be high if the concentration of anions is high but modified by the concentrations of other exchangeable cations, particularly the divalent ones.

Modification of the rhizosphere by rice roots will strongly influence these processes and hence the mobility of nutrient ions. Models of ion uptake from non-flooded soils (Bouldin 1989) indicate that the soil N supply is unlikely to limit uptake as long as the concentrations of anions in solution are high and the NH_4^+ concentration in the soil reasonable. However, as anion concentrations decrease, the cation concentrations in solution also decrease in order to maintain electrical neutrality. Consequently, NH_4^+ movement through the soil may become uptake-limiting. In most flooded soils, HCO_3^- is the dominant anion as long as the pH is above about 5.5. However, in an acidified rhizosphere, HCO_3^- concentrations will be low. Modifying factors are the concentration of Fe^{2+} (if it is high then most of the NH_4^+ will be excluded from the exchange complex and hence maintained in solution), the cation sorption capacity of ferric hydroxide formed in Fe^{2+} oxidation and the effect of acidity on cation sorption. The extent of Fe^{2+} oxidation in the rhizosphere by root O_2 release is thus crucial in the N nutrition of the plant. Oxidation reduces the pH and the Fe^{2+} concentration, and hence may adversely affect NH_4^+ mobility. In addition, acidification may reduce the rates of microbially-mediated processes affecting the N supply, such as mineralization of organic N.

On the other hand, the supply of other nutrient ions may be enhanced. In many soils, acidification increases the availability of sparingly soluble nutrients such as P and Zn, even at low pH. Thus, Fe^{2+} oxidation in the rhizosphere may increase P and Zn mobility, although this will be offset to some extent by their sorption on ferric hydroxide precipitated in the oxidation reaction. The tolerance of lowland rice to toxic acids such as Al^{3+} is a further consideration, but it is well known that upland varieties are tolerant of very high levels of Al^{3+}.

Perspectives

The model demonstrates that the soil as 'seen' by the rice root is entirely different from the bulk soil although it is the latter whose properties we normally measure. The model furthermore provides a means of assessing quantitatively how root and soil characteristics influence nutrient uptake in this highly complex system. In addition to the changes in soil conditions with distance from the root surface, the model shows that there are large differences along the root length. We need to know how the nutrient uptake capacity of the root varies along the root length – averages over the whole root system will not suffice – and how this changes as the root ages. Related to this, we need a better description of the root system: the rate at which new roots are formed and grow, their dimensions and geometry. It is hoped that with this knowledge, we will be able to breed rice varieties with root systems tailored to particular conditions, and we will be able to better match agronomic management with varietal and soil properties.

Acknowledgement

I thank Prof. Dave Bouldin for stimulating discussions.

Appendix

Root and soil parameter values.

Root (based on Luxmoore et al. 1970; Armstrong and Beckett 1987):
For z' (distance from the tip) = 0–2 cm: $\Theta_G = 0.075$, $\lambda = 10^{-2}$ dm s^{-1}, $q_s = 10^{-6}$ dm s^{-1}; for $z' = 5$ cm to $z = 0$ cm, $\Theta_G = 0.75$, $\lambda = 10^{-7}$ dm s^{-1}, $q_s = 10^{-9}$ dm s^{-1}; for $z' = 2$–5 cm, Θ_G and q_s vary linearly, λ varies logarithmically; the respiration rate at $z' = 0$ dm is 10^{-5} mol dm^{-3} (root tissue) s^{-1} and it decreases as $(z')^2$ to a minimum of 3×10^{-7} mol dm^{-3} (root tissue) s^{-1} at $z' = 10$cm; $f_G = 1$; $r_{\text{outer stele}} = 0.25$ mm; $r_{\text{inner wall}} = 0.47$ mm; $r_0 = 0.5$ mm; rooting density = 105 dm^{-2}; initial root length = 2 cm; root elongation rate = 1 cm d^{-1}.
Soil (based on Kirk et al. 1990):
Total Fe = 0.05 mol kg^{-1}; Fe^{2+} sorption Freundlich parameter $a = 1.5$ mol kg^{-1}, parameter m = 0.5; CO_2 pressure = 0.5 kPa; $k_{ox} = 0.3$ dm^3 mol^{-1} s^{-1}; $k_{red} = 10^{-6}$ s^{-1}; $\Theta_L = 0.5$; $f_L = 0.5$; $\varrho = 1.0$ kg dm^{-3}.

References

Abe J, Nemoto K, Hu D X, Morita S (1990) A nonparametric test on differences in growth direction of rice primary roots. Jpn. J. Crop Sci. 59:572–575.

Ahmad A R, Nye P H (1990) Coupled diffusion and oxidation of ferrous iron in soils. I. Kinetics of oxygenation of ferrous iron in soil suspension. J. Soil Sci. 44:395–409.

Armstrong W, Beckett P M (1987) Internal aeration and the development of stelar anoxia in submerged roots. New Phytol. 105: 221–245.

Beckett P M, Armstrong W, Justin S H F W, Armstrong J (1988) On the relative importance of convective and diffusive gas flows in plant aeration. New Phytol. 110:463–468.

Bouldin D R (1966) Speculations on the oxidation-reduction status of the rhizosphere of rice roots in submerged soils. Pages 128–139 in IAEA Technical Report No. 655. International Atomic Energy Authority, Vienna.

Bouldin D R (1989) A multiple ion uptake model. J. Soil Sci. 40:309–319.

Drenth H, Meijboom F, ten Berge H F M (1991) Effects of growth medium on porosity and branching of rice roots (Oryza sativa L.). Pages 162–175 in Penning de Vries F W T, van Laar H H, Kropff M J (Eds.) Simulation and systems analysis for rice production (SARP). Pudoc, Wageningen, 369 p.

Drew M C, Saker L R (1986) Ion transport to the xylem in aerenchymatous roots of Zea mays L. J. Exp. Bot. 37: 22–33.

International Rice Research Institute (1991) Pages 203–205 in Program Report for 1990, International Rice Research Institute, Manila, Philippines.

Kirk G J D (1990) Diffusion of inorganic materials in soil. Philos. Trans. R. Soc. Lond. B Bio. Sci. 329:331–342.

Kirk G J D, Ahmad A R, Nye P H (1990) Coupled diffusion and oxidation of ferrous iron in soils. II. A model of the diffusion and reaction of Fe^{2+}, H^{+} and HCO_3^- in soils and a sensitivity analysis of the model. J. Soil Sci. 44:411–431.

Luxmoore R J, Stolzy L H, Letey J (1970) Oxygen diffusion in the soil-plant system. Agron. J. 62:317–332.

Nye P H (1972) The measurement and mechanism of ion diffusion in soils. VIII. A theory for the propagation of pH changes in soils. J. Soil Sci. 23:82–92.

Nye P H (1979) Diffusion of ions and uncharged solute in soils and soil clays. Adv. Agron. 31:225–272.

Nye P H (1981) Changes of pH across the rhizosphere induced by roots. Plant Soil 61:7–26.

Penning de Vries F W T (1991) Improving yields: designing and testing VHYVs. Pages 13–26 in IRRI Res. Pap. Ser. 151. International Rice Research Institute, Manila, Philippines.

Yoshida S (1981) Fundamentals of Rice Crop Science. International Rice Research Institute, Manila, Philippines. 269 p.

Adjustment of nitrogen inputs in response to a seasonal forecast in a region of high climatic risk

B.A. KEATING[1], R.L. McCOWN[2] and B.M. WAFULA[3]

[1] *CSIRO, Division of Tropical Crops and Pastures, 306 Carmody Rd St Lucia, 4067 Australia*

[2] *Agricultural Production Systems Research Unit, P.O. Box 102 Toowoomba, Queensland, 4350 Australia*

[3] *Kenyan Agricultural Research Institute, P.O. Box 340 Machakos, Kenya*

Key words: CERES-Maize, fertilizer application, financial return, Kenya. maize, management strategy, risk, semi-arid zone, simulation, water- and nitrogen limited yield, weather forecast

Abstract

Responses to inputs such as nitrogen (N) fertilizer can vary dramatically from one season to the next in association with the rainfall variability in semi-arid climates. Traditional agricultural experimentation has examined fixed strategies of fertilizer management, but farmers frequently make tactical adjustments to their management in the light of what they perceive as information relevant to the prospects for the forthcoming crop. In this paper, we use a crop simulation model to examine the value of changing N fertilizer application rates in line with predictions of likely response to fertilizer based on the date of season onset. The analysis, using the CERES–Maize model, is centred on maize production in Machakos, a semi-arid district in eastern Kenya.

The timing of sowing relative to defined onset criteria was examined, and the value of minimizing delays in sowing after onset quantified. Assuming sowing takes place at onset, a set strategy based on 45 kg N ha^{-1} maximizes returns, but nil or negative returns were associated with N fertilizer inputs in 40 % of seasons.

The date of season onset was found to be a useful predictor of response to fertilizer inputs. A conditional strategy in which N fertilizer application was adjusted in relation to onset-date, resulted in only a small increase in expected gross margin. As far as production was concerned, the use of some fertilizer, irrespective of seasonal outlook, was the highest priority.

The impact of tactical adjustments in fertilizer use on the associated risks were assessed using a number of criteria. The results indicate that conditional strategies reduced the risk of negative gross margin and may be valued by extremely risk-averse farmers contemplating fertilizer use.

Introduction

Agricultural experimentation has traditionally compared farming strategies consisting of actions that are fixed over time. The yield of one cultivar is compared with another; one planting date against another; different rates of fertilizer are used in an attempt to find an optimum, and so on. While farmers may have a relatively fixed domain within which they operate, many of the decisions they make are conditional in nature. Their strategies of farming include many decision nodes in a decision tree from which various actions arise. The majority of decisions are made in the light of additional information in the form of current or preceding seasonal, management or economic events or states. The term "tactics" has been used to describe opportunistic changes to a

F.W.T. Penning de Vries et al. (eds.), Systems Approaches for Agricultural Development, 233–252.

general strategy to take advantage of short-term conditions (Connor and Loomis 1991). In the more general context of decision theory, tactics represent choices from the set of possible actions that are available at decision nodes in the decision tree. A fixed strategy is a special and simpler case of the general decision problem–that is one in which there is no possibility of obtaining further information (Anderson et al. 1977). While there is no theoretical need for such a distinction, we refer to set strategies ($S_{1 \text{ to } n}$) and conditional strategies ($C_{1 \text{ to } n}$) as we examine prospects for nitrogen fertilizer use in this paper.

Research on conditional strategies has been constrained by methodological problems. Such strategies assume greatest importance in climates of high variability and field experiments would need to run for a large number of years to adequately sample this variability. Traditionally, uncontrolled climatic variation has been lumped with all other uncontrolled variables. Advances in the simulation of crop growth hold promise for both the design and evaluation of farming strategies in which management decisions are conditional on other events or states. The assessment of a wide range of both set and conditional management practices over long periods of historical weather information through crop or cropping system models is possibly the only realistic means by which they could be evaluated.

Classes of forecast

Application of a tactic to farm management requires that some information or forecast relevant to the prospects of a forthcoming crop is available at a time when the farmer still has the chance of responding with revised management actions. This information can be directly relevant to a forthcoming crop (e.g. soil water at sowing) or be a adequate surrogate for such information (e.g. rainfall over a fallow period).

Some examples of information sources upon which forecasts have been based include:

- Climate information e.g. onset date of rains, rainfall quantities (Stewart and Faught 1984), long term weather forecasts such as the Southern Oscillation Index (Hammer et al. 1991).
- Soil information e.g. soil water content in relation to decisions on fallowing (Fischer and Armstrong 1987).
- Past crop or fertilizer management history.
- Current crop performance e.g. tissue analysis, sap tests, tiller number in relation to within-season decision making on the need for additional nitrogen fertilizer in wheat (Van Herwaarden et al. 1989).
- Past crop performance e.g. past grain protein levels in relation to decisions on nitrogen fertilization of wheat (Woodruff 1987).

Conditional strategies in traditional farming systems

The generalised notion of the decision problem, with one or more sources of information on system state providing forecasts of future system performance, is equally applicable to intensive agricultural production and to traditional subsistence farming systems. This paper is focused on the latter and our views have been strongly shaped by experiences gained in working with the predominantly subsistence farmers of a semi-arid region in eastern Kenya. In such systems, farmers have a range of tactics available to them as a season unfolds, and their perceptions of its quality develop (O'Leary 1984). Many farmers attempt to have at least a portion of their crop land planted to maize before the rains commence. The rate at which planting proceeds on the remaining land will be influenced by the timing and intensity of the opening rains (Ockwell et al. 1991). If the rains start late and, in the farmer's mind, do not hold promise for a good season, farmers will start to plant more drought tolerant crops like sorghum.

Objective of this paper

A strategy for making nitrogen fertilizer application conditional on the timing and extent of the early seasonal rains had been proposed for a region in semi-arid eastern Kenya by Stewart and Faught (1984). This was referred to as "Response Farming" and the details of the scheme have been evaluated by Wafula (1989) and McCown et al. (1991). In the current paper, we revisit response farming, but only insofar as it provides an example of a conditional management strategy. Much of the detail previously covered will not be repeated. Instead, we shall focus on the potential for simulation to assess conditional strategies of crop management.

This paper outlines the modeling approach employed and the work needed to develop a capability to realistically simulate conditional management strategies. In this region, farmers plant largely in response to what they consider to be the onset of the rainy season. This varies greatly from year to year. Hence, we initially examine simulations with plantings made at 'onset', and for various delays after onset, to identify an optimum planting strategy. The general prospects for nitrogen fertilizer use in the region are examined in terms of set strategies, and then the benefits of conditional strategies with fertilizer inputs linked to forecasts of season potential are assessed.

Methods for evaluating strategies for N management

The work was conducted in a semi-arid region of eastern Kenya where maize is the staple crop. The bi-modal rainfall regime allows two crops to be sown each year, but the risks of drought are high with rainfall less than 250 mm expected in 40 % of seasons. Little fertilizer is used (Rukandema 1984) and when adequate rain is received, yields generally remain low because of N deficiency (Jaetzold and Schmidt 1983).

Rapid population growth and degrading land resources in this region have created intense interest in more productive farming systems (Lynam 1978; Lele 1989). The high risk of poor returns to fertilizer inputs in drought years is often quoted as major deterrents to their use. Our analysis (Keating et al. 1991; McCown and Keating 1992) supports the view expounded by Ruthenberg (1980) that productive systems are unlikely to be achieved in such situations without a major injection of nutrients in the form of mineral fertilizers.

Model development and performance

The CERES–Maize model (Jones and Kiniry 1986) was chosen as an appropriate model. This model dealt with the major N transformations in the soil and the important crop growth and development processes in response to environment and management. The input requirements of the model were feasible in our situation.

We commenced this work with version 1 of CERES–Maize in 1985. The model was extensively evaluated at a number of sites in semi-arid eastern Kenya over the period 1985 to 1989. While performance of the original model was reasonable, a number of revisions were made to deal with problems encountered during its application in Kenya. In addition, a number of enhancements were made to allow for more realistic simulation of both fixed and tactical management options (Keating et al. 1991).

The severity of the water deficits encountered in the region under study were so great that crops actually died. The original model would not simulate crop death, but allowed severely stressed crops to remain in 'suspended animation'. If rain was received later in the vegetative growth period, the simulated crops recommenced growth and low, but significant, yields could be achieved. In reality, such crops were dead and the farmer would have considered re-sowing on the late rain. Routines were introduced which killed crops in response to an accumulated index of water deficit during the early- to mid-vegetative growth period.

Silking was found to be delayed by severe water or nitrogen stress, and changes were made to the model to simulate such delays. A number of other changes were made which we felt improved model integrity or had conceptual advantages. Some of these changes have also been addressed in version 2 of CERES–Maize.

Planting date was an input in the original model, fixed for any particular crop being simulated. This was unrealistic in this region where farmers plant in response to what they perceive as the onset of the rainy season. Routines were introduced which allow the user to define criteria for season onset in terms of the length, pattern and quantity of rain needed to initiate a planting opportunity. Related routines allow for replant options should a crop emerge, but fail to survive during an onset window.

Management information such as plant population and fertilizer rate were also fixed inputs for a particular crop being simulated in the original model. Enhancements were made which allowed these inputs to be conditional on the timing of onset of the season. For instance, if the rains started and sowing took place before a nominated date, high plant populations and fertilizer N could be set. If the rains started late, the simulation could be set up to use low plant populations and not apply fertilizer. Opportunities were also made for within-season management (fertilizer side-dressings, thinning) to be conditional on the timing and quantity of early-season rain.

The model validation dataset contained information from 159 crop/treatment combinations, with yields ranging from 0 to 8000 kg ha^{-1} in response to variation in sowing date, water, nitrogen, plant population and climatic conditions. The root mean squared deviation between predicted and observed grain yield was 689 kg ha^{-1}. The line of best fit was close to the 1:1 line (slope (s.e.) = 0.94 (0.03) and intercept (s.e.) = 249 (103)) and coefficient of determination (r^2) was 0.88. Further information on model performance is given by Keating et al. (1991).

Standard methods and assumptions for this paper

The standard inputs used and assumptions made throughout the simulation study have been described in detail elsewhere (Keating et al. 1991; McCown et al. 1991). Briefly, all simulations were conducted using daily rainfall data for the National Dryland Farming Research Centre, Katumani, Machakos, Kenya (lat. 1°35' S ; long. 37°14' E ; altitude 1601 m). In general, conditions selected are those thought to be typical of current practice or recommendations. Two crops per year were simulated over the 1957 to 1988 period. The short rains (SR) occur from October to January and the long rains (LR) from March to July.

Onset of the long rains season was deemed to occur when 40 mm of rain was recorded within an 8 day period, with no more than one contiguous dry day. The onset rule for the short rains was similar but based on 30 mm instead of 40 mm. Onset periods or 'windows' were defined from calendar days 289 to 327 and 38 to 106 for the SR and LR respectively. Unless specified otherwise, sowing was assumed to take place immediately season onset was detected within the window. If onset was not detected in any particular season, the crop was assumed to have been sown into dry soil at the end of the onset window. These onset criteria are based on the agroclimatic analysis of Stewart and Faught (1984) but are to some degree arbitrary and bound to be specific to regions. Nevertheless, the concepts of planting windows and minimum rain needed to initiate planting activity are consistent with farmer behaviour in this region (Ockwell et al. 1991) and are likely to be more generally applicable.

The maize cultivar, Katumani Composite B, was simulated throughout this study. The standard soil profile assumed was that of a chromic luvisol which is typical of the region. This soil has an organic carbon content of 0.8% in the surface layer, an initial mineral-N content of 54 kg ha^{-1} and a potential available water content of 173 mm over its 130 cm depth. Each season was modelled independently of other seasons with reinitialization of input parameters at the start of the onset window.

Grain yields were used to compare alternative agronomic strategies when no major input costs were involved (e.g. study of delay in planting after onset). The performance of alternative set and conditional strategies involving different input levels (e.g. studies involving different rates of fertilizer) were

Table 1. Nitrogen fertilizer and plant population levels used in the simulation study of (a) fixed strategies, (b) tactics conditional on season onset.

Set strategy	Plant population (10^3 plants ha^{-1})	N rate ($kg\ ha^{-1}$)
(a) Set strategies		
S_1	22	0
S_2	27	15
S_3	33	30
S_4	37	45
S_5	44	60
S_6	55	80

Strategy	Predictor (z_i) onset date	Plant population (10^3 plants ha^{-1})	N rate ($kg\ ha^{-1}$)
(b) Conditional strategies			
C_1-medium	Early	33	30
inputs	Late	33	0
C_2-high	Early	44	60
inputs	Late	33	30

compared using gross margin per hectare. The assumptions made in terms of prices of inputs and outputs are given in McCown et al. (1991). Monetary values are in Kenyan shillings (Kshs) and as a guide, 100 Kshs is equivalent to four US dollars.

Variable costs included seed, fertilizer (30 Kshs kg^{-1} N) and harvest costs. The price assumed for nitrogen is twice the purchase price to allow for variable costs of transport, application and additional weeding costs. A constant sale price of 3 Kshs kg^{-1} for maize grain was assumed.

Strategies examined

In this study, sowing was assumed to take place at onset or to be delayed by periods of up to 25 days from onset. The analysis was conducted with moderate levels of N fertilizer and plant population (S_4 in Table 1a). Other inputs were those described earlier as standard for the simulations.

Rates of N fertilizer (ranging from 0 to 80 kg N ha^{-1}) applied at sowing as Calcium Ammonium Nitrate were examined. Other studies have shown that plant population needs to be varied to match nitrogen supply if optimum production is to be achieved (Keating et al. 1991). Hence, these N rates were combined with plant populations ranging from 22000 to 55000 plants ha^{-1} (Table 1a).

Seasonal onset date was used as the predictor of seasonal potential. Seasons in which onset occurred before 18 March (calendar day 77) and 2 November (calendar day 306) for the long rains and short rains respectively were classified as early. Seasons starting after these dates within the defined onset windows were said to be late. These definitions of early and late onset were those developed by Stewart and Faught (1984). Two levels of management were evaluated, each with its own tactics for early and late onset (Table 1b). No low input level was considered since tactics are only relevant when at least some inputs are in use. The tactics evaluated can be thought of as a reduction in N fertilizer rate and plant population when a late onset forecasts a poor season.

Conceptual framework for the analysis of tactical decision-making

The conceptual framework for the analysis of decisions based on Bayesian statistical theory is well developed (Anderson et al. 1977). Bayes theory allows the probabilities of different outcomes (states) to be calculated, conditional on other events. The notion of a conditional strategy as used in this paper is essentially the decision problem where there is the possibility of gathering further information from an 'experiment'. The concepts needed for the analysis of the situation where information is available for tactical decisions are incorporated in the equation:

$$P(\theta_i | z_k) = P(\theta_i) * P(z_k | \theta_i) / \text{Sum}\{P(\theta_i) * P(z_k | \theta_i)\}$$

Where;

$P(\theta_i | z_k)$ – The posterior probabilities of θ_i given z_k e.g. the probability of a particular state (e.g. good season) after observing a particular forecast (e.g. late start). – -Relevant to conditional strategies.

$P(\theta_i)$ – The prior probability of the state i occurring, i.e. the assessment of the state's chance of occurrence based on historical weather data. – The only probability relevant to set strategies.

$P(z_k | \theta_i)$ – The likelihood of forecast z_k given state i, e.g. the chance of observing an early start to the season (forecast) given that a good season (state) will prevail.

$P(\theta_i)*P(z_k | \theta_i)$ – The product of prior probability and likelihood is referred to as the joint probability.

θ_i – The ith event or state e.g. good season vs. bad season; good response to fertilizer vs. poor response.

z_k – The kth prediction or forecast arising out of an 'experiment' and providing additional information about the probabilities of the states e.g. early start to the rains, late start; soil dry at planting or fully wet at planting.

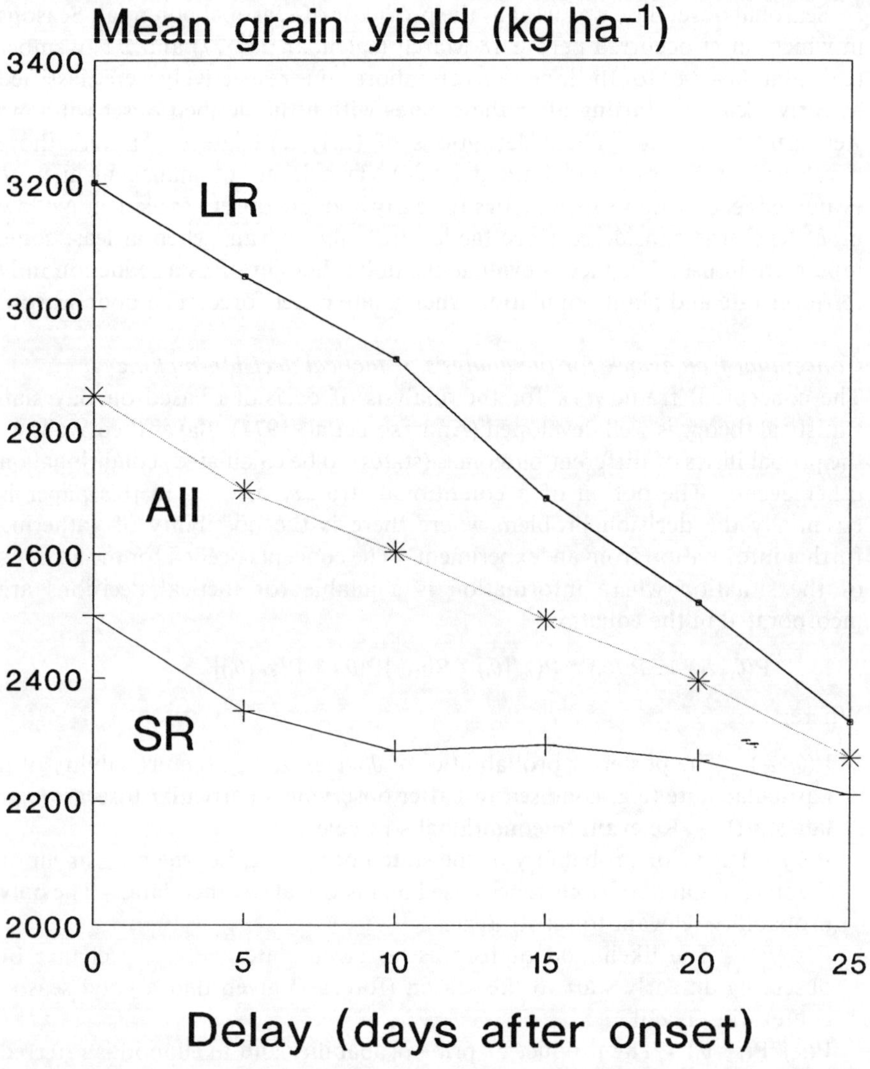

Figure 1. Effects of delay after onset in planting on the mean grain yields at Katumani in the long rains and short rains, 1957 to 1988.

Results of the simulation study

Planting strategies

It is not sensible to consider an optimal calendar date for planting in this environment, given the large variation in dates at which the seasons start. We examined the performance of crops simulated as having been planted according to some onset criteria, in comparison with crops for which some delay occurs between the time when those criteria are satisfied and planting.

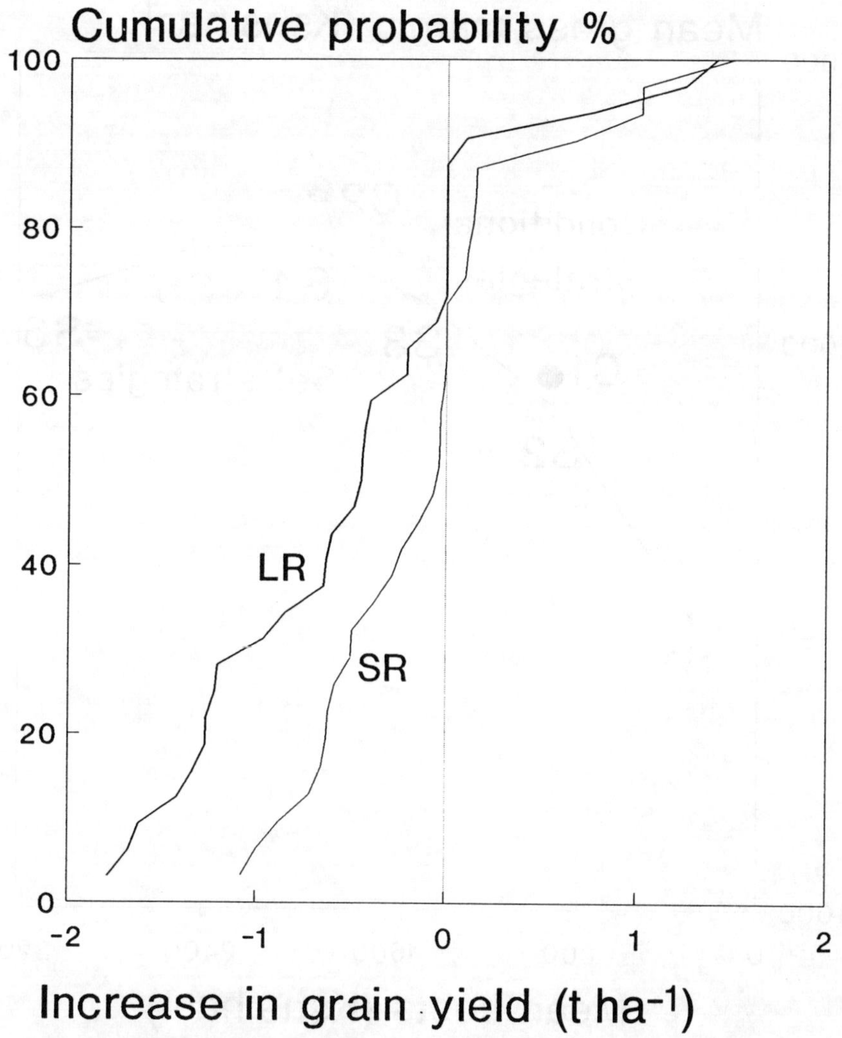

Figure 2. Cumulative distribution function for loss of grain yield associated with a 20 day delay in planting after onset for the long rains and short rains.

Mean grain yield simulated over the 63 seasons examined at Katumani declined from 2900 to 2300 kg ha^{-1} as planting was delayed 0 to 25 days after onset (Figure 1). The losses associated with delayed planting were generally greater in the long rains than in the short rains. On average, losses of 23 kg (0.8 %) of grain yield per day delay in planting were simulated over both seasons, rising to 35 kg (1.1%) per day delay in the long rains. Variation in response from season to season was great and, while losses were generally recorded, some crops benefited from delays in planting. This occurred either in situations where out-of-season rain was recorded in the January–February short dry period or

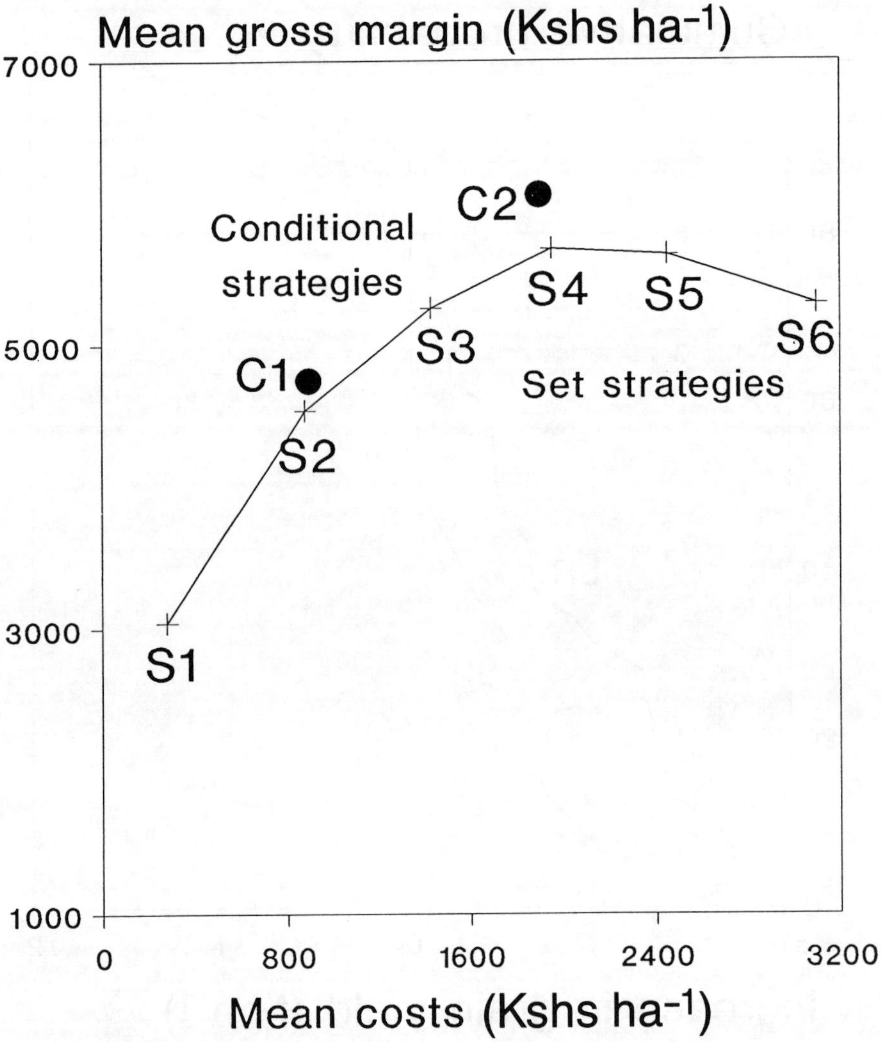

Figure 3. Average gross margin over the 1957 to 1988 period at Katumani associated with a range of fixed and conditional strategies.

when excessive nitrogen losses were limited by delayed planting in a small number of extremely wet seasons. A 20–day delay in planting after onset was estimated to lead to yield losses in 70 % of long-rains seasons and 60 % of short-rains seasons (Figure 2).

While the consequences of delays in planting will be influenced by both the definition of onset selected and interactions with other management variables such as nitrogen supply, the generally appreciated value of 'early planting' in this region is supported by this analysis (Dowker 1964). Losses associated with delays in planting can be attributed to inefficient use of both nitrogen and water

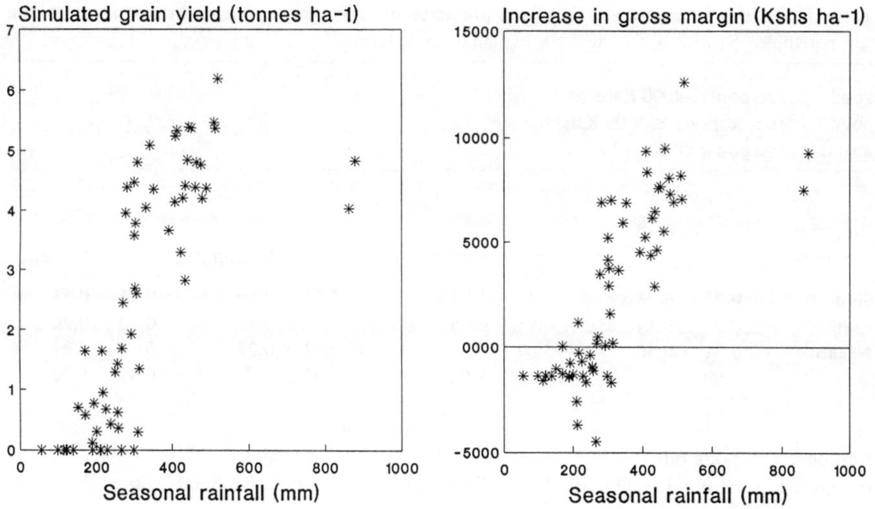

Figure 4. Relationships between seasonal rainfall (defined above) and simulated grain yield (left) and simulated response to inputs (S_4–S_1) for the long and short rains, 1957 to 1988.

resources and hence other factors which influence nitrogen and water supply or demand (e.g. fertilization, plant population, genotype) will influence the outcome of delays in planting. In the remainder of the paper, the assumption was made that planting would proceed at the optimal time, i.e. as soon after onset as possible.

Strategies of N use

Long term average grain yields increased from 1106 to 2794 kg ha^{-1} as the input level in the fixed strategies increased from S_1 to S_6. Mean gross margin were maximized (5255 to 5673 Kshs ha^{-1}) at N rates between 30 and 45 kg N ha^{-1} and plant populations between 33000 and 37000 plants ha^{-1} (S_3 and S_4 – Figure 3).

Variability in simulated response to N was extreme, ranging from positive increases in gross margin of 13000 Kshs ha^{-1} (above the crops receiving no fertilizer) in some seasons to losses of 3000 Kshs ha^{-1} associated with high rates of fertilizer use in other seasons. In general, both grain yields simulated and response to added N were strongly related to seasonal rainfall levels (Figure 4).

Tactical adjustments based on climate information

Performance of the predictor

Stewart and Faught (1984) have shown that potentially useful relationships exist in this region between the date of season onset and seasonal rainfall. McCown et al. (1991) used Bayesian statistics to update seasonal rainfall probabilities based on historical prior probabilities using a predictor (onset

Table 2. Performance of onset date as a predictor of class of response to inputs based on the comparison of S_4 with S_1 (see text for explanation of terms).

good	response>4000 Kshs ha^{-1}
poor	$0 \leq$ response ≤ 4000 Kshs ha^{-1}
negative	response<0 Kshs ha^{-1}

	Early onset							
Response to inputs	Long rains				Short rains			
	Prior	Likelihood	Joint	Posterior	Prior	Likelihood	Joint	Posterior
Good	0.47	0.73	0.34	0.74	0.32	0.80	0.26	0.53
Poor	0.22	0.14	0.03	0.06	0.20	0.50	0.10	0.20
Negative	0.31	0.30	0.09	0.20	0.48	0.27	0.13	0.27
	1.00		0.46	1.00	1.00		0.49	1.00
	Late onset							
Response to inputs	Long rains				Short rains			
	Prior	Likelihood	Joint	Posterior	Prior	Likelihood	Joint	Posterior
Good	0.47	0.27	0.13	0.24	0.32	0.20	0.06	0.12
Poor	0.22	0.86	0.20	0.36	0.20	0.50	0.10	0.20
Negative	0.31	0.70	0.22	0.40	0.48	0.73	0.35	0.69
	1.00		0.55	1.00	1.00		0.51	1.00

date). A similar approach has been taken here, except that we examined both prior and posterior probabilities of response to N inputs, rather than of seasonal rainfall. While this restricts the analysis to the N fertilization issue, it has the advantage of eliminating the scatter that we see in the relationship between response to inputs and seasonal rainfall (Figure 4). It also means that onset-date effects on factors other than rainfall, (e.g. effects on temperatures and radiation during grain filling) are captured in the analysis.

In the period studied, 47 % of long rains seasons started early, 53 % late, using the criteria outlined earlier as proposed by Stewart and Faught (1984). The likelihood of seasons with a good yield response to inputs (as assessed by the incremental yield of S_4 over S_1) starting early, is shown in Table 2. Other combinations of response and season-onset date (predictor) are also shown. Bayesian probability theory has been used to calculate posterior probabilities. For the LR, the prior probability of a good yield response to inputs is 47 % in the absence of any information concerning season onset date. The corresponding conditional or posterior probability of a good response to inputs given an early onset is 74 %. Similarly, while there was a 31 % prior probability of a poor response to inputs in the long rains, this is increased to 40 % given a late onset to the long rains.

Onset-date also changes the probabilities of obtaining responses to inputs in the short rains (Table 2). The probability of obtaining a good response to inputs is increased from 32 to 53 % if onset is known to be early. Poor responses to inputs are expected in 48 % of short rains seasons in general, but this probability is increased to 69 % if onset is late.

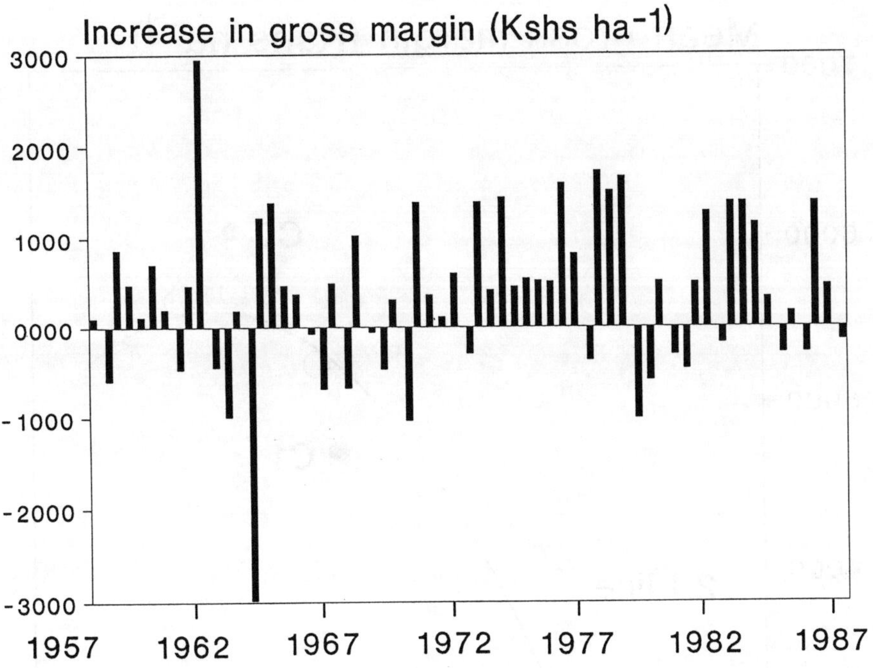

Figure 5. Variability over time of the increase in gross margin in response to conditional management (C_2 minus S_4).

It is clear that the onset-date of the rains has a major effect on the probability of obtaining a response to nitrogen fertilizer and associated inputs. A comparison of tactical management with fixed strategies considering both average returns and risks is now examined to assess the economic value of this shift in probabilities.

Average benefits

The mean gross margins for the two conditional strategies examined are compared with fixed strategies in Figure 3. For the same input cost, averaged over the period studied, tactics which link fertilizer use to onset-dates result in gains in the expected gross margin. However, the size of these gains is small in comparison with the large impact of the unconditional use of fertilizer.

Measures of risk

While the overall impact of the tactics examined on average profitability was small, reductions in the risk associated with fertilizer use also need to be assessed.

The high input conditional strategy (C_2) provided benefits over and above a comparable fixed strategy (S_4) in 67% of seasons simulated, but had a negative impact in the remainder of seasons (Figure 5). Such negative effects arose mostly (81%) from situations when an early onset was indicative of a good

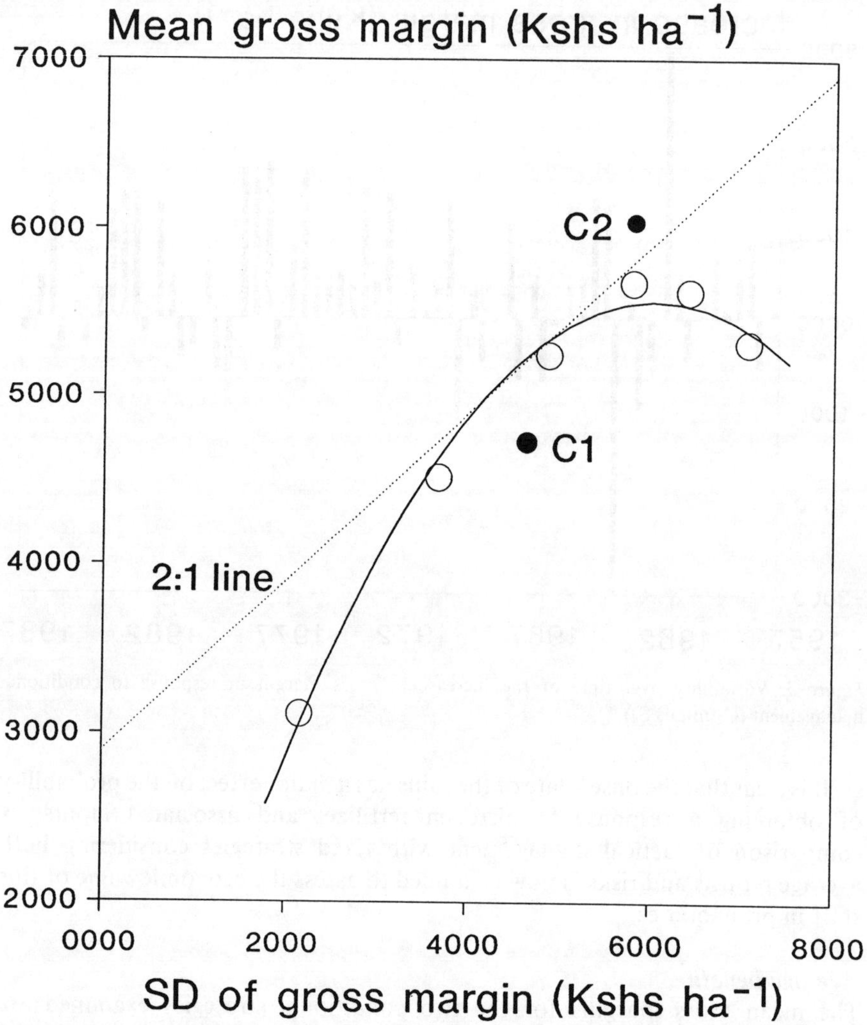

Figure 6. The outcome of various set open symbols (S_1, S_2, S_3, S_4, S_5, S_6) and conditional (C_1, C_2) strategies, specified in Table 1, plotted in mean-standard deviation space.

season and inputs were increased accordingly, but subsequent response to these additional inputs was poor.

Efficiency frontiers in Mean (E) – Standard Deviation (SD) space
McCown et al. (1991) have compared conditional strategies with fixed strategies in terms of E–SD space. The technique portrays production in terms of the long term average gross margin (E) and risk in terms of the standard deviation of gross margin (SD) over the historical period simulated (Figure 6). Compared to a corresponding set strategy (S_4), the high input conditional strategy (C_2)

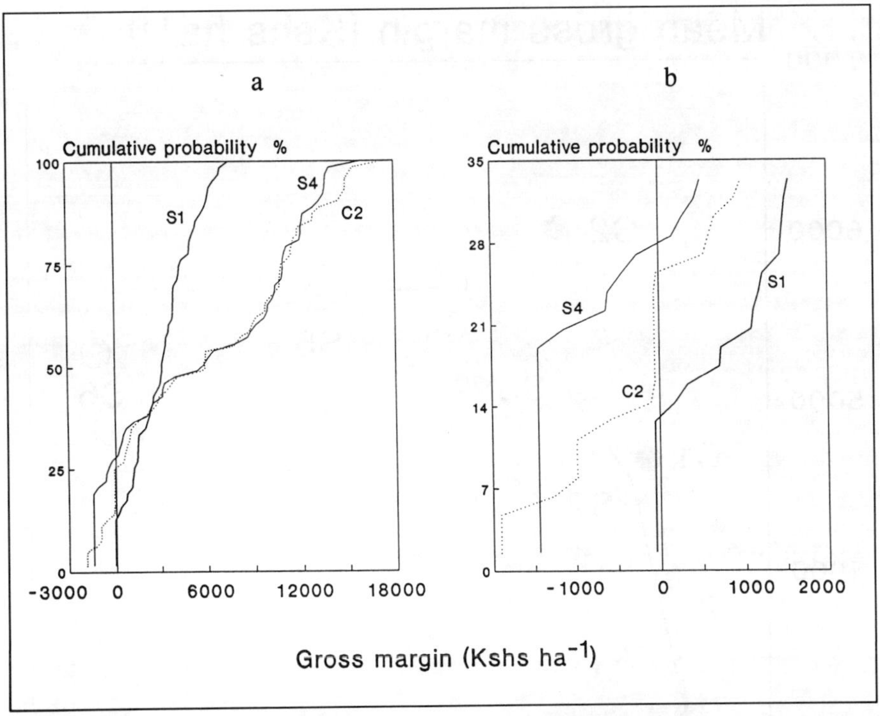

Figure 7. Cumulative probabilities of gross margin for fixed (S_1, S_4) and conditional (C_2) strategies of N fertilization. Details of strategies are given in Table 1. (a) Full range of gross margin. (b) Lower 33% of the range in gross margin.

resulted in higher mean returns with the same or slightly reduced risk, insofar as standard deviation is an adequate measure of risk. The strategy using moderate input levels conditional on onset-date (C_1) fell below the efficiency frontier generated by fixed strategies and is of no further interest. A 2:1 rule of thumb has been suggested (Ryan 1984) as a first approximation to the attitudes of farmers on small-holdings to incurring added risk in conjunction with increased gross margin, i.e. such farmers would not be averse to using inputs or technologies provided they did not increase the standard deviation of the gross margin more than twice the increase in mean gross margin. Such a rule would suggest that the S_2, S_3 and S_4 fixed strategies and the C_2 conditional strategy are realistic options for risk averse farmers. The E–SD plot also highlights the large gains in efficiency achievable through the use of inputs (S_4 vs. S_1) and conversely, the small benefits of tactics associated with their conditional use (e.g. comparing C_2 with S_4).

Stochastic dominance analysis

The cumulative distribution functions (CDF) for the gross margin (Figure 7a) compare the low-input (S_1) and high-input (S_4) fixed strategies with a high-

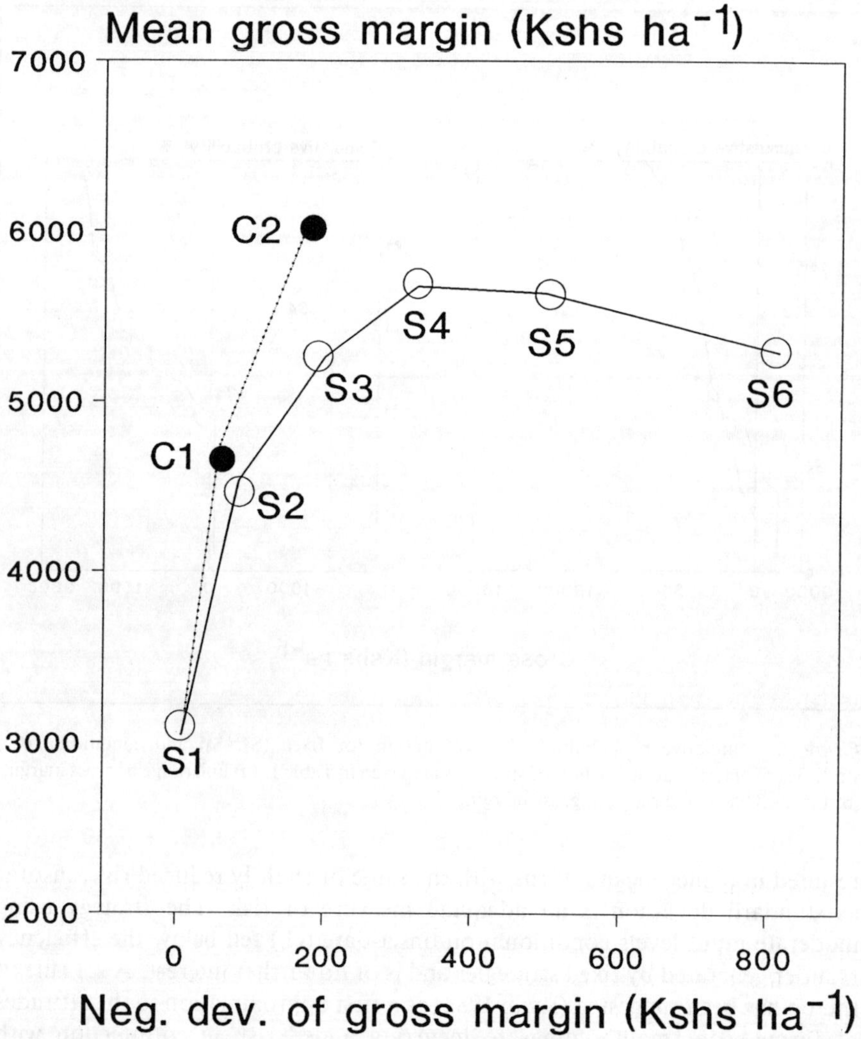

Figure 8. The value of various set (S_1, S_2, S_3, S_4, S_5, S_6) and conditional (C_1, C_2) strategies, specified in Table 1, plotted in mean-negative deviation space. The negative deviation refers to outcomes with a gross margin less than 0 Kshs ha^{-1}.

input conditional strategy (C_2). If we apply constant 'Pratt-Arrow' risk aversion coefficients (r) and stochastic dominance analysis with respect to a function (Meyer 1977; Goh et al. 1989) , the C_2 strategy dominates S_4 at all levels of risk aversion and dominates S_1 if $r > 0.00035$. Units for r are in Kshs^{-1} and the outcome space is gross margin for maize (Kshs ha^{-1}).

Stochastic dominance analysis has been used to compare risky prospects, but is limited in the absence of detailed information on farmer attitudes. We do not have an estimate of risk aversion coefficients for farmers in this region, but an earlier comparison (Keating et al. 1991) based on a study of Indian subsistence farmers (Bailey and Boisvert 1989), suggested that such a coefficient might apply to the most risk averse 8 % of farmers in that study.

A closer examination of the lower third of the probabilities for gross margin (Figure 7b) highlights the large increase in risk of a negative gross margin associated with use of fertilizer inputs (S_4 compared to S_1) and the significant reduction in this risk that was achieved with a conditional strategy using season-onset as a predictor (C_2 vs. S_4).

Safety first: Mean (E)–negative deviation (ND) space

While quantifying the production objectives of resource poor farmers is a demanding and often controversial exercise, the desire of such farmers to achieve some threshold production level needed for survival is easily envisaged. In the case of decisions concerning fertilizer inputs, the desire of farmers not to lose money (i.e. not to record a negative gross margin) can be viewed as a requirement for financial survival or 'safety-first' goal setting. Strategies can be assessed in terms of such goals by plotting the expected returns against the probability-weighted sum of deviations below some target, in this case, below a gross margin of zero (Figure 8). Such a plot in mean-negative deviation space (E–ND) (Parton, 1992) has obvious parallels to E–SD space considered earlier (Figure 6). E–SD space uses all variability in returns as an indicator of risk (i.e. deviations both up and down) while E–ND space considers only the down-side risks. While conditional strategies of fertilizer use raised expected returns slightly, it had little impact on risk as assessed in the E–SD plot (Figure 7). Tactics linked to an onset-date forecast did however have a substantial benefit in reducing negative deviations (Figure 8) and may be attractive to farmers pursuing strong 'safety first' goals.

The implicit utility function in the E–ND space as presented here is one that places a value on reducing losses that is proportional to the size of the reduction, but that places no extra value on profit over and above the target level. This is obviously an extremely risk-averse position to take, but the plot does highlight the potential attraction of conditional strategies of fertilizer use for highly risk-averse individuals.

Discussion

The date of the start of the rains at Katumani can be a useful predictor of potential yield, and hence of the capacity of crops to respond to inputs. Adjustment of N input levels and plant populations to better match the season potential is a logical response with a sound biological basis. How much value to place on the forecast is more difficult to assess. In terms of average returns, its value is small relative to the large benefits from using fertilizer irrespective of a forecast. In terms of minimisation of risks, it can be of value, substantially reducing the number of occasions when fertilizer is purchased and rainfall is insufficient to obtain a return in the year of application.

We have not considered the residual value of fertilizer applied in such situations, but other experimental (Watiki and Keating unpubl. data 1988) and simulation (Keating et al. 1991) studies suggest that it can be significant in dry years. Information concerning past fertilizer management and prior rainfall data could serve as forecasts of the likely value of residual fertilizer. Conditional strategies could be developed that adjust current fertilizer management in response to both future seasonal prospects and hence nutrient demand (as in this study), and to nutrient supply as influenced by past management and weather.

This and earlier studies in the region highlight the potential economic value of fertilizers and yet we observe few farmers using fertilizers. An analysis of possible reasons why more farmers don't use fertilizer has been given by McCown et al. (1991).

It is clear from the simulation results that anything less than 20 to 30 years would not provide an adequate picture of the variability in net benefits associated with a particular tactic. Experimental evaluation of the conditional management strategies (examined in this paper using simulation) would not have been feasible. The models we use remain fairly crude tools in the evaluation of the decisions that confront farmers. Considerable local adaptation was needed in this study to deal realistically with the crop system of interest. Some of this adaptation was of a technical nature, such as the recalibration of a function or correction of an error of logic. The most important changes in this work were, however, those that broadened the scope of the model to deal with aspects of the system we felt had to be addressed if realistic simulations were to be achieved. Changes included such issues as the ability to simulate planting in response to weather rather than as a fixed input, and the ability to simulate the death of crops severely stressed during the establishment and early periods of vegetative growth. The solutions developed on problems such as crop death were based on limited data and further research is needed. Despite the enhancements made to the model in this study, we are still working essentially with a crop model and as such, fairly inadequate tools in the assessment of matters of a cropping system or farming system nature.

Acknowledgements

Our work in Kenya was part of a collaborative project between the Australian Centre for International Agricultural Research (ACIAR) and the Kenyan Agricultural Research Institute (KARI). The authors acknowledge the many Kenyan and Australian members of the project team for their valuable contributions.

References

Anderson J R, Dillon J L, Hardaker J B (1977) Agricultural Decision Analysis. The Iowa State University Press, Ames Iowa. 344 p.

Bailey E, Boisvert R N (1989) A comparison of risk efficiency criteria in evaluating groundnut performance in drought-prone areas. Australian Journal of Agricultural Economics 33:153–169.

Connor D J, Loomis R S (1991) Strategies and tactics for water limited agriculture in low rainfall Mediterranean climates. Pages 441–465 in Acevedo E, Fereres E, Gimenez C, Srivastava J P (Eds.) Improvement and management of winter cereals under temperature, drought and salinity stresses. Instituto Nacional de Investigaciones Agrarias, Ministerio de Agricultura, Pesca y Alimentacion, Madrid.

Dowker B D (1964) A note on the reduction in yield of Taboran maize by late planting. East Afr. Agric. For. J. 30:33–34.

Fischer R A, Armstrong J S (1987) Strategies and tactics with short fallowing. Proceedings of the 4th Australian Agronomy Conference, Australian Society of Agronomy, Parkville, Australia, 300 p.

Goh S, Shih C-C, Cochran M J, Raskin R (1989) A generalized stochastic dominance program for the IBM PC. Southern Journal of Agricultural Economics 21:175–182.

Hammer G L, McKeon G M, Clewett, J F, Woodruff D R (1991) Usefulness of seasonal climate forecasts in crop and pasture management. Pages 15–23 in Proc. Conf. on Agricultural Meteorology, The University of Melbourne, 17–19 July 1991, The Bureau of Meteorology.

Jaetzold R, Schmidt H (1983) Farm management handbook of Kenya. Natural conditions and farm management information. Part C, East Kenya. Ministry of Agriculture, Kenya.

Jones C A and Kiniry J R (1986) CERES–Maize. A simulation model of maize growth and development. Texas A&M University Press. College station, 194 p.

Keating B A, Godwin D G and Watiki J M (1991) Optimising nitrogen inputs in response to climatic risk. Pages 329–357 in Muchow R C, Bellamy J A (Eds.) Climatic risk in crop production-models and management for the Semi-arid Tropics and Subtropics. CAB International, Wallingford.

Lele U J (1989) Managing agricultural development in Africa: Lessons of a quarter century. MARDIA Discussion Paper 2, The World Bank, Washington, 40 p.

Lynam J K (1978) An analysis of population growth, technical change, and risk in peasant, semi-arid farming systems: A case study of Machakos District, Kenya. PhD Thesis Stanford University, California, 266 p.

Meyer J (1977) Choice among distributions. Journal Economic Theory 14:326–336.

McCown R L, Wafula B M, Mohammed L, Ryan J G and Hargreaves J N G (1991) Assessing the value of a seasonal rainfall predictor to agronomic decisions: The case of response farming in Kenya. Pages 383–409 in Muchow R C, Bellamy J A (Eds.) Climatic risk in crop production-models and management for the Semi-arid Tropics and Subtropics. CAB International, Wallingford.

McCown R L and Keating B A (1992) Looking forward: Finding a path for sustainable development. Pages 126–132 in Proceedings of Dryland Farming Symposium, Nairobi, December 1990, Australian Centre for International Agricultural Research, Canberra, Australia.

Ockwell A P, Parton K A, Nguluu S N, Muhammed L (1992) Relationships between the farm household and adoption of improved technologies in the semi-arid tropics of eastern Kenya. Journal Farming Systems Research and Extension (in press).

O'Leary M F (1984) The Kitui Akamba: Economic and social change in semi-arid Kenya. Heinemann Educational Books, Nairobi, 139 p.

Parton K A (1992) Socio-economic modeling of decision making with respect to choice of technology by resource-poor farmers. Pages 110–118 in Proceedings of Dryland Farming Symposium, Nairobi, December 1990, Australian Centre for International Agricultural Research, Canberra.

Rukandema M (1984) Farming systems of semi-arid eastern Kenya. A comparison. East Afr. Agric. For. J. 44:422–435.

Ryan J G (1984) Efficiency and equity considerations in the design of agricultural technology in developing countries. Australian Journal Agricultural Economy 28:109–135.

Ruthenberg H (1980) Farming systems in the tropics. 3rd edition, Clarendon Press, Oxford. 424 p.

Stewart J I, Faught W A (1984) Response farming of maize and beans at Katumani, Machakos District, Kenya: Recommendations, yield expectations, and economic benefits. East. Afr. Agric. For. J. 44:29–51.

Van Herwaarden A F, Angus J F, Fischer R A, Irving G C J (1989) Tests to predict yield response of wheat to top-dressed nitrogen. Proc. 5th Australian Agronomy Conference, Australian Society of Agronomy, Parkville, 551 p.

Wafula B M (1989) Evaluation of the concepts and methods of Response Farming using crop growth simulation models. M.Ag. Sci. Thesis, Melbourne University.

Woodruff D R (1987) WHEATMAN. Project Report Q087014, Queensland Department of Primary Industries, Brisbane, Australia.

Maize modeling in Malawi: a tool for soil fertility research and development

UPENDRA SINGH[1], PHILIP K. THORNTON[1], ALEX R. SAKA[2], J. BARRY DENT[3]

[1] *International Fertilizer Development Center (IFDC), P.O. Box 2040, Muscle Shoals, Alabama 35662, U.S.A.*
[2] *Department of Agricultural Research, Ministry of Agriculture, Lilongwe, Malawi*
[3] *University of Edinburgh, Edinburgh, United Kingdom*

Key words: CERES-Maize, fertilizer application, financial return, maize, Malawi, management strategy, water-nitrogen limited yield

Abstract
The increasing population pressures on land and the limited land that can be brought into agricultural production have necessitated the development of crop production technologies that are practicable, stable, environmentally sound, and adoptable by resource-poor farmers. A maize modeling initiative, set within this general framework, is underway in Malawi. Activities include the following: Establishing a natural resources data base for Malawi; matching appropriate maize cultivars for the agroecological zones of the country under varying management inputs; and updating site- and season-specific fertilizer recommendations in the primary maize-growing agroecological zones of the country. Field trials carried out for a number of sites in Malawi have been used to validate a simulation model of the growth, development, and yield of maize. In this paper, results from the simulation work are discussed, and the model is used to illustrate weather- and soil-related variability in maize production. The way in which the modeling component can be integrated into the soil fertility research agenda of the Department of Agricultural Research in Malawi is outlined, and future prospects for model application are discussed.

Introduction

Malawi is a landlocked country located in southeastern Africa. It is approximately 900 km long from north to south, and it varies in width from 80 to 160 km. Its total area is 11.8 million ha, of which 20% is covered by water. Approximately 5.3 million ha of this land area is suitable for agriculture; the remainder includes natural and planted forests and national parks. Nearly all of Malawi receives adequate rainfall for rainfed agriculture. Over 60% of the country receives 760–1000 mm y^{-1}, and some areas receive up to 1500 mm y^{-1}. The principal rainfall pattern involves a concentration in a single rainy season from November to April. Approximately 85% of rain falls between December and March.

Over 85% of Malawi's 8 million people reside in rural areas and derive their livelihood from small landholdings of less than 2 ha per farm family. The current average population density of 85 persons per square km and the population growth rate of 3.7% per annum are quite high by African standards.

F.W.T. Penning de Vries et al. (eds.), Systems Approaches for Agricultural Development, 253–273.

The population is likely to rise to over 11 million by the year 2000, creating substantial pressures on the limited land resource.

Agriculture in Malawi contributes well over 40% of the gross national product and accounts for 90% of foreign exchange earnings. It is based on smallholder farming on tribal land and estate farming on leasehold land. The estates contribute 70% of the export trade. However, smallholders contribute more than 85% of total agricultural production in the country. The primary crop is maize; of a total of 1.8 million ha cropped in Malawi, some 65% is planted to maize. In 1989, it was estimated that local varieties were grown on approximately 90% of this area (Williams and Allgood 1990).

To achieve high crop yields on land that is now continuously cropped, farmers must apply inorganic fertilizers, use improved varieties, and take advantage of new crop husbandry practices. These have largely been taken up by medium-scale farmers and estate owners but not by smallholders–they are constrained primarily by lack of cash to purchase the inputs that accompany increased yield levels. Although Malawi has remained self-sufficient in food production since independence in 1964, average crop yields for smallholders have generally remained low (less than 1 t ha^{-1} for maize, for example). Low or declining crop yields are caused by a multitude of factors, such as poor soil fertility, soil erosion, poor soil physical conditions, and inadequate or poorly distributed rainfall. Other, equally important factors include the use of unimproved varieties, uncontrolled pests and diseases, and suboptimal agronomic practices.

With increasing population pressures on land, and limited land that can be brought into agricultural production, there is a clear need to develop appropriate crop production technologies that are practicable, stable, environmentally sound, and adoptable by resource-poor farmers. These systems could be based on increased fertilizer use efficiency, improved varieties, soil and water conservation strategies, and the use of indigenous and organic fertilizer materials as alternative sources of nutrients. It is within this framework that the maize modeling work described below is developed.

Objectives of the maize modeling initiative

The major objectives of this work are as follows: (1) To build a data base for soil and weather variables for the major maize-producing areas of Malawi. (2) To validate and calibrate the simulation model CERES–Maize for representative cultivars in three study areas. (3) To predict maize production and its variability between sites and between years in the study locations and to extrapolate the results to other regions of the country. (4) To illustrate how the maize model can be used to enhance the efficiency of the traditional approach to agricultural research based on long-term field experimentation.

The following section describes field experiments conducted at a number of sites to generate minimum data sets (IBSNAT, 1990) with which to validate

CERES–Maize for Malawian conditions and characterize commonly grown maize varieties.

Field experimentation

Field trials were conducted during the 1989–1990 wet seasons at three research stations in Central Malawi. Some salient characteristics of these sites are listed in Table 1. At each location, a completely randomized block design involving 3 replicates and 12 treatments was used. The treatments at each site were composed of two varieties, three nitrogen levels (0, 60, and 240 kg ha^{-1} as urea), and two phosphorus levels (0 and 150 kg ha^{-1} as single superphosphate). The most commonly grown local variety was planted for each site. At Mwimba, the hybrid MH16 was planted; at Chitedze the hybrid SR52 was grown; and at Chitala a dent hybrid, NSCM–41, was used. To prevent yield reductions from inadequate levels of nutrients to which the model is not yet sensitive, blanket nutrient applications were used prior to sowing: Potassium, zinc, sulfur, boron, and copper were applied to each plot at Mwimba and Chitala, and sulfur and potassium were applied at Chitedze. Treatment details are set out in Table 2. Full minimum data sets were collected from each experiment (IBSNAT, 1988). Prior to planting the field trials, soils at the three sites were characterized in terms of the physical and chemical properties specified by IBSNAT (1990).

Weather information

There is a weather station at Chitedze Research Station for recording the data required for running CERES–Maize: Solar radiation, maximum and minimum temperatures, and rainfall. For the trials executed during 1989–1990 at Chitala and Mwimba Research Stations, rainfall and maximum and minimum temperatures were recorded, and these data were supplemented by records obtained from nearby weather stations–for Chitala, from the township of Salima, and for Mwimba, from the township of Kasungu. Daily sunshine hours were used to estimate solar radiation data whenever necessary. In late 1990, automated portable weather stations were installed at Chitala and at Mwimba Research Stations. Long-term daily averages (by month) for solar radiation, maximum and minimum temperature, and monthly average rainfall for Chitedze, Chitala, and Mwimba Research Stations are shown in Figure 1.

Performance of the model

Model description

The CERES–Maize model allows the quantitative determination of growth and yield of maize (Jones and Kiniry 1986; Ritchie et al. 1989). The growth of the

Table 1. Research station experimental sites, 1989–1990.

Research Station	Area	Coordinates	Elevation (m.a.s.l.)	Soil	Rainfall, (mm) Oct. '89–May '90	Rainfall range, (mm)	Rainfall variability
Mwimba	Medium Altitude Plain	33°22'E, 13°17'S	1050	Weathered ferralitic latosol, Kamphuru series (light textured sandy loam)	1000	750–1050	medium
Chitedze	Medium Altitude Plain	33°38'E, 13°59'S	1100	Ferric luvisol, medium textured sandy clay (Oxic Rhodustalf)	1068	800–1100	medium
Chitala	Salima Lakeshore Plain (low altitude)	34°17'E,13°41'S	600	Ferroginous soil, medium-textured sandy clay	729	700–950	high

Table 2. Treatment details, 1989–1990 experiments.

Expt. number	Site	Sowing date	Plant population m^{-2}	Treatments: Varieties	Treatments: Fertilizer, kg ha^{-1}	Number of Treatments	Number of Replicates	Date N fertilizer applied	Blanket fertilization[a] kg ha^{-1}
1	Chitedze	28 Dec.	3.7	Local, SR52	0, 60, 240 N; 0, 150 P	12	3	Feb. 2	82 K_2O, 35 S
2	Chitala	14 Dec.	3.7	Local, NSCM–41	0, 60, 240 N; 0, 150 P	12	3	Jan. 23	82 K_2O, 35 S, 15 Zn, 2 Cu, 2B
3	Mwimba	20 Nov.	3.7	Local, MH16	0, 60, 240 N;0, 150 P	12	3	Jan. 12	82 K_2O, 35 S, 15 Zn, 2 Cu, 2B
4	Mwimba	20 Nov.	3.7	MH16	0, 60, 120, 180, 240 N; split applications of 30 kg at each application; 0, 30, 60 P single application	15	3	Weekly from Dec. 4	82 K_2O, 35 S, 15 Zn, 2 Cu, 2 B

[a] Blanket- and P fertilizations were carried out in one application on the following dates: Experiment 1, December 28; experiment 2, December 14; experiment 3 and 4, December 7.

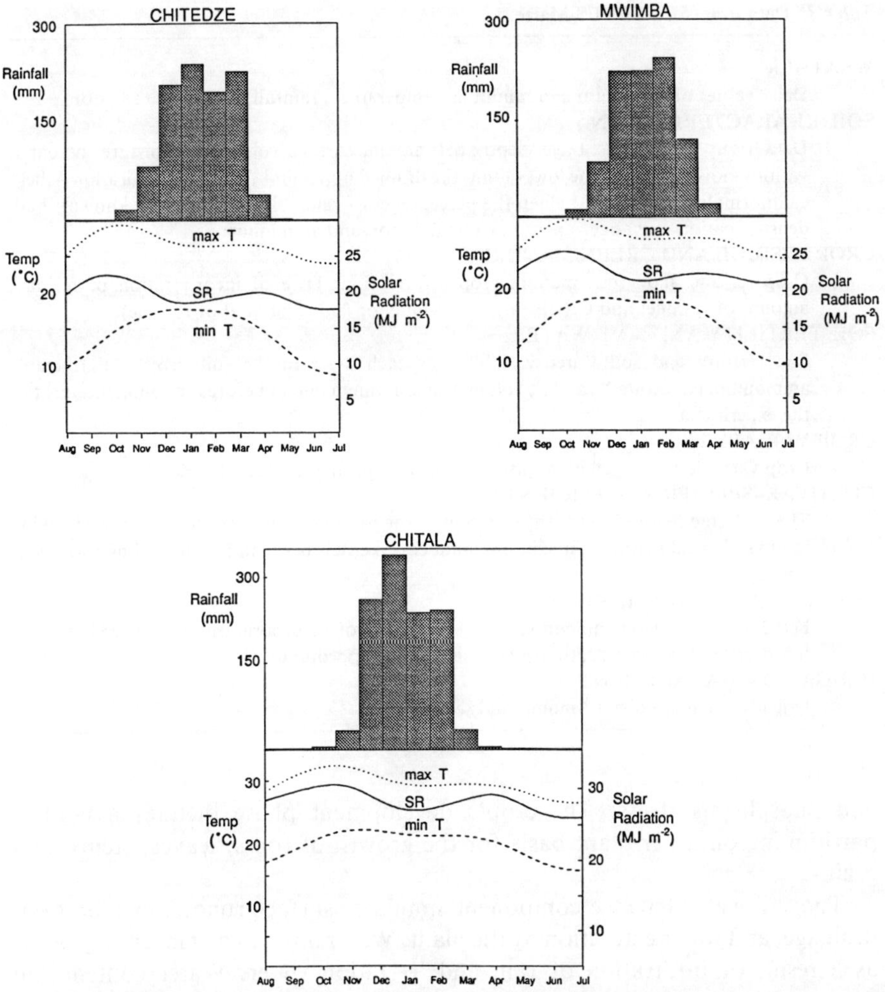

Figure 1. Long-term monthly meteorological data for rainfall, maximum temperature (max T), minimum temperature (min T), and solar radiation (SR).

crop is simulated with a daily time step from sowing to maturity on the basis of physiological processes as determined by the crop's response to soil and aerial environmental conditions. A detailed description of the crop model can be found in the references cited above and in Singh et al. (1989).

Briefly, phasic development in CERES–Maize quantifies the physiological age of the plant and describes the duration of nine growth stages. Potential growth is dependent on photosynthetically active radiation and its interception as influenced by leaf area index (LAI), row spacing, plant population, and photosynthetic conversion efficiency of the crop. Actual biomass production is constrained by suboptimal temperatures, soil water deficit, nitrogen deficiency,

Table 3. Data inputs for CERES-Maize.

WEATHER	
	Daily values of maximum and minimum temperature, rainfall, and solar radiation.
SOIL CHARACTERIZATION	
	Data for up to 15 layers: Layer depth; pH; organic carbon; volumetric moisture content at various moisture levels (the lower limit, the drained upper limit, and field saturation – these can normally be estimated when the percentages of sand, silt, and clay are known); bulk density; cation exchange capcity; extractable iron and aluminium
CROP RESIDUE AND GREEN MANURE	
	Crop residue and green manure crop information: Date of incorporation of residue; amount of residue; and C:N:P ratio of residue (or % N and % P of residue).
SOIL FERTILITY AND SOIL WATER	
	Soil fertility and soil water variables for each layer in the soil profile: Extractable ammonium N, nitrate N, and P; volumetric soil water content before the commencement of the experiment.
CROP VARIABLES	
	Crop variables: emerged plant population; row spacing; and seeding depth.
CULTIVAR–SPECIFIC COEFFICIENTS	
	Thermal time (base 8°C) for the duration of the juvenile phase and the duration of silking to physiological maturity; maximum number of kernels per plant; kernel filling rate under optimum conditions.
FERTILIZER MANAGEMENT	
	Fertilizer data: Dates, amounts, sources, method of incorporation and placement of all fertilizer applications; depth of placement where appropriate.
IRRIGATION MANAGEMENT	
	Irrigation management: Amount and schedule.

and phosphorus stress. The crop's development phase dictates assimilate partitioning on a per-plant basis for the growth of roots, leaves, stems, and grains.

The soil water balance component simulates surface runoff, evaporation, drainage, and water extraction by the plant. Water input in any layer may occur as a result of infiltration of rain and irrigation water. Water content can decrease because of soil evaporation, root absorption, or flow to an adjoining soil layer.

The nitrogen submodel simulates the processes of turnover of organic matter with the associated mineralization and/or immobilization of N, nitrification, denitrification, hydrolysis of urea, and ammonia volatilization for a variety of N sources, including chemical fertilizers, green manure, and other organic N sources. Nitrate and urea movement is simulated as a function of water flow through a layer. The nitrogen submodel also simulates the uptake and use of N by maize. A phosphorus submodel is under development at the International Fertilizer Development Center (IFDC); this simulates the processes of adsorption and desorption of P, organic P turnover, and the dissolution of rock and fertilizer phosphate. The model simulates the effects of both N and P deficiency on photosynthesis, leaf expansion, tillering, senescence, assimilate partitioning, and, in severe cases, delay in phenological development.

The input data required to run the model are shown in Table 3. The model produces outputs concerning water balance, soil and plant N balance, soil and plant P, and crop growth and development.

Model calibration

Prior to evaluating the model, the genetic coefficients for the varieties used were estimated. These were determined by using the observed silking and maturity dates and the observed kernel numbers per ear for each of the varieties in the treatment with the highest rate of nitrogen application. The coefficients were adjusted until there was a match between the observed and simulated dates of silking and maturity. The values thus obtained were then used for all other treatments. In general, local varieties have longer growth duration and higher stover production (low harvest index) than the hybrids.

At the Chitedze and Mwimba sites, the same local variety was grown. The genetic coefficients generated for the local variety at Chitedze were thus validated against field data from Mwimba. Likewise, coefficients for MH16 derived from Experiment 3 at Mwimba were used to validate the hybrid against field observations from Experiment 4 (Table 2). For both cases where it was possible to derive independent values of the coefficients, model predictions of emergence, silking, and maturity dates were accurate to within 2 days of observed results at the associated site. The hybrids SR52 and MH16 were found to be identical in terms of the genetic coefficients required for running CERES–Maize, except for the higher yield potential of MH16. The local and hybrid varieties grown at Chitala had substantially longer growth durations.

To validate CERES–Maize, the model was run using the same soil, weather, and management conditions as for the nitrogen and variety treatments of Experiments 1 to 4 (Table 2). Simulated outcomes were then compared with the measured results from these experiments. As there was no response to P fertilization at any of the sites, these treatments were not used for model evaluation purposes.

The model's performance for the different varieties and nitrogen rates is shown in Figure 2. At Chitedze Research Station on fertile soils, the local variety (Local–1) showed no significant response to nitrogen application. The model successfully simulated a similar response. The observed and simulated grain yield responses for hybrid SR52 were also closely matched, and they showed no significant response to nitrogen fertilizer beyond 60 kg N ha^{-1}. However, at Mwimba on a sandy soil with low inherent soil fertility, both the locally grown variety (Local–1) and the high-yielding hybrid MH16 showed marked response to nitrogen fertilizer application. This effect was accurately captured by the model, as shown from the simulated results in Figure 2. At Chitala Research Station, the two maize varieties (Local–2 and the hybrid NSCM–41) behaved differently. The local variety showed no response to nitrogen beyond 60 kg ha^{-1}, whereas NSCM–41 responded to nitrogen application at all rates. The simulated results were again in accord with the observed data. All of the simulated data points lie within 20% of the measured

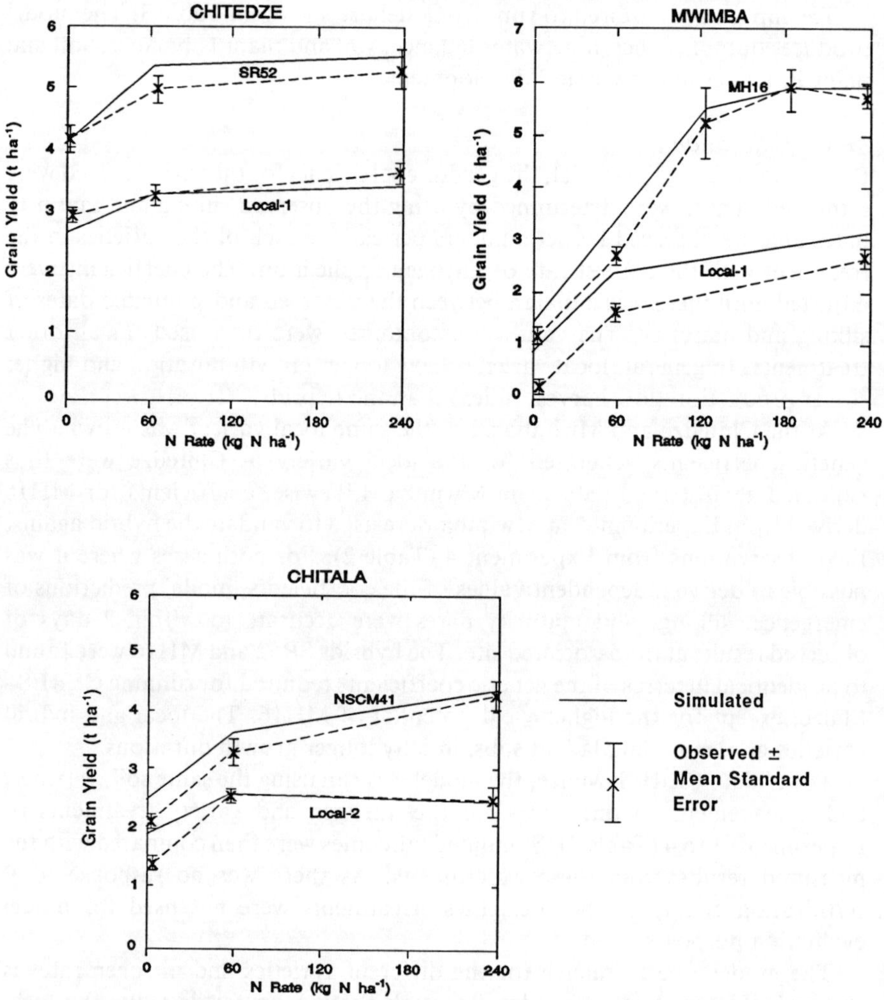

Figure 2. Comparison of observed and simulated grain yield over a range of nitrogen rates for local and hybrid maize.

values for each treatment.

The model was also validated against a nitrogen management experiment conducted at Mwimba Research Station during the 1987–1988 season. The experimental details were similar to Experiment 4 (Table 2), except there was no blanket fertilization. Maize hybrid SR52 was planted on December 12. The comparison between observed and simulated results over four rates of nitrogen application, presented in Figure 3, shows that the model overpredicted yields at high nitrogen rates. The lower observed maize grain yields at high rates of nitrogen application may be attributed largely to deficiencies of nutrients other than nitrogen and phosphorus. This illustrates an important point–experiments

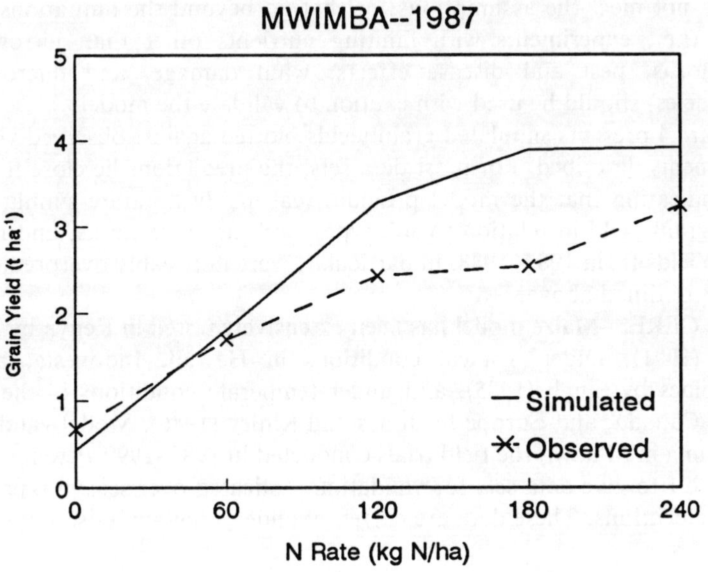

Figure 3. Performance of the CERES–Maize model over a range of nitrogen rates for grain yield with maize cultivar, SR52.

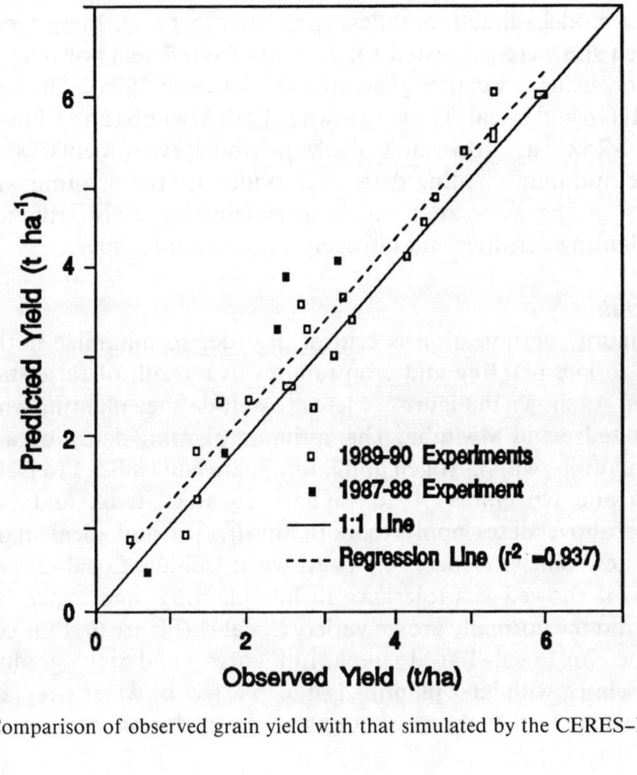

Figure 4. Comparison of observed grain yield with that simulated by the CERES–Maize.

that do not meet the assumptions, or that go beyond the limitations, of the model (i.e., experiments with limiting nutrients other than nitrogen and phosphorus, pest and disease effects, wind damage, and micronutrient deficiencies) should be used with caution to validate the model.

Figure 4 presents simulated grain yields plotted against observed yields for experiments described. For most data sets, the predictions lie close to the 1:1 line, indicating that the model provides reasonably accurate simulations of maize grain yield in relation to soil types, soil moisture status, and nitrogen rates. Yields from 1987–1988, in particular, were noticeably overpredicted for reasons outlined above.

The CERES–Maize model has been extensively tested in Kenya by Keating et al. (1991), under tropical conditions in Hawaii, Indonesia, and the Philippines by Singh (1985), and under temperate conditions in the United States, Canada, and Europe by Jones and Kiniry (1986). Model validation is continuing in Malawi; the field trials conducted in 1989–1990 were repeated in 1990–1991 to give data sets for validation replicated over season type for the various locations. These data are currently undergoing analysis.

Model experimentation

All simulations were conducted using the same soil, plant, and management inputs as for model validation, unless specified otherwise. Long-term climatic data for each site were generated for 25 years from 7 years of daily historical weather data, using a weather generator (Richardson 1985). The same maize varieties, MH16 and Local–1, were grown at both Mwimba and Chitedze (in the field study SR52 was grown at Chitedze). Model experiments were used to evaluate the optimum planting dates and to identify the planting window for each variety at the associated site. Experiments were also run to identify optimum planting densities and nitrogen fertilization regimes.

Planting window

Planting window optimization is critical in order to minimize both drought impact due to late planting and crop failures as a result of false starts to the rainy season. As shown in Figure 5, there are well-defined planting windows for maize at Chitedze and Mwimba. The optimum planting dates, obtained from 25-year simulations with nitrogen nonlimiting, are November 1 to December 15 at Mwimba and November 30 to January 15 at Chitedze and Chitala. In general, the above dates apply to both the hybrid and local maize at the associated sites. When the local variety grown at Chitala, Local–2, was planted at Mwimba, it showed less tolerance to late planting, and hence to drought stress, than did the normally grown variety Local–1 (Figure 6), thus confirming the preference for Local–1 at Mwimba. In Figures 5 and 6, the gradual decline in yield associated with later planting is due to effect of water stress on growth and yield, whereas the abrupt changes in yield during early planting are

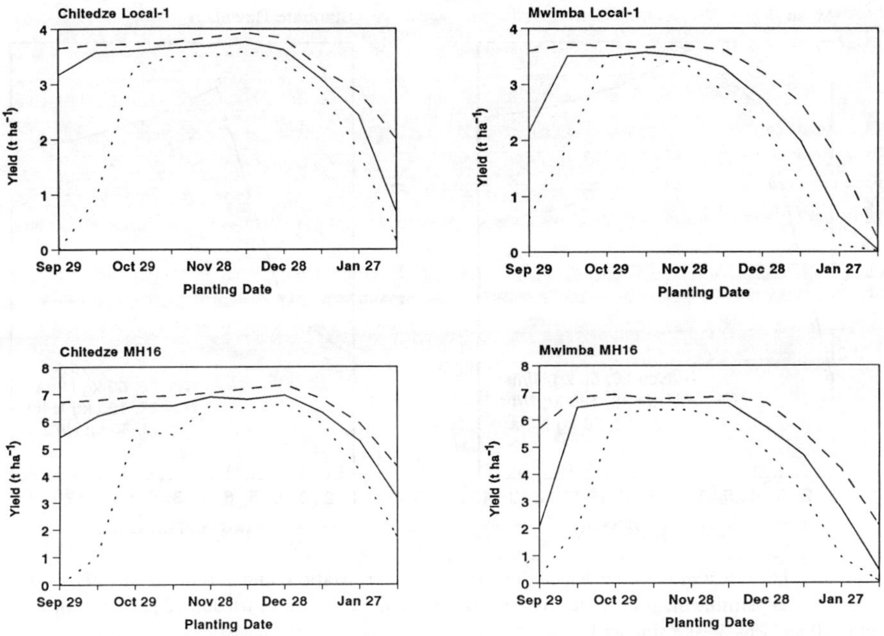

Figure 5. Simulated grain yield response to planting date as shown by median or 50th percentile ...., 90th percentile ------------, and 10th percentile ________.

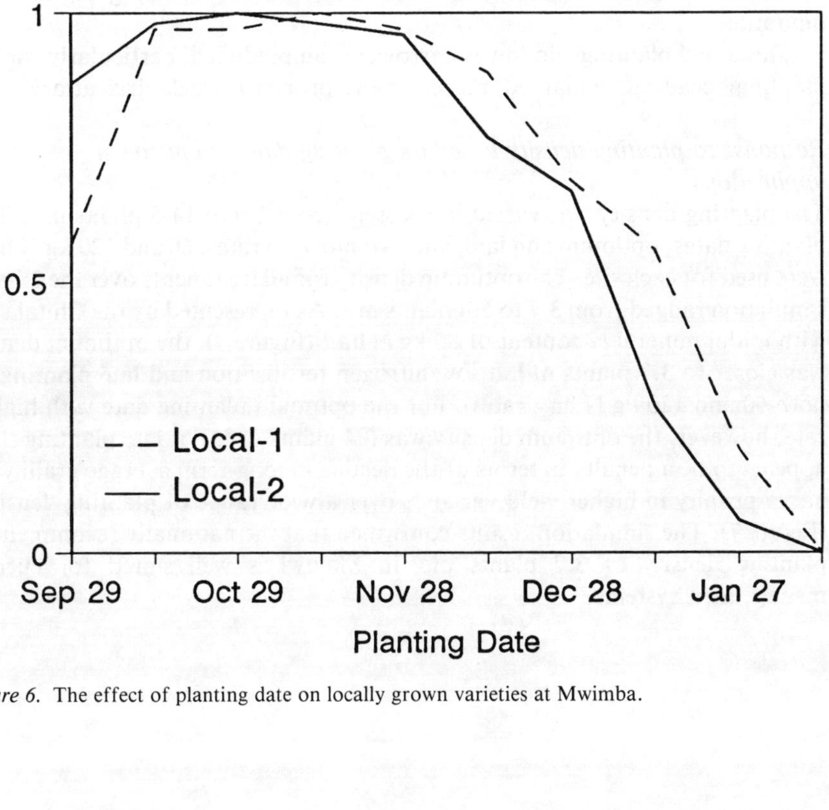

Figure 6. The effect of planting date on locally grown varieties at Mwimba.

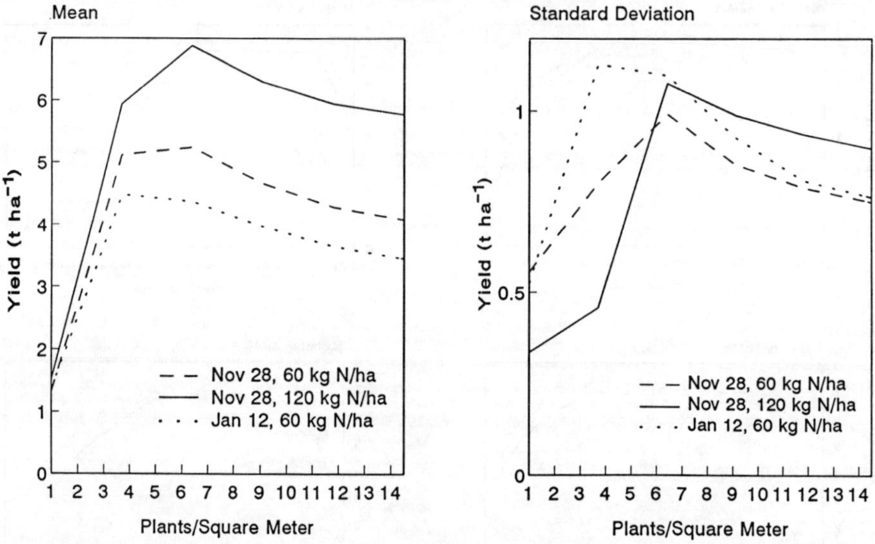

Figure 7. The effect of planting density, two rates of N fertilization, and optimum (Nov 28) versus late (Jan 12) planting on grain yield. (Note: Planting density responses for Jan 12 planting with 60 and 120 kg N ha^{-1} were similar.)

attributed to poor establishment and crop failures resulting from water limitation.

The actual planting window is narrower than predicted, particularly for late planting because late planted maize is more prone to streak virus attack.

Response to planting density based on planting date and nitrogen applications

The planting density was varied in six steps from 1.0 to 14.5 plants m^{-2}. Two planting dates, optimum and late, and two nitrogen rates, 60 and 120 kg N ha^{-1}, were used for each site. The optimum density for all treatments over the 25-year simulation ranged from 3.7 to 6.4 plants m^{-2}. As represented by the Chitala site with initial mineral N content of 80 kg N ha^{-1} (Figure 7), the optimum density was closer to 3.7 plants m^{-2} at low nitrogen fertilization and late planting (at both 60 and 120 kg N ha^{-1} rates). For the optimum planting date with high N rate, however, the optimum density was 6.4 plants m^{-2}. For late planting there appears to be a penalty in terms of the decline in long-term average grain yield and a penalty in higher yield variance over a wide range of planting densities (Figure 7). The simulation results confirmed that the nationally recommended planting density of 3.7 plants m^{-2} in Malawi is well suited for current management systems.

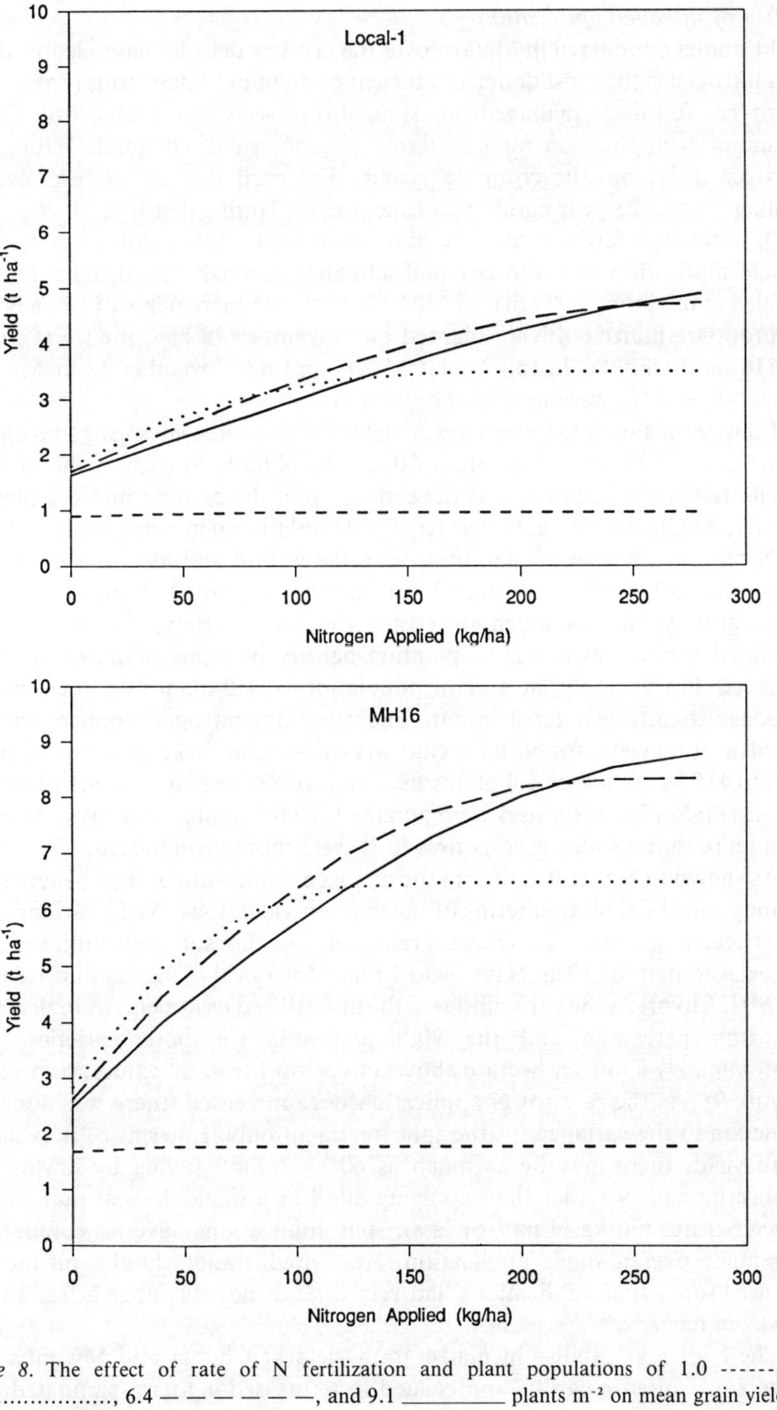

Figure 8. The effect of rate of N fertilization and plant populations of 1.0 ---------, 3.7 ………………, 6.4 ——— ———, and 9.1 ________ plants m^{-2} on mean grain yield of Local-1 and MH16 maize cultivars.

Effect of nitrogen application

Field studies conducted in Malawi over the last few decades have clearly shown that nitrogen is the most deficient nutrient element in Malawi soils (Saka 1987). Nitrogen fertilizer management simulations were conducted to identify optimum N application rates in terms of grain yield, economic return, and nitrogen loss from the cropping system. For each site, the N response was evaluated for a 25-year period as a function of planting density (1.0–9.1 plants m^{-2}), initial soil fertility, and the number of split applications of N fertilizer (single application versus three equal split applications). The nitrogen fertilizer applied as urea ranged from 0 to 270 kg N ha^{-1} with increments of 30 kg N ha^{-1}. Appropriate maize cultivars planted on November 28 at Chitedze (Local–1, MH16) and Chitala (Local–2, NSCM–41) and on November 13 at Mwimba (Local–1, MH16) were used in the simulation.

Long-term simulated mean grain yields for the three sites ranged from less than 0.5 t ha^{-1} to 9 t ha^{-1} at rates of 0–270 kg N ha^{-1}. At each of the sites, the mean grain yield response was dependent upon the cultivar and the planting density. At Chitedze on a moderately fertile soil (initial mineral N content of 60 kg N ha^{-1} at the time of planting), both the hybrid and local maize varieties responded to N application and planting density (Figure 8). Although the long-term mean yield was much lower for the local variety, the two varieties exhibited similar responses to planting density in terms of mean yield and variance. For example, at a plant population of 1.0 plant m^{-2}, there was no response to nitrogen fertilization. The optimum nitrogen application rates based on grain yield, for both the cultivars varied from 80 kg N ha^{-1} at 3.7 plants m^{-2}, to 150 kg N ha^{-1} at 6.4 plants m^{-2}, and to 200 kg N ha^{-1} at 9.1 plants m^{-2} (Figure 8). Similar responses were obtained from Chitala, whereas at Mwimba on a more sandy soil, the responses to N were more pronounced.

As shown in Figure 9, splitting the nitrogen application is also beneficial on a sandy soil (Mwimba) in terms of mean grain yield, lower yield variance, and lower leaching losses. The mean grain yield on this soil with initial mineral nitrogen content of 70 kg N ha^{-1} was 1 t ha^{-1} for Local–1 variety and 1.8 t ha^{-1} for MH16 hybrid. The variabilities in the unfertilized treatment, as indicated by the 10th percentile and the 90th percentile for both varieties, were approximately a ton per hectare above and below the mean values, respectively (Figure 9). As the rates of N application were increased, there was a marked reduction in the variance for the split treatment only. In terms of maximizing grain yield, there may be as much as 60 kg N ha^{-1} saving by having split applications of N rather than applying all N in a single dose at planting. At lower N rates (30 kg N ha^{-1} or less), split applications have no comparative advantage over a single application. At Chitedze and Chitala, on medium textured soils, split applications had very little or no advantage at rates up to 150 kg N ha^{-1}.

The higher variability in maize grain yield at Chitala and Mwimba with respect to Chitedze can be appreciated in terms of long-term simulated yield (Figure 10). To minimize yield variability, the simulation study used locally

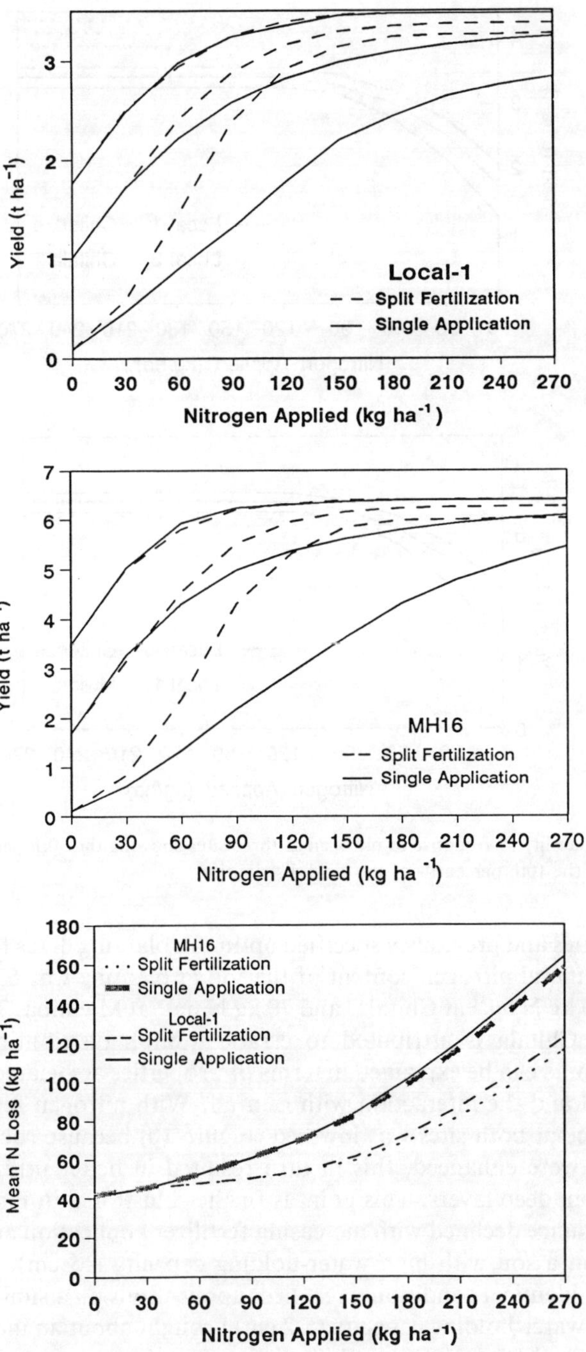

Figure 9. The effect of split versus single N application on mean grain yield, yield variability (shown by the 90th and 10th percentiles) and mean N loss.

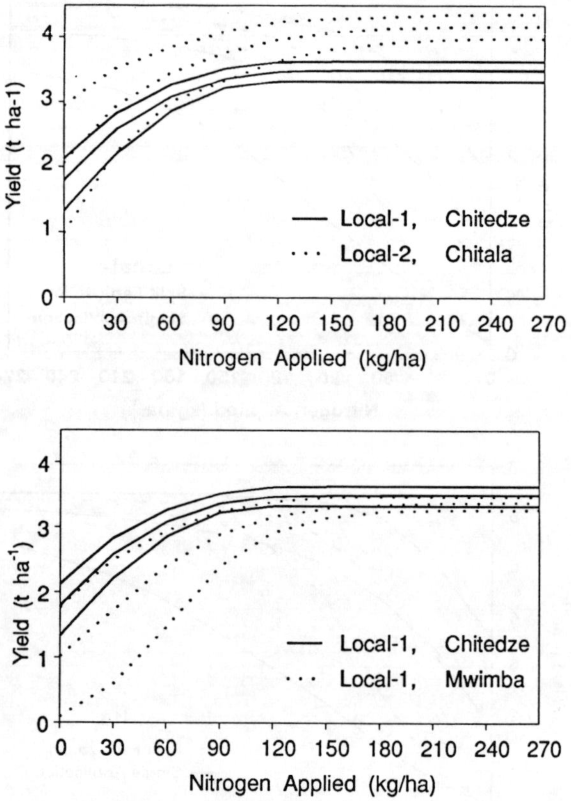

Figure 10. Variability of response to nitrogen at three sites showing the 90th percentile, the 50th percentile, and the 10th percentile of the yield distribution.

grown varieties and previously specified optimum planting dates for each of the sites. The mineral nitrogen content of the soil at planting was 60 kg N ha^{-1} at Chitedze, 80 kg N ha^{-1} at Chitala, and 70 kg N ha^{-1} at Mwimba. The high yield variance at Chitala is attributed to erratic rainfall distribution, while the Mwimba results can be explained in terms of properties associated with coarse-textured soil and the interaction with rainfall. With nitrogen fertilization the yield variance at both sites was lowered (Figure 10) because root growth and distribution were enhanced; this in turn resulted in better utilization of soil moisture from deep layers. This point is further illustrated in Figure 11 where the yield variance declined with increasing fertilizer application and grain yield at Chitala on a soil with high water-holding capacity (15 cm). On the other hand, under identical conditions, fertilizer applications on a similar soil with a much lower water-holding capacity (5.2 cm) brought about an increase in yield variability. The higher yield variability is attributed to faster growth of young plants with fertilization and drought stress during grain filling.

On soils of higher water-holding capacity, a reduction in yield variability can

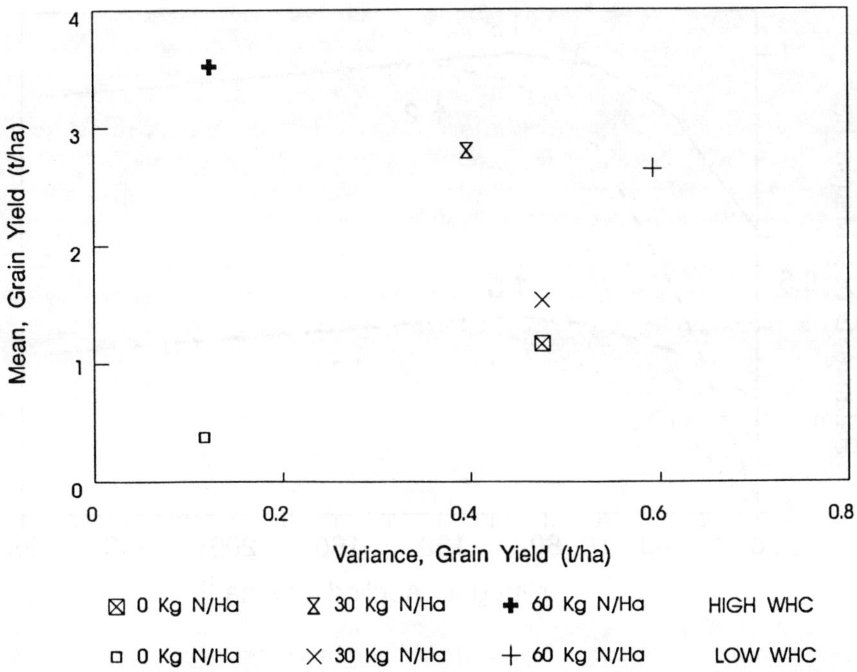

Figure 11. Effect of soil water holding capacity (WHC) on the yield response to N fertilizer at Chitala.

be obtained with application of 40–80 kg N ha^{-1}. Using long-term simulations, it was found that the above amounts of N could be supplied by 2.0–4.5 t ha^{-1} of green manure incorporated at the time of planting; however, it may not be feasible to obtain such quantities of green manure. At all three sites, the nitrogen losses were similar for green manure and for inorganic fertilization.

Throughout the nitrogen management simulations, the N response, N loss, and total N uptake were similar for the hybrids and the local varieties, although the grain yield for the latter was usually lower. The local varieties had higher stover weight. If grain yield was the only criterion, then the hybrids would be superior to local maize. However, farmer preference is also dictated by grain quality and amount of fodder.

An economic analysis of nitrogen fertilizer use at Chitedze is shown in Figure 12. Gross margins per hectare are plotted, in relative terms, against N application rate for the two varieties MH16 and Local–1 at two levels of initial soil fertility. Prices and costs pertaining to 1991 were used (in this analysis, account was taken of the economic value of the grain only). An analysis using the Mean-Gini stochastic efficiency criterion (Fawcett and Thornton 1990) led to the identification of the optimum input level (for each case, the stochastically efficient input level coincides with the profit-maximizing level for this simple analysis). Note the shift in the optimum input level for each variety as a move is made from a low- to a high-fertility soil.

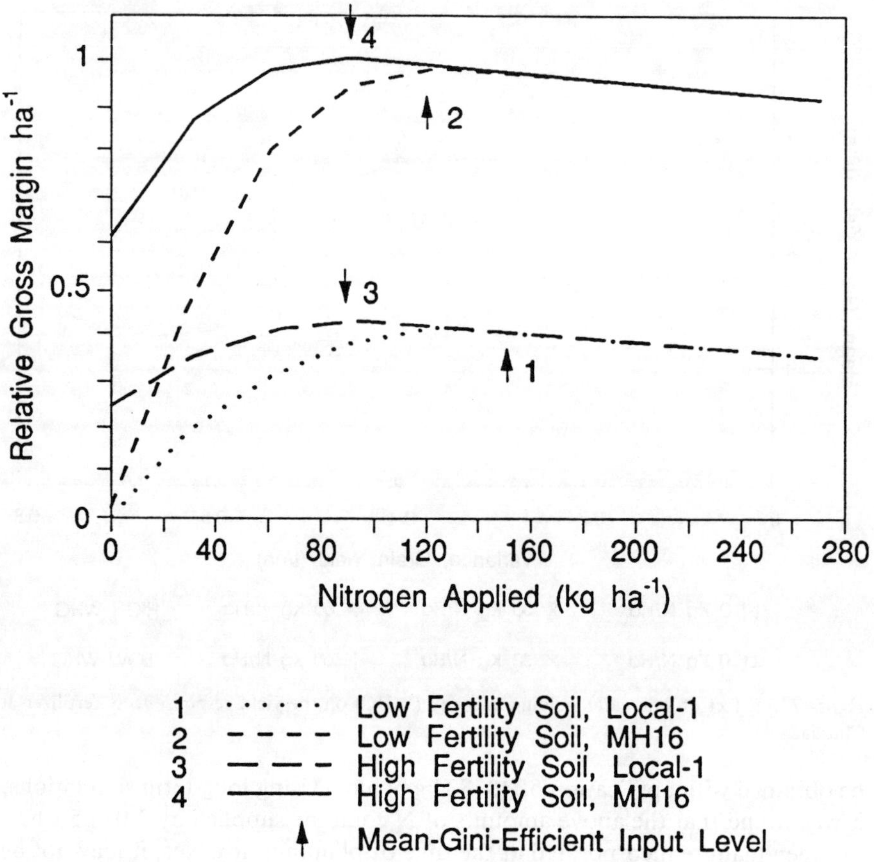

Figure 12. Relative gross margins for nitrogen fertilizer use at Chitedze, 1991 costs and prices.

Future work and prospects

The field trials carried out in 1989–1990 on the experimental stations were repeated during 1990–1991 to provide model data sets for validation over two season types. In addition, a small number of on-farm experiments were carried out, managed by collaborating farmers according to their own individual practices. In subsequent seasons, the model will be tested at a number of other sites in the country, particularly in areas that are more marginal for maize production in terms of soil P availability, soil acidity, and climatic conditions.

In terms of model application, the variability in the onset, distribution, and quantity of rainfall in Malawi indicates that optimum crop management practices will vary from year to year, depending on the actual weather experienced. It is planned to use CERES–Maize to define season types, each with specific crop management recommendations. The possibility of early identification of season type, perhaps in terms of timing of the onset of the rains and other long-term variables, should allow simple management recommendations to be derived not only for specific locations and for specific long-term weather conditions but also for the weather patterns that are most likely to occur in any particular season. Soils and climate data from the major maize-growing ecologies of the country are currently being collated into spatial data bases for manipulation within ARC/INFO, a Geographic Information System (GIS). In addition, the links between the GIS and CERES–Maize are being automated to allow regional simulation analyses to be performed for the major maize-producing areas of Malawi. This will allow year-to-year and site-to-site variation in model outputs such as production response, fertilizer efficiency and economic returns to be quantified and mapped (see Singh et al. 1991, for example).

The modeling work also has a larger role to play within the framework of the Soil Fertility Research Program in Malawi. Modeling and information technology will be used in a number of activities, including the following:
(1) To conduct soil surveying, characterization, and classification to update previous agroecological work and to facilitate land resource analyses.
(2) To assist in the agronomic evaluation of indigenous fertilizer materials, such as rock phosphates, as alternative sources of nutrients and to identify the soil types on which their use may be beneficial.
(3) To investigate the use of organic manure and crop residues to help identify low-input maize cropping management practices.
(4) To identify priority research areas concerned with the fate and transport of nutrients applied to croplands and to act as a medium whereby research results can be collated and analyzed.
(5) To identify long-term management practices that are sustainable and that minimize damage to the environment.

The CERES models are increasingly being used to address aspects of environmental quality and sustainability, in particular, with regard to ways of minimizing nutrient losses, and the effect of management practices on the long-term biological viability of cropping sequences and rotations. It is hoped that the current project will constitute an important input to these areas of model research, development, and application.

Conclusions

The results presented suggest that CERES–Maize, having been calibrated and evaluated using field experiments, performs adequately for the locations tested in Malawi. It was demonstrated that useful information on planting windows, plant population, fertilizer regimes, and variety selection can be gleaned by carrying out simulation experiments replicated over many different season types. This constitutes a preliminary step only, however, and future work will extend the applicability and usefulness of the model.

Such activities represent a considerable research investment. An integrated approach to soil fertility research makes special demands on the institutions involved and necessitates training research personnel in the use of the required methodologies and techniques. Simulation modeling has an important role to play in any integrated approach because soil fertility cuts across commodities, disciplines, and regions. It is hoped that work on maize modeling in Malawi will expand over the next few years so that real and tangible benefits to the research and development process can be demonstrated, and thence to the smallholder farmer who forms the core of the agricultural sector in so many countries of the world.

Acknowledgements

The financial assistance of the Rockefeller Foundation, New York, and the all-round help of Dr. Malcolm Blackie of the Rockefeller Foundation office in Lilongwe, in conducting this work are gratefully acknowledged. The authors also acknowledge the contribution of L. Chisenga, Department of Agricultural Research, Chitedze, in managing the field trials.

References

Fawcett R H, Thornton P K (1990) Mean-Gini dominance in decision analysis. IMA Journal of Mathematics Applied in Business and Industry 6:309–317.

International Benchmark Sites Network for Agrotechnology Transfer (1988) Experimental design and data collection procedures for IBSNAT, 3rd edition, revised, Technical Report 1, IBSNAT, Department of Agronomy and Soil Science, University of Hawaii, Honolulu. 73 p.

International Benchmark Sites Network for Agrotechnology Transfer (1990) Field and laboratory methods for the collection of the IBSNAT minimum data set. Technical Report 2, IBSNAT, Department of Agronomy and Soil Science, University of Hawaii, Honolulu; and Supplement 1. 67 p.

Jones C A, Kiniry J R (1986) CERES–Maize. A simulation model of maize growth and development. Texas A&M University Press, College Station, Texas, USA, 194 p.

Keating B A, Godwin D C, Watiki J M (1991) Optimising nitrogen inputs in response to climatic risk. Pages 329–358 in Muchow R C, Bellamy J A (Eds.) Climate risk in crop production: Models and management for the semi-arid tropics and subtropics. CAB International, Wallingford, Oxfordshire, UK.

Richardson C W (1985) Weather simulation for crop management models. Trans. Am. Soc. Agric. Eng. 28 (5):1602–1606.

Ritchie J T, Singh U, Godwin D C, Hunt L A (1989) A user's guide to CERES–Maize V2.10. International Fertilizer Development Center, Muscle Shoals, Alabama, USA.

Saka A R (1987) A review of soil fertility and fertilizer recommendations in Malawi, First Meeting of the East and South-Eastern African Fertilizer Management and Evaluation Network, Nairobi, Kenya, May 1987.

Singh U (1985) A crop growth model for predicting corn performance in the tropics. Ph.D dissertation, Agronomy and Soil Science Department, University of Hawaii, University Microfilms International, Ann Arbor, Michigan, USA.

Singh U, Godwin D C, Humphries C G, Ritchie J T (1989) A computer model to predict the growth

and development of cereals. Pages 668–675 in Proceedings of the 1989 Summer Computer Simulation Conference, Austin, Texas, USA.
Singh U, Brink J E, Christianson C B (1991) Application of the CERES–Sorghum model and geographic information systems in the Indian semi-arid tropics. International Fertilizer Development Center, Muscle Shoals, Alabama, USA (in press).
Williams L B, Allgood J H (1990) Fertilizer situation and markets in Malawi. Paper Series P–12, International Fertilizer Development Center, Muscle Shoals, Alabama, USA, 25 p.

SESSION 4

Crop production: biological constraints

Pest damage relations at the field level

K. J. BOOTE[1], W. D. BATCHELOR[1], J. W. JONES[1],
H. PINNSCHMIDT[2] and G. BOURGEOIS[3]

[1] *University of Florida, Gainesville, Florida 32611, U.S.A.*
[2] *International Rice Research Institute, P.O. Box 933, 1099 Manila, Philippines*
[3] *Agriculture Canada, Quebec, Canada*

Key words: CERES-Rice, competition, defoliation, IBSNAT, multiple pests, peanut, pest damage categories, pest scouting, PNUTGRO, rice, seed damage, simulation, soybean, SOYGRO, weed

Abstract
In view of large yield losses to various biotic pests, there is a critical need to adapt crop models to account for pest effects, either by direct coupling with mechanistic pest simulators, or by input of observed pest damage or numbers. Pest effects can be coupled to crop models by modifying crop state variables (mass or numbers of various tissues), rate variables (photosynthesis, water flow, senescence), or inputs (water, light, nutrients). An approach is presented whereby scouting data on observed pest damage is input via generic pest coupling modules in order to predict yield reduction from pests. Examples of this approach are presented for simulating the effects of defoliating insects, seed-feeding insects, and rootknot nematode on soybean growth using the SOYGRO model. A similar approach with the CERES–Rice model is used to simulate effects of leaf blast and other pests on rice growth and yield. Effects of leafspot disease on peanut growth and yield are illustrated with both the generic pest coupling approach and a mechanistic disease simulator coupled to the PNUTGRO model. A systematic approach to input pest damage effects into crop models offers potential to account for yield loss from pests and to determine action thresholds that are dynamic and different per season or region.

Accounting for pest effects in crop models

Most existing crop models predict crop growth in response to climatic, soil-related, and cultural management factors, but fail to account for effects of various biotic organisms. In view of considerable yield losses to various pests, there is a critical need to modify crop models to account for such pest effects, either by direct coupling with mechanistic pest simulators, or by inputting damage based on observed pest damage or numbers.

The point of pest coupling in crop models

Coupling of pest effects to crop simulation models can be done either at the level of inputs (water, light, nutrients), rate processes (photosynthesis, transpiration-water uptake, senescence), or state variables (numbers of organs or mass of various tissues). For example, crop carbon (C) balance is modelled in most growth simulators with a state variable approach: (i) C is assimilated by photosynthesis (rate process), (ii) C is lost to respiration, senescence, and

F.W.T. Penning de Vries et al. (eds.), Systems Approaches for Agricultural Development, 277–296.

abscission (rate processes), and (iii) C is stored in various tissues (state variables). Rate processes and state variables for plant nitrogen (N) and water balances are similar, but further depend on processes and state variables dealing with N and water in the soil.

Categories of pest damage

Pests can be classified into various categories depending on the type of damage caused to the crop. Boote et al. (1983) proposed the following classes of pest damage: stand reducers, photosynthetic rate reducers, leaf senescence accelerators, light stealers, assimilate sappers, tissue consumers, and turgor reducers. A given pest or pathogen may fall into several categories and its damage may be coupled at more than one point to the crop growth simulation model. These various types of pest categories will be briefly reviewed here.

Assimilate sapper

These pests (pathogens, nematodes, or sucking insects) remove soluble assimilates from plant cells. Biotrophic fungi, for example, produce haustoria which invade host cells and actively transport nutrients from the host cytoplasm across the haustorial plasma membrane (Gay and Manners 1981). Certain nematodes feed directly from the phloem tissue in the root and the plant is left with less assimilate for its own tissue growth. Sucking insects, such as whitefly (*Trialeurodes abutiloneus*) and aphid (*Aphis craccivora*), remove C- and N-containing assimilates from foliage. The critical coupling factor is predicting the rate of assimilate removal as a function of pest numbers and activity.

Tissue consumer

Tissue consumers differ from assimilate sappers by feeding after the plant has already converted the assimilate to tissue. In this case, the plant has already incurred respiration costs to synthesize its tissue. Tissue-consuming pests may feed on any number of plant tissues including roots, stems, pods, seeds, and leaves (state variables in crop models). Foliar-feeding insects are an obvious example of a tissue-consuming pest which removes leaf mass and leaf area, often without reducing effectiveness of adjacent leaf tissue (Poston et al. 1976; Ingram et al. 1981). The direct effect is the loss of photosynthetic tissue, with subsequent indirect effects on crop photosynthesis. Necrotrophic pathogens are a less obvious example which kill cells in advance of their growth and use the contents of the dead cells for their own growth and respiration. The critical coupling parameters for 'tissue consuming' pests are the rate of tissue consumption, tissue type, and timing of consumption (Rudd 1980; Wilkerson et al. 1983).

Stand reducers

A stand reducer is any pest that causes the loss of whole plants or tillers. Typical stand reducers are soil-borne fungi or insects that cause seedling death. The consequence is a loss of plant biomass and number of plants. To predict the consequences of damage on yield, one needs to know the timing of loss, the number of plants lost, and the approximate geometric distribution of remaining plants. Crop models generally have 'built-in' compensatory mechanisms that allow tillering and other responses to the additional available light. Striped stem borer (*Chilo supressalis*) is a pest that causes loss of tillers in rice (*Oryza sativa* L.). Observed and simulated yield loss to this pest is a function of crop age because early tiller loss is compensated by tiller replacement, whereas damage to mid-life cycle tillers causes failure of panicle formation, and even later damage affects only the grain growth rate (Rubia and Penning de Vries 1990).

Photosynthetic rate reducer

A number of pathogens directly affect the rate of photosynthesis of the host tissue, even without directly consuming the tissue. These have a direct effect on the photosynthetic process of remaining tissue as summarized by Buchanan et al. (1981). Fungi, bacteria, and viruses may reduce numbers of chloroplasts per unit leaf area or alter chloroplast ultrastructure, electron transport carriers, and electron transport reactions of photosynthesis (Magyarosy and Buchanan 1975; Magyarosy and Malkin 1978; Buchanan et al. 1981). Some pests, while being primary tissue-consumers or assimilate sappers, also may produce or induce a messenger metabolite that causes a remote effect on adjacent tissue. To simulate damage from photosynthetic rate reducers, one must be able to quantify the effect of a given level of pest infection upon the photosynthetic light response curve of intact leaves. If the light response curve is defined by quantum efficiency (Q_E) and light-saturated leaf photosynthesis rate (P_{max}), then one must describe pathogen effects on these two parameters. Typically, P_{max} will be reduced by chloroplast ultrastructural changes; however, Q_E is also frequently reduced.

Leaf senescence accelerators

Some pathogens, in addition to being tissue-consumers or assimilate sappers, enhance senescence and abscission of leaves. Senescence accelerators cause premature abscission of leaf and petiole mass and thereby reduce light interception and photosynthesis. *Cercospora* spp. on peanut (*Arachis hypogaea* L.) are typical senescence accelerators. They reduce crop-photosynthesis more from the reduction of leaf area index (LAI) and light interception, than from their effect on single leaf photosynthesis (Bourgeois 1989). To simulate this effect, the rates of leaf senescence induced by given levels of pathogen infection must be quantified and added to the normal rates of senescence caused by self-shading, drought stress, nutrient deficiency, and crop aging.

Light stealers

The primary effect of weeds is assumed to be a 'light stealing' effect by absorbing photosynthetically active radiation that could have been used for crop assimilation. In this case, light is an input to the crop simulation model.

Two simple approaches are to assume the weed leaf area is either completely above the crop (overstory) or it is equally competitive with the crop LAI. In a more complex analysis, one could simulate the vertical distribution of the weed leaf area versus the crop leaf area as they both change during the season, and compute the light intercepted by both. Such an approach would obviously require a layered canopy light interception model and the ability to predict the time course of increase in height and leaf area density of the crop and weed. Several groups have integrated dynamic growth models of crops and weeds to study effects of competition for light and water on crop growth and yield. Spitters and Aerts (1983) incorporated the effect of vertical distribution of leaf area and roots to study competition by weeds. Starting with a similar vertical weed-crop competition model, Kropff (1988) demonstrated the importance of weed relative leaf cover and other simple relationships. Wilkerson et al. (1990) integrated weed and soybean models, taking into account the horizontal non-homogeneity of weeds in a crop and demonstrated good predictions of weed and soybean growth.

Pathogens, in some cases, may act as light stealers. For example, necrotic lesions absorb light. This light is not available to foliage that may be shaded by the necrotic lesions. Thus, necrotic lesions not only constitute a loss of photosynthesizing tissue; they also remain in the canopy for light interception. The importance of such lesions in reducing photosynthesis depends on whether the lesions are predominantly on lower or upper leaves. Rabbinge and Rijsdijk (1981) simulated leaf shading as one factor associated with powdery mildew (*Erysiphe graminis*) effects on growth and yield of wheat (*Triticum aestivum* L.).

Turgor reducers

Turgor reducing pests may start out as assimilate sappers and/or tissue consumers, but additionally act to reduce water transfer through the root xylem. Nematodes and root-rot pathogens can act as turgor reducers because of their subsequent effects on plant water balance (Trudgill et al. 1975; Ayres 1981; Hornby and Fitt 1981). Vascular wilt pathogens such as *Pseudomonas*, *Fusarium*, and *Verticillium* act as turgor reducers when they block flow in the xylem. The turgor reducing effect is related to the extent to which these pests feed on roots, block phloem movement to root growing tips, reduce root growth, and reduce water and nutrient uptake. The coupling parameters for turgor-reducing pests are difficult to quantify, but include rate of feeding, secondary tissue death, and root-stem conductivity to waterflow as a function of various levels of these pests. It is readily apparent that adequate simulation of turgor-reducing pests will require a crop model that includes C balance, soil-plant water balance, and root growth. Secondary compensation effects such as increased C allocation to root or nematode growth are additional important concerns.

Another category of turgor reducers are pests that affect stomatal and non-stomatal resistance to leaf transpiration. Ayers (1981) in his review, described how rusts, powdery mildew, and some virus diseases affect guard cell function, resulting in greater resistance to CO_2 uptake by well-watered plants, but insufficient stomatal control to fully restrict water loss during severe drought stress.

Problems in categorizing type of pest damage

It is easy to hypothesize pest damage categories, but more difficult to allocate pests to given categories. It is even more difficult to quantify the amount of damage per unit or number of pests, yet that is precisely what must be done in order to couple pest damage to crop simulation models for purposes of predicting yield reduction. Damage from a given pest need not be restricted to one category or coupled at only one point to a crop model.

A generic approach for coupling pest damage to crop models

We will illustrate two approaches for coupling pest effects: first, using inputs of 'scouting data' on pest damage into a generic crop-pest coupling model, and secondly, using a mechanistic simulation of the pest population, with concurrent coupling to the crop model.

Recent efforts in the International Benchmark Sites Network for Agrotechnology Transfer (IBSNAT) project have focused on developing a generic pest coupling approach which allows the input of scouting information on pest damage. Over the past 10 years, it has become obvious that the present IBSNAT crop models are capable of predicting yield only in situations that are relatively pest-free and not suffering from nutrient deficiencies (other than N). This is not surprising since the IBSNAT models were not designed to simulate stresses from pests, diseases, and nutrient deficiencies other than N. Thus, simulated yields represent potential yield as limited by climate and soil-water holding characteristics. It has also become obvious that mechanistic simulators of pest population dynamics, while being excellent tools to understand the pests, require difficult-to-obtain regional scale inputs. For example, there is a low probability of success for an insect population dynamics model that predicts, *a priori*, insect populations during the season from an initial pest input and seasonal weather. The difficulty is not necessarily with the model, but with seasonal influxes of migrating insects dependent on numerous factors, and on highly variable mortality factors. With some scouting input, such pest simulators could be 'feedback corrected' so as to allow improved predictions of pest populations for the next few weeks.

Alternatively, the scouting inputs could be used as direct input into the crop models. This approach has been adopted recently in the IBSNAT project by Batchelor et al. (1991), whereby scouting data on pest numbers or damage to the crop is input into pest coupling modules in the IBSNAT crop models in order to predict yield reduction from pests. The pest coupling routine requires that we know the appropriate coupling points and the relationships of pest

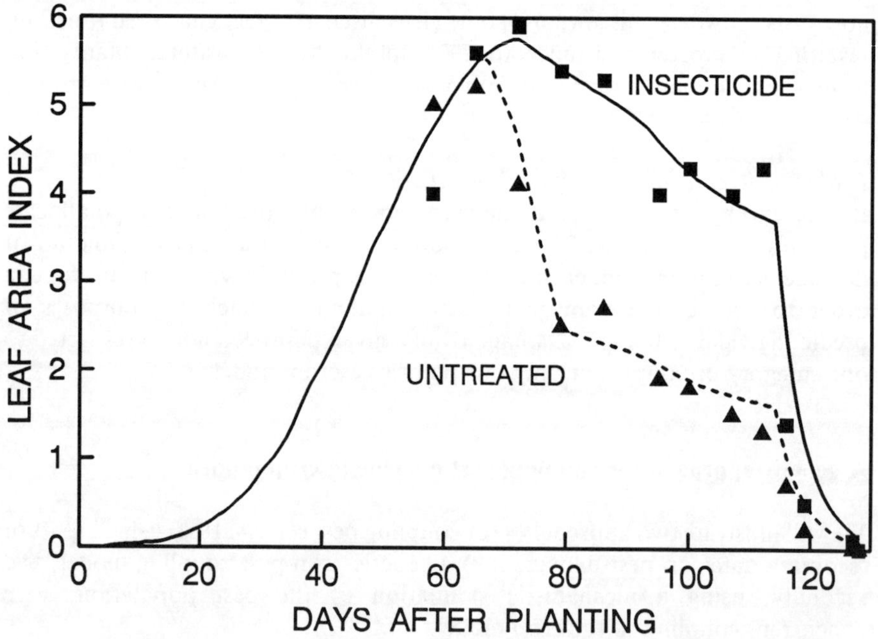

Figure 1. Leaf area index for Bragg soybean at Quincy, Florida, in 1979, for insecticide-treated and untreated plots.

damage to crop model processes or state variables. With insect defoliation, for example, the primary coupling point may be loss of leaf area; however, leaf mass and leaf protein are also reduced. The user enters data into a pest progress file that describes the level of damage or insect population observed on respective dates. The input for leaf feeding insects could be: (i) fraction defoliation, (ii) leaf mass (g m^{-2}) removed, (iii) leaf area (m^2 m^{-2}) removed, or (iv) numbers of insects of a given instar per unit land area. If inputs are given as insect populations, the pest coupling routine converts insect populations to feeding rates on designated coupling points by using coefficients from the pest parameter file. This file defines the feeding rates and coupling points of individual pests derived from values reported in the literature or by entomologists. During execution, the crop model uses linear interpolation to compute daily damage from the observed damage or insect population level on the reported scouting dates. When the model is run, the appropriate state variables of leaf mass, leaf area, and leaf protein are changed. The model is run with and without the pest damage to evaluate the net effect of the pest damage on final yield.

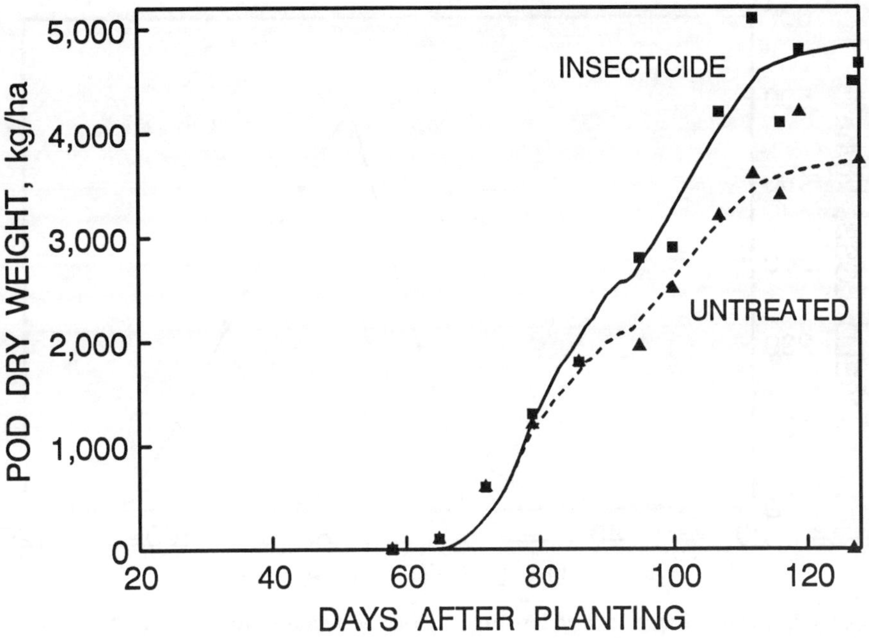

Figure 2. Pod mass over time for Bragg soybean at Quincy, Florida, in 1979, for insecticide-treated and untreated plots.

Example simulations of pest effects using coupled models

Generic coupling of insect defoliation – input of damage

The generic pest coupling approach will first be illustrated for the SOYGRO crop growth model using data from a 1979 field study of natural defoliation of soybean by velvetbean caterpillar (*Anticarsia gemmatalis* Hubner) (Ingram et al. 1981). Active defoliation in the untreated plot occurred from 65 to 79 days, during the critical period of pod formation. Field-measured LAI of insecticide-treated and untreated (defoliated) plots (Figure 1) were used to create a damage file giving the daily leaf area removed. Using this coupling point, the soybean crop model correctly predicted the decreased growth of pods and seeds and the resultant loss in yield attributed to the insect pests (Figure 2). This simple example shows that we can mimic the yield-reducing effects of defoliating pests merely by their effect on LAI.

Yield loss assessments are easily made with this generic pest coupling approach. For example, if the SOYGRO crop model is run with, and without a given pest damage, one can evaluate the seed yield loss caused by failure to control pests at a given growth stage. Figure 3 illustrates soybean yield reduction simulated by SOYGRO when 25% cumulative defoliation over a 10-day period is applied at different weeks during the season. It is obvious that

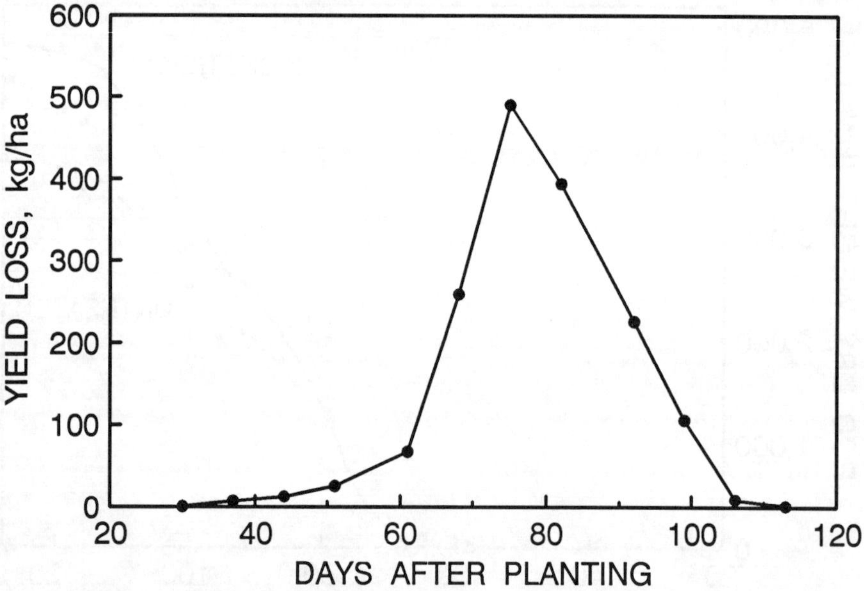

Figure 3. Simulated soybean yield loss from 25% defoliation over 10 days applied at different weekly intervals during the growing season.

yield loss is greater during the middle of the soybean life cycle, and less with early or late season infestations. This approach has excellent potential to assist in making pest management decisions. It allows for natural 'stage specific' thresholds to be determined based on the growth dynamics of the crop, and provides stage-specific thresholds that account for past-climatic and soil-related effects on crop growth.

Generic coupling – input of insect numbers for a seed-feeding pest

A more sophisticated approach to model pest effects on yield is to input scouting information on pest populations, coupled with expert knowledge of pest feeding rates. This approach was used by Batchelor et al. (1989a; 1989b) to develop the SMARTSOY expert system as an aid for soybean insect pest management. This approach uses 'expert knowledge' from entomologists regarding expected feeding rates and plant part preferences for specific pest types during specific instar phases. Data on pest populations is used to estimate the amount of defoliation and seeds damaged by various insect populations if left untreated and to determine the period of time over which the damage would occur.

The above approach of inputting pest numbers is used here to demonstrate the stage-dependent yield losses caused by corn earworm (*Heliothis zea*) (6.5 larvae m^{-1} of row) on soybean (Figure 4). For purposes of this simulation, corn earworm was assumed to feed only on pods and seeds. Corn earworm

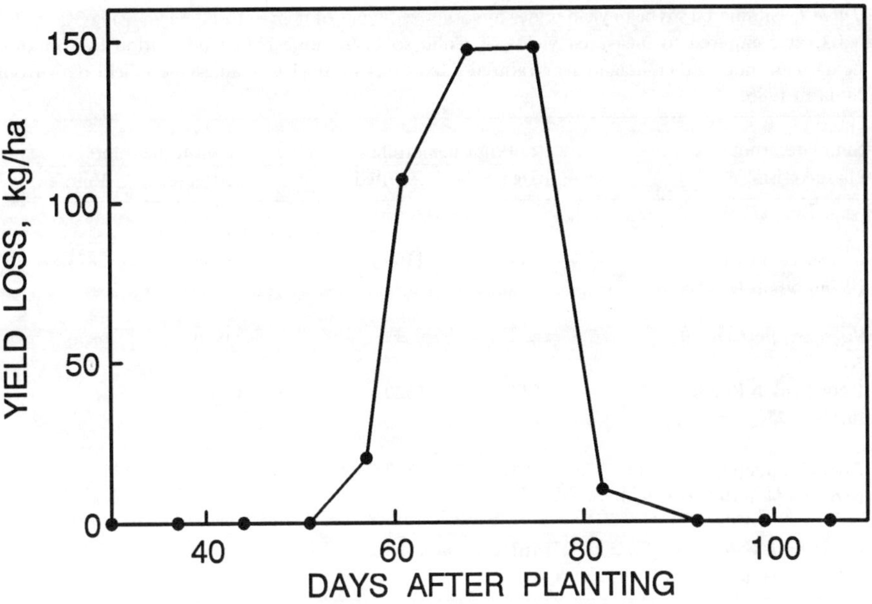

Figure 4. Simulated yield loss of Bragg soybean in Mitchell County, GA, as affected by 6.5 corn earworm m^{-1} of row introduced at different times after planting (SMARTSOY simulation).

preferentially feeds on small pods, small seeds, larger seeds/foliage in that order of approximate preference. Simulations reveal a critical window for damage that is related to setting of pods and seeds. Corn earworm damage causes pod and seed addition to be prolonged, fewer seeds to be set, seed growth to be delayed, and final yield to be reduced.

Generic coupling of a pest with multiple types of damage

We anticipate the need to handle multiple damage from a given pest. If the same insect (e.g. corn earworm) is feeding on leaves, pods and seeds, then the user should define the damage to each structure and input that into the damage file. However, if the input is number of larva of a given instar, then an interface module allocates feeding damage to seeds, pods, and other plant parts based on availability of those plant components and based on entomological 'expert knowledge' on feeding rates dependent on instar, pod age, and preference for plant parts. Internal to the model, the appropriate changes are made to compute number and mass of pods and seeds lost or damaged.

Generic coupling to SOYGRO: rootknot nematode as an assimilate sapper

Assimilate sappers are a special category that we would like to illustrate with an example of rootknot nematode *(Meloidogyne arenaria)* on soybean. All pests

Table 1. Simulated soybean yield under hypothesized types of nematode damage, using SOYGRO V5.3, as compared to measured yields of Cobb soybean subjected to infestation by rootknot nematode under either irrigated or rainfed conditions in 1985 at Gainesville (Yield data from Stanton 1986).

Simulated root characteristic	Rootknot nematode		Uninfested plots	
	Irrigated	Rainfed	Irrigated	Rainfed
	seed yield, kg ha^{-1}			
Measured Yield (From Stanton 1986)	1766	1109	3566	2873
Standard Pest-free Run	---	---	3570	2820
Root Length Reduced (6500 to 2500 cm g^{-1})	1520	1320		
Root Conductivity Reduced ($4.3 \cdot 10^{-5}$ to $1.0 \cdot 10^{-9}$)	2260	750		
Direct Sink Effect (32% to Nematode/gall)	1610	1280		

obtain their nutrition from the crop by either direct consumption of tissue or by feeding on available soluble assimilates (with little tissue damage). Our experience with rootknot nematode on soybean (Stanton 1986) leads us to conclude that the rootknot nematode/gall complex is a major assimilate drain on the plant, and one that appears to have first priority for assimilates over all other plant tissues including seeds. Plant organs appear to share the remaining assimilate according to 'normal' plant allocation rules. Competition from this nematode does not necessarily reduce root growth per se because plant water status, stomatal conductance, and single leaf photosynthesis were insignificantly reduced (Stanton 1986). Other nematodes may have different effects however.

Rootknot nematode as an assimilate sapper on soybean will be illustrated here based on two experiments on Cobb soybean at Gainesville, Florida (FL). and Marianna, FL. In her experiments, Stanton (1986) established differential nematode infestations and observed substantial yield loss due to rootknot nematode in irrigated and non-irrigated treatments (Table 1). We attempted to simulate nematode effects on soybean growth from either: (i) direct damage to roots via reduced root length, or (ii) direct damage to roots via reduced root conductivity for water uptake, or (iii) direct competition by nematodes and galls for assimilate. We initially made minor adjustments in the SOYGRO model to correctly simulate yield in the non-infected soybeans (irrigated and non-irrigated treatments in Table 1). Adjustments were primarily to set the ratio of root length to root mass and the shape of the rooting profile in order to fit the observed root length density profiles for the irrigated and non-irrigated treatments of non-infected soybean. Next, we evaluated three

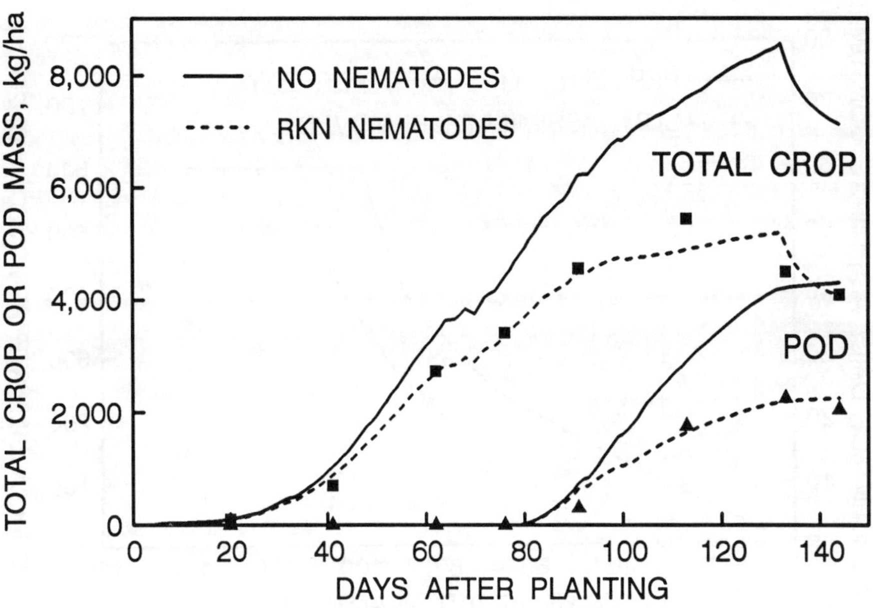

Figure 5. Dry matter accumulation in aboveground biomass and pod mass for Cobb soybean with rootknot nematode injury at Marianna, Florida, in 1988. Simulation of a pest-free crop is included for comparison.

hypotheses for nematode damage. The assumption of direct damage via reduced root length, was accomplished by reducing the ratio of root length to root mass in the SOYGRO model. This ratio had to be reduced from the original 6500 to 2500 cm g^{-1} in order to come close to the simulated yield. This scenario not only failed to mimic differences between rainfed and irrigated treatments of infected plots, but it produced a much lower simulated root length density than was measured. The hypothesis that nematodes reduced root conductivity to water uptake, was tested by reducing a conductivity function in the water uptake per unit root length. The conductivity value had to be reduced from $4.3 \cdot 10^{-5}$ to $1.0 \cdot 10^{-9}$ in order to mimic the correct average yield reduction from rootknot nematode, yet with this low conductivity,the simulation was much too sensitive to water deficit as shown by the inability to predict the difference in yield in irrigated and rainfed treatments of the infested soybeans. The assumptions of reduced root length or loss in root conductivity to water flow resulted in simulated severe water deficit. This was not confirmed by measurements of water potential and stomatal conductance. Furthermore, these simulations gave incorrect patterns of dry matter accumulation. The third hypothesis of an assimilate-sapping effect was tested by allowing a fixed fraction of daily crop photosynthesis to be sequestered by the nematode/gall complex. Allowing approximately 32% of the current photosynthate to be diverted to nematodes/galls resulted in a good fit to seasonal growth and yield of the rainfed and irrigated treatments without overly-severe water deficit

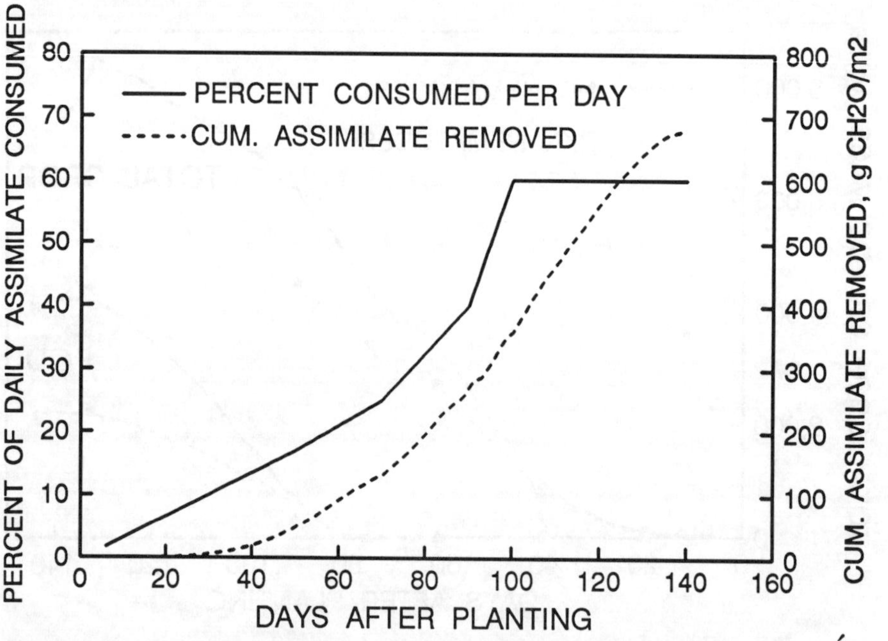

Figure 6. Hypothetical function describing rootknot nematode damage to Cobb soybean at Marianna, Florida, in 1988: 1) as percent of daily assimilates (glucose) removed, and 2) as cumulative assimilate removed over the growing season.

(Table 1). Moreover, this assumption was supported by much larger root:shoot ratio for nematode-infested plants (0.75 versus 0.27 for uninfested plants at 97 days after planting) observed by Stanton (1986). Minor effects on root conductivity to water flow were observed late in the season by Stanton.

Rootknot nematode infestation and galling were visually observed by mid-season in an 'on-farm' study of Cobb soybean on a producer field near Marianna, FL. We did not have a good control treatment available; however, the SOYGRO model correctly predicted the growth and yield of Cobb for the 'no nematode' treatment of Stanton (1986) and for a 1981 experiment at Gainesville. We developed an assimilate sapping function (fraction of assimilates removed per day) that resulted in the correct accumulation of aboveground dry matter (Figure 5). When calibrated to aboveground dry matter in this way, the seed yield was correctly simulated. The nematode-assimilate sapping function (fraction of assimilates removed per day and cumulative assimilates removed) is shown in Figure 6. The model-simulated root:shoot ratio with or without nematodes was 0.66 and 0.24, respectively, at 97 days assuming that assimilates were converted to gall mass (and added to root mass) using 1.7 g glucose per g dry matter. This compares well with Stanton's measurements.

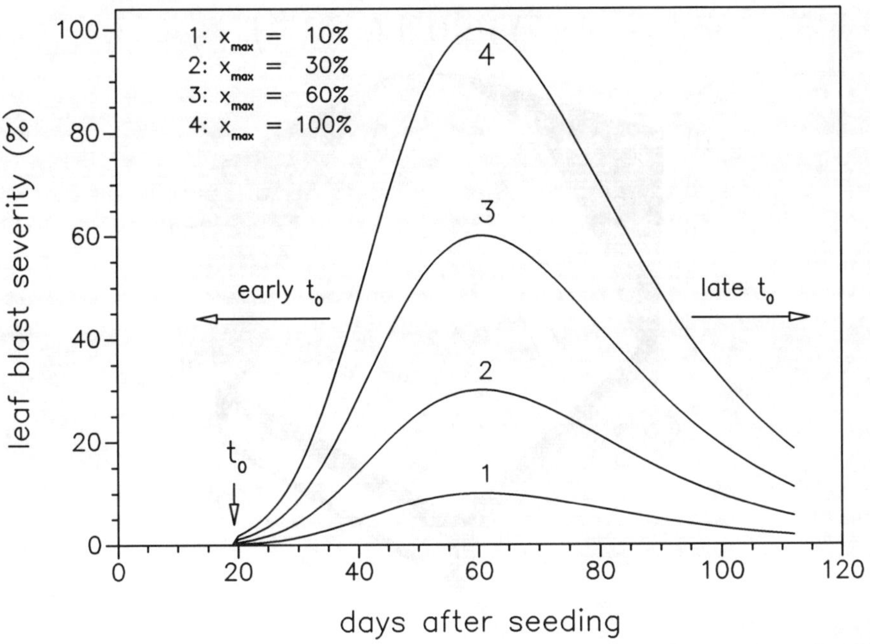

Figure 7. Example of disease progress curves used for simulating the effect of leaf blast on rice (X_{max} = maximum disease severity (%), and t_o = disease onset time in days after seeding).

Generic coupling of leaf blast effects on the CERES–Rice model

The same generic pest coupling approach has been used with the CERES–Rice model by Teng et al. (1990; 1991) and Pinnschmidt et al. (1990). Simulation of effects of leaf blast (*Pyricularia oryzae* Cavara) on rice growth and yield will be illustrated based upon the hypothetical leaf blast progress curves shown in Figure 7. The curves represent a series of leaf blast scenarios, with different maximum intensities, X_{max} (10, 30, 60 and 100%). Leaf blast is assumed to have two major effects on the rice plant: (i) reduction in the amount of light interception/photosynthesis due to leaf spots that occupy otherwise productive leaf area, and (ii) reduction of photosynthesis through toxic effects as described by Bastiaans (1991).

The combined effects of disease onset time (to) and increasing disease intensity are shown in Figure 8 and illustrate that predicted yield loss is greater with earlier onset of leaf blast and with greater disease intensity. Early disease onset caused lower LAI, lower biomass production, and lower yield than later disease onset.

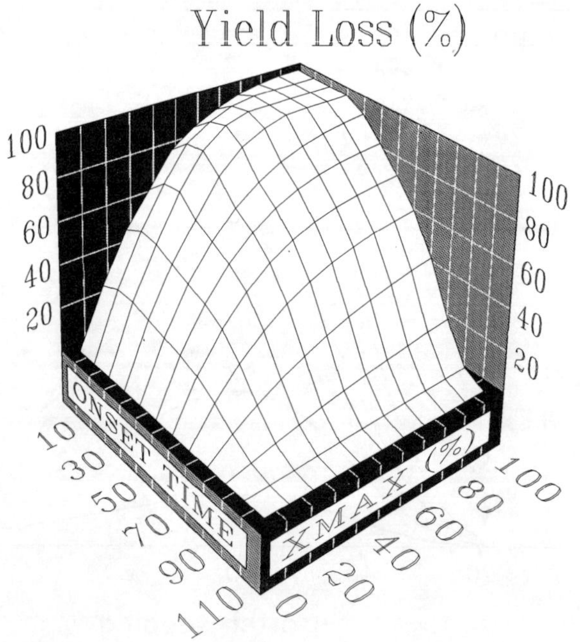

Figure 8. Simulated yield loss due to leaf blast in rice as affected by onset time (t_o, days after seeding) and maximum severity level of the disease.

Coupling of multiple pests (Leaf Blast, Leaffolder, and Weeds) on rice

The generic pest coupling approach was used to hypothetically evaluate multiple combinations of three pests on rice using CERES–Rice. Figure 9 shows progress curves for leaf blast severity, defoliation by leaffolder (*Cnaphalocrocis medinalis*), and weed density, each at low or high intensities. Onset time for leaf blast was 20 days, defoliation by leaffolders was assumed to begin at 50 days, and weed competition at 30 days after seeding. The effects of rice blast were modelled as described previously. Defoliation is assumed to reduce green leaf area per plant. Weeds were assumed to compete only for light, assuming a well-irrigated, well-fertilized crop. Weeds were assumed to be uniformly distributed in the crop and to have the same height and light extinction traits.

Simulated combinations of these three pests usually led to a less-than-additive effect on yield loss, compared to summing-up the single pest effects (Table 2). As expected, increasing pest infestation levels resulted in greater yield losses. Leaf blast and weed scenarios, as single effects or in combination, caused considerable yield loss and caused greater loss than did defoliation from leaffolder.

Table 2. Simulated yield loss due to combined infestations of leaf blast, defoliators, and weeds in rice. Pest onset times for leaf blast, weed competition, and defoliation were 20, 30, and 50 days after seeding, respectively.

Final weed	maximum leaf blast severity: 0%			10%			30%		
Leaf area	final defoliation:								
Index	0%	15%	66%	0%	15%	66%	0%	15%	66%
(m^2m^{-2})	simulated yield loss (%)								
0.0	0.0	3.7	21.5	30.6	35.0	51.1	73.4	74.2	81.5
0.7	10.0	12.6	26.6	39.3	42.2	54.6	76.4	77.1	83.1
4.4	42.5	42.6	46.6	63.5	63.8	67.3	100.0	100.0	88.7

Effect of Cercospora leafspot on peanut – mechanistic disease simulator

In the southeastern USA, late leafspot induced by *Cercosporidium personatum* (Berk and Curt.) commonly reaches epidemic proportions on peanut *(Arachis hypogaea* L.) and may reduce yield as much as 50% if not controlled with fungicides (Gorbet et al. 1982). Bourgeois (1989) developed a mechanistic leafspot disease simulator called LATE SPOT and coupled it to a peanut crop growth simulator (PNUTGRO; Boote et al. 1989). PNUTGRO was modified by Bourgeois (1989) to predict individual leaf cohorts so as to allow infection on individual age classes of leaves. The LATE SPOT model predicts the processes of pathogen infection, lesion expansion, lesion progression through its stages, and subsequent sporulation as a function of climatic parameters, pest state variables, and crop state variables. During the growing season, most of the spore release is from sporulating lesions in the field and is thus auto-generating, but the start of the infection cycle depends on old peanut residue in the field and a low level of external spore load that depends on temperature and humidity. In LATE SPOT–PNUTGRO, individual leaf senescence is accelerated as disease progresses on individual leaflets (the senescence function was calibrated to field data). Despite observed disease effects on single leaf photosynthesis, the majority of the reduction in canopy assimilation appeared to be caused by leafspot-induced defoliation. Thus, canopy photosynthetic reduction was computed from the reduction in total healthy leaf area.

The combined LATE SPOT–PNUTGRO model was calibrated to one year (1986) at Gainesville, and subsequently tested and shown to work well for three other years in Florida (Bourgeois 1989). Disease progress (percent necrosis) and resultant LAI simulations are shown for Florunner peanut grown at Marianna, FL in 1983 (Pixley et al. 1990). Disease progress curves (Figure 10) and LAI loss (Figure 11) are reasonably well-predicted considering that there was no calibration of disease parameters for this new site and different year. Figure 12

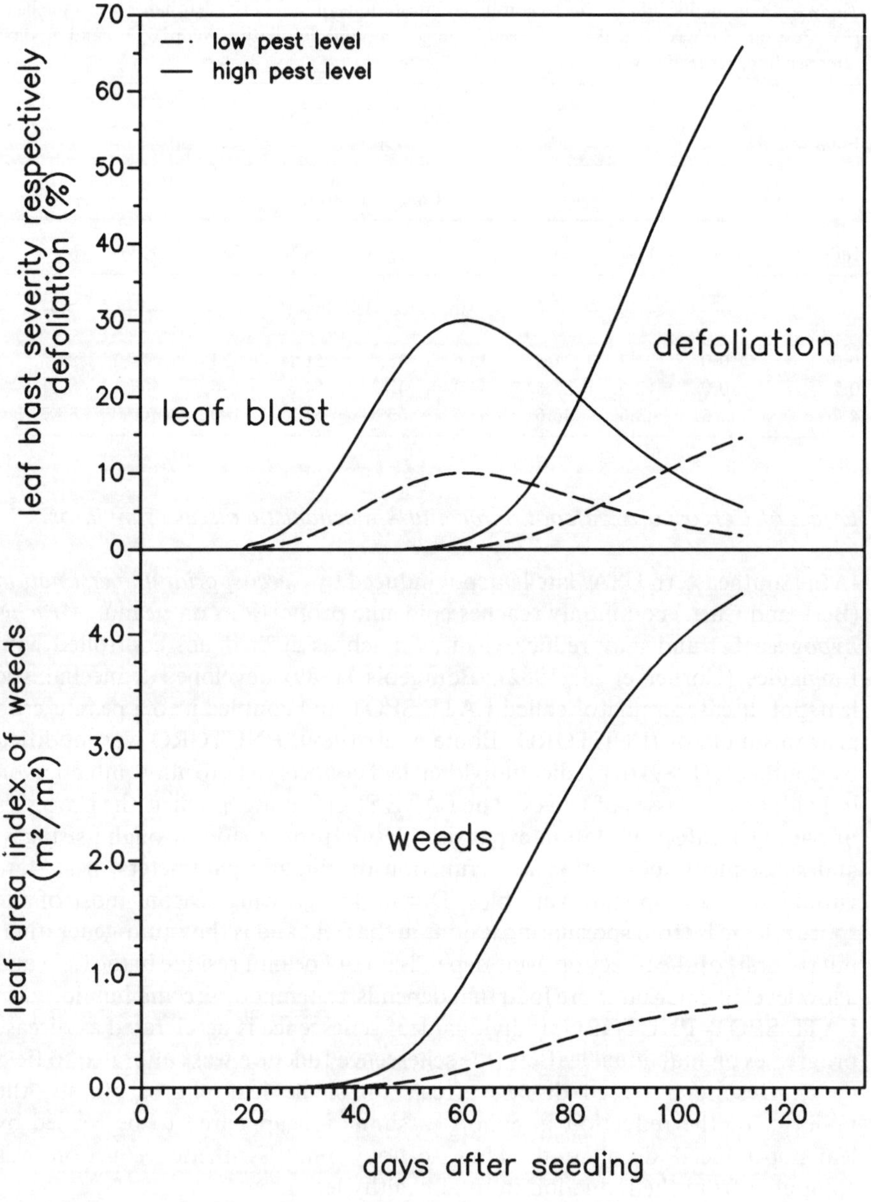

Figure 9. Pest progress curves used to simulate multiple pest effects on rice.

illustrates the resulting pod mass accumulation curves. We do not have a pod abscission routine in the LATE SPOT–PNUTGRO simulator, so simulated yields represent the potential produced and would be reduced with pod abscission as harvest date is delayed.

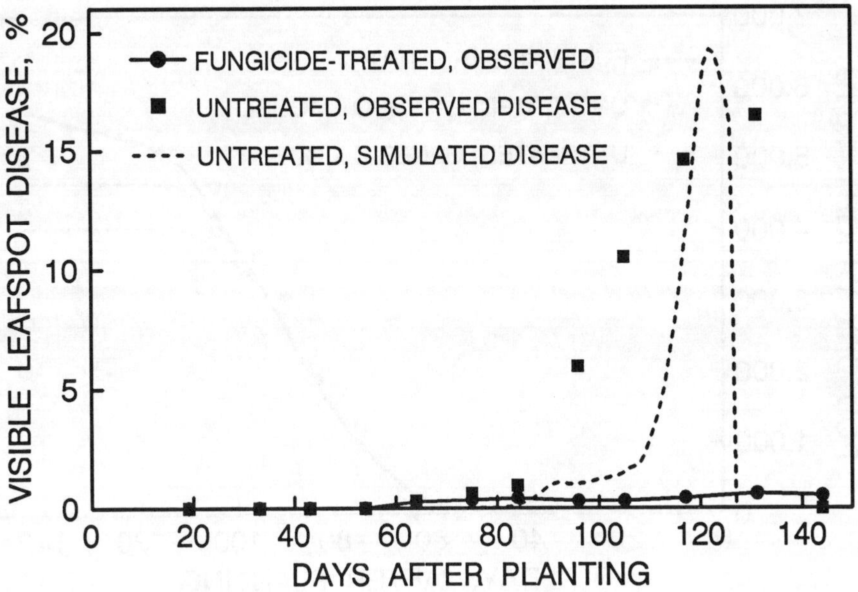

Figure 10. Predicted and observed leafspot disease progress curves on unsprayed Florunner peanut grown at Marianna, Florida, in 1983. Observed disease progress on fungicide-treated crop is also shown, but is < 1%.

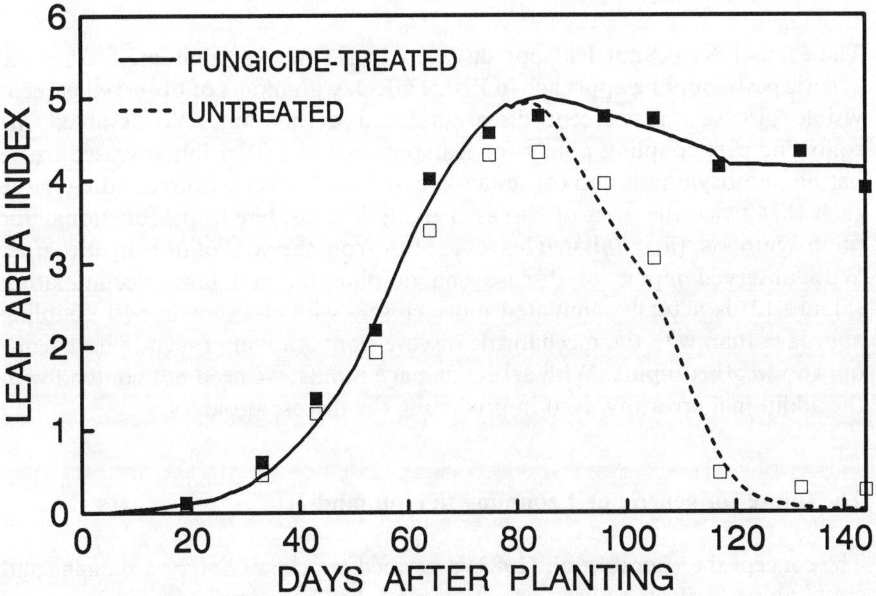

Figure 11. Predicted and observed leaf area index of fungicide-treated and unsprayed Florunner peanut grown at Marianna, Florida, in 1983.

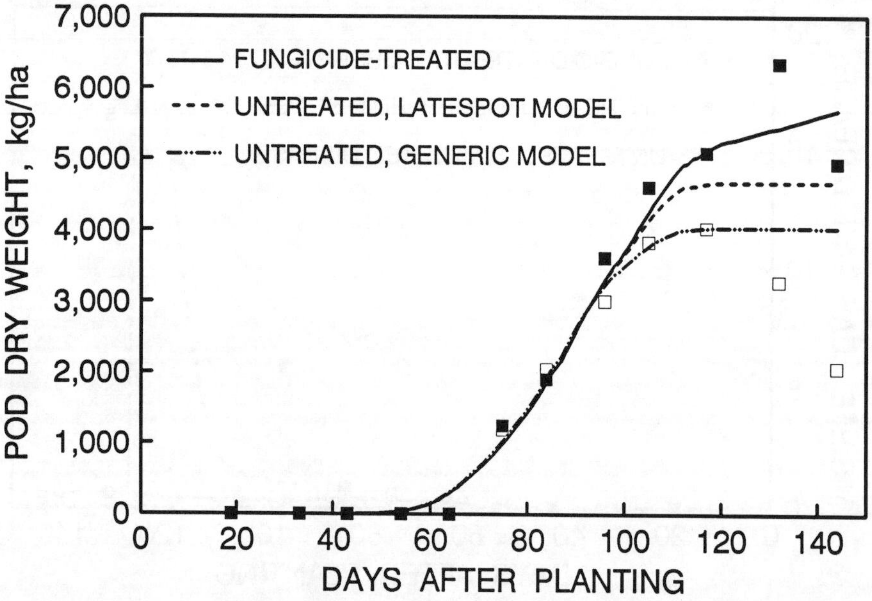

Figure 12. Predicted and observed pod dry matter accumulation for fungicide-treated and unsprayed Florunner peanut grown at Marianna, Florida, in 1983. Additional dashed lines illustrate simulation by the generic pest coupling approach with PNUTGRO.

Generic coupling of leafspot on peanut – input of disease ratings

The same 1983 peanut leafspot disease experiment was simulated using the generic pest-coupling approach in PNUTGRO, with inputs of observed percent visible disease and percent disease-induced defoliation. We assumed the following pest coupling points for leafspot diseases: (i) visibly diseased tissue has no photosynthesis and causes a toxic or 'halo' effect on surrounding tissue such that twice the area of diseased tissue is considered non-functional for photosynthesis, (ii) defoliated leaves are lost from the pool of healthy leaf area. With observed inputs of disease and defoliation, pod mass accumulation (Figure 12) is actually simulated more closely with the generic pest coupling approach than with the mechanistic disease simulation approach that depends only on weather inputs. With actual damage inputs, we need not contend with the additional error involved in predicting the disease progress.

The future for generic pest coupling to crop models

The concept that Boote et al. (1983) described for classifying pest damage and coupling pest effects to crop growth models has been implemented in several IBSNAT crop models and has been shown to be valid and useful. The resulting models can be used for pest loss assessment,for projecting effects of damage or pests on future yield, and for determining sensitivities of crops to timing and intensities of damage in specific environments. The approach appears to be generic since it allows coupling of damage resulting from many pests (even where research data are limited and mechanistic pest simulators are not yet developed for a given pest). The approach is further generic since the pest coupling modules can be easily developed or transferred for other crop models. Lastly, pest models are needed, but they may need to be location dependent because of the range of pests and difficulty in initializing models and in quantifying movement, reproduction, and mortality. Our approach is compatible with future linkage of pest simulators since the coupling points should be the same and the scouting inputs should serve as data for developing pest simulators.

References

Ayres P G (1981). Effects of disease on plant water relations. Pages 131–148 in Ayres P G (Ed.) Effects of disease on the physiology of the growing plant. Cambridge University Press, Cambridge, England.

Bastiaans L (1991). The ratio between virtual and visual lesion size as a measure to describe reduction in leaf photosynthesis of rice due to leaf blast. Phytopathology 81;611–615.

Batchelor W D, McClendon R W, Jones J W, Adams D B (1989a). An expert simulation system for soybean insect pest management. Trans. ASAE 32;335–342.

Batchelor W D, McClendon R W, Adams D B, Jones J W (1989b). Evaluation of SMARTSOY: An expert simulation system for insect pest management. Agric. Syst. 31;67–81.

Batchelor W D, Jones J W, Boote K J, Pinnschmidt H (1991). Extending the use of crop models to study pest damage. Paper No. 91-4502. American Society of Agricultural Engineers., St. Joseph, MI 49085, USA.

Boote K J, Jones J W, Hoogenboom G, Wilkerson G G, Jagtap S S (1989). PNUTGRO V1.02, Peanut Crop Growth Simulation Model, User's Guide. Fl. Agric. Exp. Sta., Journal No. 8420. Univ. of Florida, Gainesville, USA. 76 p.

Boote K J, Jones J W, Mishoe J W, Berger R D (1983). Coupling pests to crop growth simulators to predict yield reductions. Phytopathology 73;1581–1587.

Bourgeois G (1989). Interrelationships between late leafspot disease and Florunner peanut: A modeling approach. Ph.D. dissertation, Univ. of Florida, Gainesville, USA. 219 p.

Buchanan B B, Hutcheson S W, Magyarosy A C, Montalbini P (1981). Photosynthesis in healthy and diseased plants. Pages 13–28 in Ayres P G (Ed.) Effects of disease on the physiology of the growing plant. Cambridge University Press, Cambridge, England.

Gay J L, Manners J M (1981). Transport of host assimilates to the pathogen. Pages 85–100 in Ayres P G (Ed.) Effects of disease on the physiology of the growing plant. Cambridge University Press, Cambridge, England.

Gorbet D W, Shokes F M , Jackson L F (1982). Control of peanut leafspot with a combination of resistance and fungicide treatment. Peanut Sci. 9;87–90.

Hornby D, Fitt B D L (1981). Effects of root-infecting fungi on structure and function of cereal roots. Pages 101–130 in Ayres PG (Ed.) Effects of disease on the physiology of the growing plant. Cambridge University Press, Cambridge, England.

Ingram K T, Herzog D C, Boote K J, Jones J W, Barfield C S (1981). Effects of defoliating pests on soybean canopy CO_2 exchange and reproductive growth. Crop Sci. 21;961–968.

Kropff, M J (1988). Modelling the effects of weeds on crop production. Weed Res. 28;465–471.

Magyarosy A C, Buchanan B B (1975). Effect of bacterial infiltration on photosynthesis of bean leaves. Phytopathology 65;777–780.

Magyarosy A C, Malkin R (1978). Effect of powdery mildew infection of sugar beet on the content of electron carriers in chloroplasts. Physiol. Plant Pathol. 13;183–188.

Pinnschmidt H, Teng P S, Yuen J E (1990). Pest effects on crop growth and yield. Pages 26–29 in Teng P, Yuen J (Eds.) Proc.workshop on modeling pest-crop interactions. Univ. of Hawaii at Manoa, Honolulu, HI, January 7–10, 1990. IBSNAT project, Univ. of Hawaii, Honolulu, USA.

Pixley K V, Boote K J, Shokes F M, Gorbet D W (1990). Disease progression and leaf area dynamics of four peanut genotypes differing in resistance to late leafspot. Crop Sci. 30;789–796.

Poston F L, Pedigo L P, Pearce R B, Hammond R B (1976). Effects of artificial and insect defoliation on soybean net photosynthesis. J. Econ. Entomol. 69;109–111.

Rabbinge R, Rijsdijk F H (1981). Disease and crop physiology: A modeler's point of view. Pages 201–220 in Ayres P G (Ed.) Effects of disease on the physiology of the growing plant. Cambridge University Press, Cambridge, England.

Rubia E G, Penning de Vries F W T (1990). Simulation of yield reduction caused by stem borers in rice. J. Plant Prot. Trop. 7;87–102.

Rudd W G (1980). Simulation of insect damage to soybeans. Pages 547–555 in Corbin F T (Ed.) Proceedings World Soybean Research Conference II, Raleigh, NC., Westview Press, Boulder, CO., USA.

Stanton M A (1986). Effects of root-knot nematodes (*Meloidogyne* spp.). on growth and yield of 'Cobb' soybean (*Glycine max* (L.) Merrill). M.S. Thesis, Univ. of Florida, Gainesville, USA. 120 p.

Spitters C J T, Aerts R (1983). Simulation of competition for light and water in crop-weed associations. Aspects Applied Biol. 4;467–483.

Teng P S, Calvero S, Pinnschmidt H (1990). Simulation of rice pathosystems and disease losses. Paper presented at the IRRI Thursday Seminar Simulation Series, Dec. 6, 1990. IRRI, Los Baños, Philippines.

Teng P S, Klein-Gebbinck H, Pinnschmidt H (1991). An analysis of the blast pathosystem to guide modeling and forecasting. Pages 1–31 in IRRI: Selected papers from the International Rice Research Conference, 27–31 Aug. 1990, Seoul, Korea. IRRI, Los Baños, Laguna, Philippines.

Trudgill D L, Evans K, Parrott D M (1975). Effects of potato cyst nematodes on potato plants. II. Effects on haulm size, concentration of nutrients in haulm tissue and tuber yield of a nematode resistant and a nematode susceptible potato variety. Nematologia 21;183–191.

Wilkerson G G, Jones J W, Boote K J, Ingram K T, Mishoe J W (1983). Modeling soybean growth for crop management. Trans. of ASAE 26;63–73.

Wilkerson G G, Jones J W, Coble H D, Gunsolus J L (1990). SOYWEED: A simulation model of soybean and common cocklebur growth and competition. Agron. J. 82;1003–1010.

Quantification of components contributing to rate-reducing resistance in a plant virus pathosystem*

F. W. NUTTER, Jr.
Department of Plant Pathology, Iowa State University, Ames, IA. 50011, U.S.A.

Key words: aphid, breeding, Integrated Pest Management, pepper, rate reducing resistance, resistance components, tobacco etch virus, virus epidemiology, weed hosts

Abstract
In fungal pathosystems, it has been shown that rate-reducing resistance can be separated into components to mechanistically explain why epidemics are fast or slow, as affected by host genotype. Components of resistance to plant viruses may be quantified in a similar fashion but with several subcomponents added to account for alternative weed hosts and/or infected volunteers that are of primary importance to initiate epidemics and for vectors which are essential for transmission. Epidemiological principles and operational definitions were developed and utilized to identify and quantify components contributing to rate- reducing resistance in the*Capsicum annuum* L./weed hosts/tobacco etch virus (TEV)/aphid pathosystem. Pepper genotypes which could be infected with TEV, but delayed symptom appearance and slowed the rate of virus antigen accumulation over time, were also found to have a rate-reducing affect on TEV epidemics in the field compared to fully susceptible genotypes. Rate-reducing genotypes were found to have lower receptivities, represent smaller lesions (epidemiologically), had longer latent periods, and produced fewer viruliferous aphids over time, compared to susceptible pepper genotypes. Insect vector(s) and weed hosts were found to be important subcomponents of the pathosystem and these have also been quantified and related to the development of TEV epidemics in time and space. Substantial yield benefits were realized when rate-reducing resistance was deployed to control TEV. The use of resistant cultivars to reduce the apparent infection rate should increase the effectiveness of disease control tactics that reduce initial inoculum.

Introduction

Resistance that reduces the rate of plant disease epidemics caused by fungi has been shown to play an important role in Integrated Pest Management (IPM) programs, but this strategy has not been fully utilized as a means to control plant virus epidemics (Evered and Harnett 1987; Ponz and Bruening 1986). Slow mildewing and slow rusting are terms that were coined to describe host genotypes that possess this form of resistance to fungal pathogens (Kuhn et al. 1978; Shaner 1973). An analogous type of resistance related to plant viruses would be found in hosts which restrict virus replication and/or translocation within plants (Debokx et al. 1978; Gray et al. 1986; Kegler and Meyer 1987; Kuhn et al. 1989). Numerous examples of this type of virus restriction are known, but have rarely been studied with regard to their epidemiological effects

* Journal Paper No. J-14826 of the Iowa Agriculture and Home Economics Experiment station, Ames, Iowa.

F. W. T. Penning de Vries et al. (eds.), Systems Approaches for Agricultural Development, 297–308.

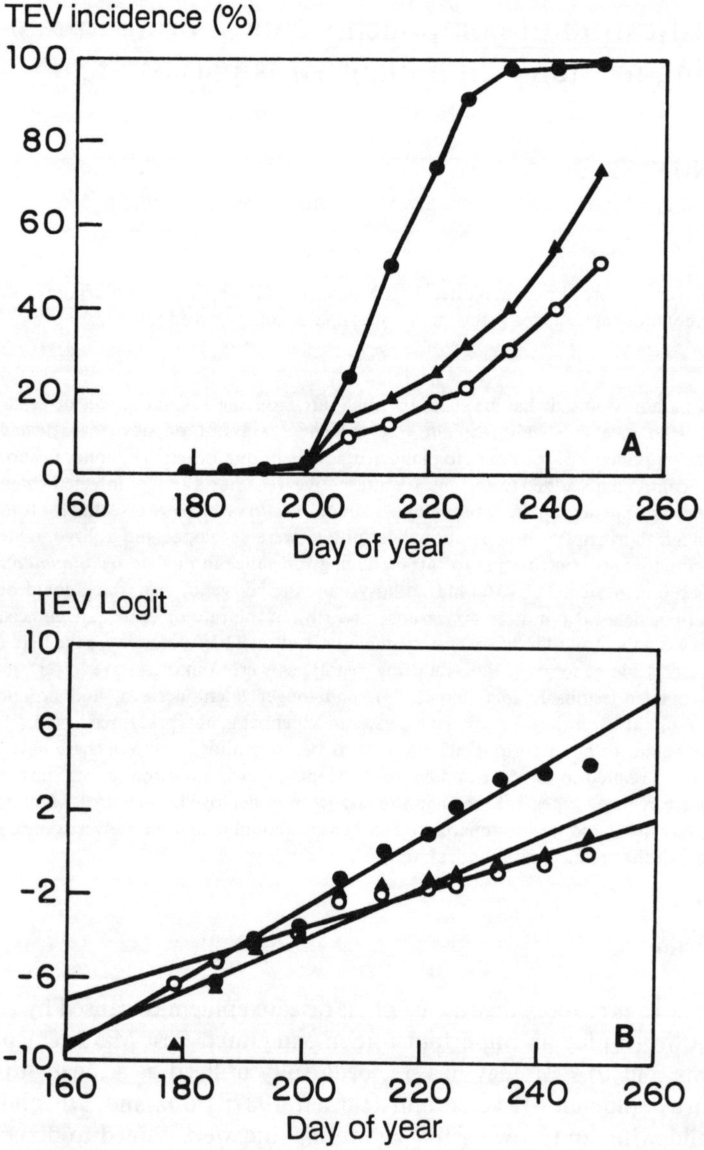

Figure 1. A. Tobacco etch virus (TEV) disease progress curves for Yolo Wonder B, Tambel 2, and Asgrow XPH-5021 during the 1986 growing season. B. Logit lines from TEV disease progress curves in 1986.

in the field. It is suspected that virus restriction is controlled polygenically. Polygenic resistance has several advantages over the monogenically controlled resistance that is currently being utilized in breeding programs as a strategy to control viral diseases. Monogenic-based resistance has been shown to 'break down' or fail when new viral (Wyatt and Kuhn 1980) or fungal strains develop

(Zadoks and Schein 1979). However, polygenic resistance in fungal pathosystems has been shown to remain stable and effective against all pathogen races or strains (Parlevliet 1979; Zadoks and Schein 1979).

Tobacco etch virus (TEV), a member of the potyvirus group which is transmitted nonpersistently by aphids, causes severe epidemics in bell pepper (*Capsicum annuum* L.) throughout the southeastern United States (Benner et al. 1985; Nutter et al. 1989; Padgett et al. 1990) as well as in Arizona, California, Texas and Mexico (Laird et al. 1964; Villalon 1985). In North Carolina, disease incidence is reported to approach 75% each season (Main and Gurtz 1988). Since 1984, detailed surveys have shown that TEV incidence exceeds 90% annually by the time of harvest in all pepper fields in Northeast Georgia (Benner et al. 1985; Nutter et al. 1989.). Early plant infection reduces fruit set as well as fruit size and weight, whereas late season infection has little effect on fruit set and fruit weight, and losses are usually less than 5% (Nutter et al. 1989). Thus, management strategies that delay TEV infection have the potential to greatly limit yield losses experienced by growers.

Resistance that reduces the rate of plant virus disease epidemics

Kuhn et al. (1989) identified multiple levels of resistance to TEV in bell pepper genotypes. Two genotypes, Tambel 2 (Villalon 1985) and Asgrow XPH-5021, were shown to have a moderate level of resistance, as indicated by a delay in symptom appearance and a slower rate of virus antigen accumulation compared to the susceptible genotype Yolo Wonder B. Although both of these moderately resistant genotypes could be infected and virus antigen levels eventually reached levels comparable to susceptible pepper genotypes, it was hypothesized that pepper genotypes with moderate levels of resistance would reduce the rate of TEV epidemics at the population level (in the field). To test this hypothesis, the effect of host resistance on the development of TEV epidemics was monitored in three pepper genotypes (Padgett et al. 1990). Rates of disease progress were reduced by half in the genotypes possessing moderate levels of resistance (Tambel 2 and Asgrow XPH-5021) compared to susceptible Yolo Wonder B (Fig. 1). Tambel 2 and Asgrow delayed the time to reach 50% disease incidence by 25 and 34 days, respectively, compared to Yolo Wonder B. This delay resulted in pepper fruit yields that were 24% higher in the rate-reducing varieties compared to Yolo Wonder B (Padgett el al 1990).

Once methods to quantify rate-reducing pepper genotypes in the field were developed, the next goal was to identify and quantify resistance components contributing to the rate-reducing phenotype. In fungal pathosystems, Parlevliet (1979) showed that rate-reducing resistance can be separated into components to mechanistically explain why epidemics are fast or slow as affected by host genotype. Components of resistance to viruses could be quantified in a similar fashion, but with several subcomponents added to account for vectors which are essential for transmission. Therefore, the next objective was to develop and utilize epidemiological principles and operational definitions to identify and quantify rate-reducing components in the TEV/pepper pathosystem.

Operational definitions for quantifying components of resistance contributing to rate-reducing resistance

Receptivity

Receptivity can be defined as the proportion of plants in a population that become infected when host genotypes are challenged using a defined inoculation technique. Plant genotypes may differ in the amount of infection that occurs even though each genotype receives the same amount of inoculum. Relative receptivities among host genotypes can be determined once an acceptable inoculation method has been developed. When dilute sap inoculum was used in the TEV/pepper system, mechanical inoculation revealed differences between the susceptible and moderately resistant genotypes (Kuhn et al. 1989) (Table 1). Furthermore, similar significant differences in receptivity among these same genotypes were detected when single viruliferous aphids were used to inoculate test plants (Table 1). In our system, test plants were assayed by ELIZA (Clark and Adams 1977) approximately 21 days after aphid inoculations were performed.

Table 1. Receptivity (%) of three bell pepper genotypes mechanically and aphid inoculated with tobacco etch virus.

Genotype	Mechanical	Aphid
Yolo Wonder B	100 a[z]	78 a
Tambel 2	67 b	43 b
Asgrow XPH-5021	60 b	45 b

[z] Means followed by the same letter within each column are not significantly different using the Waller-Duncan K-ratio test ($P<0.05$). Receptivity is the percentage of plants within a genotype, that become infected after plants are mechanically innoculated (mechanical) or exposed to a single viruliferous aphid (aphid).

Lesion size

Assessment of disease incidence is the most common method used to quantify disease intensity in plant pathosystems involving systemically infected plants. Plants are judged as being either infected (+) or not infected (-). When a population of plants is assessed in this manner, the incidence of diseased plants is obtained by counting the number of infected plants and then dividing by the total number of plants assayed or observed. Disease assessments in plant virus systems are usually qualitative (+ or –) and it is assumed that plant virus is present systemically throughout the plant (roots, stems, leaves, fruit, etc.). For this reason, a systemically infected plant is considered to be, epidemiologically, a single lesion. The concept of lesion size has not been utilized in plant virus pathosystems as a means to select for plant genotypes that reduce the rate of virus spread in the field.

The concept of individual, systemically infected plants being considered a single lesion can be logically extended to provide a quantitative measurement with regard to lesion size. For example, there are six weed species in Georgia that are known to serve as alternative hosts for TEV, but these six weed species vary greatly in size. For example, a Jimsonweed plant (*Datura stramonium* L.) systemically infected by TEV, would represent a larger lesion (epidemiologically) than an infected annual ground cherry plant (*Physalis angulata* L.) because of the physical difference in size of the plants. The leaf area available for aphids to acquire the virus is much greater with Jimsonweed assuming that aphids are equally attracted to both weed species (which they are) and that the virus titre is similar in both species (also true).

The concept of lesion size can also be applied in cases where a plant genotype, through some resistance mechanism, slows or impedes virus movement within the plant. This would be analogous to a slower rate of lesion expansion as occurs with fungal pathogens that cause leaf spots or blotches. A host genotype that restricts virus movement would result in a slower rate of 'lesion expansion' and therefore would represent (epidemiologically-speaking) a smaller lesion relative to susceptible genotypes in which virus movement is not impeded. In the TEV/bell pepper pathosystem, lesion size was quantified by sampling single leaves from several branch terminals of a pepper plant over time and testing these by enzyme-linked immunosorbent assay (ELIZA). The number of branch terminals that were positive for TEV divided by the number of branch terminals tested/plant gave a quantitative measurement in regard to the leaf area (lesion size) available for virus acquisition by aphids. Using this operational definition for lesion size, the proportion of terminals positive for TEV in the resistant genotype Tambel 2 was 0.56 compared to 0.96 for the susceptible genotype Yolo Wonder B. The lesion size of Tambel 2 relative to Yolo Wonder B was $0.56/0.96 = 0.58$ and represents the probability of aphids acquiring TEV from Tambel 2 relative to Yolo Wonder B.

Latent period

A general definition for latent period is the time from infection to the time an infected plant (or lesion) produces new infection units (such as spores or viruliferous aphids). This definition is not very practical because the precise time infection occurs is difficult, if not impossible, to measure. A more practical definition for measuring the latent period of plant viruses (in the host) would be the time from inoculation to the time infection efficiency exceeds 10% using the appropriate virus vector. For our experiments, 10% represents a percentage that was three times the percentage of noninoculated (control)

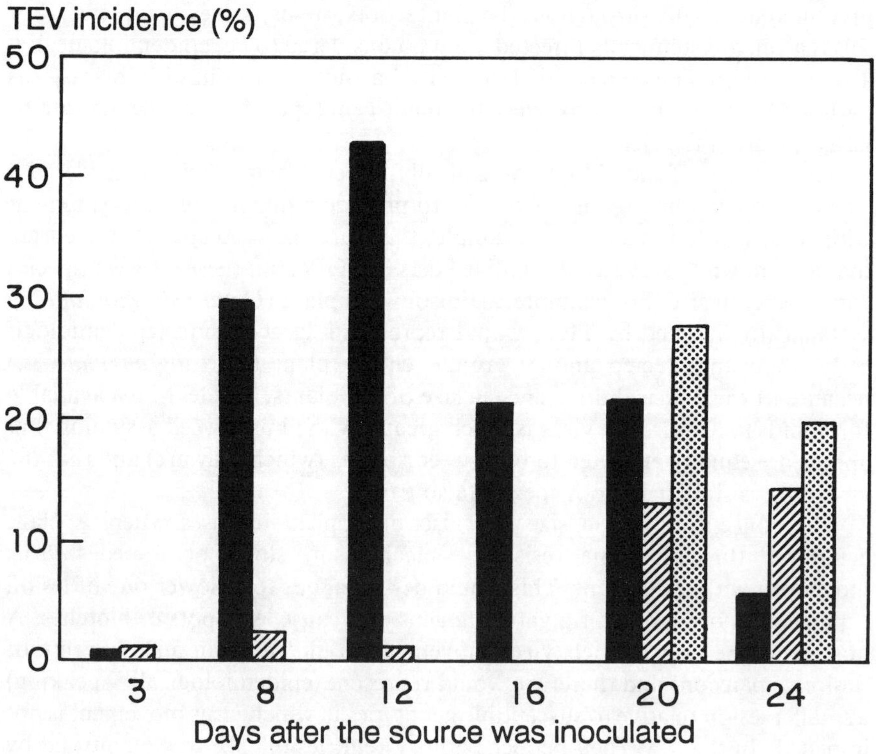

Figure 2. Latent period of three pepper genotypes infected with tobacco etch virus (TEV) as determined by TEV incidence in susceptible target plants. Individual aphids were observed to probe on inoculated source genotypes (Yolo Wonder, Tambel 2, Asgrow) before transferring single aphids to susceptible target plants 3, 8, 12, 16, 20, and 24 days after source genotypes were inoculated mechanically with TEV.

plants that became infected with TEV during the course of the greenhouse experiments. Such a method was developed and used in the TEV/bell pepper pathosystem to quantify the length of the latent period for TEV in bell pepper genotypes. Aphids were starved for 30 minutes and then allowed a specified period (120 seconds) to probe a host genotype (source plant) that had already been mechanically inoculated. After probing, single aphids were placed on healthy target plants susceptible to TEV and allowed to feed for eight to ten hours. Infection of target plants would only occur if the mechanically-inoculated source plants were infectious (i.e., aphids could acquire TEV). This process was repeated every three to four days after source plants were mechanically inoculated to determine when Yolo Wonder B, Tambel 2, and Asgrow could serve as efficient sources for virus acquisition by aphids. Infection efficiency, the number of infected target plants divided by the total number of plants receiving single aphid transfers, was determined three weeks after each date that aphids were transferred from source plants to target plants.

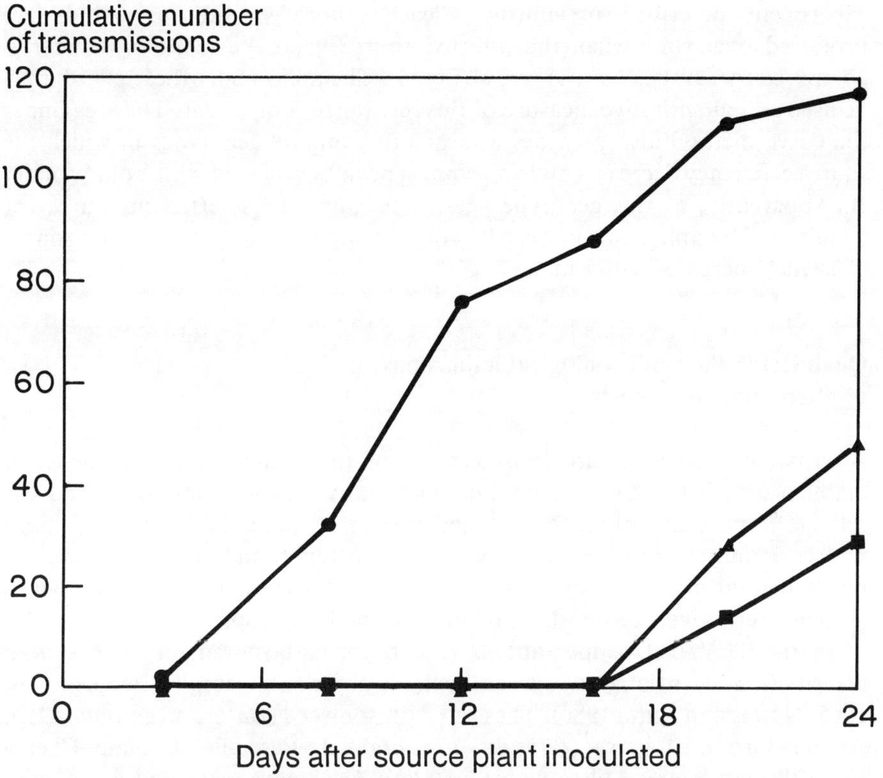

Figure 3. Cumulative number of successful aphid transmissions of tobacco etch virus (TEV) as affected by host genotype. Data are derived from cumulative TEV incidence values for Yolo Wonder B, Tambel 2, and Asgrow as shown in Figure 2.

Latent period was defined as the time from mechanical inoculation of the source genotype to the time infection efficiency (using aphids) exceeded 10%. Based on this operational definition, the latent period for Yolo Wonder B was greater than three days but less than eight days whereas the latent periods for both Tambel 2 and Asgrow were greater than 16 days but less than 20 days (Fig. 2). Therefore, Yolo Wonder B can serve as a potential source for secondary spread of TEV much sooner than infected Tambel 2 and Asgrow plants.

Capacity to produce viruliferous aphids

This component is analogous to the number of spores produced per lesion (over time) in fungal pathosystems. In the TEV/pepper pathosystem, we are interested in the relative number of viruliferous aphids being produced over time as affected by a virus-infected host genotype. We can obtain a quantitative measure of this component by plotting the cumulative number of successful TEV transmissions over time using data obtained from the latent period experiment described previously. Clearly, more viruliferous aphids are produced over time when the infected source genotype is Yolo Wonder B compared to Tambel 2 and Asgrow (Fig. 3). The areas under these curves may be used as a quantitative measure of this resistance component. The area under the curve method also provides a means to compare genotypes in which the source efficiency decreases after reaching a peak as occurred with Yolo Wonder B. Apparently, as this genotype ages, it becomes a less efficient source for acquisition by aphids, as opposed to other pepper genotypes in which source efficiency increases with time.

Quantifying the epidemiological importance of alternative weed hosts

Alternative weed hosts are considered to be important sub- components of many plant virus pathosystems but quantitative information regarding the relative importance of different weed species as potential sources of plant viruses is largely lacking. Since weeds are often found within, as well as bordering agronomic crops, it is important to know the relative receptivities among alternative weed hosts in relation to the host crop.

In the TEV/bell pepper/aphid/weed hosts pathosystem, TEV has been identified in six solanaceous weed hosts in northeast Georgia (Benner et al. 1985; Nutter and Kuhn 1989). These are: Jimsonweed (*Datura stramonium* L.), Annual Ground Cherry (*Physalis angulata* L.), Perennial Ground Cherry (*Physalis virginiana* Mill), Apple of Peru (*Nicandra physalodes* L.), and Horsenettle (*Solanum carolinense* L.). Of these six species, only Horsenettle and Perennial Ground Cherry are perennial species, the other four are annuals. The relative importance of these six species was determined by comparing receptivities and by comparing relative efficiency of these weeds as sources for aphids to acquire TEV.

Relative receptivities of alternative weed hosts

Relative receptivities for the six solanceous weeds plus bell pepper were compared by determining the number of viruliferous aphids required (per plant) to obtain a 50% probability of infection by TEV. In other words, the less receptive a weed or host genotype is, the more viruliferous aphids needed per plant to obtain a 50% probability of plant infection. Using this operational definition for receptivity, Annual Ground Cherry was the most receptive weed species while Black Nightshade, Horsenettle and Perennial Ground Cherry were the least receptive (Table 2). Perennial Ground Cherry and Horsenettle remain important epidemiologically because, once infected, TEV resides within these alternative hosts during intercrop periods. More that 90% of the Perennial Ground Cherry plants sampled in the field tested positive for TEV (using ELIZA) at the time peppers were being transplanted to the field. TEV

Table 2. Number of viruliferous aphids required per plant to achieve a 50% probability of infection by Tobacco Etch Virus.

Plant species	Aphid number[z]
Annual Ground Cherry	1.0
Jimsonweed	1.3
Apple of Peru	3.3
Pepper	>4.0
Black Nightshade	>6.0
Horsnettle (Perennial)	>6.0
Perennial Ground Cherry	>6.0

[z] Aphids were allowed a 30 sec acquisition period on pepper (cv Yolo Wonder B) and were then transferred to individual test plants. Test plants received 1, 2, 3, 4, or 6 viruliferous aphids per plant.

incidence in Horsenettle ranged from 40 to 70% over a 5 year period (Nutter and Kuhn 1989).

Relative source efficiences of alternative weed hosts

The relative source efficiencies of weed host genotypes and bell pepper were determined in greenhouse experiments. Each weed species (and pepper cv Yolo Wonder B) were inoculated mechanically. Each inoculated source plant represented a replication and served as a source for 20 Yolo Wonder B (target) plants. There were four replications per source genotype. A single aphid was placed on each target plant after aphids were allowed a standard acquisition period on a TEV-infected source plant. Aphid inoculations of target plants were conducted 4, 8, 12, 16, and 20 days after the source plants were inoculated mechanically. Target plants were observed visually for TEV symptoms and the presence of TEV in target plants was confirmed by ELIZA. Since target plants (Yolo Wonder B) were fully suceptible, differences in disease incidence among target plants was related to the efficiency with which aphids could acquire TEV from the source genotypes. Source efficiencies for weed hosts were determined relative to pepper cv. Yolo Wonder B (Fig. 4). Source efficiencies were higher for Jimsonweed and Annual Ground Cherry relative to pepper. This would indicate that these two annual species, when present within pepper fields, may contribute to a higher rate of within field spread of the virus relative to infected pepper plants. Source efficiency of Apple of Peru was similar to that of pepper while Perennial Ground Cherry and Horsenettle were less efficient sources than pepper. Black nightshade could not be tested because only a small percentage of this population could be infected either mechanically or with aphids.

Summary

Traditionally, plant breeders have adopted the approach of selecting plants that yield well after exposure to pathogen infection. Selected genotypes are often

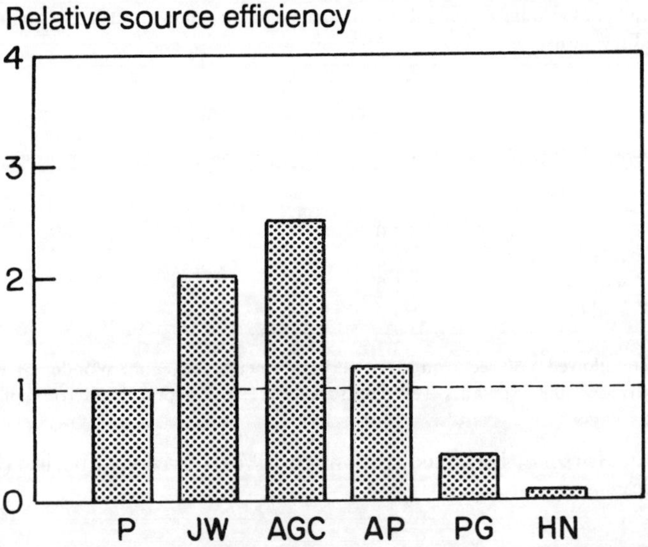

Figure 4. Relative source efficiency of five solanaceous weed hosts for acquisition of tobacco etch virus by aphids relative to bell pepper cv. Yolo Wonder B. P = Bell Pepper, JW = Jimsonweed, AGC = Annual Ground Cherry, AP = Apple of Peru, PG = Perennial Ground Cherry, HN = Horsenettle.

referred to simply as 'resistant' with no regard to the underlying biochemical mechanisms of resistance. Furthermore, only limited attention has been given to the epidemiological consequence of combining virus-resistant cultivars with other disease control tactics. Resistance to viruses has been classified into several categories based upon the response of plant genotypes exposed to infection in laboratory – greenhouse studies. These categories are:

(i) extreme resistance,
(ii) partial resistance to infection,
(iii) resistance to the spread of the virus within the host plant,
(iv) resistance to virus replication within the plant.

These resistance categories have been discussed in a 1986 review article (Ponz and Bruening 1986) and in a 1987 symposium (Evered and Harnett 1987). The epidemiological significance of these types of resistance has been described qualitatively for several virus pathosystems but the effect of these types of resistance, individually and in combination, on reducing the rate of plant-to-plant spread in host populations in the field has rarely been studied.

Epidemiological components that contribute to rate reducing resistance in the TEV pathosystem can be defined as follows: resistance to infection (reduced receptivity or efficiency), resistance to spread of the virus within the host (lesion size), resistance that increases the time from inoculation to the time the host can produce viruliferous aphids (latent period) and the capacity of the host to produce viruliferous aphids over time. A simulation model to test different scenarios regarding the elimination of TEV weed hosts reservoirs, a reduction in aphid vectors and/or the deployment of pepper genotypes differing in the expression of components contributing to rate-reducing resistance is nearly complete. Early simulations have shown that plant genotypes with reduced receptivities to TEV can greatly reduce the rate with which plants become infected in the field. This is particularly true when the primary source of virus inoculum is from weed hosts which border the field as opposed to infected weed (or host) sources within the field. An influx of viruliferous aphids would result in fewer successful plant infections over time compared to highly receptive host genotypes, such as Yolo Wonder B. If infected weed and/or host plants within the field are the primary source of virus inoculum, then genotypes with longer latent periods will have a greater impact on reducing the rate of infection.

The use of rate-reducing genotypes is consistent with the IPM concept that some damage to individual plants in a population may be economically acceptable (Zadoks and Schein 1979). The development and use of plant virus simulation models will provide a better understanding of the epidemiological effects that resistance components have on reducing the rate of plant virus epidemics. This will provide plant breeders with better screening methods to develop resistant cultivars and will provide a knowledge base as to how to best integrate resistance with other disease control tactics.

References

Benner C P, Kuhn C W, Demski J W, Dobson J W, Colditz P, Nutter F W Jr. (1985) Identification and incidence of pepper viruses in northeastern Georgia. Plant Dis. 69;999–1001.

Clark M F, Adams A N (1977) Characteristics of the microplate methods of enzyme-linked immunosorbent assay for the detection of plant viruses. J. Gen. Virol. 34;475–483.

DeBokx J A, Van Hoof H A, Pirone P G M (1978) Relation between concentration of potato virus Y and its availability to *Myzus persicae*. Neth. J. Plant Pathol. 84;95–100.

Evered D, Harnett S (Eds.) (1987) Plant Resistance to Viruses. Ciba Foundation Symposium 133. Wiley and Sons, Chichester, UK. 215 p.

Gray S M, Moyer J W, Kennedy G G, Campbell C L (1986) Virus-suppression and aphid resistance effects on spatial and temporal spread of watermelon mosaic virus 2. Phytopathology 76;1254–1259.

Kegler H, Meyer U (1987) Characterization and evaluation of quantitative virus resistance in plants. Arch. Phytopathol. Pflanzenschutz 5;343–348.

Kuhn C W, Nutter F W Jr., Padgett G B (1989) Multiple levels of resistance to tobacco etch virus in pepper. Phytopathology 79;814-818.

Kuhn R C, Ohm H W, Shaner G E (1978) Slow leaf-rusting resistance in wheat against twenty-two isolates of *Puccinia recondita*. Phytopathology 68;651–656.

Laird E F, Jr., Desjardins P R, Dickson R C (1964) Tobacco etch virus and potato virus Y from pepper in southern California. Plant Dis. Rep. 48;772–776.

Main C E, Gurtz S K (Eds.) (1988) Estimates of Crop Losses in North Carolina Due to Plant Diseases and Nematodes. Department of Plant Pathology Special Publication No. 7, North Carolina State University, Raleigh, NC., USA. 209 p.

Nutter F W, Jr., Kuhn C W (1989) Epidemiological importance of six solanaceous weed hosts in the tobacco etch virus/bell pepper pathosystem. Phytopathology 79;375.

Nutter F W, Jr., Kuhn C W, All J N (1989) Models to estimate yield losses in bell pepper caused by tobacco etch virus epidemics. Phytopathology 79;1213.

Padgett G B, Nutter F W Jr., Kuhn C W, All J N (1990) Quantification of disease resistance that reduces the rate of tobacco etch virus epidemics in bell pepper. Phytopathology 80;451-455.

Parlevliet J E (1979) Components of resistance that reduce the rate of epidemic development. Annu. Rev. Phytopathol. 17;203–222.

Ponz F, Bruening G 1986 Mechanisms of resistance to plant viruses. Annu. Rev. Phytopathol. 24;355- 381.

Shaner G (1973) Evaluation of slow-mildewing resistance of Knox wheat in the field. Phytopathology 63;867–872.

Villalon B (1985) 'Tambel-2'- a new multiple virus resistant bell pepper. Texas Agric. Exp. Stn. Leaflet L-2172. Texas, USA.

Wyatt S D, Kuhn C W (1980) Derivation of a new strain of cowpea chlorotic mottle virus from resistant cowpeas. J. Gen. Virol. 49;289–296.

Zadoks J C, Schein R D (1979) Epidemiology and Plant Management. Oxford University Press, New York, USA. 427 p.

The rice leaf blast simulation model EPIBLAST

CHANG KYU KIM AND CHOONG HOE KIM

Agricultural Sciences Institute, Rural Development Administration, Suweon 441–707, Korea

Key words: disease forecast, EPIBLAST, inoculum, Pyricularia orizae, rainfall, spore trap, sporulation, temperature, yield loss

Abstract

Rice blast, caused by *Pyricularia oryzae*, is one of the biggest constraints to stable rice production in most of the temperate regions and some tropical and subtropical countries. To minimize yield loss by blast disease, large efforts have been made to solve the blast problem. However, it is very difficult to eliminate blast from fields due to the complexity of the life cycle of *P. oryzae*. In Korea, we experimented and collected field data of the blast fungus to study in epidemiology and developed a leaf blast simulation model EPIBLAST. It is devised for quantitative forecasting of the incidence of the leaf blast disease. In EPIBLAST, temperature, relative humidity, rainfall, dew period, and wind velocity are the meteorological input variables. Healthy, diseased and dead leaf area are plant physiological state variables. Inoculum potential, sporulation, conidia release and dispersal, penetration, and incubation period are by epidemiological processes. The accuracy of EPIBLAST predictions was field tested during the 1991 cropping season. EPIBLAST predicts peak of the leaf blast epidemic as early as middle of July. It can also predict similar disease progress patterns with direct observation, but some fluctuations were observed due to sensitivity of EPIBLAST to minute weather changes. It needs further revision.

Introduction

Rice, as the most widely cultivated food crop in the world, has a very important role in supporting a rapid increasing world population. Rice is grown within a broad latitude band, from as far north as Hokkaido in Japan to as far south as Australia (Brady 1976). As rice production increases with ever more intensive agricultural practices, threats such as diseases and insects as well as unfavourable environmental conditions that affect rice productivity become more and more important.

Rice blast disease, caused by *Pyricularia oryzae*, is of major economic importance in the rice-growing world. It has been reported to occur in some 60 countries (Parthasarathy and Ou 1965). Yield losses due to blast are tremendous in different regions and countries, but exact figures for yield losses due to blast are few. In 1953, an epidemic year, the loss in Japan was about 0.693 Mt; from 1953 to 1960, annual losses varied from 1.4 to 7.3%, with an average of 3% (Goto 1965). Yield losses in the Philippines in 1962 and 1963 were estimated to reach more than 90% in some areas, with 50–60% losses occurring in Bicol and Leyte Provinces (Nuque et al. 1979). Outbreaks in the Philippines in 1969 and 1970 also resulted in considerable yield losses (Nuque et al. 1979). In Korea, yield losses were reported to be 4.2% in 1978 and 3.9% in

F.W.T. Penning de Vries et al. (eds.), Systems Approaches for Agricultural Development, 309–321.

Table 1. Sporulation of *Pyricularia oryzae* from natural leaf blast lesions in Icheon, Korea, 1991.

Date	Lesion size (mm)	Lesion type	No. released[1]	No. remained[2]
July 5	6.2	Acute	360	8720
9	8.9	Chronic	3400	24400
	6.7	Acute	4880	48080
10	8.8	Chronic	14800	205700
	8.3	Chronic	2080	48760
11	15.9	Chronic	2620	58860
	4.4	Acute	300	1620
12	14.2	Chronic	5700	19520
	6.4	Acute	1940	13520
24	28.0	Chronic	200	140

[1] Total number of conidia collected by the KY-type spore trap (Kim and Yoshino 1987).
[2] Total number of conidia remained on both sides of lesion. Remaining conidia are thoroughly scraped with a 1 ml hypodermic syringe needle, and counted using a hemacytometer.
Crop growth stages during the experiment were maximum tillering to panicle initiation and the period was rainy season. Rainfall amounts were: 5, 9 and 12 July: 0 mm; 10 July: 0.5; 11 July: 1.4 and 24 July: 15.5 mm.

1980 (Rural Development Administration 1989).

One characteristic of *P. oryzae* is its sensitivity to minute temperature changes, despite the wide temperature range from 8 to 36 °C that can occur during its life cycle in different regions. Rice plants are also very sensitive to temperature changes. As a result, blast epidemics have become a serious problem in temperate countries such as Korea and Japan, especially in years when cold damage is prevalent. Blast epidemics also have been reported in tropical regions, such as in the Philippines and Thailand, where the annual mean temperature is almost 30 °C.

The final goal of studying epidemiology is to precisely forecast disease incidence and thus to minimize yield damage from the disease. It is quite desirable to incorporate all known factors into a simulation model of rice leaf blast to predict damage to rice more precisely. However, we cannot expect that one successfully developed model can be widely adaptable in other places where varieties, cultural practices and environments are quite different. Thus researchers are trying to develop their own models. Effective modern plant disease forecasting relies heavily on epidemiology. During the last 20 years, computer simulation models have been developed for potato late blight (Stephan and Gutsche 1980; Gurevich et al. 1979; Waggoner 1968), wheat and barley rusts (Teng et al. 1980; Shrum 1975; Zadoks 1968), and southern corn leaf blight (Massie 1973; Waggoner et al. 1972). These models are all developed by epidemiological basis. Two leaf blast simulation models – BLASTAM (Koshimizu 1982) and BLASTL (Hashimoto et al. 1984) – have been developed in Japan. The rice sheath blight simulation model BLIGHTAS was developed by Ijiri and Hashiba (1986).

Table 2. Comparison of sporulation between natural leaf blast lesions and conidia and conidiophore removed lesions in Icheon, Korea, 1991.

Date	Conidia removed lesion					Natural lesion				
	Size (mm)	Type	Rel[1]	Adax[2]	Abax[3]	Size	Type	Rel	Adax	Abax
July 19	5.9	Acute	320	560	2060	8.6	Chronic	8240	2220	63250
	7.3	Chronic	160	540	580					
23	6.5	Acute	360	8240	12480	7.0	Acute	560	3620	24280
24	28.0	Chronic	200	100	40	18.3	Chronic	2240	2840	34420
25	9.8	Chronic	160	840	15520	10.5	Chronic	4020	3440	56020

[1] Total number of conidia collected by the KY-type spore trap (Kim and Yoshino 1987).
[2] Total number of conidia remained on adaxial surface of lesion. Remained conidia are thoroughly scraped with a 1 ml hypodermic syringe needle and counted using a hemacytometer.
[3] Total number of conidia remained on abaxial surface of lesion. Remained conidia are thoroughly scraped with a 1 ml hypodermic syringe needle and counted using a hemacytometer.
[4] Total sporulation amount derived from released and remained during 1700–0900 h.
Crop growth stage during the experiment was panicle initiation and the period was rainy season.

Background of EPIBLAST

Plant disease forecasting primarily aims to predict either time of probable outbreaks or rate of increases in disease intensity. With an accurate disease forecasting system and appropriate disease management technology, farmers can avoid excessive and untimely application of agrochemicals. The forecasting program benefits the farmers: optimal use of labour, materials and money results in a better harvest.

During the leaf blast season under temperate conditions, the fungus completes one life cycle within one week. Each phase of the life cycle, i.e. sporulation, germination and/or penetration, requires a different range of temperature and relative humidity. In Korea, the *P. oryzae* season usually start in mid-June and normally ends in early August. Within that season, life cycles is repeated 8 to 11 times, resulting in a tremendous inoculum population. To develop a quantitative leaf blast simulation model, we initiated research on the unknown factors, such as on the sporulation and conidia release phase of *P. oryzae* during one rice cropping season under natural conditions. All the experiments and measurements were done at Icheon, 60 km east of Suweon.

The amount of sporulation from a discrete lesion during one night in the field varies depending upon lesion type, age, size and weather conditions from a few hundred to few hundred thousand (Table 1). Sporulation amount was also different between natural lesions (pre-existing conidiophores and conidia not removed) and the amount from removed lesions (Table 2). In most cases, more spores were formed on abaxial surface of leaves.

Beside the sporulation potential, the conidia release phase is very important for production of inoculum. The Kim and Yoshino type (KY-type) spore trap

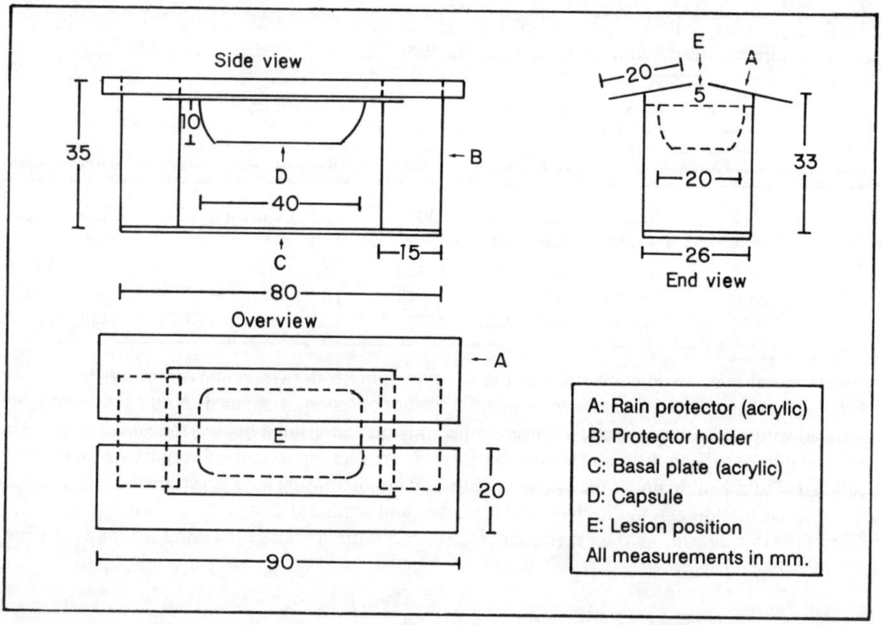

Figure 1. The KY type spore trap, Korea, 1985.

devised in 1985 for measuring conidia release under field condition (Kim and Yoshino 1987) is shown in Figure 1. A leaf with discrete lesion is fixed on the acrylic rain protector using double-sided adhesive tape without detaching the leaf from the plant. A basin-shaped acrylic capsule 40 mm long is placed just beneath the lesion. Figure 2 and 3 illustrate spore catches in different devices: the KY-type spore trap for measurement of conidia release; the rotation sampler (Suzuki 1969a) for measurement of dynamic conidia dispersal and a horizontal spore trap for measurement of static conidia dispersal.

Conidia release patterns can vary with weather conditions, type and age of lesions, as we have measured since 1987. Under clear day conditions on 9–10 July 1987 (Figure 2), conidia release measured by the KY-type spore trap and conidia dispersal measured by other two traps exhibited one typical peak at 0600–0700 h. Suzuki (1969b) and Kato (1976) reported one peak at 0200–0300 h. However, under continuous cloudy conditions 14–15 July 1987, there were 2–3 conidia release and dispersal peaks between 0100 and 0800 h (Figure 3).

Conidia release and dispersal patterns from 3 and 7 d old lesions were studied during the 1988 rainy season. On 13–14 July, it rained from 2400 to 1000 h. Conidia were released during the entire period, with two distinct peaks between 0100 and 0700 h (Figure 4). The first peak (0100–0300 h) is thought to be caused by the production of conidia from pre-existing conidiophores; the second peak (0700 h) was due to conidia produced from conidiophores formed on the evening of 13 July. The peak came immediately after rainfall exceeding 16 mm h^{-1}. More conidia were released from the older lesions. We also

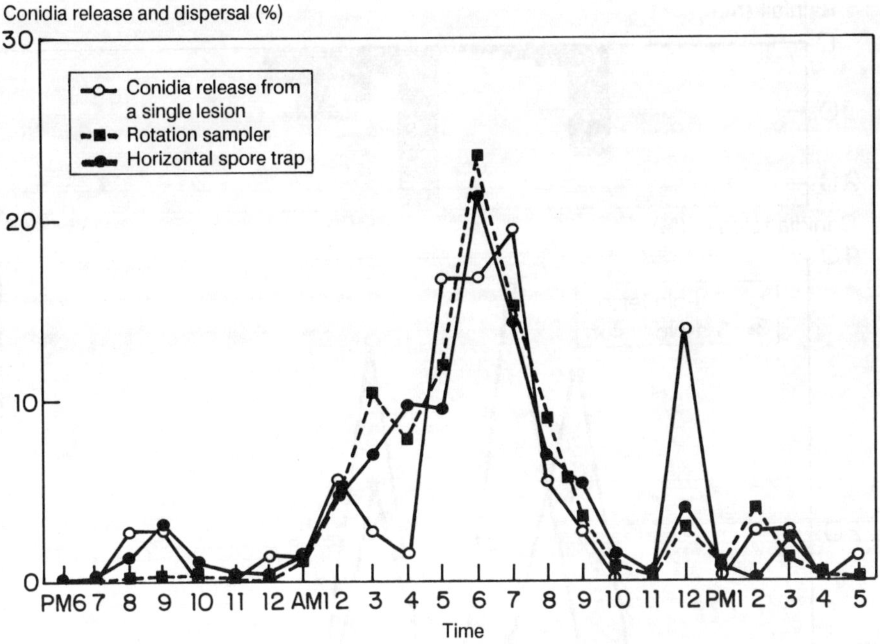

Figure 2. P. oryzae conidia release and dispersal pattern, measured hourly by different types of spore trap under clear day condition in Icheon, Korea, 9–10 July, 1987.

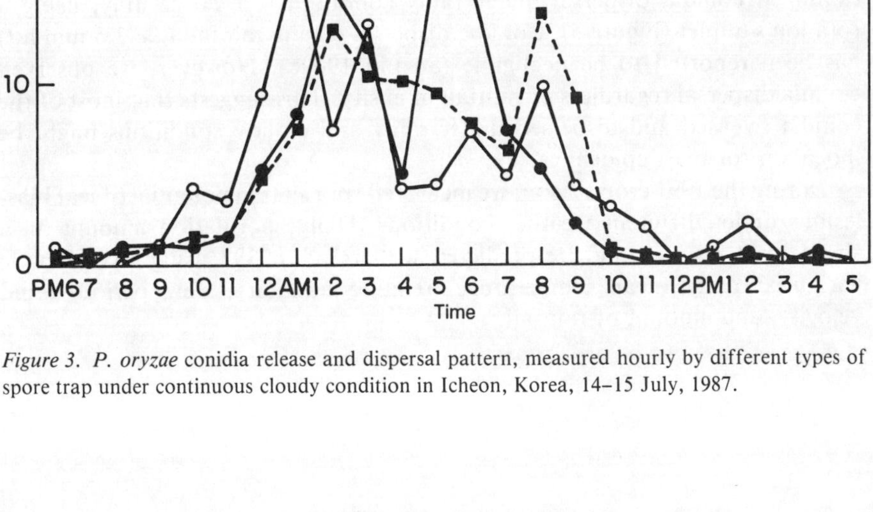

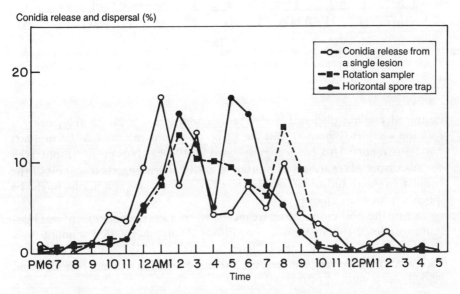

Figure 3. P. oryzae conidia release and dispersal pattern, measured hourly by different types of spore trap under continuous cloudy condition in Icheon, Korea, 14–15 July, 1987.

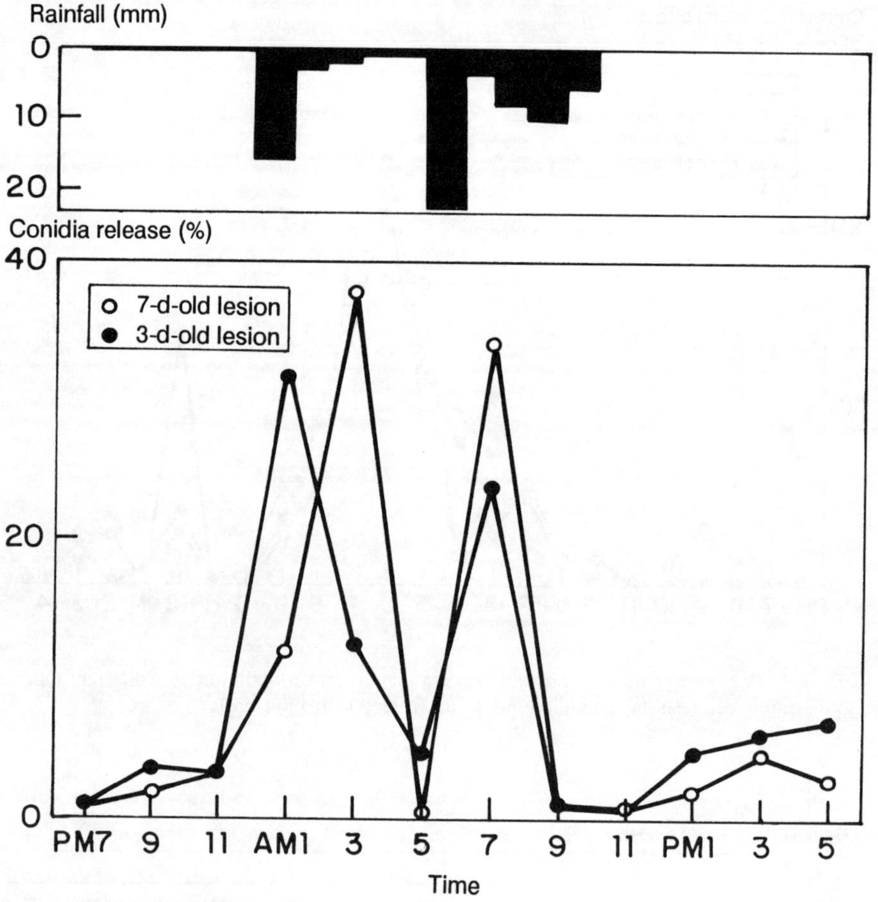

Figure 4. P. oryzae conidia release pattern under rainy conditions in Icheon, Korea, 13–14 July, 1988.

examined conidia dispersal under rainy conditions on 21–22 July, using a rotation sampler (Figure 5). Conidia dispersal during rainfall (> 3.5 mm h^{-1}) has been reported to be negligible (Suzuki 1969b). However, we observed conidia dispersal regardless of rainfall intensity. This suggests that most of the conidia released and dispersed under rainy and cloudy conditions might be inoculum for blast epidemics.

During the 1991 crop season, we measured sporulation potential of leaf blast lesions under different weather conditions. Daily sporulation amount on a single blast lesion could be predicted with r^2 of 0.7397 using five climatic variables: daily mean temperature, relative humidity, dew period, wind velocity, and amount of rainfall.

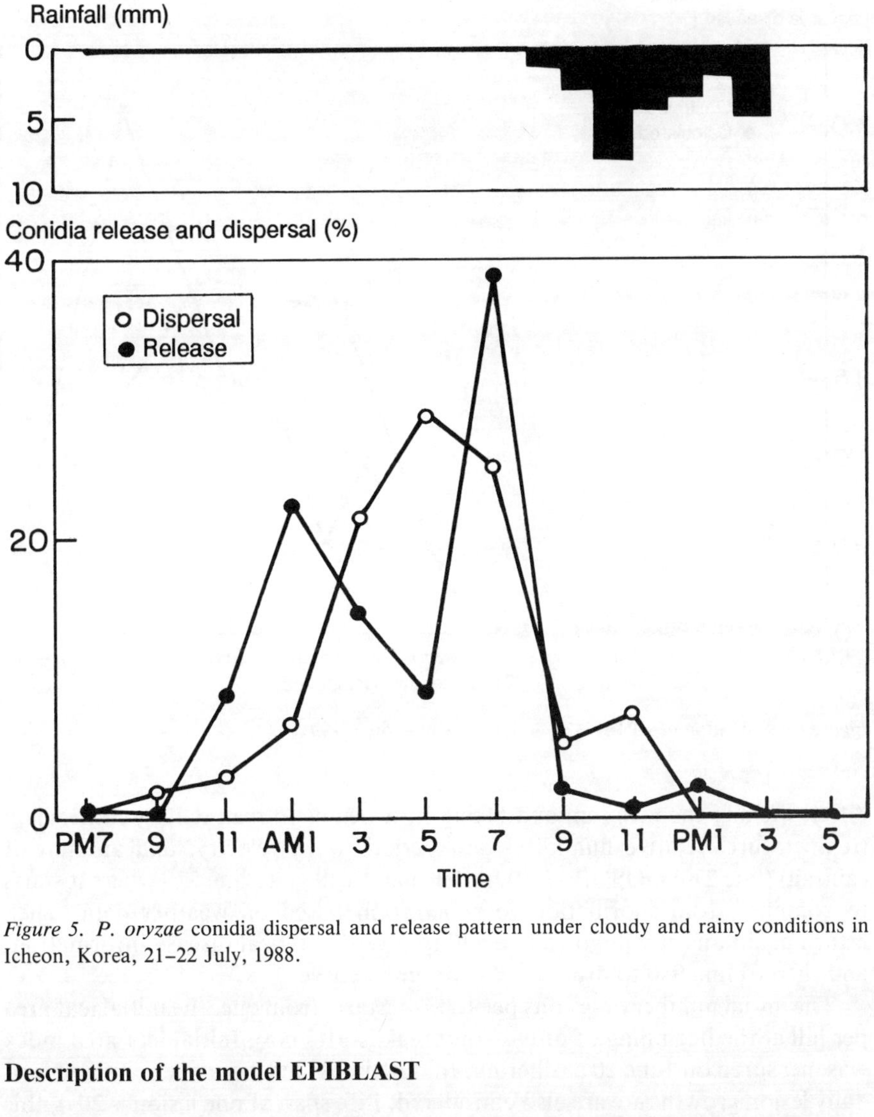

Figure 5. P. oryzae conidia dispersal and release pattern under cloudy and rainy conditions in Icheon, Korea, 21–22 July, 1988.

Description of the model EPIBLAST

On the basis of such spore release data, we developed a dynamic simulation model EPIBLAST for quantitative forecasting of incidence of rice leaf blast. The model is composed of one main program with three subroutines, and is discussed below.

The program is written in BASIC. It consists of three parts: the main program (line 10 to line 540), the disease progress program that uses weather information (line 570 to line 960), and a graphics program (line 980 to line 1340). The influence of weather on behaviour of spores of blast fungus for disease development is illustrated step by step in the disease progress reaction.

The order of operation in EPIBLAST is as follows :

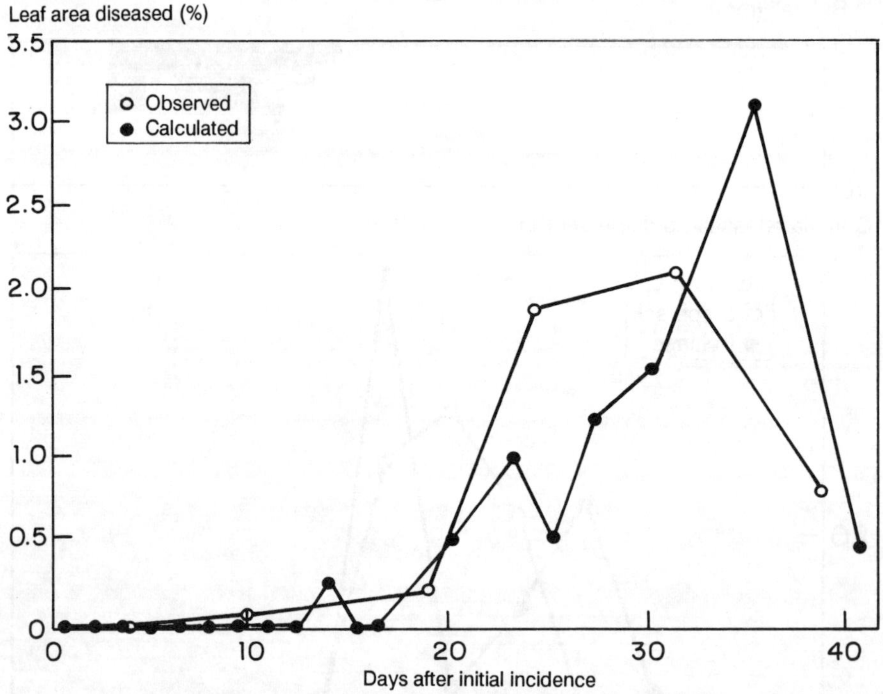

Figure 6. Validation of EPIBLAST in Icheon site, Korea, 1991.

After opening the program and listing it, it begins to read daily weather data (temperature, relative humidity, dew period, wind velocity, and amount of rainfall) (line 230: GOSUB 1350). After that it will go to line 570 where it starts to compute from sporulation to penetration based on weather data. Then return again and it will go to line 240 to give theoretical disease information, and then to line 980 to draw disease progress curve.

The initial number of lesions per 1000 m^2 starts from one, the initial leaf area per hill at the beginning of disease outbreak is 31.7 cm^2. Initial leaf area index was measured on June 20 at tillering. Initial diseased leaf area per unit area and daily lesion growth rate are also considered. Life span of one lesion is 20 d; this corresponds to the observed leaf senescence rate. The relationship between the incubation period of a leaf blast lesion (y) and temperature (x, °C) is described by Yoshino's equation (1979):

$$y = 0.6x + 20.8$$

Sporulation potential, conidial release, deposit of released spore on leaf surface, survival of deposited spore, germination, appressorium formation, and penetration into the epidermal cells, respectively, are calculated in relation to these weather factors.

Results

The model was validated during 1991 blast season at the Icheon site, Korea and the result is shown in Figure 6. Direct observations were made six times starting 28 June until 2 August with intervals of seven days. The model started running from 21 June. The model suggests that the epidemic started on 10 July; start of epidemic by direct observation was on 12 July. This means that the start of leaf blast epidemic almost coincided by the two methods and we can consider the application of this EPIBLAST model in the forecasting sites in the near future.

Discussion

We have tried to formulate a simulation model for rice leaf blast forecasting based on epidemiological data with special emphasis on the behaviour of blast fungus under the natural conditions; i.e. sporulation and conidial release. Of course, once a rice plant is infected by blast fungus – especially under high nitrogen level – the plant is totally destroyed and sometimes even heading is impossible. High plant density and a high level of nitrogen result in vigorous vegetative growth and a dense leaf canopy which leads to a favourable micro-climate for blast disease development. At the same time each weather factor: temperature, relative humidity, dew period, wind velocity and rainfall, is correlated with behaviour of *P. oryzae*, creating a complex pest behaviour pattern. Due to these multiple dependencies, the model reacts very sensitive to minute weather changes.

In the 1992 blast season, we will try to validate and test EPIBLAST in some districts other than Icheon site, based on data from weekly weather forecasts and past weather data. Meanwhile, the model needs further revision to improve the fit with observed disease progress. We are currently attempting to modify the model for developing a panicle blast prediction model by adding some different variables before and after heading time.

Appendix

Appendix 1. List of variables in the program EPIBLAST

APPR : number of germinated conidia with appressorium formation
DEFL : defoliation: leaf loss due to rice leaf blast ($cm^2 hill^{-1}$)
DEP : number of released conidia deposited on leaf
DEW : dew period – leaf wetness ($h\ d^{-1}$)
DIS : disease – diseased leaf area ($cm^2 hill^{-1}$)
GER : germination of survived conidia
INAR : initial diseased leaf area at the end of June ($cm^2\ hill^{-1}$)
INDI : number of leaf blast lesions per 10000 m^2
LAT : leaf area with latent infection – incubation period ($cm^2\ hill^{-1}$)
LFAR : leaf area ($cm^2\ hill^{-1}$)

NLFAR : net leaf area available for infections (cm^2 hill^{-1})
NTS : total number of possible penetration sites
PDIS : percent diseased leaf area
PENT : number of penetrating spores
RAIN : daily rainfall amount (mm)
RDEF : rate of defoliation (cm^2 cm^{-2} d^{-1})
REL : number of conidia released from lesions
RH : daily mean relative humidity
RLB : rice leaf blast
RPT : fraction – % penetration of blast fungus into rice leaf tissue
SPOR : sporulation – number of spore formation from lesions
SRCE : source – estimated initial leaf area per plant at June 20 (cm^2 hill^{-1})
SURV : number of surviving deposited conidia
TEM : daily mean temperature (°C)
UNLG : unit of leaf growth (cm^2 hill^{-1} d^{-1})
UNLS : unit leaf senescence (cm^2 hill^{-1} d^{-1})
VLAT : velocity of latent to appearance – from penetration to visual lesion appearance (cm^2 cm^{-2} d^{-1})
VLS : velocity of natural leaf senescence (cm^2 cm^{-2} d^{-1})
VPEN : velocity of penetration (number of penetrating spores per number of spores with appressorium per day)
WINV : daily mean wind velocity (m s^{-1})

Appendix 2. Listing of the program EPIBLAST

```
10 REM THIS IS A SIMPLE MODEL 'EPIBLAST' FOR LEAF BLAST PROGRESSION
20 DIM RLB(40,2)
30 INPUT "INITIAL NUMBER OF LESIONS PER 10a;INDI
40 INPUT "NUMBER OF DAYS";N
50 OPEN "0",#1,"A:BLAST.DAT"
60 FOR I=1 TO N
70 DAY=I
80 SRCE=31.7
90 REM SRCE:SOURCE
100 INAR=INDI*3.14*0.08^2/24000
110 REM INAR:INITIAL DISEASED LEAF AREA
120 REM INDI:NUMBER OF LESIONS/10a
130 UNLG=0.23*DAY-0.00324*DAY^2
140 REM UNLG:UNIT LEAF GROWTH
150 LFAR=SRCE+UNLG
160 REM LFAR:LEAF AREA
170 IF DAY<21 THEN VLS=0 ELSE VLS=1/20
180 REM VLS:VELOCITY OF LEAF SENESCENCE
190 UNLS=VLS*LFAR
200 REM UNLS:UNIT LEAF SENESCENCE
210 NLFAR=LFAR-UNLS
220 REM NLFAR:NET LEAF AREA AVAILABLE FOR INFECTIONS
230 GOSUB 1350
240 LAT=NLFAR*RPT
250 REM LAT:LATENT, RPT:RATE OF PENETRATION
260 VLAT=1/(-0.6*TEM+20.8)
270 REM VLAT:VELOCITY OF LATENT TO APPEARANCE
280 DIS=LAT*VLAT
```

```
290 REM DIS:DISEASE
300 RDEF = 1/20
310 REM RDEF:RATE OF DEFOLIATION
320 DEFL = DIS*RDEF
330 REM DEFL:DEFOLIATION
340 PDIS = 100*(DIS-DEFL)/NLFAR
350 REM PDIS:PERCENT DISEASE
360 PRINT PDIS;”P”
370 ‘RLB(I,1) = DAY
380 ‘RLB(I,2) = PDIS
390 WRITE #1,DAY,PDIS
400 NEXT I
410 CLOSE #1
420 ‘PRINT DAY; “ “;PDIS
430 OPEN “I”,#1,”A:BLAST.DAT”
440 FOR I = 1 TO N
450 FOR J = 1 TO 2
460 INPUT #1,RLB(I,J)
470 NEXT J
480 NEXT I
490 CLOSE #1
500 CLS
510 PRINT “DAY”;” PDIS (%)”;”
520 FOR I = 1 TO N
530 PRINT USING “–# –.–––#”;RLB(I,1);RLB(I,2)
540 NEXT I
550 GOSUB 980
560 END
570 IF DAY = 1 THEN SPOR = INAR*10^7*(974.81-17.17*TEM + 35.08*DEW-8.060001*RH)
580 REM SPOR:SPORULATION, TEM:DAILY MEAN TEMPERATURE(C)
590 REM DEW:DEW PERIODS(HR), RH:DAILY MEAN RELATIVE HUMIDITY
600 IF DAY = <2 THEN SPOR = DIS*10^7*(974.81-17.17*TEM + 35.08*DEW-8.060001*RH)
610 IF TEM<20 AND TEM = <22 THEN SPOR = SPOR*2
620 IF TEM>25 THEN SPOR = SPOR/10^2.3
630 IF WINV = <2 THEN GOTO 650 ELSE GOTO 720
640 REM WINV:DAILY MEAN WIND VELOCITY (M/SEC)
650 IF RH> = 96 THEN REL = 0.2*SPOR
660 REM REL:RELEASE
670 IF RH = >90 AND RH<96 THEN REL = 0.15*SPOR
680 IF RH = >85 AND RH<90 THEN REL = 0.1*SPOR
690 IF RH = >80 AND RH<85 THEN REL = 0.07*SPOR
700 IF RH<80 THEN REL = 0.03*SPOR
710 GOTO 750
720 IF RH = >96 THEN REL = 0.1*SPOR
730 IF RH = >90 AND RH<96 THEN REL = 0.07*SPOR
740 IF RH>90 THEN REL = 0.03*SPOR
750 IF WINV<1 AND RAIN<3 THEN DEP = 1/100*REL ELSE DEP = 1/200*REL
760 REM RAIN:RAINFALL AMOUNT(MM)
770 REM DEP:DEPOSITION
780 IF DEW>4 THEN SURV = DEP ELSE SURV = 0
790 REM SURV:SURVIVAL
800 IF DEW>4 THEN GERM = SURV ELSE GERM = 0
810 REM GERM:GERMINATION
```

```
820 IF DEW = >18 THEN APPR = GERM*0.8
830 REM APPR:APPRESSORIUM FORMATION
840 IF DEW = >15 AND DEW<18 THEN APPR = GERM*0.7
850 IF DEW = >12 AND DEW<15 THEN APPR = GERM*0.6
860 IF DEW = >8 AND DEW<12 THEN APPR = GERM*0.5
870 IF DEW = >6 AND DEW<8 THEN APPR = GERM*0.4
880 IF DEW = >4 AND DEW<6 THEN APPR = GERM*0.2
890 VPEN = (-14.569 + 0.985*DEW + 0.494*TEM)/100
900 REM VPEN:VELOCITY OF PENETRATION
910 PENT = APPR*VPEN
920 REM PENT:PENETRATION
930 NTS = NLFAR*10^3.89
940 REM NTS:NUMBER OF TOTAL SITES
950 RPT = PENT/NTS
960 REM RPT:RATE OF PENETRATION
970 RETURN
980 DIM D%(2)
990 OPEN "I",#1,"A:BLAST.DAT"
1000 FOR I = 1 TO 2
1010 FOR J = 1 TO 2
1020 INPUT #1,RLB(I,J)
1030 NEXT J
1040 NEXT I
1050 CLOSE #1
1060 CLS
1070 SCREEN 2
1080 LOCATE 3,30,0
1090 PRINT "DISEASE PROGRESS CURVE"
1100 LOCATE 6,6,0
1110 PRINT "% DISEASE"
1120 H = 14
1130 V = 8
1140 LINE(6*H,6*V)-(6*H,16*V + 4),1
1150 LINE(6*H,16*V + 4)-(41*H,16*V + 4),1
1160 FOR I = 0 TO 10 STEP 2
1170 LOCATE 17-I,7,0
1180 PRINT USING "-#";I*10
1190 IF I = O THEN GOTO 1240
1200 LINE (5*H + 4,16*V + 4-I*V)-(6*H,16*V + 4-I*V),1
1210 FOR J = 0 TO 41-7
1220 LINE((6 + J)*H,16*V + 4-I*V)-((6 + J)*H + 4,16*V + 4-I*V),1
1230 NEXT J
1240 NEXT I
1250 LOCATE 20,35,0
1260 PRINT "TIME IN DAYS"
1270 FOR I = 1 TO 31 STEP 2
1280 LOCATE 18,9 + I*2,0
1290 PRINT USING "-";I
1300 NEXT I
1310 LOCATE 23,5,1
1320 IF I = 0 THEN GOTO 1340
1330 LOCATE 22,1,1
1340 RETURN
```

```
1350 READ DAY, TEM, RH, DEW, WINV, RAIN
1360 DATA 1, 23.3, 88, 11, 0.6, 6.5
1370 GOTO 570
```

References

Brady N C (1976) Foreword. In Climate and Rice. International Rice Research Institute, P.O.Box 933, Manila, Philippines.

Goto K (1965) Estimating losses from rice blast in Japan. Pages 295–202 in The rice blast disease. The John Hopkins Press, Baltimore, Maryland, USA.

Gurevich B E, Filippow A V, Tverskoi D L (1979) Forecasting the development of harmfulness of potato late blight (*Phytophthora infestans*) under different meteorological conditions on the basis of simulation model 'Epiphtora'. Mikol. Fitopathol. 13:309–314.

Hashimoto A, Hirano K, Matsumoto K (1984) Studies on the forecasting of rice leaf blast development by application of the computer simulation [In Japanese, English summary]. Special Bull. of the Fukushima Prefecture Agric. Exp. Stn. Jpn. 2:1–104.

Ijiri T, Hashiba T (1986) Computerized forecasting system for rice sheath blight disease(BLIGHTAS) [In Japanese]. J. Plant Prot. 40:42–45.

Kato H (1976) Some topics in a disease cycle of rice blast and climatic factors. Pages 417–425 in Climate and Rice. International Rice Research Institute, P.O.Box 933, Manila, Philippines.

Kim C K, Yoshino R (1987) Epidemiological studies of rice blast disease caused by *Pyricularia oryzae* Cavara. I. Measurement of the amount of spores released from a single lesion. Korean J. Plantpathol. 3:120–123.

Koshimizu Y (1982) A forecasting method for leaf blast outbreak by the use of AMeDAS data. [In Japanese]. Kongetsu No Noyaku 26(1–4):46–53, 64–67, 74–83, 68–78.

Massie L B (1973) Modeling and simulation of southern corn leaf blight disease caused by race T of *Helminthosporium maydis* Nisik. & Miyake. Ph.D thesis. Penn. St. Univ., USA. 93 p.

Nuque F, Bandong J, Estrada B, Crill P (1979). Fungal disease of rice. Rice Production Training Series, Slide-tape instructional unit WDC-4. IRRI, P.O.Box 933, Manila, Philippines.

Parthasarathy N, Ou S H (1965). International approach to the problem of blast. Pages 1–5 in The rice blast disease. The John Hopkins Press, Baltimore, Maryland, USA.

Rural Development Administration (RDA) (1989). Crop Protection Report for 1989 [In Korean].Suweon, Korea. 209 p.

Shrum R (1975) Simulation of wheat stripe rust (*Puccinia striiformis* West) using EPIDEMIC, a flexible plant disease simulator. Penn. State Univ. Agric. Exp. Stn. Prog. Rep. 347:1–81.

Stephan S, Gutshe V (1980) An algorithmic model to simulate Phytophthora epidemics (SIMPHYT). Arch. Phytopathol. Pflanzenschutz 16:183–192.

Suzuki H (1969a) Studies on the behavior of rice blast fungus spore and application to outbreak forecast of rice blast disease [In Japanese, English summary]. Bull. Hokuriku Agr. Exp. Stn. 10:1–118.

Suzuki H (1969b) Interrelationship between the occurrence of the rice blast disease and the meteorological conditions [In Japanese]. J. Agr. Met. Japan. 24:211–218.

Teng P S, Blackie M J, Close R C (1980) Simulation of the barley leaf rust epidemic: structure and validation of BARSIM–1. Agric. Syst. 5:55–73.

Waggoner P E (1968) Weather and the rise and fall of fungi. Pages 45–66 in Biometeorology. W.R.Lowry, ed., Oregon State Univ. Press, USA.

Waggoner P E, Horsfall J G, Lukens R J (1972) EPIMAY, a simulator of southern corn leafblight, Conn. Agric. Exp. Stn. Bull. 729:1–84.

Zadoks J C (1968) Reflections on resistance. Page 223 in Abstracts of papers. First International Congr. Plant Pathol. Working. Gresham, UK.

SESSION 5

Farming systems

Potential for systems simulation in farming systems research

J.B. DENT
Institute of Ecology and Resource Management,
University of Edinburgh, West Mains Road, Edinburgh EH9 3JG, Scotland

Key words: CERES-Wheat, farm management, farm model, farmer participation, Farming Systems Research, household decision model, IBSNAT, institutional constraints, political constraints, simulation, socio-economic data, minimum data set

Abstract
Attention is drawn to the benefits and limitations of Farming Systems Research (FSR) as a means of developing and extending new technology. The value of crop modelling to the FSR process is outlined. Emphasis is placed on the fact that the driving force for farming system change is social and cultural. The point is made that if a modelling approach is to assist the process of technology development, adoption and change then work must include but extend beyond the use of crop models in the research phase of FSR. The format and advantages of a whole farm model are presented. Lack of socio-cultural data limits traditional approach to modelling and consequently a rules based framework is discussed. Such an Expert Systems approach permits the use of qualitative data necessary to express socio-cultural conditions. The concept of a Minimum Data Set for socio-cultural data is presented in relation to a generic whole farm model.

Farming Systems Research

Targeting of research and translation of findings into technology packages that are relevant to the needs of local farmers and therefore will be adopted are major issues of concern in many countries. Some of the reasons for delay between identification and adoption of technology are:

- much applied research is not directly relevant to the immediate needs of farms in the district and in any case usually does not produce the kind of results that are useful for decision making at farm household level. (This is partly because the objectives of farm families are different to those assumed by researchers and extension personnel);
- there are risks involved in adoption of new technology: some of them perceived as being great enough to discourage farmers;
- there are often management problems associated with new technology that will prevent adoption without an active extension programme focussing on these problems;
- farmers operate with physical, economic and social constraints which are not well appreciated.

Collinson (1988) suggests that farmers themselves should be part of the process of technology choice and development, and that their needs and problems should be the basis of establishing research priorities.

F.W.T. Penning de Vries et al. (eds.), Systems Approaches for Agricultural Development, 325–339.

In so far as these elements are inherent in Farming Systems Research (FSR) he views them as being the innovating strengths and of course they do address some of the issues affecting the efficiency of technology identification, development and adoption.

The process of FSR is not strictly a sequential series of events but there are a number of definable phases: appraisal, when the current production systems in an area are described and analysed and a resource audit of farm household is made; experimentation, during which potential improvements to existing production systems are developed; evaluation, when promising technology is tested in farmers fields and demonstrated to local farmers; and finally extension to the target group of farmers. The whole process is obviously dynamic in nature and there are strong feed-back elements between the phases (Hildebrand 1990).

Farmer involvement is a characteristic of FSR and is seen as an important advantage over traditional Research and Development (R&D) processes in agriculture. Farmers themselves are experimenters and as Hildebrand (1990) emphasises their degree of involvement in FSR has been controversial, "with social scientists urging more and biological scientists resisting the loss of control associated with too much farmer participation". However, farmer involvement in conceptualising the experimental programme is most often beneficial and must be seen as a strength of FSR. More crucially, the demonstration work in farmers fields, testing technology and making the results explicit to local farmers, is a lynch-pin in the whole approach. It is at this stage where the holistic characteristic of FSR becomes obvious. The effect of technology, which may relate to a single enterprise is viewed against the background of the organisation and operation of the whole farm. Conflicting resource demands, trade-offs between enterprise and socio-economic constraints and attitudes may all be involved in the overall assessment of a technology package. Only in the fields of collaborating farmers will these issues become clear to neighbouring farmers who are considering the technology of their own farms.

The philosophy of FSR is sound and the general advantages in directing technology development and adoption appear powerful. In practice however, problems have arisen; some of a fundamental nature and some institutional and political.

Determination of the experimental programme

Involving farmers in the research priorities has already been mentioned as a method of focussing experimentation but this involvement can also prove limiting. There is a case for the development of simple technology that relates to existing methods in the region because the adoption process is likely to be much quicker (Wake et al. 1988). Local farmers are most likely to indicate research priorities leading to such technology but in spite of rapid adoption economic progress may be slow. Changes in infrastructure support (subsidies credit, an intensive extension programme, etc.) may change the prospects for adoption of a more comprehensive technology.

Limitation of the value of experimental results and demonstrations
This point has been discussed by many others (e.g. Lightfoot 1987). The issue relates to the fact that field experimentation and the demonstration of technology in farmers' fields is critically dependent upon seasonal climate and within this the sequence of weather. In field trials, weather sequence/treatment interactions are usually strong; in demonstration, the timing of whole farm resource (e.g. labour, water etc.) demands is weather sequence dominated. The consequence is that work needs to be repeated over a sample of seasons to determine the natural variability of results. Furthermore, in demonstration work there are carry-over effects from one season to the next, not only in terms of soil nutrient and moisture status but more importantly in relation to the economic status of the farm.

Put simply, technology developed from experimentation carried out in one or two good or average years, may fail in demonstration in a sequence of poor years either because the physical conditions were inadequate or because the farmer did not have the resources following one poor year to maintain the technology into a second.

In development, time is always limiting and replication of experimentation or demonstration over many years will never be a viable option. Consequently, the FSR process may lead to inappropriate technology and inevitable meagre adoption rates. The ability of field experimentation ever to produce results that are comprehensive enough to have relevance to household decision-making must be questioned (Nix, pers. comm. 1991) since no measure of climate-related risk can be generated. The experimental phase to be of any value must be a long one, certainly over several seasons, but in practice it can never be long enough to expose the technology to external variables such as rainfall and temperature.

In general then, the results of experiments are heavily dependent on the climate sequence but also on the specific soil type and the numerous management factors under the control of the investigator – e.g. plant population, timing of cultivations and sowing and pest control. The critical issue is whether the results from such experiments have much relevance to the way technology packages based on them will operate in the fields of farmers, in a different place, year, and soil-type, where farming operations are limited by cultural and economic conditions neither in existence or perhaps perceived at the research station.

Institutional and political constraints
Collinson (1988) points out that the political, institutional and professional infrastructures in developing economies are 'top-down' in nature. As a result 'grassroot views' are rarely weighed in the decision making process at national, regional or district level. These underlying weaknesses clearly work against the philosophy and practical application of FSR. Because of budgetary constraints morale in national research institutions is generally low and this creates a climate where scientists cling to whatever status they have and stay within 'safe' disciplinary boundaries. Collinson (1988) concludes that these factors have

permitted "only a muted impact from FSR" as far as Africa is concerned.

Systems and simulation

The concepts involved in the systems approach have been well documented in the past (Rountree 1977; Spedding 1979). The fundamental issue that coincides with the concepts of FSR is that an isolated study of one or a number of components (sub-systems) that make up the system is inadequate to permit complete understanding. Components interact and it is these interactions that give the system identity and organisational integrity (Doyle 1990). In farming systems, interactions are not only or even mainly of a biological type; important economic and socio-cultural links between components are involved. It is evident that farming systems are dynamic in nature and also subject to uncertain elements that rightly may be considered exogenous to the system. Such elements impact upon all three component types of the farm system (biological, economic and socio-cultural) creating uncertain behaviour of farming families in response, for example, to exposure to new technology, the availability of new opportunities created by markets or to provision of credit and uncertain outcome of any management strategy selected by the farm household.

Logically, the study of farming systems must progress in balance: clearly, the weakest area of understanding will be the one which limits knowledge of the whole. Research programmes that create this balance are desirable but unfortunately have not been the reality. The fact is that research funding has been (and is) concentrated in the biological disciplines because these have (erroneously) been considered the foundation of agriculture. Neglect of the economic and socio-cultural components in research programmes has been responsible for enormous deficiency in understanding farming systems. Policy makers and planners of land use and food throughout the world are currently greatly hindered in their work because of this and consequentially farmers and the population in general are achieving less welfare than they might otherwise have had. And the unsatisfactory distribution of research effort goes on as national agriculture and food research programmes continue to be overwhelmingly dominated by biological science.

Farming systems research and simulation

The use of mathematical modelling techniques has become integrated with the study of systems. The reasons for this have been explained fully elsewhere (e.g. Doyle 1990) but perhaps the most persuasive in the context of farming systems is the fact that experimentation in the real world is expensive, time consuming and there are severe problems in controlling variables exogenous to the experiment. This is particularly the case in field experimentation which is

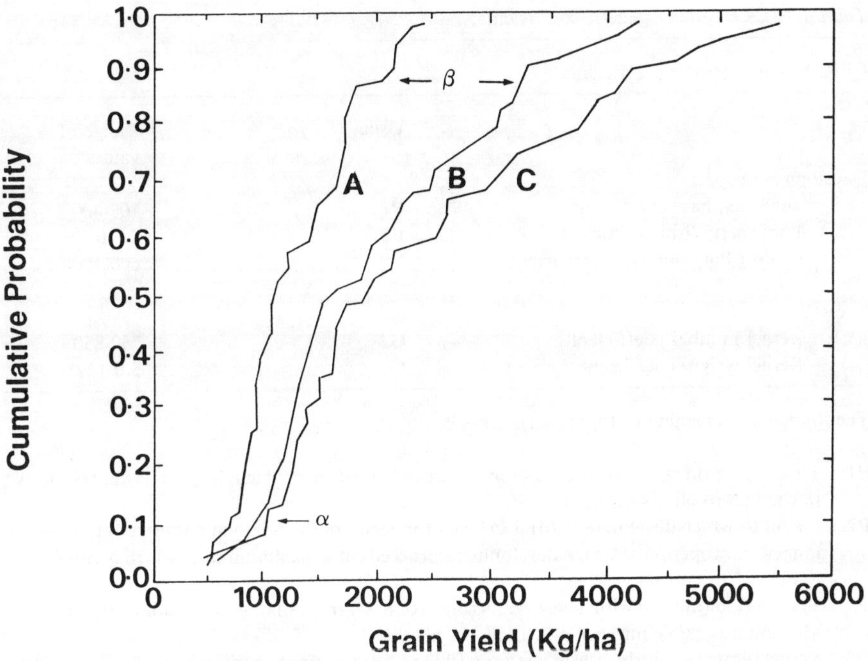

Figure 1. Cumulative probability functions for simulated grain yield response to three fertilizer strategies at Dubbo, Australia.
(A = 0 kg N ha^{-1}; B = 30 kg N ha^{-1}; C = 60 kg N ha^{-1}.)
(Figure from Godwin et al. 1989.)

consequentially a poor framework for expanding knowledge. As a result, the concept has been discussed of using locally validated detailed models of farming sub-systems as the preferred medium for applied research. (Baldwin and Hanigan 1990; Bywater 1990; Dent and Edwards-Jones, 1991).

It is well appreciated that it is now possible to create, for example, a model that will simulate the growth, development and final yield of a crop in a specific locality with tolerable precision. Assuming that confidence has been established in the capability of the model to produce outcomes similar to those experienced in the real world, then the model can become the medium whereby alternative management strategies can be assessed (experimentation). Using site specific, historical or simulated weather time series, frequency distributions of the seasonal outcomes from each strategy quickly may be established.

The work of Godwin et al. (1989) provides a useful example of the capability of the CERES Wheat model in relation to fertilizer application in Australia. Simulation of the same variety in the same location, subject to three nitrogen fertilizer levels, reveals large variations in yield from season to season according to the weather. These authors used simulated climate data for the locality over a period of 50 years. Figure 1 expresses the simulated yield data for each fertilizer level expressed in the form of cumulative distribution functions. The

Table 1. CERES-Maize genetic coefficients.

CERES-Maize genetic coefficients		
	Usual	Range of values
Development aspects		
Juvenile phase coefficient	P1	100-400
Photoperiodism coefficient	P2	0-4.0
Grain filling duration coefficient	P5	600-1000
Growth aspects		
Kernel number coefficient	G2	750-850
Kernel weight coefficient	G3	5.0-12.0

The 'biological' meaning of these coefficients is:

P1: Time period (in °C·d with a base temperature of 8°C) during which the plant is not responsive to changes in photoperiod.
P2: Extent to which development (d) is delayed for each hour increase in photoperiod above the longest photoperiod at which development proceeds at a maximum rate (which is considered to be 12.5 h).
P5: Time period (in °C·d with a base temperature of 8°C) from silking to physiological maturity.
G2: Maximum possible number of kernels per plant.
G3: Kernel filling rate during the linear grain filling stage and under optimum conditions (mg d^{-1}).

(Data from Ritchie et al. 1989).

range of variation for each fertilizer treatment is illustrated and this range increases as the fertilizer applied increases (compare α with β). However, the high fertilizer treatment will, in all years, out yield the zero fertilizer option. Such data can not be provided by way of field trials which may over a two or three year trial period only indicate the means for a small sample of climate years and some view of experimental error from the plots. What can be achieved for simple treatments like nitrogen fertilizer level can equally easily be done by way of the model for much more complex strategy-type 'combination treatments'. For example a strategy can be defined as a specific cultivar of wheat sown at a given seed rate, fertilizer level and date of sowing. Variations on this strategy can be compared over many years of representative simulated climate. This kind of research opportunity presents a flexibility not available to research restricted only to field trials. Consider further the exploration of cropping sequences: biological and economic performance of a particular sequence is dependent on a range of factors such as actual date of sowing, length and date of commencement of the growing season, weather sequence, ability to harvest on time and hence to prepare subsequent seed beds. Cropping sequences with alternative management rules can be explored to locate stability of output and economic achievement.Equally useful within Farming Systems Research is to direct aspects of technical research: both field and laboratory based. An example may be taken from the IBSNAT crop models where

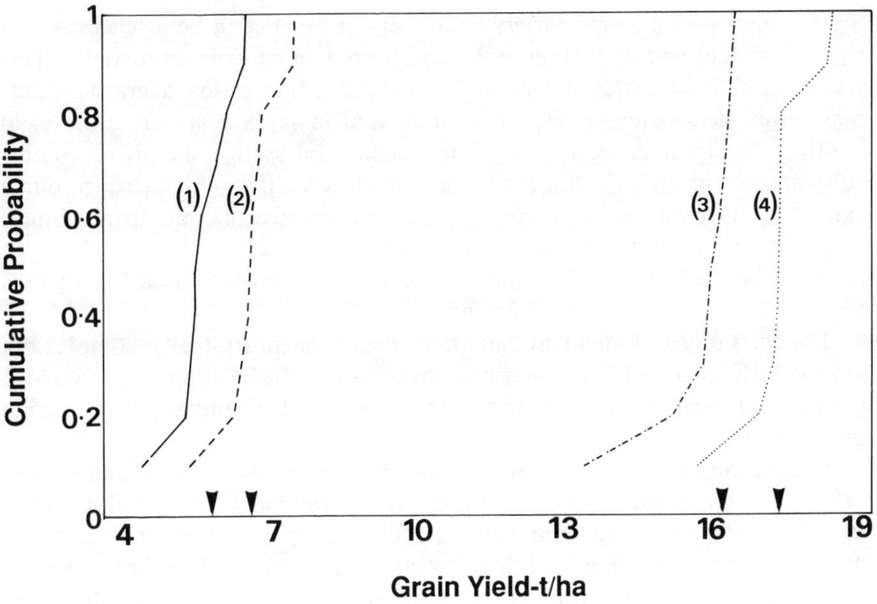

Figure 2. Cumulative probability functions for four hypothetical varieties of Maize 'grown' in Gainesville, Florida.

cultivars are characterised by a small number of " genetic coefficients" (Hunt et al. 1989). For the CERES-Maize model these coefficients are shown in Table 1.

Any particular variety is represented by a number from a range for each coefficient. For any particular location defined by soil type and climate a number of hypothetical cultivars can be defined – perhaps by making a factorial selection across the range or partial range for each coefficient. Figure 2 demonstrates the output from 4 'treatments' of such a procedure for maize 'grown' in Gainesville, Florida.

The hypothetical cultivars all have a very low coefficient for the juvenile stage and for photo periodism and a high coefficient for grain filling duration; they are differentiated by a factorial arrangement of high and low coefficient for Kernel number and a high and low coefficient for Kernel filling rate. All cultivars have the same fertilizer treatment, fit an identical position in the cropping year and are assumed to have identical seed rates. Over 50 years of simulation it can be seen that 'cultivar 4' is preferred over the others. While there may be discussion over the precise meaning of the genetic coefficients, this type of procedure provides the plant breeder with insights into the characteristics best suited to the locality in question: in this case measured by yield and variance of yield. Potential for expressing such results in terms of economic parameters clearly exists: gross margin, for example, might be another simple but useful indicator. Searches may be made of current cultivars or guidance for breeding work given, bearing in mind of course essential characteristics such as resistance to relevant pests.

Dominance of a single variety is unlikely in practice to be as clear as that shown for 'cultivar 4' in Figure 2. An alternative cultivar with similar pest resistance characteristics may provide for a stable though low average output, may never yield below a certain amount, and in say 3 years in 10 out-yield 'cultivar 4'. This may be, in the appropriate farming system, a more acceptable cultivar than the high yielding type represented by 'cultivar 4'. Cash or resource carry over from one season to another is important to economic sustainability. Dependable performance may be crucial to survival and if this is associated with low average yields it will still be preferable to a cultivar which occasionally fails to provide for minimum requirements.

The decision to adopt a new cultivar or management strategy is complex but at least with yield frequency distributions for a specific location generated by a crop model there is more information on which to formulate an extension programme.

Integration of such a model into the R&D process of a research station extends the scope and efficiency of the agricultural scientist. He will still need to carry out field trials to validate the model in the local situation and certainly he would wish to provide field demonstrations to farmers. But the simulation work provides more extensive data, more suitable for economic analysis and hence more valuable to the needs of farmer decision-making.

The previous discussion has been concerned to stress the potential advantages to the research phase of FSR of involving simple biological modelling as an adjunct to field research. Such models are able to strengthen the arm of the agricultural research scientist by improving relevance of results, timelines and cost-effectiveness. This is important in FSR but it does not, directly, address the key area integrating research and adoption. Farmers usually operate multi-enterprise units and the output from simple commodity based research will always be inadequate to meet the full needs of technology development. Technology adoption for farmers represents a process of change – usually gradual. As current systems fail to meet family demands then the type, scale, structure or operation of enterprises and how they integrate will be examined. Family demands are dynamic and multi faceted and will, over time, modify the level of satisfaction with the current system and methods. Opportunities for technology options within this process of change have to fit within economic and socio-cultural values. Even comprehensive results leading to the determination of yield dominant strategies within an enterprise is a relatively small part of the tapestry of factors influencing farm change by 'improved' technology.

Certainly biological understanding is important but if FSR has taught us anything it is that the essential driving force for change is social and cultural in nature even though its manifestation may be partly technological and partly economic (Simmonds 1987). So, those who believe that modelling work can assist the process of FSR must aim their sights a good deal higher than encouragement of the use of crop models in the research phase.

Regression models based on empirical data for a specific location which

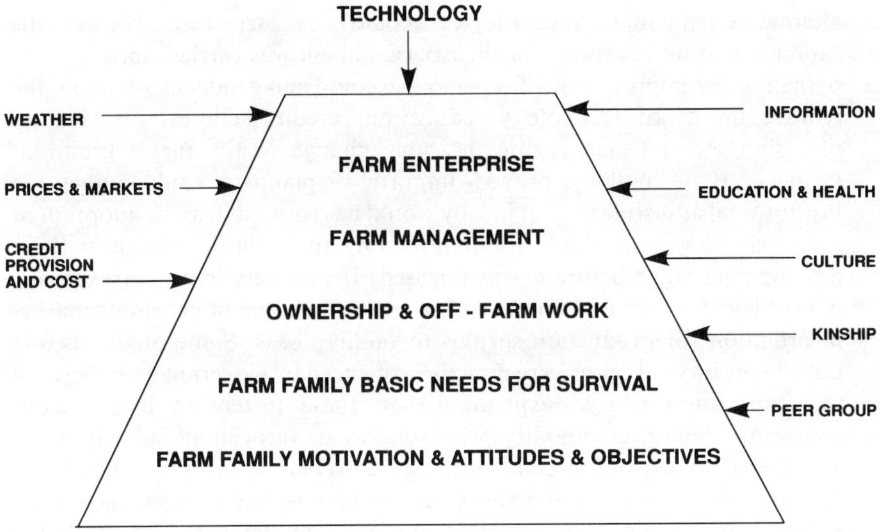

Figure 3. Components for a whole-farm model.

relate the rate of adoption, say, to the age of the farmer, the size of holding and whether the farmers are recent arrivals in the area (Polson and Spencer 1991) help to understand local circumstances but are somewhat equivalent to the early crop growth models (Crowther and Yates 1942) developed before mechanistic simulation of the growth and development process was contemplated. They are predictive for the precise local circumstances to which the data apply: they are in no way explanatory of the processes nor the interactions between farm decisions, farm family behaviour and the socio-cultural network in which rural people live and work. A more acceptable model might, at this stage, be described as in Figure 3 in which the farm/farm-family components are presented within the box. Those factors on the right hand side of the box influence behaviour by acting on socio-cultural components, while on the left hand side, economic and physical forces are described which might be expected to have a more immediate impact on decision-making.

By way of such a model a farm family representative of a target group of farm families could be 'exposed' singly or in combination to new circumstances that might for example include alternative prospective new technologies in the various enterprises, alternative market scenarios and credit opportunities and alternative levels of extension support. With such a model the FSR worker would be armed with the same kind of enhanced capability as the agricultural scientist is with a crop or livestock biological model. Similar types of advantages would accrue. They would include:

1. adoption rates for specific technology given certain market, credit and extension conditions could be predicted together with the likely impact on family economic sustainability as well as total farm output.

2. alternative possible technologies could be screened before the experimentation necessary for their development was carried out.
3. perhaps more importantly, FSR personnel could make judgements about the relative merits of technology, education, credit facilities, information provision etc. in encouraging desirable change in the target group of farmers. If technology proved important, planners could judge the institutional adjustments that ideally would be required to assist adoption of that technology and think about providing these during the technology development stage before it was released. If not, resources, for example, may be best directed into improving extension activities or providing market information for production surplus to family needs. Some of the factors outside the box in Figure 3 are directly influenced by Government policy and the implication is that expenditure on these (extension information, education, health, commodity price support or farm input subsidy) is an alternative to expenditure on agricultural research for new technology development at least in the short-run. Achieving the correct balance of expenditure at regional or district level is crucially important not solely for the limited objective of enhancing adoption of relevant technology but for the general goal of improving welfare.

Not only could research and adoption be coordinated in this manner but the whole aspect of rural development would be inter-linked with the technical R&D process. This, of course, is exactly what FSR aspires to and sometimes has achieved. Potentially, a model of the type outlined could greatly assist this process and therefore it is worth examining the necessary characteristics of such a model.

1. It would need to be a model of a farm/farm-family which is representative of a group. Classification could only be achieved by way of rural survey to designate socio-agro-climatic groups (Carter 1990). Consequently, the model would be generic in form so that it could be applied to farms representative of various groups without structural modification. The model would then receive additional data from representative farms in order to particularise it for the specific target group. As with the IBSNAT crop models the question of a Minimum Data Set becomes relevant: this would be the minimum set of socio-cultural and economic data required to set-up the generic model to run in any given circumstance. All the usual riders about validation of generic models obviously apply.
2. There is an obvious requirement to have appropriate biological models for major farm enterprises. It is doubtful that much reduction in the biological detail presently encompassed in say, the IBSNAT crop models would be acceptable because of the nature of the technology that will need to be explored. Reorganisation of such models will be inevitable to permit farmers fields to be simulated in parallel and to permit independent cropping sequences in individual fields with a variable fallow period between crops. These are required so that decisions related to allocation of resources (labour, equipment, credit) between competing enterprises can be taken on

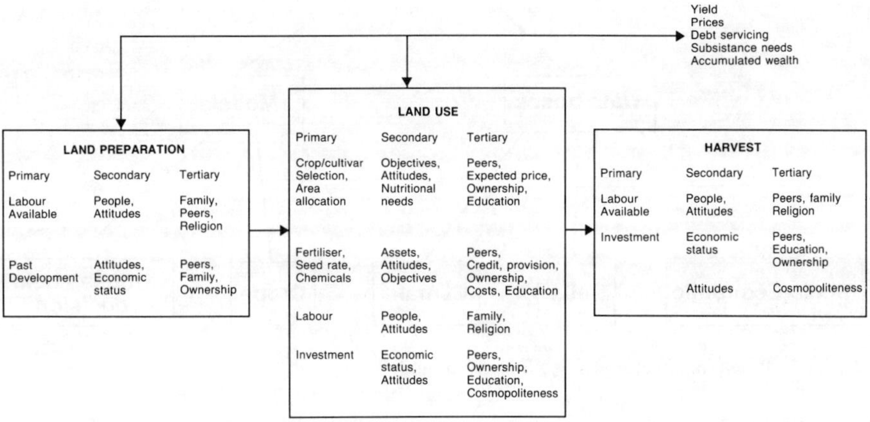

Figure 4. A schematic framework showing factors influencing key farm management decisions.

a seasonal basis. Some improvements in current biological models will be necessary, particularly in relation to the ability to simulate pest/disease incidence and development and also the competitive forces of intercropping. Work is progressing in both these areas.

3. The model must incorporate components dealing with the resource availability on the farm and the socio-cultural elements of the farm family. These components perhaps might define traditional behaviour, the level of innovativeness, the resources available, and the tenure arrangements. These are broadly defined because there is a fundamental lack of understanding of these social, economic and cultural sub-systems (Gasson 1973).

In seeking to pursue the objective of creating a functional generic model with the above characteristics, two interlinked lines of investigation are necessary: one to develop the basic framework; the other to consider data provision.

A framework

A schematic potential framework is presented in Figure 4 in which a preliminary attempt is made to distinguish between 'primary', 'secondary' and 'tertiary' elements involved in the major farm management decisions. Figure 4 is meant to be illustratory in nature rather than comprehensive but it is clear that the central decision relates to land use: this sets the whole pattern of farming and to some extent the costs and output from the farm. The types of crop grown, the area allocated to each, and the cultivar selected will be related to the pressures from peer groups, cultural norms, religion and tradition which shape the family objectives, attitudes and relationships. The dynamics of the family farm system is represented in Figure 4 by indicating that the ability to create a timely seed bed is dependent on investment in equipment and/or motive power which, in turn,

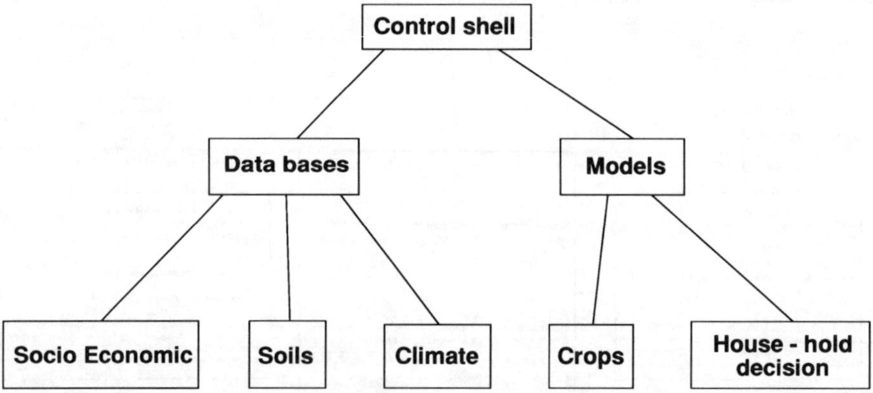

Figure 5. Potential structure for a whole-farm model.

is related to immediate and long-term past physical and financial performance as well as for example attitude to and provision of credit.

The general structure of Figure 4 is a long way from a functional computer model. The framework development to some extent is a matter of applying past experience but it is likely to improve as the question of data is examined. Even from a preliminary view it is obvious that the kind of framework that might eventually be created cannot bear much structural resemblance to the crop simulation models previously discussed. So many of the mechanisms and relationships simply have not been explored, never mind placed in to a quantitative relationship.

Because of this, it seems likely that any basic structure will involve rules-based algorithms, relying on experiential data to express many of the behavioural characteristics. A first concept might be that the overall model will take the form of a shell which will call, as appropriate, enterprise simulation models (crop models for example), databases and an expert system, reflecting the farm household decision process. The rules base approach of the expert system permits a formal structure while at the same time handling both quantitative and qualitative data. Weis and Kulikowski (1984) have defined an expert system as a computer structure which incorporates expert human reasoning that should reach similar conclusions to the expert given the same data sets. So, expert systems can make use of databases and models just as a human expert can. In circumstances of data scarcity rules may be formulated within the expert system in a manner similar to the expert using 'rules of thumb' when data are limited.

Data

The likely main components of the whole farm model are illustrated in Figure 5. Three data-bases are shown, two of which (for climate and soil) are already well known. The socio-economic data base is not yet explored but will have endogenous and exogenous elements: endogenous data will include land tenure

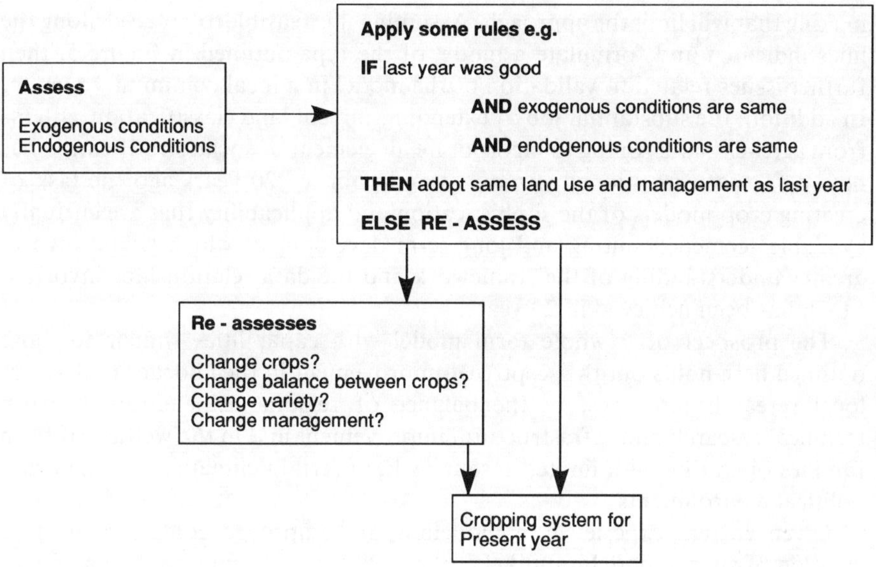

Figure 6. Initial rules for household decision model.
(Figure from Edwards-Jones 1991.)

arrangements, economic status and resources, family size, age and education and a range of cultural factors including traditional values and beliefs; exogenous factors will include credit provision and cost, market prices for inputs and products, and availability of extension services. These data would be managed within the rules based household decision model.

The model might be initiated in the manner shown in Figure 6 indicating that if immediate past experience with the current system has been satisfactory (meets nutritional and cash needs of the family) and if the exogenous and endogenous conditions appear unchanged then it is likely the farmer will continue with existing land use and management. Otherwise he will consider options. Rules for this type of consideration are likely to be complex and extensive. A start has been made, however, to define a suitable set of rules and these are displayed in the presentation by McGregor et al. (1991) and Sharma et al. (1990) have recently applied a similar format to represent smallholder decision making in agro-forestry systems in India.

Conclusion

To some extent the integration of systems simulation with FSR is a speculative issue. Development of the type of models that might be required to bring integration about will be a difficult process. It is unlikely that either computer hardware or relevant software will be limitations. Rather, it is a matter of lack of full understanding of the socio-cultural complex of farm household decision making that will limit the approach. Assuming it is feasible to proceed along the lines indicated and formulate a model of the type outlined in Figure 5, then further issues related to validation of the model in a local community remain. In addition, the substantial job of extending current land classification systems from agro-climatic zoning to incorporate an element of social stratification has not so far been mentioned. From a stand-point of 20 years ago the task of creating crop models of the sophistication and applicability that are currently available seemed daunting and long term development which would involve greater understanding of the framework and the data relationships involved. Much has been achieved in 20 years.

The prospect of a whole farm model with capabilities similar to those outlined here holds out the sceptre of improvement in the allocation of scarce local research resources, in the balance of regional expenditure between technical research and infrastructural improvement and in the welfare of farm families operating with limited resources in uncertain climatic, economic and political environments.

Given current experience in modelling and improved computer facilities creation of such a model is unlikely to take 20 years, even given the paucity of formal socio-cultural data.

References

Baldwin R L, Hanigan M D (1990) Biological and physiological systems: animal sciences. Pages 1–21 in Jones J G W, Street P R (Eds.) Systems Theory Applied to Agriculture and the Food Chain, Elsevier Applied Science, London, UK.

Biggs S D (1985) A farming systems approach: Some unanswered questions. Agric. Administration 18:1–12.

Bywater A C (1990) Exploitation of the Systems Approach in Technical Design of Agricultural Enterprises. Pages 61–88 in Jones J G W, Street P R (Eds.) Systems Theory Applied to Agriculture in the Food Chain, Elsevier Applied Science, London, UK.

Carter S E (1990) A survey method to characterise spatial variation for rural development projects. Agric. Syst. 34:237–258.

Collinson M (1988) The development of African Farming Systems: Some personal views. Agric. Administration & Extension 29:7–22.

Crowther E M, Yates F (1942) Fertilizer policy in wartime. The fertilizer requirements of arable crops. Emp. J. Exp. Agric. 9:77–98.

Dent J B, Edwards-Jones G (1991) The context of modelling in the future: the changing nature of R&D funding in future. Aspects of Applied Biology 26:183–193.

Doyle C J (1990) Application of Systems theory to farm planning and control: Modelling resource allocation. Pages 89–112 in Jones J G W, Street P R (Eds.) Systems Theory Applied to Agriculture and the Food Chain, Elsevier Applied Science, London, UK.

Gasson R (1973) Goals & values of farmers. J. Agric. Economics XXIV:521–38.

Godwin D C, Thornton P K, Jones J W, Singh U, Jagtap S S, Richie, J T (1989) Using IBSNAT's DSSAT in Strategy Evaluation. Proc. IBSNAT Symposium, 81st Ann. Meeting of Amer. Soc. Agronomy, Las Vegas: 59–71. USA.

Hildebrand P E (1990) Farming systems research-extension. Pages 131–143 in Jones J G W, Street P R (Eds.) Systems Theory Applied to Agriculture and The Food Chain, Elsevier Applied Science, London, UK.

Hunt L A, Jones J W, Richie J T, Teng P S (1989) Genetic coefficients for the IBSNAT crop models, Proceedings of IBSNAT Symposium, 81st Ann. Meeting of Amer. Soc. Agronomy, Las Vegas: 15–30.

Lightfoot C (1987) Indigenous research and on-farm trials. Agric. Administration & Extension 24:79–89.

Polson R A, Spencer D S C (1991) The technology adoption process in subsistence agriculture: The case of casava. Southwest Nigeria, Agricultural Systems 36:65–78.

Rhoades R E, Booth R H (1982) Farmer back to farmer: A model for generating acceptable agricultural technology. Agric. Administration 11:127–38.

Richie J T, Singh U, Godwin D C, Hunt L A (1989) A users guide to CERES Maize: Version 2.10, International Fertilizer Development Centre, Muscle Shoals, USA.

Rountree J H (1977) Systems thinking – Some fundamental aspects. Agric. Syst. 2:247–54.

Sharma R A, McGregor M J, Blyth J F (1990) The socio-economic evaluation of social forestry in Orissa (India). Int. Tree Crops J. (Submitted).

Simmond N W (1981) Farming Systems Research – A Review. World Bank Technical Paper No.43. Washinton, USA.

Spedding C R W (1979) An Introduction to Agricultural Systems, Elsevier Applied Science Publishers, London, UK.

Wake J L, Kiker C F, Hildebrand P E (1988) Systematic learning of agricultural technologies. Agric. Syst. 27:179–93.

Weiss S M, Kulikowski C A (1984) A practical guide to designing Expert Systems. Rowman and Allan, New York, USA.

Making farming systems analysis a more objective and quantitative research tool

L. STROOSNIJDER[1] and T. VAN RHEENEN[2]

[1] *Department of Irrigation and Soil and Water Conservation Wageningen Agricultural University, Nieuwe Kanaal 11, 6709 PA Wageningen, The Netherlands*

[2] *Interdisciplinary Research Training Project (INRES), Malang Brawijaya University, Kotak Pos 176, Malang, Indonesia*

Key words: development scenario, farm household survey, Farming Systems Analysis, land evaluation, multiple goal linear programming, optimization, socio-economic data

Abstract

Farming Systems Analysis (FSA) should lead to insight in the functioning of farming systems and help to develop various alternative scenarios for development.

New as well as old developments in research technologies and farm dynamics are used to improve FSA. Technical options are considered through a Quantified Land Evaluation (QLE) approach distinguishing three levels of inputs. With optimal inputs, a constraint free environment is assumed so that potential production is achieved. At the second level, water limited conditions are assumed with optimal nutrient supply, and at the third level, water as well as nutrient constraints may occur.

Economic conditions at the farm are investigated with the aid of a Farm Household Survey (FHS). This survey also yields a wealth of agronomic, animal husbandry and sociological information. Decision making processes are studied, giving more insight into the factors which contribute towards certain conclusions made by the farmer.

Development possibilities that fulfill different aims at the farm level under various constraints are explored making use of Interactive Multiple Goal Linear Programming (IMGLP) providing feasible development scenarios and their trade-offs.

The methods reported in this paper are being developed and tested in the limestone area of East Java by an interdisciplinary team of Indonesian and Dutch scientists.

Introduction

There are probably few research approaches in the agricultural world today that have received so much attention and popularity amongst scientists, particularly those working in the developing world, as the farming systems approach. This can be concluded from the enormous amount of literature that has appeared since the mid 1970s. A systems approach implies "studying the system as an entity made up of all its components and their interrelationships, together with relationships between the system and its environment. Such a study may be undertaken by perturbing the real system itself (e.g. via farmer managed trials or by pre- versus post adoption studies of new technology) but more generally it is carried out via models (e.g., experiments, researcher and/or farmer managed on-farm trials, unit farms, linear programming and other mathematical simulations) which to varying degree simulate the real system" (TAC

F.W.T. Penning de Vries et al. (eds.), Systems Approaches for Agricultural Development, 341–353.

1978). Over the years, different variants of systems approaches in agricultural research have been published. A specific group of approaches characterized with key-words like 'farmer's orientated' or 'bottom-up' is commonly referred to as farming systems research and development (FSR&D). Shaner et al. (1982) define FSR&D as "an approach to agricultural research and development that (i) views the whole farm as a system, and (ii) focuses on the interdependencies among components under the control of members of the farm household and how these components interact with the physical, biological, and socio-economic factors not under the households' control. The approach involves (i) the diagnostic phase: selecting target areas and farmers, identifying problems and opportunities; (ii) the development phase: designing and executing on-farm research; and (iii) the implementation phase: evaluating and implementing the results. In the process, opportunities for improving public policies and support systems affecting the target farmers are also considered". Expectations were high in the 1970s. However, looking back it becomes clear that FSR&D has not always been able to live up to these expectations.

Farming systems analysis (FSA) is the initial and crucial stage of FSR&D and comprises the above step (i) and partly (ii). FSA is the understanding of the structures and functions of farming systems, the analysis of constraints on agricultural production at farm level and ways to translate this understanding into adaptive research programmes (Fresco 1988). In other words, FSA is a tool that may be used to set a research agenda. The basic steps in FSA are (i) diagnosis: the analysis of farming systems and the identification of constraints; and (ii) design: the step from diagnosis to research, both on- and off-station.

The next section describes how, with the aid of new and old developments in research techniques and farm dynamics, an attempt is made to improve the methodologies used for FSA. With the aid of knowledge gained from soil, crop and livestock sciences, development options will be explored at farm level. The feasibility of these options will then be examined. This analysis will take into account the constraints that exist at farm level, and not the constraints that exist at levels higher than the farm level. An Interdisciplinary Research Training project (INRES) is developing and testing these improvements and executes an FSA study in the limestone area south of Malang on the eastern part of the island of Java, Indonesia. The research team comprises seven staff members of the Malang University representing five disciplines and two Dutch scientists with support of interdisciplinary task groups of the Brawijaya University in Malang, Wageningen Agricultural University and the State University of Leiden. It is envisaged that the methodologies being developed will overcome some of the shortcomings of FSA. These shortcomings will be discussed in the section 'Critique of farming systems analysis'. When it appears successful, FSA will become more than a tool for cropping and livestock systems optimization to which it has evolved at present despite its definition. The last section of this paper will present some preliminary conclusions.

INRES FSA methodology

This section describes the techniques used for the development of a new FSA methodology. Quantification of cropping and livestock subsystems is needed and this will be described in the following sub-section where also attention will be given to the interactions of these components in mixed systems. The next sub-section will focus on the quantification of the socio-economic components in a farming system and the last sub-section will present the technique used to combine technical and socio-economic information. This integration, as developed in on-the-job research in the INRES-project, goes into a direction similar to a theoretical framework being developed for the FAO called LEFSA (Fresco et al. 1990).

Analysis of the technical components of farming systems

For the methodology, as it is being developed at the INRES project, the following information will be required from bio-technical disciplines: (i) potential and attainable production levels for various product groups on well-defined land units in the considered region; product groups are represented by: (a) single cropping, e.g. cassava, maize, and (b) intercropping, e.g. cassava/maize and perennials; (ii) the technologies related to the attainable and potential yields; variation in agronomic methods, such as fertilizer use, soil and water conservation, pest and disease control, etc.; (iii) analysis of the reasons for the yield gaps between potential and attainable and between attainable and actual yields for the various product groups; (iv) worked-out concepts of the ways to sustain production potential: concepts of soil erosion and degradation, ways to maintain structure and depth of soils; (v) per livestock type, actual and potential technologies with related inputs and outputs.

To provide the information, the bio-technical disciplines will combine disciplinary knowledge in an extended Quantified Land Evaluation (QLE). For more details on quantified land evaluation, see Driessen (1986; 1988), Van Diepen et al. (1989) and Van Lanen (1991). In this quantified land evaluation approach, a farming system is defined as a combination of different land use systems, practised by a household on the basis of decisions made in response to physical factors, its own priorities, and external incentives.

A land use system is a combination of a land utilization type and a land unit. A land unit is an area that can be considered homogenous with regard to the defined land utilization type. A land utilization type is a collection of 'key attributes', i.e. biological and technical aspects of the production environment that are relevant to the production capacity of the land. Examples are: crop(s) grown, animals kept, utilization of inputs like implements, labour, fertilizer, etc..

A land unit can be characterized from a 'supply' and a 'demand' point of view. From the supply view a land unit can be described by a number of relevant characteristics which form together the land quality. The demand point of view is determined by the land utilization type (i.e. the crops that one wishes to grow and the cropping techniques one wishes to use) and includes both physical and non-physical requirements which form together the land use requirements.

Quantified Land Evaluation matches the land qualities with the requirements, not, as was and still is common practice in most land evaluation methods, through a simple rating system, but by using dynamic crop modelling. Only then can an optimal matching between the varying (in space and time) requirements with similarly varying qualities be achieved. In this analysis the socio-economic attributes are tentatively considered exogenous and invariate.

Production potential is obtained by modelling production in three steps: (i) unconstrained (potential) production without water and nutrient stress in a pest and diseases free environment, (ii) water-limited production and (iii) water and nutrient-limited production. Modelling needs input data which are obtained from an intensive survey on representative farms with regards to land utilization and land utilization types with corresponding land qualities and land utilization requirements.

The above quantified land evaluation analysis will generate the following output:

1. a data base on land units with their land qualities;
2. a data base on land utilization types (key attributes) with corresponding land utilization requirements;
3. estimates of crop production in selected land utilization systems with specified (set) activities and inputs for the three hierarchical input levels;
4. estimates of animal production on selected farms with specified (set) activities and inputs.

Multiple land utilization systems (more than one crop on a land unit at one time) will be handled by combining single – land utilization system analysis taking into account effects exerted on the crops by each other. In order to be able to handle mixed farming systems an interfacing module, which describes the feedback mechanisms between on-farm primary and secondary production, will be used to link models for plant production with those for animal production. Use will be made of crop growth simulation models (e.g. Spitters et al. 1989) and livestock simulation models (Kingwell and Pannell 1987; Udo and Brouwer 1991). The cropping and livestock component of a farming system affect each other, both on the output as well as the input side and the interaction will be established by linking the crop growth and livestock simulation models.

Finally, the bio-technical disciplines will provide the data needed in the integration phase of the new methodology being developed, i.e. input data in a comprehensive farming system model. A major need is a number of development scenarios based on the previous analysis of constraint solving technologies. Development scenarios are options for development. These scenarios consist of sets of activities, inputs and production potentials to be used in input–output matrixes below.

Analysis of the socio-economic components of farming systems

As Byerlee et al. (1982) point out it is farmers, not fields, that make the decisions, and therefore socio-economic criteria may be just as important as land qualities and key attributes in determining farmers' activities. This implies that special attention will have to be given to both the socio-economic environment as well as to the decision making processes taking place within a household. The socio-economic environment can be split into three levels, the micro or household level, the meso or sectoral and regional level and the macro or national level. Variables that are exogenous at the micro level (prices) may be endogenous at the macro level (Erenstein et al. 1991). For the methodologies being developed knowledge at all three levels is required.

The contribution of the socio-economic disciplines to the development of the methodology will be in providing information on the following topics: (i) a detailed input – output analysis of the farm activities being practised at present in the region and their constraints; (ii) a summary of the most important changes that have taken place in the past five years; (iii) the reasons why farmers introduced technological innovations; (iv) the way farmers received information about possible innovations (extension services, farmer meetings, radio); (v) the objectives of the farmers and their priorities (profit maximization, risk minimization); (vi) the ways in which the farmers' social and cultural environment (norms and values) influences their activities; (vii) the activities that are gender-specific; (viii) information on the educational level of the household; (ix) organisations of which the farmers are members; (x) the gap between potential, attainable and actual production levels for the activities of the farmer and her or his household; and the socio-economic and technical explanation for these gaps; (xi) an indication of the stakeholders in agricultural development; (xii) contribution towards defining options for development; and (xiii) indications of demographic trends.

To gain a greater understanding of the socio-economic structures of the micro level, INRES conducted a Farm Household Survey (FHS), in which input – output data was collected for 36 selected farms, for both on-farm and off-farm activities. It would have been desirable to have had a larger sample than 36 farms. However, due to logistic constraints this was not feasible. Conducting surveys in order to gain information concerning rural households has been done very often; however, such detailed surveys as undertaken in this project are exceptional. For the development of the new methodology, a FHS as the one conducted by INRES is essential (for more details on the FHS, see Appendix). One may wonder whether it is necessary to collect data in such great detail and indeed Byerlee et al. (1982) argue against it. The authors, however, are convinced that if the aim is to formulate changes in technology or the environment in which the household systems are located, it is necessary to have detailed and accurate knowledge of the systems being studied. Not only quantitatively but also qualitatively knowledge of the farming systems increases tremendously with the data collecting system used by INRES.

Socio-economic data at the meso and macro level was collected with which projections can be made using simple models which take into account past trends, elasticities and base level data.

Decision making processes within the household were also studied. These studies were done using in-depth interviews with nucleus farms (selected farms for in-depth studies) mainly by looking back at the last 5 years and considering important decisions that were made within the household as well as decisions that were made during the FHS. Special attention was given to decisions that involved technological innovations and for this purpose use was made of several existing theories on decision making (Barlett 1980, Huijsman 1986, Van Dusseldorp 1991). An attempt was made to determine the reasons for the household to reach certain decisions, where did they get their information from, which decisions are farmers likely to take considering their objectives, and which cultural factors determined these decisions? Information gained from the FHS and the decision making analysis together with estimates on demographic developments are to be used to judge, select and adapt the scenarios proposed by the technical disciplines, as mentioned before. This is a challenging task since socio-economic disciplines hitherto often limit their analysis to socio-economic changes in the rural society based on a comparison of the present situation with that of the past.

Development scenarios are to be established at the farm level, in combination with the farmers, as well as for the region, reflecting policy goals. It is not yet clear whether the village (council) is an independent stakeholder in agricultural development which has to be taken into account.

Linking the bio-technical and socio-economic components of farming systems

To link bio-technical and socio-economic components of farming systems Interactive Multiple Goal Linear Programming (IMGLP) is used. This mathematical programming technique is more suitable than econometric modelling techniques to explore development options, taking into account technological innovations and risk acceptance at farm level. With IMGLP various goals can be taken into account, as well as their trade-offs be illustrated. IMGLP was developed by Spronk and Veeneklaas (1983) and its value for agricultural development is described by Van Keulen (this volume). For illustrations of its application on a regional basis, one is referred to De Wit et al. (1988), Veeneklaas (1990) and Van Keulen and Veeneklaas (this volume). However, IMGLP has so far not been used at the household level, taking into account all the activities being practised at present and data generated by quantified land evaluation as inputs.

Development scenarios identified above will be used for further study. From these scenarios certain goals can be established. An input – output matrix will be put together containing all existing and alternative activities using generated data. IMGLP involves a number of iteration cycles. During the first cycle the lower bounds of all the determined goals are set at their minimum requirements. The user will obtain a feasible solution that satisfies these minimum requirements. Each goal is then maximized on its own, with the lower bounds of the other goals defined as minimum restrictions. After the first cycle a situation may be reached where for each of the goals no better value may be obtained than the one calculated, and a value less favourable than the minimum goal restriction generated will be unacceptable. In continuing cycles, one or more goals may be tightened and the iteration cycles will be repeated for the other goals. The choice of the goal restriction and the degree in which they are tightened will depend on the user and on her or his specific interest. In the course of tightening goal restrictions the solution space will be narrowed until it will not be possible to improve on any of the goals without sacrificing on any of the other. The opportunity cost of one goal can then be expressed in terms of the other goals. This provides the various stakeholders with a clear insight in the trade-offs between the different (and often conflicting) goals.

The interesting aspect of IMGLP actually is that there is interaction not only between the user and the model, but also between the user and the stakeholders in agricultural development. For the purposes of the INRES project two main groups of stakeholders can be distinguished: farmers and policy makers. Within these two main groups there will be sub-groups, for example, small farmers and large farmers, policy makers responsible for the agricultural sector and policy makers responsible for the financial situation of the region's administrative institutions. Representatives of these groups will be identified and with them the user of the IMGLP will interact. IMGLP has been applied for regional development in Mali and for evaluation of economic policy in the Netherlands. The following 10 points describe how at various levels the decision makers can be included in IMGLP.

1. A vector of minimum goal values will be presented to the decision makers (farmers, policy makers), together with a set of potential improvements within the set of feasible solutions, of these minimum goal values.
2. The decision makers are asked to indicate whether or not they find the solutions meeting minimum requirements satisfactory.
3. If not, the decision makers are asked to give an indication which minimum goal values will have to be increased.
4. On the basis of a the new vector of minimum goal values, a new set of the potential improvements of these values is calculated and presented to the decision maker.
5. As a result of the new vector of minimum goal values, there will be a shift in the indicated minimum goal values. The question will arise for the decision maker: is this shift outweighed by the shift in the potential values of the other goal variables?
6. If the shift is unacceptable, the decision maker gets the opportunity to revise her or his earlier wishes with respect to the changed minimum goal value.
7. If the shift is acceptable the decision maker can continue to raise any of the other or even the same minimum goal value.

8. A reduction will take place in the set of feasible solutions, and the decision maker will have to decide whether or not to continue or to stop.
9. When the decision maker decides to stop, she or he can select a suitable solution from the set of solutions satisfying the minimum conditions. Each time a set of Pareto optimal solutions has been produced. One refers to a 'Pareto optimal solution' when it is impossible to improve the welfare of one individual or group of individuals without reducing the welfare of another individual or group of individuals, see also Alocilja and Ritchie (this volume).
10. If the decision maker so wishes, a set of feasible solutions satisfying the minimum conditions on the goal variables can be subjected to a second analysis.

IMGLP will show how the farming systems could look like in 'n' number of years seen against the background of the development scenarios of the various stakeholders in rural household development. For each of the development scenarios it is conceivable to assume that the type of activities the household will be practising in 'n' years will be different. An indication will be given of the various development pathways. One of the goals of INRES, i.e. to incorporate sustainability in long-term development, is supposed to be achieved in this way. It is realized that when going from the household level to the regional level, aggregation biases will occur, because not all household are alike. Hazell et al. (1986) state that ideally a model should be constructed for every individual farm, and all individual models linked together form a sector model. This is of course not feasible and it will be necessary to work on the basis of representative household groups. The effects of an optimal regional situation on variables, which are exogenous at farm level, but need not be at regional level may also become subject of further research at the INRES project. However, at present attention will be focused at the farm level.

Critique of farming systems analysis

Methodologies used in FSA are documented in Byerlee and Collinson (1980), Conway (1985) and Collinson (1987). The diagnostic phase usually includes a study of background information, an informal survey (rapid rural appraisal/sondeo) and a formal verification survey. Collinson (1987) mentions that the output of a good diagnosis will include: (i) the identification of problems for which experiments may be done with a priority ranking; (ii) assessment of the extent in which certain technological innovations are suitable with the system and a system – wide cost/benefit analysis for each innovation; (iii) a description of the characteristics of target group farmers and farms as a basis for the choice of representative locations where on-farm experiments may be executed; (iv) a description of current husbandry practises for setting the levels of non-experimental variables for experiments to be done by scientists and for evaluating farmer management; (v) considering the farmers' circumstances, the identification of realistic treatments for the experiments; and (vi) an assessment of the possible ways in which the farmers will judge the results gained from the experiments.

FSA as it has been practised over the years has been subjected to criticism and problems. The methodologies described above will, hopefully, overcome the following criticisms.

1. FSA can be vulnerable to subjectivity. Strong emphasis is laid on the participation of the farmer in determining the main constraints to be solved, i.e. a bottom-up approach. In practise, however, FSA can be vulnerable to subjectivity, i.e. it may be scientist top-down biased. This can be the case when the scientist perceives the problems of the farmer and decides the priority for problem solving, often not considering the interaction between the various activities being practised by the farmer.
2. FSA is mainly farmer orientated. It should be, but it should not only be farmer orientated. The farming systems lie in a region and the region will be administered by policy makers. These policy makers also have certain development scenarios for the region. The instruments they may chose to use (subsidies, taxes, infrastructure, etc.) will influence the 'operational space' of the farming systems. In a region, for example, where the main crop being cultivated is cassava, it is conceivable that policy makers will want to stimulate the cultivation of cash crops (e.g. coffee). Subsidies may be given to farmers to grow the cash crop, while at the same time measures may be taken to discourage farmers from growing cassava. In this case research resources may be better spent analyzing the transition possibilities from cassava to coffee. While in most countries it will be the farmer who eventually decides which crop will be grown, at the same time the farmer will also be influenced by his environment and by policy makers.
3. FSA has mainly been crop oriented. Norman (1978) mentions that FSA is somewhat a misnomer. He notes that to date research has been mainly confined to crop production processes and that the approach has rarely been applied to livestock processes. He continues to state that the other areas generally omitted from consideration to date are off-farm enterprises and a more holistic systems approach, which goes beyond the farm gate and attempts to endogenize, for example, the marketing process.
4. FSA has suffered from institutional problems (Collinson 1982; Moscardi et al. 1983; Gilbert et al. 1980). It is argued that recommendations may be rejected because they are inappropriate to the institutional setting for which they were designed. Programmes would become more realistic, appropriate and acceptable if they took account of the capabilities, resources and past activities of the host institutions. Heinemann and Briggs (1985) further stresses that only with the active and constructive support of the local staff and farmers can there be a self-sustaining problem-solving research system. At the INRES project the Indonesian staff from a local university with a strong commitment to provide a scientific basis for rural development play a dominant role in the execution of the programme.

5. FSA is confronted with time conflicts. As discussed in Norman (1978), FSA is confronted with time conflicts in two ways. Firstly, in the FSR approach a conflict exists between short-run private gains and long-run social costs. If only the farmer is allowed to indicate the constraints in his system, these will tend to be biased towards the former, which could exacerbate the latter. The linear programming methodology described will enable the user to also take into account the long-run social costs. Secondly, there is inevitably a time lag in the recognition of a problem, the finding of a relevant solution and its adoption by farmers (Norman 1978). The use of multiperiod linear programming can be of aid in simulating the time gap and making the options for development more realistic.
6. FSA has been too qualitative. This has made it a difficult tool for policy makers and scientists to accurately assess the problems in a region. Determining the order in which problems would have to be addressed has been also obscure as a result of this.
7. FSA has concentrated insufficiently on gender differentiation. Numerous studies have pointed out that many household activities are gender specific. Consequently this will have an impact on the type of activities as well as on the extent in which they are practised by members of the household. Certain solutions proposed on the basis of FSA may therefore not be feasible as they are not in conformity with the realities of on-farm circumstances.
8. There has been no unification of FSA methods. In the literature one comes across many different descriptions of how FSA should be conducted. However, each approach will be different. If standardization of these methods were possible, this would reduce costs for future FSA studies.

Conclusions

The methodology being developed by INRES makes use of and integrates various research techniques. These are quantified land evaluation, crop and livestock simulation models, FHS data collection techniques and decision analysis techniques. Development scenarios for the farming systems will be established together with representatives of the main groups of stakeholders in rural agricultural development. From the development scenarios certain development goals can be established. With IMGLP options for development can be investigated.

The present methodology aims at overcoming part of the criticism concerning FSA. Emphasis will be on farmers and policy makers as the main decision makers. The methodology will give due attention to all groups of stakeholders in rural development, and it will be suitable for gender differentiated analysis without losing its broad perspective of which so many woman orientated studies suffer. Both on-farm and off-farm activities will be subjected to analysis. Considering the cooperative nature of the INRES project and the active and constructive support that has been received from the

Indonesian staff, the implementation problems mentioned by Heinemann and Briggs (1985) could be overcome. In order to make the exercise more cost effective it is of crucial importance that the present experience lead to a clear framework for future studies. Such a framework should be accompanied by user's manuals, software and a standardized data collection system.

The authors do not have the illusion that methodologies being developed at the farm level are a panacea to rural development. Studies at the regional level are also required. In the study attention will primarily be focused at the farm level, where variables which are determined at the regional or macro level (e.g. prices, subsidies, taxes, etc.) will be exogenous. It will therefore not be correct to aggregate farm level optimizations to the regional level without taking into consideration the behaviour of variables that are endogenous at higher levels. An aggregation from the farm to the regional level should also take into account the aggregation bias. However, if some of the problems in the previous section are solved, FSA will have been made a more objective and quantitative research tool.

Still pressing however are institutional and organizational issues that the final outcome of the INRES project will face. The impact of the methodology being developed will remain limited unless it is part of a larger long term rural development effort, so that non-agricultural, non-experimental variables (such as prices, marketing, input supply etc) can also be tackled effectively.

Appendix

The data collection system is based on the Farm Management Package (FARMAP) of the FAO (1988). However, the system was adapted in such a way that use could be made of a spreadsheet for data entry which is widely available and easy to handle. For a complete description, the reader is referred to Moll (1990). The FHS used by the INRES project monitors all the activities that are done by or for members of the Farm Household. This means that the data gathered allow a gender specific analysis. The activities can be divided into two main categories, farm activities and off-farm activities. Farm activities are all activities (agricultural and non-agricultural) practised on the farm managed by the members of the household whereas activities practised outside the farm are categorized as off-farm activities. INRES defined members of the household as those people who lived under the same roof of the head household for more than 6 months per year in total and each member was given a specific number. The land managed by the farm household was divided into parcels, which are defined as one unit of land with more or less the same characteristics, and with the same tenure status. A parcel may be divided into two or more plots if the crops or crop combinations grown on them differ. Thereby, parcels refer to the actual land use. Whenever an operation is performed by a member of the household or is performed for the benefit of the household the information is recorded in DBase datafields.

Example:
Household number 112 grows maize on plot 2 which is situated in parcel 1. The farmer has spent 16 hours weeding on this plot, in period 3 of the survey. The data will be entered as follows:
Datafield 1: 112 (code of the household)
Datafield 2: 10 (code for the activity 'maize')

Datafield 3: 1 (code for the parcel)
Datafield 4: 2 (code for the plot)
Datafield 5: 1 (code for the type input – output)
Datafield 6: 41 (code for the type of operation)
Datafield 7: 3 (code for the period)
Datafield 8: 16 (quantity: hours, kilograms, etc.)
Datafield 9: 1 (code given to the source of the input if it came from the household)
Datafield 10: not applicable (price/value)
Datafield 11: 2 (cash/kind)

Enumerators interviewed the farmers every 6 days, during one year and at the same time independent measurements were also taken, in order to check the information quantitatively.

References

Alocilja E C, Ritchie J T (1992) Multicriteria optimalization for a sustainable agriculture. Pages 383–397 in Penning de Vries F W T, Teng P S, Metselaar K (Eds.) Systems Approaches for Agricultural Development, Proceedings of the International Symposium on Systems Approaches for Agricultural Development, 2–6 December 1991, Bangkok, Thailand (this volume).

Barlett P G (ed.) (1980) Agricultural Decision Making: Anthropological contribution to rural development. Academic Press, New York Inc, USA. 378 p.

Byerlee D, Collinson M P (1980) Planning Technologies appropriate to Farmers: Concepts and Procedure. CIMNYT, El Batan, Mexico. 71 p.

Byerlee D, Harrington L, Winkelman D L (1982) Farming Systems Research: Issues in Research Strategy and Technology Design. Amer. J. Agric. Econ. 64:897–904.

Collinson M (1982) Farming Systems Research in East Africa: The Experience of CIMMYT and Some National Agricultural Research Services, 1976–'81. International Development Paper No. 3, Department of Agricultural Economics, Michigan State University, USA.

Collinson M P (1987) Farming Systems Research: Procedures for Technology Development. Expl. Agric. 23:365–386.

Conway G R (1985) Agroecosystem Analysis. Agricultural Administration 20:31–55.

De Wit C T, Van Keulen H, Seligman N G, Spharim I (1988) Application of Interactive Multiple Goal Programming Techniques for Analysis and Planning of Regional Agricultural Development. Agric. Syst. 26:211–230.

Driessen P M (1986) Quantified Land Evaluation (QLE) Procedures, A New tool for Land Use Planning. Neth. J. Agric. Sci. 34:295–300.

Driessen P M (1988) The QLE Primer: A first Introduction to Quantified Land Evaluation procedures. Landbouwuniversiteit, Wageningen. 120 p.

Erenstein O, Schipper R (1991) Land Use Planning: An Application of Multilevel and Multiobjective Linear Programming Models. Department of Development Economics. Agricultural University of Wageningen, Hollandseweg 1, 6706 KN Wageningen, The Netherlands. 178 p.

FAO (1988) The FARMAP, Farm Survey Data Analysis Package. Farm Management and Production Economics Service, AGSP, FAO Rome, Italy.

Fresco L O (1988) Farming Systems Analysis, An Introduction. Tropical Crops Communication No. 13. Dept. of Tropical Crop Sciences, Wageningen Agricultural University, P.O.Box 341, 6700 AH Wageningen, The Netherlands.

Fresco L, Huizing H, Van Keulen H, Luning H, Schipper R (1989) Land Evaluation and Farming System Analysis for Land Use Planning. FAO Guidelines: Working document. 176 p. + appendices. Rome, Italy.

Gilbert E H, Norman D W, Winch F E (1980) Farming Systems Research: A Critical Appraisal.

Rural Development Paper No. 6, Department of Agricultural Economics, Michigan State University, USA.

Hazell P B R, Norton R D (1986) Mathematical Programming For Economic Analysis In Agriculture. Macmillan Publishing Company, New York, USA. 400 p.

Heinemann E, Biggs S D (1985) Farming Systems Research: An Evolutionary Approach To Implementation. Journal of Agricultural Economics. 36:59–65.

Huijsman A (1986) Choice and Uncertainty in a Semi Subsistence Economy: A Study of Decision Making in a Philippine Village. Royal Tropical Institute, Mauritskade 63, 1092 AD Amsterdam, The Netherlands. 335 p.

Kingwell R S, Pannell D J (1987) MIDAS, a bioeconomic model of a Dryland Farm System. Simulation Monograph, PUDOC Wageningen, The Netherlands. 207 p.

Moscardi E et al. (1983) Creating an On Farm Research Programme in Ecuador. Working Paper 01/83, Economics Programme, CIMMYT, El Batan, Mexico.

Moll H A J (1990) Farm Household Data Collecting System. Department of Development Economics. Agricultural University of Wageningen, Hollandseweg 1, 6706 KN Wageningen, The Netherlands.

Norman D W (1978) Farming Systems Research To Improve the Livelihood of Small Farmers. Amer. J. Agric. Econ. 60:813–818.

Shaner W W, Philipp P F, Schehl W R (1982) Farming Systems Research and Development: Guidelines for developing countries. Westview Press, Boulder, Colorado, USA.

Spitters C J T, Van Keulen H, Van Kraalingen D W G (1989) A simple and universal growth simulator: SUCROS87. Pages 147–182 in Rabbinge R, Ward S A, Van Laar H H (Eds.) Simulation and Systems Management in Crop Protection. Simulation Monographs. PUDOC, Wageningen, The Netherlands. 420 p.

Spronk J, Veeneklaas F R (1983) A Feasibility Study of Economic and Environmental Scenarios by Means of Interactive Multiple Goal Programming. Regional Sci. and Urban Econ. 13:141–160.

Technical Advisory Committee (TAC) (1978) Review Team of the Consultative Group on International Agricultural Research. Farming systems research at the International Agricultural Research Centres. The World Bank, Washington, D.C., USA.

Udo H M J, Brouwer B O (1992) Concepts and practices for Systematically Analyzing the Livestock Component of Farming Systems. In Proceedings International seminar on livestock and feed development in the Tropics, Commission of the European Communities and LUW–UNIBRAW Animal Husbandry Project, Malang, Indonesia.

Van Dusseldorp D B W M (1991) Integrated Rural Development and Interdisciplinary Research, A link Often Missing. Wageningen Agricultural University, Hollandseweg 1, 6706 KN Wageningen, The Netherlands.

Van Diepen C A, Wolf J, Van Keulen H, Rappoldt C (1989). WOFOST: a simulation model of crop production. Soil Use and Management 5:16–24.

Van Keulen H (1992) Options for agricultural development: a new quantitative approach. Pages 357–397 in Penning de Vries F W T, Teng P S, Metselaar K (Eds.) (1992) Systems Approaches for Agricultural Development, Proceedings of the International Symposium on Systems Approaches for Agricultural Development, 2–6 December 1991, Bangkok, Thailand (this volume).

Van Keulen H, Veeneklaas F R (1992). Options for agricultural development: a case study for Mali's fifth Region. Pages 369–382 in Penning de Vries F W T, Teng P S, Metselaar K (Eds.) Systems Approaches for Agricultural Development, Proceedings of the International Symposium on Systems Approaches for Agricultural Development, 2–6 December 1991, Bangkok, Thailand (this volume).

Van Lanen H A J (1991) Qualitative and Quantitative Physical Land Evaluation: An Operational Approach. PhD-thesis Wageningen Agricultural University, The Netherlands. 196 p.

Veeneklaas F R (1990) Dovetailing Technical and Economic Information. PhD-Thesis. Erasmus University Rotterdam, The Netherlands. 159 p.

Options for agricultural development: a new quantitative approach

H. VAN KEULEN
Center for Agrobiological Research (CABO-DLO),
P.O.Box 14, 6700 AA Wageningen, The Netherlands

Key words: animal husbandry, Australia, cropping system, development scenario, Egypt, fertilizer, land use planning, multiple goal linear programming, Negev, nutrient-water limited yield, optimization, QUEFTS, wheat, WOFOST

Abstract
A basis for agricultural development planning is a thorough analysis of the options for agricultural development. Such an analysis aims at the identification of the 'best' use of land. This definition implies the existence of generally accepted objectives. However, in practice different actors in the development process often have different, and at least partially conflicting objectives. To make a well-founded choice in these conditions, it is necessary to define on one hand the technical possibilities in quantitative terms, and on the other hand all possible development objectives. Confrontation of the two then should indicate to what degree the various objectives are attainable, to what degree they are conflicting and what the 'exchange value' among the objectives is.

In this paper a method is presented that can be used as a tool in such an analysis. The technical possibilities are characterized by the technical coefficients of various production techniques, derived partly from simulation models, partly from farming systems analysis. For evaluation of the degree of realization of the various objectives the technique of interactive multiple goal linear programming (IMGLP) is employed, based on definition of the regional constraints and the development objectives. Application of the technique indicates to what extent the available techniques can meet the alternative demands under various policy options and different socio-economic conditions.

Introduction

Regional agricultural development, with 'regional' referring to a specific geographical part of a national economy, can be defined as the dynamics of changes in infrastructure and technology that are pursued to improve existing agricultural systems and the welfare of the population. A technology in this context is understood as a well-defined way of converting inputs into agricultural products. In general, a wide variety of technologies is available, in terms of both the commodities and the mix of inputs used to produce them, but not all these technologies are feasible in the physical environment and the socio-economic context of a given region. The physical environment is mainly characterized by the relevant soil properties and the prevailing weather conditions, whereas the socio-economic environment is characterized by the regional capital and human resources and constraints, and the prices of inputs and products in the region. Analysis of options for agricultural development in a region, as a basis for planning should result in selection of the 'best'

F.W.T. Penning de Vries et al. (eds.), Systems Approaches for Agricultural Development, 355–365.

development option. As 'best' can only be defined in relation to well-defined criteria, this definition implies the existence of established and generally accepted objectives in regional planning. However, various actors involved, all with different interests at stake in the development process, leads to, at least partly, conflicting objectives. To do justice to all actors in this situation, and arrive at a well-founded choice, the technical possibilities in a region, as dictated by the natural resource base and the available production techniques, should be identified on the one hand and all possible development objectives on the other, and their viability should be analyzed in the context of the prevailing socio-economic situation and its possible modifications.

In this paper a method is presented that may assist in exploring the agro-technical possibilities and quantifying the necessary technical coefficients, to identify the possible degree of goal attainment, and to judge the economic viability of different development options.

Interactive multiple goal linear programming

Land use planning is an essential component in the analysis and planning of regional agricultural development. For effective planning it is necessary to quantify the technical options on the basis of the regional resources and constraints and identify the objectives for development. Emphasis on different objectives, such as for example contribution to foreign exchange, self-sufficiency in food production, risk-avoidance, achievement of parity income for the rural population, may lead to different development options, with their associated differences in choice of production techniques. Any development plan for a region must be technically feasible and it must take into account all the possible goals imposed on the region and the constraints to satisfy the various goals.

The method described here can be used to evaluate the agricultural potentials of a region and to analyse to what extent the available techniques can meet the demands under various constraints, under various policy options and under different socio-economic conditions. The input requirements and the investment needs also follow from the analysis.

The method

The method, only briefly explained here (for a full description cf. De Wit et al. 1988), is based on a linear programming approach that optimizes a mix of production techniques, subject to a set of constraints. The production techniques are defined as 'activities', each yielding certain 'outputs' and requiring certain 'inputs'. The inputs draw on resources that are limited, and may therefore be constraining application of the techniques or their level of intensity. When only one goal is pursued (optimized) the approach is straightforward. However, in real life, different goals, such as food production, export, employment, environmental protection, etc. are pursued, which at least are partially conflicting. In that situation, the choice for a certain development option becomes dependent on the relative value attached to each of the goals, which is not necessarily the same for different actors. The Interactive Multiple Goal Linear Programming technique (Veeneklaas 1990; De Wit et al. 1988) allows attainment of an optimum solution by stepwise optimization of the various objectives. Initially, the lower bounds of all the goals considered are set at their minimum values, to ascertain that feasible solutions are obtained, that satisfy all these minimum requirements simultaneously. Each of the goals is then optimized on its own, with the lower bounds on the other goals set as minimum goal restrictions. For each of the goals this first cycle yields the most favourable value that can be expected and the most unfavourable value that has to be accepted. The first cycle defines the total solution space, 'the feasible area'. However, the ideal situation where all the goals reach their maximum value simultaneously does not exist. The most satisfactory solution from the point of view of a particular 'user' may now be obtained in successive iteration cycles by tightening one or more of the goal restrictions and repeating the optimization for one or more of the other goals. The choice of the goal restrictions and the degree to which they are tightened reflect the specific interests of the user. During the stepwise maximization of various goals, under increasingly tighter restrictions on the other goals, the solution space is gradually reduced until a situation is reached where the user cannot improve on any of his goals without sacrificing on another one. In that way he becomes aware of the opportunities for exchange between the various goals in his desired solution space, i.e. he obtains the opportunity costs of one goal in terms of the other goals.

Different actors in the development process may have different objectives or attach different weights to the various goals, and may therefore end up in different corners of the solution space. In terms of analyzing options for agricultural development, this implies that in interactive contact with different interest groups (government, development agencies, local population) different desired options could evolve. The method furthermore allows to explore the possibilities for a compromise that is satisfactory to all interest groups, even though it is not ideal for anyone in particular.

Regional analysis, farming systems analysis and planning

When the method described above is applied to regional analysis and planning in the field of agriculture, the activity matrix contains 'all' existing and conceivable production techniques for a region, including those that may still be in the Research and Development (R&D) pipeline. These may include cropping activities, animal husbandry activities, and any other activities related to the agricultural sector. For production activities currently practiced in a region, the technical coefficients in the matrix, which quantify the inputs and outputs for implementing and operating each activity, may be obtained from available statistical information and from farming systems analysis. For production activities not yet practiced coefficients may be obtained from crop

growth simulation models and animal production models. The relevant production techniques should be identified through land evaluation.
The resources of the region (or constraints) include the area and the 'quality' of the various land types available, which have to be defined on the basis of land evaluation. In addition, other resources, such as the population living in the region, its demographic composition and participation rate, additional labour that may be available from outside the region, endowment of capital goods, crop rotation constraints, animal breeds and herd sizes present in the region, etc., are also included in the model. Again, most of these data will have to be derived from farming systems analysis and rural surveys, as well as from statistical sources.

In applying the model, a distinction is made between tradeables and non-tradeables. Prices are attached only to goods and services that can be traded across the border of the region, such as fertilizers, products from arable farming (e.g. grains, tubers, fibers) and from animal husbandry (e.g. meat and milk) or to persons that have an alternative employment in other sectors of the economy, as is the case with labour of the local population, if off-farm employment opportunities exist. Non-tradeables, for example labour of the local population for which no alternative employment exists, or land that can only be used for activities included in the model, or products that cannot be easily transported such as straw and organic manure, often do not have a directly observable price. However, these goods and services may have an opportunity cost, and therefore an implicit price.

The results

The analysis results in (i) identification of consistent, technically feasible development options, for what is regarded as the most satisfactory combination of all goal variables; (ii) identification of the major constraints for such developments; (iii) evaluation of the costs of greater achievement of one goal in terms of sacrifices on the other goals and the constraints, which can lead to identification of technical bottlenecks and constraints; (iv) translation of the selected combinations of goal achievement into a combination of activities, i.e. the mix of production techniques (e.g. cropping activities and animal husbandry activities), necessary to achieve the goals, the needs for investments, imports, exports and credit facilities, the labour requirements and their qualifications, etc.

The method of analysis is not an econometric one, with many (often uncertain) behavioural relations. Social constraints, like unequal accessibility to the means of production, land titles, or economic behavioural patterns are also not taken into account. The analysis therefore does not 'predict' the future development of a region, but it identifies technically feasible development options, that best attain a certain set of goals. A further analysis is necessary ('post-model'), to identify, among others, the policy measures necessary to realize the required developments (cf. Van Keulen and Veeneklaas 1992). That, however, is beyond the scope of this paper.

Technical coefficients from mechanistic crop growth models

Over the last two decades the system-analytical approach to crop ecology has resulted in the development of many crop growth simulation models, that combine the insights in the factors and processes that determine crop growth and yield in such a way, that quantitative estimates can be made of the yield potential of the main agricultural crops under a wide range of environmental conditions and major constraints can be identified (Seligman 1990; Penning de Vries et al. 1989; Van Diepen et al. 1989; De Wit and Van Keulen 1987; Van Keulen et al. 1987; Jones and O'Toole 1987; Van Keulen and Wolf 1986; Whisler et al. 1986). In first instance, comprehensive models have been developed, mainly aiming at increasing understanding of the interactions between the crop and the main growth factors, which mainly served as research tools (De Wit et al. 1978). On the basis of their results, more simplified versions, so-called 'summary models' (Penning de Vries 1982), have been developed and the range of applications has increased, among others for quantified land evaluation (Van Keulen et al. 1987; SOW 1985).

Many objections have been raised to the use of deterministic crop growth models, ranging from disenchantment with the method altogether (Monteith 1981; Passioura 1973), through the problems associated with their data requirements, the 'parameter crisis' (Burrough 1989a), the stochastic nature of the input data used (Burrough 1989b), and the fact that model results necessarily pertain to 'single events' which causes application problems in a spatially and temporally variable environment, to the complaint that the models cannot reproduce the actual situation. However, development of such models provides the opportunity (or rather, creates the necessity) to formulate consistent quantitative opinions on the behaviour of the systems under consideration, their potentials and the biophysical constraints that are operative. The consequences of alternative opinions can therefore easily be made explicit, and as such these models form a tangible basis for discussion.

In the framework of examining the options for agricultural development, deterministic crop growth models find their major application in the formulation and quantification of alternative production techniques that are not (yet) practiced in a region, but have potential applicability, in view of the prevailing agro-ecological conditions (Van Keulen 1990).

A typical example of such a deterministic crop growth model is WOFOST (Van Diepen et al. 1988), that simulates growth of an annual crop during one growing season in daily intervals, using a state variable approach. This assumes that the state of the system at any moment can be described in quantitative terms and that changes in the state can be described by mathematical equations, that contain the state of the system at that moment, and driving variables. Major physical and physiological processes such as CO_2 assimilation, respiration and phenological development are quantitatively described, and the exchange processes with the environment as CO_2 uptake, transpiration, and water uptake are incorporated. The rates of all these processes are described as a function of the state of the crop at any moment and the controlling environmental conditions.

The effects of the main yield-determining factors are evaluated following a hierarchical approach. At the highest hierarchical level the number of factors that is considered is restricted, by assuming that constraints that can feasibly be removed technically, have indeed been eliminated. At subsequently lower hierarchical levels increasingly more factors are taken into account. Hence, first potential yield is calculated, reflecting the genetic potential of the crop under the prevailing weather conditions, that determine the duration of the growth period and the length of the various phenological phases (temperature) and the rate of growth during that period (solar radiation). Such yields, that assume optimum growing conditions throughout the growth period, are achieved in agricultural practice for instance in Western Europe and in South American plantation crops. In most developing countries such yields are not aimed for, but they may serve as a yardstick against which the current situation and possible future developments can be measured.

At the next level water-limited yield is calculated, taking into account periods of water shortage and/or excess water. To quantify the soil water balance, in addition to rainfall, soil physical properties with respect to transport and storage of water are considered. These calculations provide the dynamics of water in the soil, from which moisture availability for the crop is derived. The relative availability, i.e. in relation to crop water demand, is used to quantify the effect of soil moisture conditions on carbon dioxide exchange and growth. This analysis thus quantifies the possible yield-reduction, resulting from sub-optimal soil water conditions, and the requirements for irrigation and/or drainage.

At the next hierarchical level the effects of the major plant nutrients are quantified, to arrive at nutrient-limited yield. Nutrient availability from natural sources is estimated in this approach using the QUEFTS system (Janssen et al. 1990), and translated into crop yield by assuming maximum dilution of the elements in the tissue (Van Keulen and Van Heemst 1982). These calculations also quantify the amounts of fertilizer required to attain either water-limited or potential yield.

On the basis of the results of these simulation models, target yields at the various production levels are identified. To realize these target yields, in addition to material inputs, such as irrigation water, fertilizers and crop protection agents, labour and monetary inputs are necessary. The material inputs can be directly derived from the results of the simulation models, while for the remaining ones application of the general body of knowledge is required, supplemented by information from local Farming Systems Analysis (Fresco 1988).

Table 1. Definition of three typical regions.

Socio-economic characteristic	'Negev'	'W. Australia'	'W. Desert'
Prices – outputs (\$ kg^{-1})			
Local market price, lamb	2.4	1.5	2.4
Export price of lamb	1.7	1.2	1.7
Export price of wheat	0.16	0.15	0.16
Prices – inputs			
Imported concentrates (\$ kg^{-1})	0.15	0.17	0.15
Phosphorus fertilizer (\$ kg^{-1} P)	0.40	0.60	0.40
Nitrogen fertilizer (\$ kg^{-1} N)	0.30	0.60	0.30
Hired labour (1000 \$ $person^{-1}$ y^{-1})	10.0	10.0	5.0
Regional constraints			
Permanent settlers (persons)	400	100	1000
Local market for lamb (1000 t mutton)	10	5	20

Validation

One of the major difficulties in application of the proposed method for regional development planning is validation of the model. In other words, do the results inspire enough confidence to justify their use as guidelines for regional development and research planning? As the objective is to explore the options for regional development, rather than predicting them, establishing data for model validation is generally impossible. However, some confidence in model performance can be gained by comparing model results under different environmental conditions. For example, it can be examined, whether the model can respond in a reasonable way to different socio-economic scenarios.

The case of the Mediterranean region

In order to test the performance of the type of model described, an input-output table for mixed farming (sheep husbandry, combined with rainfed cultivation of wheat and legumes) in a Mediterranean region (De Wit et al. 1988) was exposed to widely varying socio-economic conditions, to see if it could reproduce realistic technology mixes, and how it responds to smaller changes within such situations.

Three different socio-economic regions were defined that roughly reflect the situation in the northern Negev of Israel, a region in the West Australian wheat belt and the coastal strip of the Western Desert of Egypt. All three areas are semiarid with a Mediterranean-type climate. Dryland farming and extensive animal husbandry are the major agricultural practices. The population density is highest in the Egyptian region and lowest in the Australian region; the distance from markets is largest in the Australian region and smallest in Israel.

Table 2. Summary of main results of 'typical region' development features (mid-development phase).

Item	'Negev'	'W. Australia'	'W. Desert'
Cumulative consumption* (M$)	244	64	282
Wheat area (1000 ha y^{-1}):			
- Continuous wheat	9.12	0	10.91
- Wheat/fallow rotation	0	1.15	0
- Wheat/legume rotation	7.82	17.07	4.82
Total wheat area	16.94	18.21	15.73
Straw area usage (1000 ha y^{-1}):	13.03	9.25	13.32
Concentrate imports (1000 t)			
- Obligatory concentrate	69.06	0	67.9
- Concentrate, replaceable by legume	2.64	0	6.3
Fertilizer use (1000 t):			
- Nitrogen	1.06	0.33	1.24
- Phosphorus	0.53	0.68	0.46
Grain utilization (1000 t):			
- Exported from region	20.71	10.05	20.70
- Used in the region	0	7.84	0
Lamb and mutton marketing (1000 t)			
- Exported from region	4.33	5.79	0
- Used in the region	9.92	0.50	14.93

* Cumulative consumption over the 15-year development horizon.

The local market for lamb is largest in the Egyptian region. In order to keep the test conditions relatively simple and open to fairly direct analysis, model performance was checked in relation to one goal only: maximization of revenue for consumption. The socio-economic environment is defined in terms of the prices of outputs and bought inputs, and in terms of regional constraints like the number of 'permanent settlers' and the local demand for lamb and mutton. This is a highly simplified definition of a region, but should be sufficient to illustrate the effect of regional characteristics on development options and technology selection for development. A more detailed definition is given in Table 1. To simplify the comparison, the region has been assumed to comprise 50000 ha in all three cases. The results of the runs are given in Table 2. The model produces strikingly different results and technology mixes in the different socio-economic regions. Total regional income in the 'West Australian' scenario is lowest, but income per capita is the highest. The cropping system there is overwhelmingly a wheat/legume rotation whereas in the 'northern Negev' and the 'Western Desert' of Egypt it is mainly continuous grain cropping. In the 'West Australian' scenario, the use of nitrogen fertilizer is lowest while the use of phosphorus fertilizer is highest; no imported concentrate feed is given at all, while in the other regions, it is used heavily. Whatever concentrate feed is given for fat-lamb production comes from the locally produced grain.

Despite the simplistic nature of this exercise and the fact that alternative feasible technologies like wool or sheep milk production were not included among the available technologies, the results of these runs reflect current development trends in the northern Negev of Israel and in the north-western coastal zone in Egypt. The trends suggested by the model for Western Australia are much less realistic, because the main relevant activities there are based on fine wool Merino sheep in which the wool is the main product and the mutton is a by-product. These activities were not considered in the present version of the model. Including them requires data on the Australian sheep husbandry budget, that can then be translated into a vector of technical coefficients. That would then enable planners to use the model to explore innovative technologies, including some that these days are being discussed intensively by scientists and pastoralists in Australia. The technologies concern the introduction of specialized fat lamb production systems based on the Awassi breed that is particularly suitable for the live sheep export trade to the Middle East. This, as well as many other hypothetical alternative production systems could be assessed with the model, before investing the huge amount of capital involved. In the same way, this model can be adapted to other situations, such as semiarid regions in South America or the southern parts of the USA.

Conclusions

The interactive multiple goal linear programming technique can help to decide on feasible development possibilities in a region, within a wide range of technical and socio-economic conditions, and as such forms a powerful tool in regional agricultural planning. The validity of the results obtained depends largely on the accuracy of the technical coefficients in the activity matrix, on the degree of realism of the scenarios that govern the interactions between the activities, and on proper definition of the goal variables. The results of analyses with this method can be used as a basis for discussion with various interest groups in a region, and can help in making the consequences of goals and aspirations explicit. It should be emphasized that the analysis does not provide a prediction of what will happen, but a consistent picture of the technical capabilities of a region within a well-defined (socio-)economic environment. Within regional agricultural planning projects it may be applied to examine the possibilities for alternative land use plans under different conditions.

In recent years many new tools have been developed with potential applicability in regional development planning (Fresco et al. 1990; Van Diepen et al. 1991). Within the framework of this article not all of these could be treated, but some have been illustrated, indicating that a major development is the application of quantitative methods, combined with geo-referenced data bases, allowing consistent analyses of possibilities and the effects of physical and (socio-)economic conditions.

References

Burrough P A (1989a) Modelling land qualities in space and time: the role of geographical information systems. Pages 45–60 in Land qualities in space and time. Bouma J, Bregt A K (eds.) Proc. ISSS Symp., Wageningen, Pudoc, Wageningen.

Burrough P A (1989b) Matching spatial databases and quantitative models in land resource assessment. Soil Use and Management 5:3–8.

De Wit C T, Van Keulen H (1987) Modelling production of field crops and its requirements. Geoderma 40:253–265.

De Wit C T, Van Keulen H, Seligman N G, Spharim I (1988) Application of interactive multiple goal programming techniques for analysis and planning of regional agricultural development. Agric. Syst. 26:211–230.

De Wit C T et al. (1978) Simulation of assimilation, respiration and transpiration of crops. Simulation Monographs, Pudoc, Wageningen, The Netherlands. 141 p.

Fresco L O (1988) Farming systems analysis. An introduction. Tropical Crops Communication No. 13. Dept. of Tropical Crop Science, Wageningen Agricultural University, P.O.Box 341, 6700 AH Wageningen, The Netherlands.

Fresco L, Huizing H, Van Keulen H, Luning H, Schipper R (1990) Land evaluation and farming systems analysis for land use planning. FAO Guidelines: Working document. 176 p. + appendices. Rome, Italy.

Janssen B H, Guiking F C T, Van der Eijk D, Smaling E A M, Wolf J, Van Reuler H (1990) A system for quantitative evaluation of the fertility of tropical soils (QUEFTS). Geoderma 46:299–318.

Jones C A, O'Toole J C (1987) Application of crop production models in agro-ecological characterization. Pages 199–209 in Bunting A H (Ed.) Agricultural Environments. CAB International, Wallingford, Oxon, England.

Monteith J L (1981) Epilogue: Themes and variations. Plant and Soil 58:305–309.

Passioura J B (1973) Sense and nonsense in crop simulation. J. Austr. Inst. agric. Sci. 39:181–183.

Penning de Vries F W T (1982) Phases of development of models. Pages 20–25 in Penning de Vries F W T, Van Laar H H (Eds.) Simulation of plant growth and crop production. Simulation Monographs, Pudoc, Wageningen, The Netherlands.

Penning de Vries F W T, Jansen D M, Ten Berge H F M, Bakema A (1989) Simulation of ecophysiological processes of growth of several annual crops. Simulation Monograph 29, Pudoc, Wageningen, The Netherlands and IRRI, Los Baños, Philippines. 271 p.

Seligman N G (1990) The crop model record: promise or a poor show? Pages 249–263 in Rabbinge R, Goudriaan J, Van Keulen H, Penning de Vries F W T, Van Laar H H (Eds.) Theoretical production ecology: reflections and prospects. Simulation Monographs 34, Pudoc, Wageningen, the Netherlands.

Stichting Onderzoek Wereldvoedselvoorziening (SOW) (1985) Potential food production increases from fertilizer aid: A case study of Burkina Faso, Ghana and Kenya. Vols. I and II. CFWS, P.O.Box 14, Wageningen, The Netherlands.

Van Diepen C A , Rappoldt C, Wolf J, Van Keulen H (1988) CWFS crop growth simulation model WOFOST. Documentation Version 4.1, CWFS, P.O.Box 14, Wageningen, The Netherlands.

Van Diepen C A , Wolf J, Van Keulen H, Rappoldt C (1989) WOFOST, a simulation model of crop production. Soil Use and Management 5:16–24.

Van Diepen C A, Van Keulen H, Wolf J, Berkhout J A A (1991) Land evaluation: from intuition to quantification. Adv. Soil Sci. 15:139–204.

Van Keulen H (1990) A multiple goal programming basis for analysing agricultural research and development. Pages 265–276 in Rabbinge R, Goudriaan J, Van Keulen H, Penning de Vries F W T, Van Laar H H (Eds.) Theoretical production ecology: reflections and prospects. Simulation Monographs 34, Pudoc, Wageningen, The Netherlands.

Van Keulen H, Wolf J (1986) Modelling of agricultural production: Weather, soils and crops. Simulation Monographs, Pudoc, Wageningen, The Netherlands. 479 p.

Van Keulen H, Van Heemst H D J (1982) Crop response to the supply of macronutrients. Agric. Res. Rep. 916, Pudoc, Wageningen, The Netherlands. 46 p.

Van Keulen H, Berkhout J A A, Van Diepen C A, Van Heemst H D J, Janssen B H, Rappoldt C, Wolf J (1987) Quantitative land evaluation for agro-ecological characterization. Pages 185–197 in Bunting A H (Ed.) Agricultural Environments. CAB International, Wallingford, Oxon, England.

Van Keulen H, Veeneklaas F R (1992) Options for agricultural development: A case study for Mali's fifth region. Pages 369–382 in Penning de Vries F W T, Teng P S, Metselaar K (Eds.) Systems Approaches for Agricultural Development, Proceedings of the International Symposium on Systems Approaches for Agricultural Development, 2–6 December 1991, Bangkok, Thailand (this volume).

Veeneklaas F R (1990) Dovetailing technical and economic information. Ph.D.Thesis, Erasmus University of Rotterdam, The Netherlands. 159 p.

Whisler F D, Acock B, Baker D N, Fye R E, Hodges H F, Lambert J R, Lemmon H E, McKinion J M, Reddy V R (1986) Crop simulation models in agronomic systems. Adv. Agron. 40:141–208.

Options for agricultural development: a case study for Mali's fifth Region

H. VAN KEULEN[1] and F.R. VEENEKLAAS[2]

[1] *Center for Agrobiological Research (CABO-DLO), P.O.Box 14, 6700 AA Wageningen, The Netherlands*

[2] *Present address: Winand Staring Center for integrated land, Soil and Water Research (SC-DLO), P.O.Box 125, 6700 AC Wageningen, The Netherlands*

Key words: animal husbandry, arable farming, crop production technique, development scenario, fisheries, institutional constraints, land use planning, Mali, multiple goal linear programming, optimization, risk, socio-economic constraints, subtainability, yield, nutrient water limited

Abstract

Identifying options for agricultural development requires a method that allows quantification of the technical possibilities, definition of the regional constraints and identification of the development objectives. Such a method, based on the interactive multiple goal linear programming (IMGLP) technique, has been applied in the framework of land use planning for the fifth Region in Mali.

The results of the analysis show that financially the most attractive sector is animal husbandry. A high degree of food self-sufficiency can be achieved, but requires intensification of arable farming, which is not economically viable.

Application of the method in this actual situation shows it to be a useful tool for integration of knowledge from various disciplines and to be able to serve as a means of communication among scientists, planners and policy makers. Using this method thus contributes to a more efficient application of knowledge in establishing agricultural development options for regional planning.

Introduction

Regional agricultural development can be defined as the dynamics of changes in infrastructure and technologies that are pursued to improve existing agricultural systems and the welfare of the population. In general, a wide variety of technologies, i.e. well-defined ways of converting inputs into agricultural products is available, both in terms of the commodities, and in terms of the mix of inputs used to produce them. But not all of these technologies are technically feasible and/or economically viable in the physical environment and the socio-economic context of a given region. The physical environment can be characterized by the relevant soil properties and the prevailing weather conditions; the socio-economic environment is characterized by the regional capital and human resources and constraints and the prices of inputs and products in the region.

Analysis of options for agricultural development in a region, as a basis for planning aims at selection of the 'best' development option. However, in actual practice, often various actors with different interests are involved in the

F.W.T. Penning de Vries et al. (eds.), Systems Approaches for Agricultural Development, 367–380.

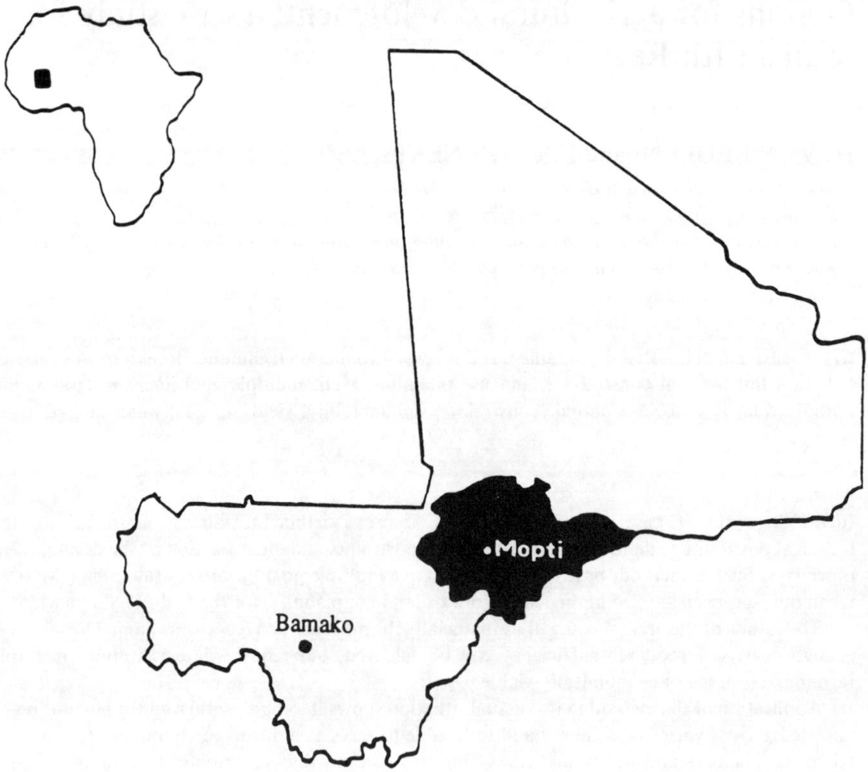

Figure 1. The location of the 5th Region of Mali.

development process, so that different, at least partly conflicting objectives exist, thus precluding the unequivocal definition of 'best'. To arrive at an acceptable choice in this situation, the technical possibilities in a region, as dictated by the natural resource base and the available production techniques, should be identified on one hand, and on the other hand an inventory of all possible development objectives or goals should be made, while taking into account the socio-economic situation.

In a previous paper (Van Keulen 1992), a method has been elaborated to explore the agro-technical possibilities and quantify the necessary technical coefficients, to identify the possible degree of goal attainment, and to judge the economic viability of different development options. In this paper, an application of that method as an aid in land use planning for one of the provinces of Mali, West-Africa, is presented.

Land use planning for the fifth Region of Mali

In the framework of the second five-year plan for the fifth Region in Mali

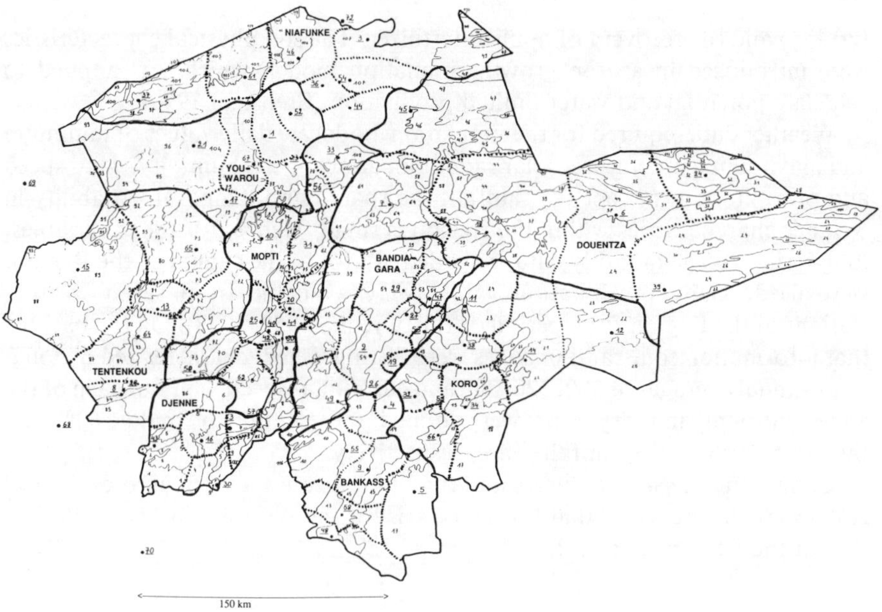

Figure 2. The Region and its eleven agro-ecological zones

(Figure 1), formulated in cooperation with the World Bank, a study on agricultural production systems in the Region was initiated. The study aimed at increasing insights in the agricultural production systems in the Region, and their major constraints, as a basis for the formulation of a land use plan, that would take into account the potentials of the natural resources and the development objectives of the various actors involved in the development process. The DLO-Center for Agrobiological Research (CABO-DLO) was requested by the Minister of Natural Resources and Animal Husbandry of Mali to assist in this study, which was considered a challenging opportunity to apply the method of interactive multiple goal linear programming (De Wit et al. 1988) in an actual situation. The results of the study have been extensively reported (Van Duivenbooden et al. 1991; Veeneklaas et al. 1991; Cissé and Van Duivenbooden 1990; Veeneklaas 1990) and in this paper only some highlights of methodology and results are presented.

Sources of data

The soil resource was defined on the basis of an extensive inventory of land resources (PIRT 1983). Seven main soil groups were distinguished on the basis of soil texture. These groups were further subdivided based on properties of the top soil, soil fertility status and the depth of the groundwater table, in a total of 18 subgroups. For each of the units distinguished, a representative set of soil physical and chemical properties was defined. The soil physical properties refer mainly to the water transport and storage characteristics, the soil chemical properties refer to the supply of plant nutrients from natural sources (soil fertility) and the recovery of applied fertilizer. The soil physical characteristics were introduced in a crop growth simulation model, which was applied to calculate potential and water-limited production (Erenstein 1990).

Weather data required for the simulation model are daily values of minimum and maximum temperature, solar radiation, atmospheric humidity, wind speed and precipitation. Except for rainfall, the spatial and temporal variability in weather characteristics within the Region is small. Hence, for those variables, long-term average monthly data, recorded at Mopti, the capital of the Region, were used. Daily precipitation values were available from seven weather stations in the Region for a period of thirty years (1959–1988). On the basis of that information four rainfall zones were distinguished, characterized by long-term annual averages of 300, 320, 382 and 485 mm, respectively. For each of the zones, 'normal' and 'dry' years were defined, representing the middle 60% and the driest 20% of the rainfall years, respectively.

Combining the soil resources and the rainfall zones, eleven agro-ecological zones were distinguished that form the basis for the agro-physical characterization of the Region (Figure 2).

Agricultural activities

Agricultural activities in the Region comprise arable farming, animal husbandry and fisheries. Each of the activities may take place anywhere in the Region, provided that the required inputs are available. For each of the activities different production techniques have been defined, characterized by a specific combination of technical coefficients, quantifying their inputs and outputs.

All agricultural activities should be sustainable, so that their production capacity in the long run is not jeopardized. Operational definitions of sustainability, therefore, had to be identified. For arable farming activities, sustainability is defined in terms of nutrient elements, i.e. the total amount of each of the macro-elements in the soil is assumed to be constant. This implies that in the absence of application of nutrients from external sources (manure or chemical fertilizer), export in agricultural products should not exceed the supply from natural sources, which is achieved by defining sufficiently long fallow periods. For animal husbandry systems, sustainability implies stable animal numbers for each of the species, based on sustainable forage production, which in addition to nutrient limitations is subject to persistency limitations, so that only a fraction of total annual production can be exploited. For fisheries, sustainability refers to a maximum quota of fish that can be caught annually.

In arable farming three cultivation types are considered: rainfed, flood retreat and irrigated or inundated. Each of these may be further sub-divided by crop species: millet, rice, sorghum, fonio, cowpea, groundnut, forage crops, shallots and 'other vegetables' (comprising tomatoes, tobacco, cassava, cabbage, etc.), although not each combination is relevant. For each crop, different technologies are defined on the basis of the following criteria: (i) application of fallow periods, (ii) oxen traction, (iii) application of farmyard

Table 1. Input-output table of millet production techniques on soil type C1. (Table continued on page 372.).

Characteristic	Extensive				Semi-intensive	Intensive
	1	2	3	4	5	6
Animal traction	–	–	+	+	+	+
Manure	–	+	–	+	+	+
Chemical fertilizer	–	–	–	–	+	+
Fallow	+	–	+	–	–	–
INPUTS [ha⁻¹ y⁻¹]						
FALLOW/MANURE/FERTILIZER						
Ratio fallow years/year cultivated*	5	–	6	–	–	–
Manure [kg DM]*	0	1 930	0	2 290	2 530	1 930
Fertilizer N [kg]*	0	0	0	0	12	96
Fertilizer P [kg]*	0	0	0	0	0	12
Fertilizer K [kg]*	0	0	0	0	0	56
Labour[a] [man days]6						
Cleaning the field	5	1	5	1	1	16
Transport and appl. of manure*	-	17.5	-	21	15.5	121
Basic dressing	-	-	-	-	1	11
Land preparation	3	3	4. + 2 At	4. + 2 At	4. + 2 At	12. + 6 At
1 Sowing	5	5	5	5	5	2. + 1 At
2 Weeding 1	15	15	10. + 2 At	10. + 2 At	10. + 2 At	10. + 2 At
2 Top dressing 1	-	-	-	-	4	4
2 Pesticide spraying 1	-	-	-	-	-	0.5
3 Weeding 2	12	12	12	12	12	12
3 Top dressing 2	-	-	-	-	-	4
3 Pesticide spraying 2	-	-	-	-	-	0.5
4 Harvesting*	5	5	6	6	5	12
6 Transport, threshing and winnowing*	16.5	16.5	13.5	13.5	19.5	46
Total	61.5	75	55.5 + 4 At	72.5 + 4 At	77. + 4 At	117. + 9 At

manure and (iv) application of chemical fertilizer. This leads to formulation of techniques at three levels of intensity: extensive, without any application of external inputs; intensive with high levels of such inputs, and semi-intensive with intermediate levels. Moreover, in intensive techniques a high degree of innovative practices is included. The degree of differentiation considered for each crop depends on its relative importance. For millet, a major crop in the Region, six different production techniques have been formulated, whereas for a minor crop like fonio (*Digitaria exilis*), only one production technique is included. Of course, for specific purposes alternative techniques can be included, provided that the technical coefficients are available. An example of an input-output table for millet production techniques (applying to a specified

Table 1. Continued.

Characteristic	Extensive				Semi-intensive	Intensive
	1	2	3	4	5	6
MONETARY INPUTS [FCFA]						
Capital charges						
Small equipment	700	700	700	700	1000	1500
Plough	-	-	1670	1670	1670	5260
Sowing machine	-	-	-	-	-	1600
Sprayer	-	-	-	-	-	1200
subtotal	700	700	2 370	2370	2670	9560
Operating costs						
Seeds	60	60	60	60	60	60
Pesticides	100	100	100	100	250	6500
subtotal	160	160	160	160	310	6560
Total	860	860	2530	2530	2980	16120
OXEN [ox]	-	-	0.33	0.33	0.33	0.75
OUTPUTS [ha^{-1} y^{-1}][b]						
Grain [kg DM]*	500	500	600	600	960	2 390
Straw [kg DM]*	1750[c]	1750[c]	1980[c]	1980[c]	2800[c]	4570[d]

At = Animal traction.

[a]) Numbers in front of operations refer to the period of the year

[b]) In a normal year in rainfall zone I (average precipitation in May-October: 530 mm).

[c]) Average N-content is 3.9 g kg^{-1}.

[d]) Average N-content is 5.1 g kg^{-1}.

*) Varies as function of activity.

soil type in a specified agro-ecological zone) is given in Table 1.

Animal husbandry systems in the Region comprise cattle, sheep, goats, camels, donkeys, horses, pigs, poultry and wild game at different degrees of importance. Different production techniques have been formulated on the basis of four criteria: (i) animal species (cattle, sheep, goats, donkeys and camels), (ii) main production objective (meat, milk, traction, transport), (iii) mobility of the animals (migrant, semi-nomadic, sedentary) and (iv) production level (low, intermediate, high). The inputs in these systems consist of required feed, specified according to season, spatial distribution and quality, labour and monetary costs. As an example the input/output table for cattle activities is given in Table 2.

The fishing activities are distinguished on the basis of the main occupation and the mobility of the households engaged in fishing: (i) households practicing fishing as main occupation, migrant (MFF); (ii) households practicing fishing as main occupation, sedentary (MSF); (iii) households practicing fishing as side activity, sedentary (SSF). The three household types vary in capital endowment

Table 2. Input/output table for cattle activities (TLU^{-1} y^{-1}); required quality of diet, intake of forage and concentrates (kg dry matter), total labour in the wet (W) and dry (D) season (man-day) and money (1000 FCFA), meat (Me, kg liveweight), milk (Mi, kg), number of animals for traction and transport (A), manure (Ma, kg dry matter).

Main prod.	Mobility	Diet	Intake Forage	Intake conc.	Labour W	Labour D	Money	Me	Mi	A	Ma
Oxen	sedentary	I	2000	-	2	8	2.3	22	0	0.55	442
Meat	semi-mobile	I	2000	-	3	8	2.3	37	0	-	298
Meat	semi-mobile	II	2000	-	3	9	3.5	56	92	-	285
Meat	migrant	I	2000	-	3	8	2.3	37	0	-	230
Meat	migrant	III	2100	-	3	9	3.5	71	219	-	222
Milk	sedentary	II	2100	-	3	9	2.3	54	165	-	444
Milk	sedentary	III	2200	-	4	10	3.5	62	376	-	445
Milk	migrant	II	2100	-	3	9	2.3	54	165	-	232
Milk	migrant	III	2200	-	4	10	3.5	62	376	-	232
Milk	semi-int.	IVc	1820	380	5	14	22.0	61	520	-	415
Milk	semi-int.	IV	2200	-	5	14	22.0	61	520	-	415

and in productivity. In contrast to the arable farming and animal husbandry activities, the available data did not allow calculation of target yields, hence the amount of fish captured and the required labour have been used as the basis for the input/output table (Table 3).

The regional constraints

Agricultural exploitation of the Region is subject to various constraints, of which the known physical and technical ones are incorporated in the model, but the institutional and/or social constraints that can be equally important cannot be taken into account directly in a formal quantitative model, because of lack of reliable quantitative insight. Hence, these must be taken into account in a post-model analysis.

The basic restriction for land is that it can be used for one purpose at a time only, thus introducing the competition for land. Moreover, not all land is suitable for agricultural exploitation, for instance due to degradation. To take that into account, each soil type in each agro-ecological zone has been assigned a so-called 'utility index', ranging from 1, if all land can be used, to 0, if none is suitable. In addition specific soil types may only be suitable for certain crops, such as land that is temporarily flooded. Finally, all land outside a radius of 6 kilometres from a permanent water point is considered suitable for pasture only. If techniques requiring a fallow period are selected, the specified area per unit cultivated land is used for fallowing.

With respect to labour, the year has been sub-divided in six periods, on the basis of the cropping calendar. The labour requirements for arable farming activities are specified in terms of these periods. For animal husbandry, only the wet and the dry season are considered separately. Where fishing is practiced as

Table 3. Input/output table for fishery activities MMF: migrant, main occupation fishing; MSF sedentary, main occupation fishing; SSF: sedentary, side activity fishing.

	Activity		
	MMF	MSF	SSF
INPUT (household^{-1} y^{-1})			
Labour (man-year)	3.62	3.62	1.81
Monetary inputs (1000 FCFA)			
depreciation equipment	182	155	31
Maintenance equipment	41	32	7
Fuel for motor boats	48	30	6
Firewood (normal/dry)	77/44	62/35	13/7
Total (normal/dry)	348/315	279/252	57/51
OUTPUT (household^{-1} y^{-1})			
Fish (ton fresh weight)			
Normal year	4.68	3.74	0.76
Dry year	2.68	2.14	0.43

the main occupation, labour is required throughout the year, whereas for the third activity labour is required during off-peak periods only. In the model the specified condition is that in each agro-ecological zone, in each of the periods labour demand should not exceed labour availability.

Animal traction is assumed to be provided solely by oxen, under the condition that in each agro-ecological zone the number of oxen pairs required should not exceed the number available, which is an output of the first animal husbandry activity in Table 2. Manure is also specified as output of animal production activities (Table 2) and used in arable cropping activities and in some agro-ecological zones as a substitute for firewood. In the model the condition is specified that per agro-ecological zone demand should not exceed the supply.

Forage availability from natural pastures is specified per agro-ecological zone, taking into account its temporal distribution (wet season and dry season) and its quality. In addition, availability and quality of crop residues and concentrates (both in the dry season only) are taken into account to estimate total feed availability. The total forage demand (differentiated in time and quality) per agro-ecological zone should not exceed the supply.

Total catch of fish in the Region is subject to an upper limit, that depends on flood level and is different, therefore for 'normal' years and 'dry' years.

Transport animals, donkeys and camels, are indispensable for daily life in the Region. However, it is difficult to assign them directly to specific agricultural activities. Therefore, the required number of transport animals is related to population density, for camels defined for the Region as a whole, for donkeys per agro-ecological zone.

The goals

The potentials of the multiple goal linear programming technique are best utilized if the number of goal variables is high and the number of constraints accordingly low. In that way, a high degree of flexibility is achieved, and the options for technically feasible development possibilities are kept as open as possible. In this study twenty variables have been defined that, in principle, can be optimized. However, most of these do not act directly as goal variables, but only serve to specify pre-set minimum or maximum levels. The goal variables can be sub-divided into a number of categories:

Physical production in a normal year, specified separately for: total millet, sorghum and fonio production; total rice production; total marketable crop production, i.e. total production minus subsistence needs; total meat production, comprising beef, mutton and goat meat; total beef production; total milk production, comprising cow, sheep and goat milk; total herd size

Monetary targets, specified separately for: total gross revenue, i.e. the total value of the marketable product minus the total monetary inputs plus the money generated by 'emigrants', that part of the base population that has left the agricultural sector; total monetary inputs in arable farming activities; total monetary inputs in animal husbandry activities; – total monetary inputs in arable farming, animal husbandry and fisheries combined.

Risks in a dry year, specified separately for: total millet, sorghum and fonio production; total rice production; total crop production; total regional grain deficit; sum of grain deficits per agro-ecological zone; total number of animals at risk.

Employment and emigration, separately specified for: total emigration; total employment.

Development scenarios

On the basis of the possible goals identified, and the constraints and relations included in the model, feasible development options for the Region can be generated. Each scenario is characterized by the goal optimized and the set of restrictions imposed on the other goals. As an example of the application of the technique, one particular goal has been optimized, total gross or monetary revenue, under two sets of goal restrictions. One set represents a more risk-taking attitude, the other is characterized by a more cautious attitude, and emphasis on avoidance of disaster in unfavourable conditions.

The more risky scenario (R-scenario) is characterized by permitted emigration of 250000 persons (one fifth of the original population of the Region); no strong demands on the level of food self-sufficiency; relatively large numbers of animals at risk in dry years.

Table 4. Maximum attainable regional gross revenue under different values of the goal restrictions.

		Maximum gross revenue (billion FCFA)
R-scenario		66.7
step 1:	emigration not exceeding 50000 (was 250000)	45.7
step 2:	total regional grain deficit not exceeding 110000 ton millet equivalents (was 150000) sum sub-regional grain deficits not exceeding 130000 (was 150000)	43.1
step 3:	total number of animals at risk not exceeding 100000 TLU (was 400000)	36.0
step 4:	rice production in normal year at least 42000 t (was 20000 t)	35.2
step 5:	monetary inputs in crop activities not exceeding 15 billion FCFA (was 20 billion)	33.7
step 6:	employment exceeding 336000 man-year (was 300000). = S-scenario	32.5

The 'safety-first' scenario (S-scenario) is characterized by a high level of food self-sufficiency; limited risks; balanced distribution of production over various agro-ecological zones; restricted emigration; high level of employment.

Results

In the R-scenario total regional gross revenue is optimized under relatively loose restrictions on the other objectives. In successive steps these restrictions are tightened one by one and the optimization repeated. The results of this procedure are given in Table 4.

Total regional monetary revenue varies from 66.7 billion FCFA (222 million U$) in the R-scenario to 32.5 billion in the S-scenario, or equivalent to a per capita income of 64000 and 26000 FCFA, respectively. The large difference in total gross revenue between the two scenarios can largely be explained by the restrictions on emigration and the number of animals at risk in dry years (Table 4). The rather low total revenue in both scenarios is mainly the result of the low profitability of arable farming, which in both cases yields negative revenues, which in addition to the unfavourable price ratios, is due to the satisfaction of the subsistence requirements for grain and the requirement of sustainable exploitation in terms of nutrient elements.

In both scenarios a number of the restrictions is binding, i.e. limiting attainment of a higher value for the optimized variable. Logically, this occurs more frequently in the S-scenario than in the R-scenario. To what extent the restrictions limit the value of the goal variable, is numerically expressed by its 'shadow price', defined as the change in the value of the goal variable at relaxation of the restriction by one unit. High shadow prices are exhibited by

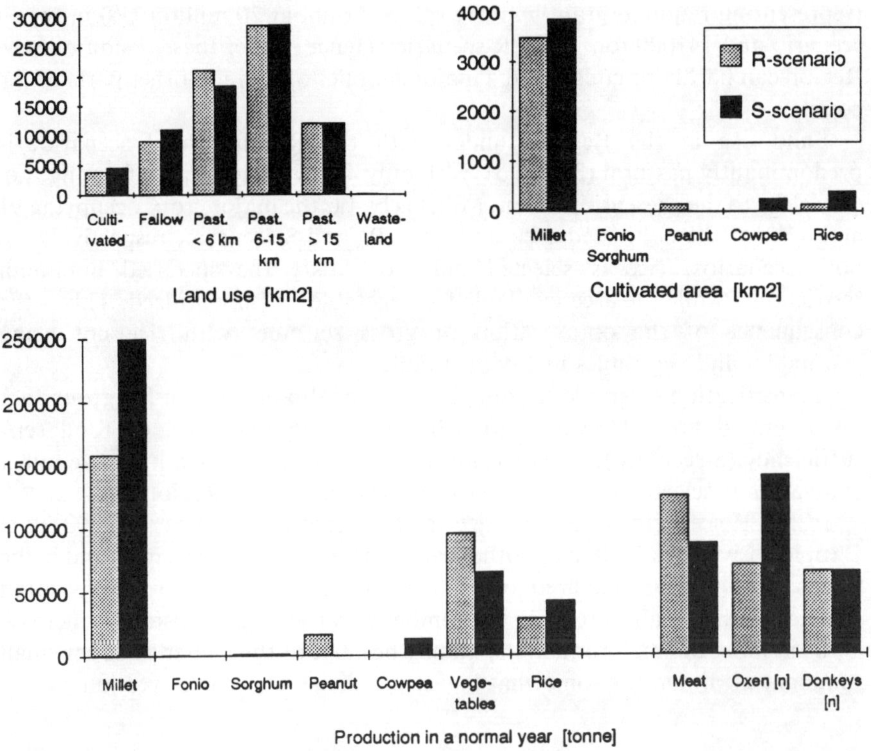

Figure 3. Land use (a) and cropping pattern on cultivated land (b) in the two scenarios

the restrictions 'animals at risk in a dry year' and 'permitted emigration': in the R-scenario the values are 18000 and 96000 FCFA, respectively, and in the S-scenario 54000 and 236000 FCFA, respectively. The high values for 'animals at risk' indicate the major role played by animal husbandry in the generation of regional revenue. For 'emigration' the direct effect is a smaller contribution from money generated outside the sector set at 75000 FCFA per man-year. However, the higher values of the shadow prices indicate that an additional effect exists, originating from the higher subsistence requirements in the absence of emigration, which is not compensated by the increased labour availability.

Other restrictions that are binding in both scenarios, albeit also with widely differing values for the shadow prices are 'rice production in dry years' and 'sum of grain deficits over agro-ecological zones'. In the S-scenario, the Region is self-sufficient in terms of food under favourable weather conditions: in normal rainfall years 349000 ton of grain (expressed in millet-equivalents) is produced, giving a surplus of 65000 ton, while sufficient animal protein is produced for a considerable export. In the R-scenario the latter is also the case, but then a grain deficit of 23000 ton exists. In dry years, maximization of gross revenue results in an overall grain deficit of 110000 ton millet equivalents (representing required grain imports valued at about 20 million U$) in the S-scenario and 141000 ton in the R-scenario. Hence, under these conditions the Region can hardly be considered a major supplier of grain to other parts of the country.

Land use in the Region, under both the R- and the S-scenario, is predominantly pastoral (Figure 3), with only 4–5% under arable farming, i.e. very close to the present situation. Millet is by far the major crop, occupying 91 and 85% of the cultivated area under the R- and S-scenario, respectively. In both scenarios, rice is selected only to satisfy the specified minimum restrictions, while the area available for vegetables is fully utilized. This is the consequence of the optimization of gross revenue, with rice not being profitable, and vegetables highly profitable.

Intensification of arable farming is hardly profitable, except for groundnut on a limited area. However, to achieve the required level of food self-sufficiency (S-scenario), intensification is necessary, and then intensive millet cultivation is selected in the southern (wetter) part of the Region.

In the R-scenario total herd size in the Region comprises 1762000 TLU (Tropical Livestock Unit, a hypothetical animal of 250 kg liveweight) and in the S-scenario 1491000, compared to 1123000 estimated in 1987. Herd composition changes, with a small increase in the number of cattle, and a substantial increase in the number of small ruminants, mainly because of the higher price for small ruminant meat and the somewhat higher efficiency of feed conversion.

Conclusions

The analysis presented in this study indicates the scope for development of the Region, under the condition of sustainability. Some of the major results may be summarized as follows:

- arable farming under these conditions is not an economically attractive activity. Its products are by far the most important source of energy in the diet of the local population, and thus serve to satisfy the subsistence needs, but they hardly contribute to the generation of income. Food production could be increased, if arable farming would be intensified, but under the prevailing socio-economic conditions this is not economically viable.
- cultivating vegetables is economically attractive, but the available area is limited due to the requirements for full irrigability. Moreover, a more thorough analysis of marketing possibilities is necessary to examine the scope for expansion.
- animal husbandry contributes by far the largest amount to Regional income, as only a limited part of its production is required for subsistence needs, while a substantial proportion of its 'inputs' (i.e. natural pastures) are 'free of charge'. It should be taken into account, however, that not all costs associated with this sector have fully been accounted for, such as the costs of drinking water, and the costs associated with exploitation of dry-season

pastures around villages.

- fisheries also is an important sector in terms of income, but because of the large number of people employed in the sector, its labour productivity is low.

The results of the study have been discussed in two workshops in Mali, one at the regional level and one at the national level. These workshops showed that national scientists and planners clearly recognized the Region as considered in this study, and that tangible steps aiming at alleviating the current constraints for development of the Region could be derived. Hence, even though the results of the multiple goal optimization model cannot, as such, be used to guide regional agricultural development planning, they clearly provide a consistent framework for a 'post-model' analysis, which takes into account conditions that could not be directly introduced in the model as 'hard' relations. Such an analysis may then lead to the formulation of policy measures aiming at the required changes in regional agricultural practice.

The method therefore provides an efficient instrument for integration of agro-technical and socio-economic information, and as such serves as a means of communication between scientists, planners and policy makers. That allows a more thorough and efficient use of the current knowledge base for agricultural development planning.

References

Cissé S, Gosseye P A (Eds.) (1990). Competition pour des ressources limitées: Le cas de la cinquième Région du Mali. Rapport 1: Ressources naturelles et populations. DLO-Centre for Agrobiological Research (CABO-DLO), Wageningen, The Netherlands and Etude sur les Systèmes de Production Rurales (ESPR), Mopti, Mali. 106 p.

De Wit C T, Van Keulen H, Seligman N G, Spharim I (1988) Application of interactive multiple goal programming techniques for analysis and planning of regional agricultural development. Agric. Syst. 26:211–230.

Erenstein O (1990) Simulation of water-limited yields of sorghum, millet and cowpea for the 5th Region of Mali in the framework of quantitative land evaluation. Report Dept. Theoretical Production Ecology, Wageningen Agricultural University, P.O.Box 430, Wageningen, The Netherlands 60 p. + annexes.

Projet Inventaires des Ressources Terrestres (1983) Les ressources terrestres au Mali, Volume I: Atlas, Volume II: Rapport technique, Volume III: Annexes. Projet Inventaires des Ressources Terrestres au Mali, Mali and USAID, Bamako, Mali, TAMS, New York, USA.

Van Duivenbooden N, Gosseye P A, Van Keulen H (eds.) (1991) Competing for limited resources: The case of the fifth Region of Mali. Report 2: Plant, livestock and fish production. DLO-Centre for Agrobiological Research (CABO-DLO), Wageningen, The Netherlands and Etude sur les Systèmes de Production Rurales (ESPR), Mopti, Mali. 206 p.

Van Keulen H (1990) A multiple goal programming basis for analysing agricultural research and development. Pages 265–276 in Rabbinge R, Goudriaan J, Van Keulen H, Penning de Vries F W T, Van Laar H H, (Eds.), Theoretical production ecology: reflections and prospects. Simulation Monographs 34, Pudoc, Wageningen, The Netherlands.

Van Keulen H (1992) Options for agricultural development: a new quantitative approach. Pages 357-367 in Penning de Vries, Teng P S, Metselaar K (Eds.), Systems Approaches for Agricultural Development, Proceedings of the International Symposium on Systems Approaches for Agricultural Development, 2-6 December 1991, Bangkok, Thailand (this volume)

Veeneklaas F R (1990) Competing for limited resources: The case of the fifth Region of Mali. Report 3. Formal description of the optimization model MALI5. DLO-Centre for Agrobiological Research (CABO-DLO), Wageningen, The Netherlands and Etude sur les Systèmes de Production Rurales (ESPR), Mopti, Mali. 64 p.
Veeneklaas F R, Cissé S, Gosseye P A, Duivenbooden N van, Van Keulen H (1991). Competing for limited resources: The case of the fifth Region of Mali. Report 4: Development Scenarios. DLO-Centre for Agrobiological Research (CABO-DLO), Wageningen, The Netherlands and Etude sur les Systèmes Production Rurales (ESPR), Mopti, Mali. 182 p.

Multicriteria optimization for a sustainable agriculture

E.C. ALOCILJA[1] and J.T. RITCHIE[2]

[1] *Agronomy, Plant Physiology and Agroecology Division, International Rice Research Institute, P.O. Box 933, 1099 Manila, Philippines.*

[2] *Department of Crop and Soil Sciences, Michigan State University, East Lansing, MI 48824, U.S.A.*

Key words: CERES-Maize, financial return, leaching, multicriteria optimization, nitrogen-water limited yield, SIMOPT2: Maize, simulation, sustainability

Abstract

The existence of multiple objectives in agriculture is a rule, rather than an exception. The issue of economic profitability is a part of the overall concern for agricultural sustainability.

The objective of this paper is to present the analytical structure, validation results, and applications of SIMOPT2:Maize, a simulation-multicriteria optimization software designed to optimize two conflicting objective functions simultaneously.

SIMOPT2:Maize is a coupling of the CERES–Maize model and the Pareto optimization algorithm. It is a tool to help identify a non-inferior fertilizer nitrogen application schedule in maize production, such that profit is maximized and nitrate leaching is minimized, subject to constraints in physical and biological resources and the natural environment.

Effects of weather factors, soil properties, genetic properties, and cultural management practices are simulated through the CERES–Maize model. Yield and amount of nitrate leaching are output of the simulation. Yield is converted into revenue and profit is calculated. Optimization is conducted to find the best trade-off fertilizer N application schedule with respect to profit and nitrate leaching.

The use of SIMOPT2:Maize has demonstrated that a set of feasible, efficient fertilizer N application schedules, which is an optimal compromise between maximum profit and minimum nitrate leaching, can be identified.

Multicriteria decision making for a sustainable agriculture

Traditionally, mathematical decision making in agriculture has followed the argument that the problem in question involves a well-defined single criterion or objective. The solutions that satisfy the constraints of the problem are ordered according to the given criterion in the search of an optimal solution. This single-criterion paradigm has served us only partially because in reality, the decision maker is engaged in the pursuit of several, and oftentimes, conflicting objectives which cannot be expressed in the same unit of measurements. In fact, the existence of multiple objectives in agriculture is the rule rather than an exception, be it by a farmer or by a policy decision maker. For example, a farmer may be interested to maximize revenue, maximize use of available labour, and minimize risk. A cabinet minister for agriculture may be interested in identifying production measures that will maximize profit for the

F.W.T. Penning de Vries et al. (eds.), Systems Approaches for Agricultural Development, 381–396.

farm producers and minimize environmental damage to the region. All these goals mentioned above are conflicting because maximizing revenue or profit may mean high risk, mechanization (less labour), or high chemical inputs (environmental pollution). In these examples, the decision maker is not interested in ranking the feasible choices according to a single criterion but rather strives to find an optimal compromise among the several objectives identified.

According to Romero and Rehman (1989), a single criterion optimization problem is a technological problem while a multicriteria optimization problem is an economic one. Single criterion optimization is technological in the sense that the optimization process involves search and measurement. For example, the problem of finding a fertilizer nitrogen (N) application that maximizes rice yield is a technological problem because it involves a search of the appropriate amount among the feasible applications. To solve this problem, the rice farmer searches from a recommended, feasible set of fertilizer applications and checks the expected yield for his/her condition. If 100 kg N of urea will result in 7 tons of rice grain per hectare (t ha^{-1}) while 60 kg N will result in 5.5 t ha^{-1}, the farmer will apply 100 kg N ha^{-1} to the field; no decision making is required. On the other hand, if the farmer is interested to find the best N application that results in maximum profit and minimum nitrate leaching, the problem becomes an economic problem since now it becomes necessary to find a compromise between the two conflicting objectives, conflicting because higher N fertilizer results in higher yield and consequently, higher profit, but may lead to higher nitrate leaching. To solve this problem, the preference of the farmer with respect to the two objectives has to be taken into consideration. As a result, the identified solution may change from one farmer to another according to individual preferences.

The issue of multicriteria optimization has been made more relevant and urgent with the worldwide concern for the environment. While food security and economic profitability continue to be major concerns in agriculture, ecological balance is emerging to be a global issue.

Many agricultural areas, lakes, and rivers have become unproductive over time due to desertification, salinization, deforestation, siltation, and accumulation of chemicals and toxic substances in the soils, surface water, and groundwater. Not only that, traces of pesticides are found in our food and high concentration of nitrates are found in our drinking water. In addition, the CO_2 concentration is increasing in the atmosphere (causing the greenhouse effect) while our minable energy resources are being depleted. In short, we are heading towards ecological disaster if we continue with our present high energy consumption, unsustainable agricultural production practices, and environmental negligence. While it is true that natural systems, such as soils, lakes, and groundwater aquifers, have capabilities to regenerate, they have 'limited capacities', that is, each of these natural systems is able to process only a fixed amount of pollutants per unit time. If the rate of pollution is higher than its rate

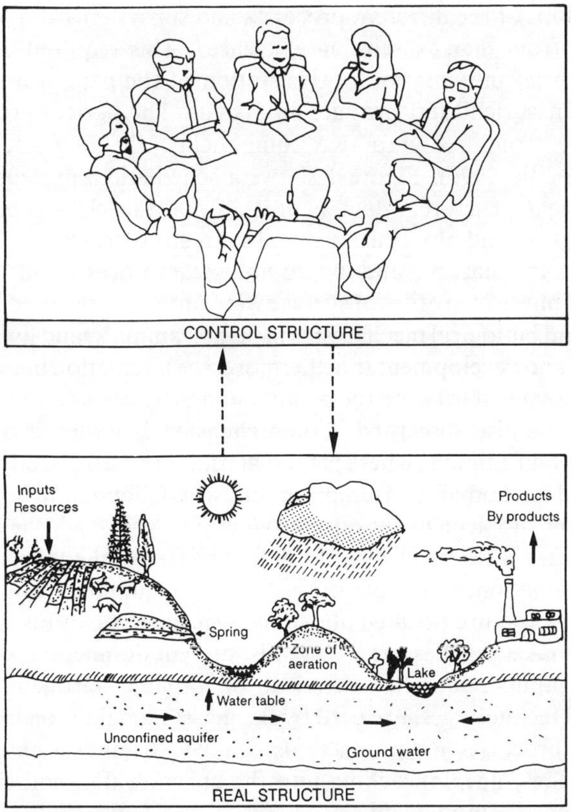

Figure 1. Schematic diagram of the *Real* and *Control* structures in agricultural system.

of regeneration or processing and the rate of discharge, then there will be a net accumulation in the natural system. When the accumulation of pollutants gets to the irreversible pollution level, the natural system deteriorates and eventually 'dies'.

Thus, there is a growing concern to adopt more sustainable agricultural production systems that depend less on chemicals and other fossil-based inputs, while at the same time increase yield, reduce production costs, improve farm profit, reduce risk, and sustain the productivity of the soil and water resources. In other words, we need to manage agriculture toward sustainability–that is, toward greater human utility and not deterioration, toward greater efficiency of resource use and not exhaustion, and toward the maintenance of ecological balance that is good both to humans and to other living species (Harwood 1990).

To analyze an integrated production system toward sustainability is a complex process because the production system is multi-level in its overall structure and highly interconnected to each of the system components at all structural levels. It requires an understanding of the vertical, horizontal, and loop relationships of the different processes and subsystems and an integration of knowledge from many disciplinary sciences. This required understanding necessitates a brief presentation in this section of the paper on the structural hierarchy of an agricultural production system. The agricultural production system can be structured into two components: the *real* and the *control* structures (Alocilja 1990). Figure 1 shows a schematic representation of this concept. The *real* structure, which is in itself multi-level, is composed of the natural, biological, and physical systems. In this structural level, it is important to understand the natural and microbiological processes in the soil, the hydrology and physical profiles of surface water and underground aquifers, and the chemical and biological mechanisms in plants, animals, and living organisms during growth and development. Furthermore, the interaction between the plant and the soil system and between the plants and pests, such as weeds, insects, and diseases, must be also discerned. Comprehension is required of the linkage between plants and animals, where plant materials are used directly or processed for animal and/or human consumption and where animal or human waste is used as fertilizer supplement for crop production. Above all, these subsystems and processes are influenced directly or indirectly by the stochastic weather factors (solar radiation, air temperature, daylength, humidity, wind speed, etc.).

The *control* structure (located above the *real* structure in Figure 1) is where the decision maker resides. It is in this human-dominated structure that management inputs, such as chemical or organic fertilizers, pesticides, irrigation, and technologies, are used by the decision maker to alter the natural and biological processes and interactions, manage the undesirable by-products (pesticide residues, nitrate leaching into the groundwater, soil erosion, etc.), and control the thermodynamic processes in the food chain. The decision maker in the *control* structure sets the objectives of agricultural production and determines how each of the system components in the *real* structure are interconnected (among such components where the decision maker has freedom to link) in order to achieve the objectives, subject to the constraints imposed by his/her environment (market, social, cultural, and other economic factors).

Multicriteria optimization with simulation models

The introduction of crop simulation models to represent the natural, biological, physical, and stochastic behaviours of the system components in agricultural production is a welcome alternative in studying the interactions of the various agricultural components in the *real* structure.

The technique of using simulation models for multicriteria optimization is the focus for the remaining part of this paper. The coupling of the CERES–Maize model and Pareto optimization algorithm (which shall be referred to as SIMOPT2:Maize) to study simultaneously the conflicting issues of maximizing profit and minimizing environmental concerns in maize production will be presented.

SIMOPT2:Maize was developed at Michigan State University in response to a major environmental concern: groundwater pollution from nitrate leaching in maize production. Shallow aquifers, especially on sandy soils, have been found to have nitrate concentration above the drinking water standard of 10 mg l^{-1}. There are obviously many ways of reducing nitrate leaching, including the fallow system, however, we have the responsibility as well to produce food for the present population. Thus, any recommended technology must satisfy at least three conditions: maintain high yield, be economically profitable, and be environmentally sustainable. SIMOPT2:Maize was therefore, designed to meet these three conditions. The result is a computer-aided tool that can help identify non-inferior fertilizer N application schedule–in terms of timing and amount–in maize production, such that profit (Φ) is maximized and nitrate leaching (Ω) is minimized, subject to constraints in resources and the natural environment.

The CERES–Maize simulation model is a major component of SIMOPT2:Maize. CERES–Maize is a crop model designed to simulate the effects of weather, cultivar, management practices, soil water, and nitrogen fertilizer on maize growth, development, and grain yield (Ritchie 1986). The effects of weeds, insects, diseases, other nutrient deficiencies and toxicities, and catastrophic weather events are not considered in this version of the model.

Problem formulation in multicriteria optimization

The specific multicriteria optimization problem is formulated as follows: find a fertilizer N schedule, in terms of timing (x_1) and amount (x_2), satisfying constraints and boundary conditions in maize production, such that profit is maximized ($f_1(\mathbf{x})$) and nitrate leaching is minimized ($f_2(\mathbf{x})$). That is, find $\mathbf{x}^*$ such that

$$\mathbf{f}(\mathbf{x}^*) = \text{opt } \mathbf{f}(\mathbf{x}) \qquad (1)$$

subject to:

$$\alpha_n \leq \mathbf{x}_n \leq \beta_n \qquad n = 1,2 \qquad (2)$$

$$g_m(\mathbf{x}) \geq 0 \qquad m = 1,2,\ldots,M \qquad (3)$$

where $\mathbf{x}^* = [x_1^*, x_2^*]^T$ denotes a vector of efficient or Pareto optimal fertilizer N schedules. The elements of the vector function $\mathbf{f} = [f_1, f_2]^T$ are the two objective functions, profit and nitrate leaching, that are to be optimized. The inequality constraints given by equations (2) and (3) define the feasible region X and represent the boundary conditions imposed on the decision variables, $\mathbf{x} = [x_1, x_2]^T$. α and β are constants representing the lower and upper boundaries, respectively, allowed on the decision variables x_1 and x_2. The constraint functions $g_m(\mathbf{x})$ are linear and non-linear functions of the variables x_1 and x_2. Any fertilizer N schedule $\mathbf{x} \in X$ defines a feasible solution and the vector function $\mathbf{f}(\mathbf{x})$ maps the set X in the set F, where F represents all possible values of the

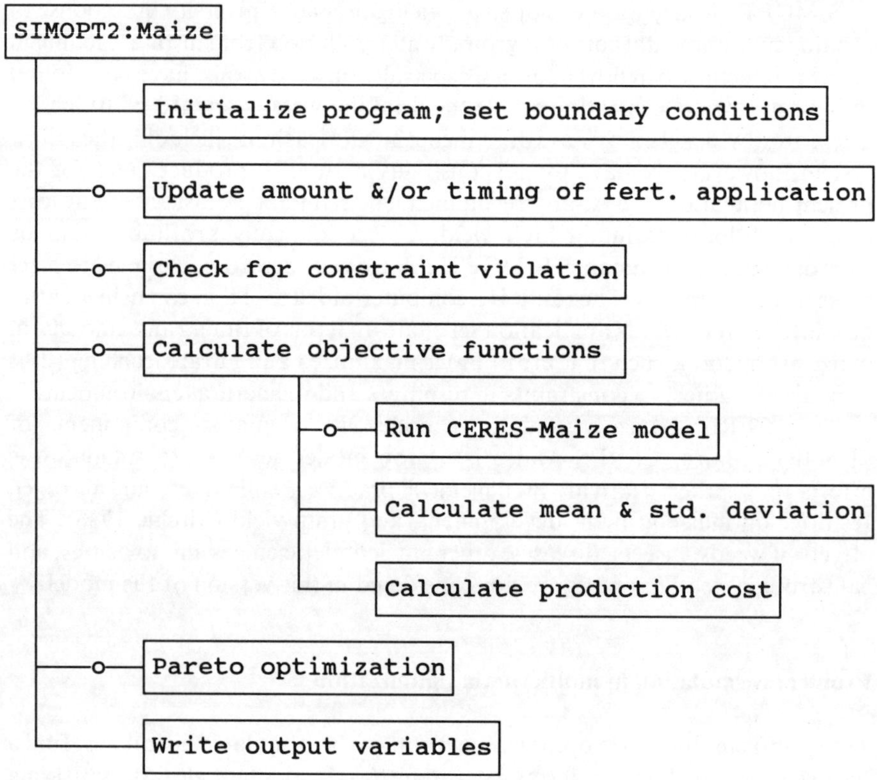

Figure 2. SIMOPT2:Maize Version 1.0 Decomposition Chart. ——— module is called only once during the optimization; ——o—— module is called more than once.

objective functions.

Profit (Φ), expressed in \$ ha^{-1}, is defined here as the difference between revenue and total cost of production. Revenue is the product of the market price (π), expressed in \$ kg^{-1}, and grain yield (Γ), expressed in kg ha^{-1}. Total cost of production (σ) is presented in Table 1. Nitrate leaching (Ω) is expressed in kg N ha^{-1} and represents accumulated nitrates below the 1.8-meter soil profile for a period of one year. Φ, Γ, σ, and Ω are all functions of $\mathbf{x}$. The two objective functions can be analytically expressed as follows:

$$\max f_1(\mathbf{x}) = \max \{\pi\Gamma(\mathbf{x}) - \sigma(\mathbf{x})\} \tag{4}$$

$$\min f_2(\mathbf{x}) = \min \{ \sum_{j=1}^{365} \Omega(\mathbf{x})_j \} \tag{5}$$

Table 1. Cost of production.

Nitrogen fertilizer per kilogram:[1]		
dry (for split application)		$ 0.40
anhydrous (for single application)		$ 0.31
Cost per application per hectare:[2]		$ 12.00
Fixed cost per hectare:[2]		$ 304.71
seed	$ 71.15	
phosphorus	$ 45.80	
potash	$ 38.38	
lime	$ 11.11	
insecticide	$ 27.16	
weed spray	$ 32.10	
repairs	$ 51.85	
gas, fuel, oil (non-irrig.)	$ 27.16	
Irrigation cost per 2.56 cm per hectare:[2]		$ 8.65
Harvesting cost per kilogram grain per hectare:[2]		$ 0.014

Sources of information:
[1] Fertilizer store by telephone call.
[2] Extension specialist at Michigan State University.

Pareto optimization

Multicriteria optimization in SIMOPT2:Maize is along the concept of Pareto optimality (Osyczka 1984), which was introduced in 1896 by Vilfredo Pareto (French et al 1983). Pareto's study provided the earliest formal recognition of the difficulty of reducing decision problems to forms involving a single objective. However, the analytical representation of Pareto optimality became popularized only in the last 20 years (beginning in the early 1970's) in the fields of Engineering, Management Science, and Operations Research (Romero and Rehman 1989).

In the problem formulation here, a non-inferior fertilizer N schedule $\mathbf{x}^* \in X$ is defined as Pareto optimum if for every $\mathbf{x} \in X$ either,

$$\bigwedge_{i \in I} (f_i(\mathbf{x}) = f_i(\mathbf{x}^*)) \text{ , } I = 1,2 \tag{6}$$

or, there is at least one $i \in I$ such that

$$f_i(\mathbf{x}) > f_i(\mathbf{x}^*) \tag{7}$$

This simply means that the fertilizer N schedule $\mathbf{x}^*$ is chosen as the optimum if none of the two objective functions, $f_1(\mathbf{x})$ and $f_2(\mathbf{x})$, can be improved without worsening at least $f_1(\mathbf{x})$ or $f_2(\mathbf{x})$, that is, where no other feasible fertilizer application can achieve the same or better performance for the two criteria under consideration and strictly better for at least one criterion. Pareto optimal solutions are also called efficient or non-inferior solutions. A set of these optimal, non-inferior fertilizer N schedules is generated to form a Pareto optimal curve.

Input requirement and algorithm of SIMOPT2:Maize

Inputs required to run SIMOPT2:Maize can be categorized into two: those that are needed to simulate maize growth and development and those required to execute the optimization routines. Input to the CERES–Maize model include daily weather data, soil characterization data, genetic coefficients, and management data. Input to execute the optimization routines are the constraint functions, boundary conditions, initial values of the decision variables, frequency of application, number of simulation runs, maximum number of iterations, and some other initialization variables.

Structured in a modular fashion, SIMOPT2:Maize is composed of the main program, 10 general subroutines, 2 user-supplied subroutines for constraints and cost functions, and the CERES–Maize subroutines. A decomposition chart of SIMOPT2:Maize is presented in Figure 2.

The algorithm of SIMOPT2:Maize starts with the opening of input and output files, initialization of decision variables x_1 and x_2, checking of boundary conditions ($\propto$ and β), and setting program control parameters. Initial values are recorded into an output file. The constraint subroutine is called to check for constraint violation. If violation occurs, x_1 and x_2 are updated by the Monti Carlo method (Osyczka 1984). If no violation, the objective functions are calculated which involves four steps. The first step is to run the CERES–Maize model according to the number of simulations as defined by the user, given x_1 (timing) and x_2 (amount of fertilizer per application). $\Gamma(\mathbf{x})$ and $\Omega(\mathbf{x})$ are outputs of this simulation process. If the simulation is for a period of one cropping season only, mean and standard deviation for $\Gamma(\mathbf{x})$ and $\Omega(\mathbf{x})$ are not calculated, otherwise, the subroutine to calculate mean and standard deviation is called. The third step is to calculate the total cost of production from $\Gamma(\mathbf{x})$ or its mean. The fourth step is to calculate $f_1(\mathbf{x})$ and $f_2(\mathbf{x})$.

The next process in the software execution is the optimization according to the Pareto optimization algorithm outlined previously. If the number of iterations is less than the maximum set by the user, x_2 is updated by the Monti Carlo method, keeping x_1 constant. Once again, the new x_2 is checked for constraint violation. If no violation, the objective functions are calculated accordingly and the optimization proceeds. This iteration continues until the maximum number of iterations has been satisfied. During the optimization, a feasible set of solutions are generated. This feasible set is then partitioned into two: the subset of feasible, inferior solutions and the subset of feasible, non-inferior solutions. The subset of feasible, non-inferior solutions comprises the set of Pareto optimal solutions $\mathbf{x}^*$. At this point, the program checks for the number of searches, that is, the number of times the user has indicated to search for new x_1 (timing of application). If the search is more than one, the program returns to update x_1 and x_2 and the whole optimization process is repeated, beginning with the checking of constraint violations. Each time x_1 is updated, a set of Pareto optimal solutions $\mathbf{x}^*$ is generated. When the maximum number of timing search is satisfied, the program conducts another optimization of all the Pareto optimal sets to identify the best set, that is, the best set of optimum N schedule from all the optimum N schedules generated. The best can be a single trade-off technology or a set of trade-off technologies which are non-inferior with respect to maximum profit (Φ_{max}) or minimum nitrate leaching (Ω_{min}). Before the program terminates, all the sets of Pareto optimal N schedules $\mathbf{x}^*$ are written into the output file.

Validation of CERES–Maize in SIMOPT2:Maize

SIMOPT2:Maize was validated using the lysimeter experiment at St. Joseph county by Werner and Ritchie (1988, unpublished data, Michigan State University). A maize fertilizer experiment was set up during the spring of 1988 in a farmer's field (Oshtemo sandy loam) at St. Joseph county, Michigan using Pioneer 3475 variety. A non-weighing lysimeter, 2 meters in area and 1.8 meters

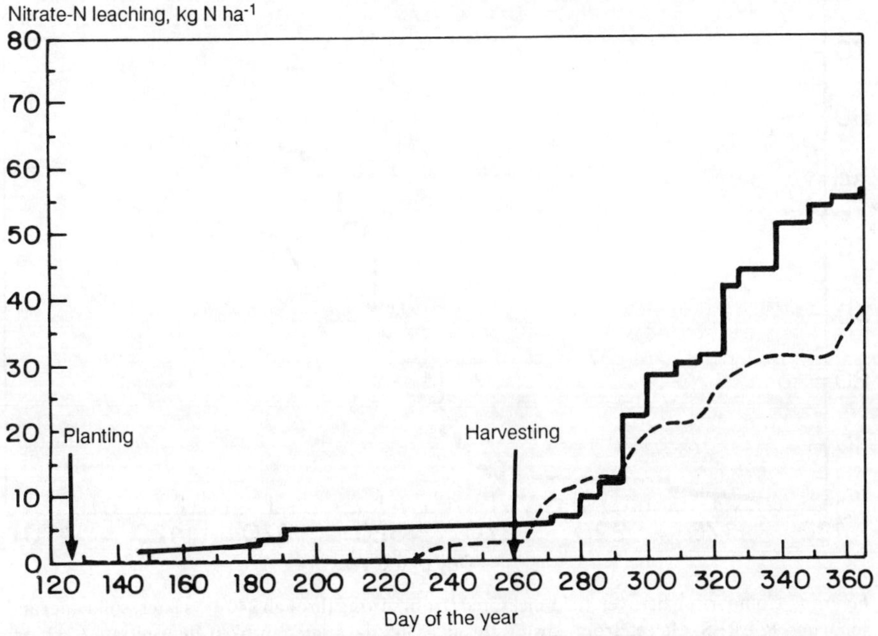

Figure 3. Comparison between field-measured (solid line) and simulated (CERES–Maize, broken line) nitrate leaching (kg N ha^{-1}) from sowing until 106 days after harvest of a maize crop sown on 5 May 1988 and harvested on 16 September 1988. Lysimeter study by A. Werner and J.T. Ritchie, Michigan State University.

deep, was earlier installed under the experimental field. The lysimeter collected the leachate at the bottom of the 1.8-meter soil profile. The leachate was analyzed for nitrate concentration and reported as kg N ha^{-1}. In this experiment, three-split fertilizer N applications were made. The first one was done during sowing (5 May 1988) with 10 kg N ha^{-1} in the form of diammonium phosphate. The second and third applications were done during shoot growth stage (17 June) and tasseling stage (8 July) with 40 and 80 kg N ha^{-1} in the form of ammonium nitrate, respectively. Irrigation was applied during the cropping season (262 mm). A total of 635.5 mm of rainfall were recorded from sowing until the end of the year. The crop was harvested on 16 September 1988. The field-measured grain yield was 9372 kg ha^{-1} and the accumulated nitrate leaching from sowing to the end of the year was 56.56 kg N ha^{-1}.

Recorded 1988 weather data, soil characterization of the production field, and management data with a 10–40–80 split application (as in the experiment) were used as input to validate CERES–Maize. The simulated output were 9177 kg ha^{-1} grain yield and 41.87 kg N ha^{-1} nitrate leaching accumulated from sowing to the end of the year. Figure 3 shows the actual and simulated incremental accumulation of nitrates. The simulated result is lower than the field measurement. A one-year simulation of the nitrate leaching is shown in Figure 4.

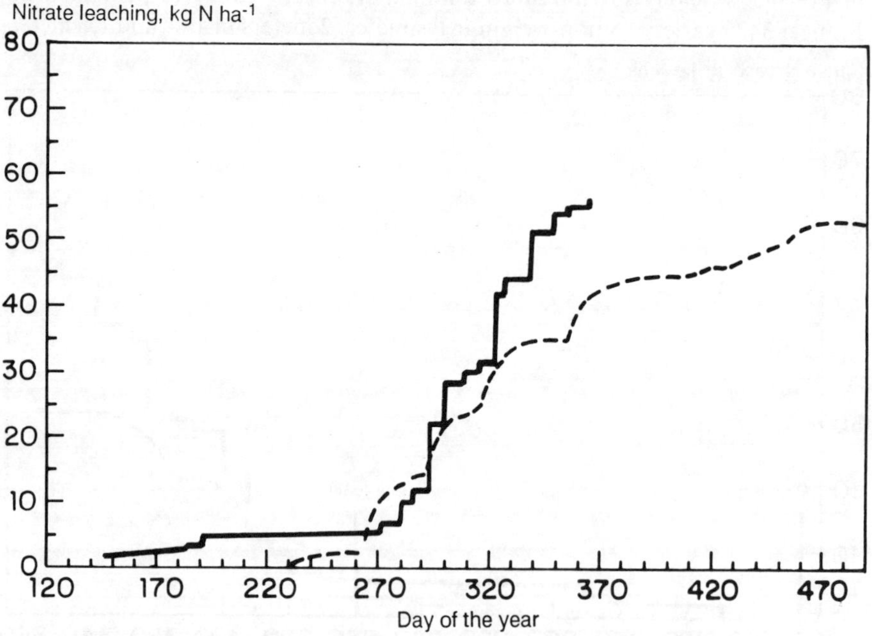

Figure 4. Comparison between field-measured (from sowing through 240 days later, solid line) and simulated (CERES–Maize, from sowing through 365 days later, broken line) nitrate leaching. Lysimeter study by A. Werner and J.T. Ritchie, Michigan State University.

Applications of SIMOPT2:Maize in production analysis

For use in production analysis, SIMOPT2:Maize was run to identify alternative three-split fertilizer applications. Given the weather and soil initial conditions of the production area, SIMOPT2:Maize was run with the following constraints: the budget for fertilizer at \$0.40 kg^{-1} was not to exceed \$200. The lower bounds for the three-split applications were given zero values while the upper boundaries were set to 75, 150, and 100 kg N ha^{-1} for the first (a), second (b), and third (c) applications, respectively. The market price of grain (π) was \$0.10 kg^{-1}. Nitrate leaching was accumulated over a year, from one sowing date to the following year's sowing date.

The constraints and boundary conditions can be expressed analytically as follows:

$$200-0.40\,x_2 \geq 0$$
$$0 \leq x_{2a} \leq 75$$
$$0 \leq x_{2b} \leq 150$$
$$0 \leq x_{2c} \leq 100$$
$$\pi = \$0.10$$
$$\Omega(x) = \sum_{j=1}^{365} \Omega(x)_j$$

Simulated yield, profit, nitrate leaching, and Pareto optimal fertilizer applications are outputs of the optimization. Figure 5 shows the curvilinear relationship between fertilizer application, expressed in kg N ha^{-1} (on the

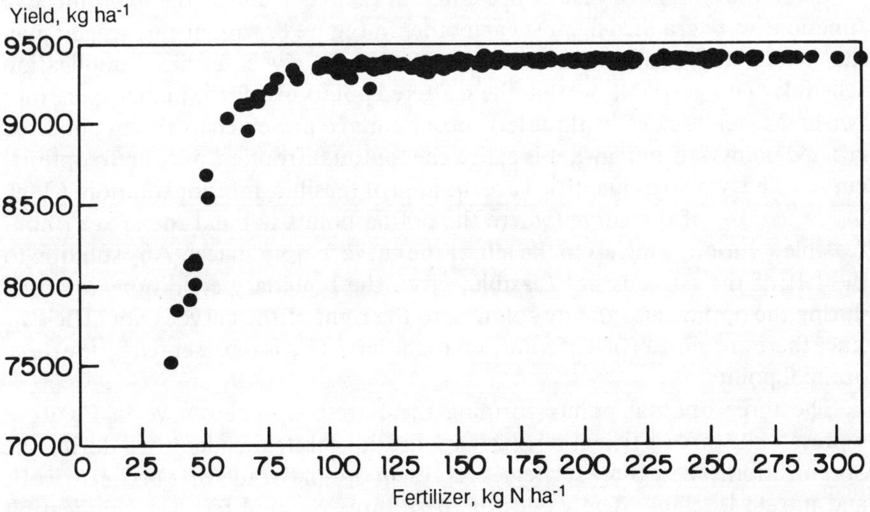

Figure 5. Relationship between simulated fertilizer 3-split application (kg N ha^{-1}) and grain yield (kg ha^{-1}).

horizontal axis) and grain yield, expressed in kg ha^{-1} (on the vertical axis). As demonstrated, the yield increases linearly with fertilizer application between 36 and 62 kg N ha^{-1}, with an estimated rate of 62 kg grain yield for every kg N. The rate of increase drastically diminishes thereafter (about 9 kg grain yield per kg N) and finally approaches zero starting at about 83–85 kg N ha^{-1} as indicated by a plateau in the yield curve.

Converting yield to profit shows an interesting result. Figure 6 shows that profit increases linearly with fertilizer application between 36 and 62 kg N ha^{-1}, at a rate of about \$5 for every kg N. The rate of gain diminishes thereafter until it reaches a maximum between 83–85 kg N ha^{-1}. After this point, the slope becomes negative with a rate of profit reduction at \$0.36 per kg N, as demonstrated by a decreasing profit curve. This phenomenon can be explained by the fact that, since increasing fertilizer does not result in increased yield between 85 to 310 kg N ha^{-1}, production cost is increased without commensurate revenue, thus profit is reduced.

The effect of fertilizer application on nitrate leaching is shown in Figure 7. The graph shows that from 36 to 180 kg N ha^{-1} of applied fertilizer, between 48 to 57 kg N ha^{-1} of nitrates are leached out below the 1.8 meter soil profile (very little variation). However, the rate of nitrate leaching starts to increase sharply and linearly with increase in fertilizer application after 180 kg N ha^{-1}, with a rate of 0.98 kg N leached per kg N applied (almost a one-to-one ratio). It is shown in Figure 5 that fertilizer application greater than 85 kg N ha^{-1} does not result in increased grain yield. Yet, the leaching is not apparent until about 180 kg N ha^{-1}. This implies that there is a build-up of N in the soil profile. However, CERES–Maize is not designed to account for long-term N dynamics, hence, simulation for subsequent years is not conducted in this study.

Given the simulated results presented in Figures 6 and 7, the two objective functions were graphed against each other in Figure 8, with nitrate leaching on the x–axis and profit on the y–axis, for every given fertilizer application schedule. This graph shows that the scattered points divide the planar space into two areas, demarcated by three left-most points (represented by three connected crossed points) forming what is called the 'optimal frontier' or 'Pareto optimal curve'. The two areas identified are the area of feasible, inferior solution, which is to the right of the curve (where the dotted points lie) and the area of non-feasible solution, which is to the left of the curve (empty space). Any solution to the left of the curve is not feasible, given the boundary conditions imposed during the optimization. Any solution to the right of the curve is not efficient, since there are more efficient solutions to the left of it, as represented by the three optimal points.

The three optimal points forming the Pareto optimal curve in Figure 8 represent the three efficient fertilizer application strategies identified during the optimization. These three strategies represent optimal trade-offs between profit and nitrate leaching. The values of these three optimal fertilizer application alternatives are shown in Table 2. Selecting one from among the three alternatives requires trade-off between the two objectives since the three

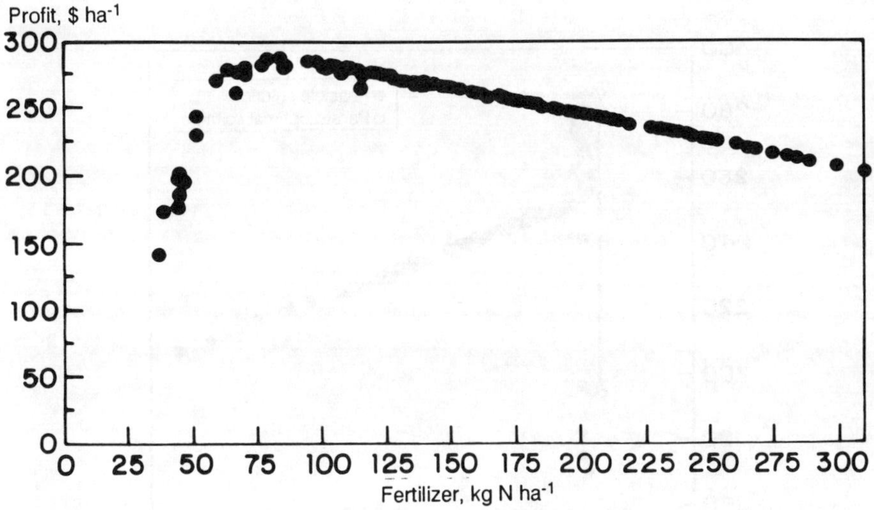

Figure 6. Relationship between simulated fertilizer 3-split application (kg N ha^{-1}) and profit (\$ ha^{-1}).

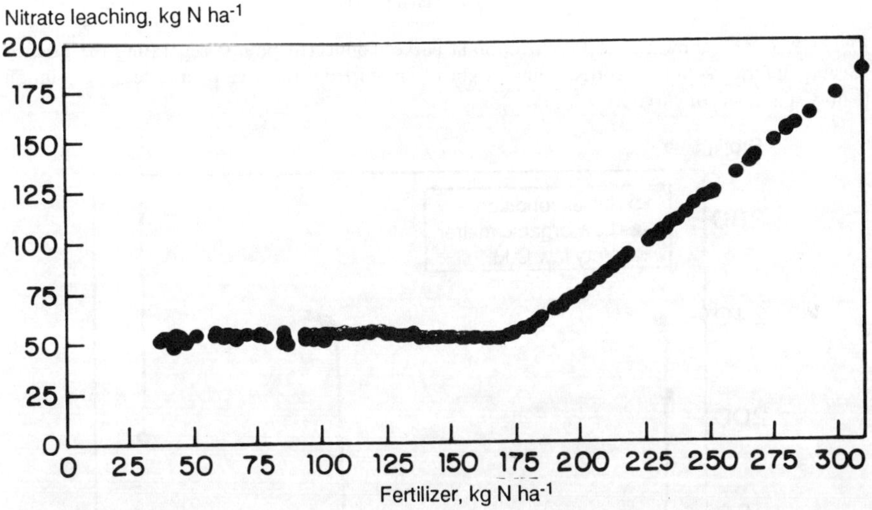

Figure 7. Relationship between simulated fertilizer 3-split application (kg N ha^{-1}) and nitrate leaching (kg N ha^{-1}).

alternatives are optimal in the Pareto sense. For example, while solution no. 2 in Table 2 has lower nitrates, it has lower profit compared to solution no. 3. To choose between solution nos. 2 and 3 requires a decision that must take into consideration the preference of the decision maker for the two objectives under consideration.

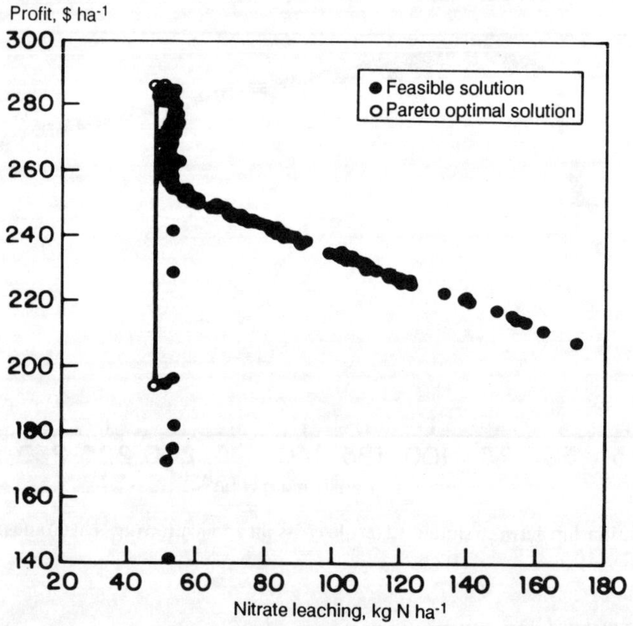

Figure 8. Graph showing the Pareto optimal curve (connected points) separating the area of feasible, inferior solution (dotted points, right of curve) from the area of non-feasible solution (empty space, left of curve).

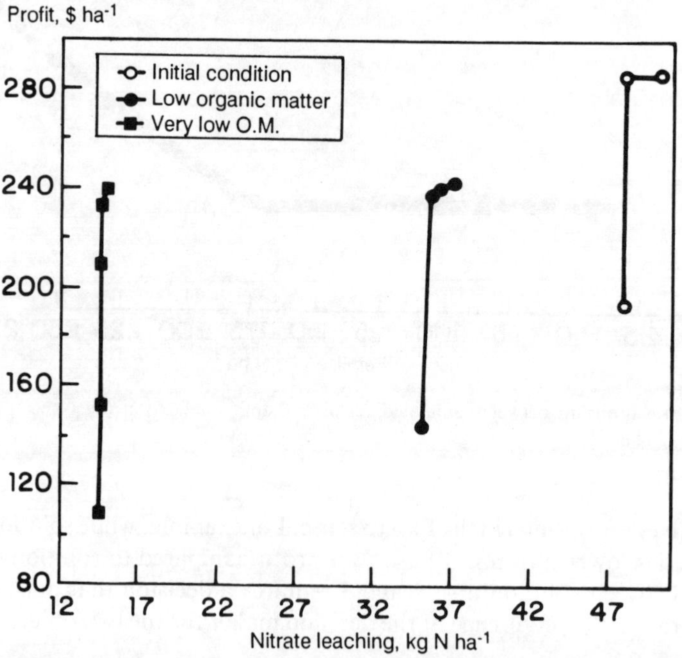

Figure 9. Three Pareto optimal curves generated from three varying initial soil conditions.

Table 2. Pareto optimal strategies.

Solution no.	Fertilizer application (kg N ha^{-1})	Grain yield (kg ha^{-1})	Profit ($ ha^{-1})	Nitrate leaching (kg N ha^{-1})
1	42	8111	193.80	47.93
2	89	9336	285.73	48.14
3	84	9317	286.06	50.23

SIMOPT2:Maize was also used to identify the soil condition that may lead to a reduced nitrate leaching. Figure 9 shows three Pareto optimal curves representing three soil conditions. The curve to the right corresponds to the initial conditions of the production area. The curves at the centre and to the left represent simulated soil conditions with low and very low soil organic carbon contents, respectively. These curves show that nitrate leaching can be reduced when soil organic carbon is low although profit decreases as well.

Conclusion

Multicriteria decision making in agriculture, particularly in dealing with the conflicting issues of economic profitability and environmental sustainability, can be facilitated by the coupling of multicriteria optimization techniques and crop simulation models. SIMOPT2:Maize is one such application. SIMOPT2:Maize promises to be a useful tool in identifying alternative non-inferior fertilizer N application schedules–in terms of timing and amount–in maize production, such that profit is maximized and nitrate leaching is minimized, subject to constraints in resources and boundary conditions imposed during the optimization.

At present, the optimization structure in SIMOPT2:Maize is designed to accommodate up to 10 objective functions, 20 decision variables, and 500 constraint functions. The program structure can be easily modified to suit a particular problem.

As the coupling is modular, the CERES–Maize model in SIMOPT2:Maize can be replaced with any crop model of similar structure, hence its potential use in other crop production analysis.

References

Alocilja E C (1990) Process network Theory. Proceedings of the International Conference on the IEEE Systems, Man, and Cybernetics. Los Angeles, California, USA.

French S, Hartley R, Thomas L C, White D J (1983) Multi-objective decision making. Academic Press, New York, USA.

Harwood R R (1990) A History of Sustainable Agriculture. Pages 3–19 in Edwards et al. (Eds.)

Sustainable Agricultural Systems. Soil and Water Conservation Society, Ankeny, Iowa, USA.
Osyczka A (1984) Multicriterion optimization in engineering with Fortran programs. John Wiley and Sons, New York, USA.
Ritchie J T (1986) The CERES–Maize Model. Pages 3–6 in Jones C A, Kinery J R (Eds.) CERES–Maize, A Simulation Model of Maize Growth and Development. Texas A&M University Press, Texas, USA.
Romero C, Rehman T (1989) Multiple Criteria Analysis For Agricultural Decisions. Elsevier, New York, USA.

Simulation of multiple cropping systems with CropSys

R. M. CALDWELL and J. W. HANSEN
Department of Agronomy and Soil Science, University of Hawaii,
1910 East West Road, Honolulu, Hawaii, U.S.A. 96822

Key words: cropping system, CROPSYS, decision support system, IBSNAT, intercropping, maize, management strategy, pests, photosynthesis, rice, risk, simulation, soybean, weed

Abstract
While the development of sustainable agriculture in the tropics often involves improvement of intercropping or double cropping systems, most decision support aids are designed for single crops. A simulation model, called CropSys, was therefore developed to evaluate management decisions for multiple cropping. CropSys builds on the work of the International Benchmark Sites Network for Agrotechnology Transfer (IBSNAT) and includes their CERES models for upland rice, maize, wheat, barley, sorghum and millet, SOYGRO and SOYNIT (which includes a model of nitrogen fixation) for soybean, and a preliminary version of SUBSTOR-Aroid for *Colocasia* and *Xanthosoma* spp. Sequential cropping capabilities were created by reprogramming the crop models within a systems hierarchical framework, using a central soil model to provide continuity in the water and nitrogen budgets across different crop cycles. Process-level submodels of competition for water, soil N and light (with hedge and solar geometry) were added for intercrops. CropSys allows simulation of any sequence of the sole crops, two-species intercrops, and fallow. These systems, and their management permutations (i.e. different planting dates, plant spacings, cultivars, irrigation and N rates), can be evaluated over the range of expected weather conditions using the stochastic strategy analysis in IBSNAT's Decision Support System for Agrotechnology Transfer.

Introduction

Multiple cropping is a salient feature of agriculture in many parts of the tropics. By growing more than one crop in a field in a year, farmers can gain a number of benefits, including better utilization of limited resources and reduction in the risk of low yields (Francis 1986).

Because of the complexities of multiple cropping, a systems approach is required for the evaluation and improvement of the practice. While use of crop simulation models would be a natural part of a systems approach, little work has been done modeling sequences and combinations of crops. This is in contrast to sole cropping, for which most of the major crops have one or more simulation model available (Joyce and Kickert 1987).

Given the value of crop simulations in support of decision making, a new model, called CropSys, was developed to simulate multiple cropping systems. Version 1.0 of CropSys incorporated submodels from CERES-Maize (Ritchie et al. 1989; Jones and Kiniry 1986) and SOYGRO (Jones et al. 1988) along with process- level models of competition for light and competition for soil moisture (Caldwell 1990a, Caldwell and Hansen 1990). Both of the single-species models were reorganized within a subroutine structure that was defined along

F.W.T. Penning de Vries et al. (eds.), Systems Approaches for Agricultural Development, 397–412.

hierarchical boundaries (Caldwell 1989). The resulting model was organized to handle a variety of multiple cropping problems, including management of mixed, row, and strip intercropping, double and triple cropping, and relay intercropping (Caldwell and Hansen 1990). CropSys version 1.0 was tested with a limited amount of field data (Caldwell 1990b).

Because of two basic limitations, CropSys version 1.0 had a narrow range of applicability in agricultural development. The species combination – maize and soybean – was of little practical interest in most parts of the tropics, and one of the key dynamics of the system was missing: nitrogen cycling, including biological nitrogen fixation.

Research was therefore undertaken to overcome these limitations. The structure of CropSys was refined using principles of object-oriented design (Caldwell et al. 1991) to simplify the inclusion of additional species models (Hansen 1991). The new structure was then utilized to link in a modified soybean model, SOYNIT, that included submodels for biological nitrogen fixation and physiological sensitivity to internal nitrogen status. Additional crop models were also included. The objective of this paper is to introduce the new capabilities of CropSys version 2.0, with background information on the approach used. A brief description is given for some of the model algorithms, and example simulations are illustrated.

Problem identification

The nature of plant interactions, especially competition, historically has been an important focus of ecology (Grace and Tilman 1990; Harper 1977). Throughout the development of the science, agriculture has provided a context for basic ecological work (Clement et al. 1929) and has contributed to the development of theory (Donald 1963) and quantitative approaches (Willey and Heath 1969; De Wit 1960). One result of the interaction between ecology and agronomy has been a rich set of simple models that relate plant productivity in mixtures to the elements of plant spacing (i.e. Chapter 10 of Vandermeer 1989).

While this body of theory relates directly to the design of intercropping systems, it is often inadequate for our task in agricultural development. Major hurdles persist beyond pursuing the empirical research required by the simple models. The effect of plant spacing on intercrop components changes with management (i.e. fertilizer, irrigation, weed control, cultivar selection) and the interaction of management and plant spacing itself changes from site to site (Russell and Caldwell 1989).

Even if the multiple interactions could be quantified for a given growing season, the task in agricultural development is to test and recommend management packages that will be useful in the future, under the range of environmental conditions (both biophysical and socioeconomic) farmers will face. We have a special interest in intercropping precisely for this reason. Intercropping can be less risky to farmers than sole cropping (Lynam et al. 1986)

Risk assessment, necessary for a proper comparison of multiple cropping and sole cropping, cannot be accomplished by field experimentation within the context of agricultural development projects. Too many crop cycles are required. Computer models are required for the kinds of risk assessments needed to guide those who set development policy and priorities for field work. Simulation is the only effective means of screening the large number of probable environmental sequences a farmer might face.

Approach

Given the need for computer simulation in risk assessment, a project was initiated to develop a generalized, process-level intercrop model. While the initial focus was on intercrops, the capacity to simulate sequential crops was solved as a matter of course. The resulting model was named CropSys.

Based on known management-by-genotype-by-environment interactions and the perceived use of the model in decision support, sensitivity to the following factors was designed into CropSys: (i) soil chemical and physical properties, varying by layer; (ii) daily weather conditions; (iii) tillage and residue management; (iv) irrigation; (v) nitrogen fertilization; (vi) planting date, permitting relay intercrops with any degree of overlap in time, as well as permitting sequential crops; (vii) plant spacing, giving users control over population densities, row widths, and row arrangement, including the variety of patterns used in strip intercrops; and (viii) cultivar differences for phenology and stature (height and compactness).

Approach to model building

Given the number of species important in multiple cropping systems, we concluded that it was not feasible to write and validate a separate model for each species combination (Caldwell 1990c). CropSys needed to be generalized to handle any combination or sequence of the species implemented. While simple models of competition can be generalized for large numbers of species (i.e. Caldwell 1989), CropSys was limited to simulating any two-way combination of species. This made practical a detailed process-level treatment of resource competition.

It was not feasible for the research project to build, with its own resources, a number of the components required for the application: (i) soils databases; (ii) cultivar databases; (iii) weather databases; (iv) models of fundamental plant physiology for each species (i.e. phenology, photosynthesis, dry matter partitioning); and (v) a user interface for the application, including a database management system, graphics programs, and risk assessment software. Though these elements are essential for simulations used in decision support, their development would have been impossible within the limits of our funding and expertise.

IBSNAT files and models

Most of the elements required from outside the project were available through the International Benchmark Sites Network for Agrotechnology Transfer (IBSNAT). The data, models, and software developed by IBSNAT and contained within their Decision Support System for Agrotechnology Transfer (DSSAT) were selected as the starting point for CropSys. DSSAT file formats (IBSNAT 1986) were adopted so that CropSys could use input files residing in DSSAT and could create output files to be graphed and analyzed by DSSAT programs.

Of particular interest were the programs and files used for strategy analysis. A stochastic weather generator provided in DSSAT, called WGEN, was placed inside CropSys, allowing CropSys to create daily weather that mimics the variety of sequences found in historical records. By simulating a strategy repeatedly with different weather sequences, CropSys can create an output file conforming to DSSAT's strategy evaluation program. While the program found in DSSAT 2.10 was written for sole crops, CropSys was programmed to integrate across crop cycles (with the length of the time period set by the user) and species, letting the user define any set of management options as the basis for a strategy. Maize economic equivalents were used to provide a common basis for integrating across species (i.e. the maize economic equivalent for a soybean yield would be the soybean yield times the ratio of the soybean commodity price over the maize commodity price; gross return for the intercrop would be the sum of the equivalent yields times the maize commodity price). This allowed the strategy evaluation program in DSSAT, using maize cost data, to properly calculate economic return for the strategy.

Six IBSNAT crop models were incorporated within CropSys version 2.0: CERES-Maize version 2.1, SOYGRO version 5.41 and SOYNIT version 5.99 for soybean, CERES-Rice version 2.1, SUBSTOR-Aroid version 1.0 for *Colocasia*, *Xanthosoma*, *Alocasia*, and *Cyrtosperma* spp., and Generic CERES version 1.99. Generic CERES contained code for CERES-Maize, CERES-Sorghum, CERES-Millet, CERES-Wheat, and CERES-Barley.

A goal in model development was to recreate the IBSNAT models within CropSys. When simulating a single crop, CropSys was designed to provide results identical to the IBSNAT counterpart. Validation of the sole crop models was thereby maintained, and a mechanism was provided for verifying the integrity of programming.

After reviewing the literature we concluded that, at present, there was insufficient process-level data on interspecific interactions (competition for water and nutrients, especially) to justify selecting a particular modeling approach from among the various possibilities. In lieu of that knowledge base, we referred back to the most basic theory of competition, expounded by Clements *et al.* (1929), that "competition is a purely physical process". At the surfaces of leaves and roots, their is no physical distinction between interspecific and intraspecific competition. We therefore hypothesized that *the representation of plant physiology and interplant (intraspecific) competition within valid crop models captures the basic interaction of crops with their immediate physical environment. That information is transferable to intercrop systems and can be used to predict plant response to interspecific competition.* The task of modeling interactions between species then became coding, by analogy, the basic physical relationships contained in the sole-crop models.

The IBSNAT models set the level of detail in CropSys. Spatial variability within layers of the soil was not considered and plant-to-plant variation within a species was not included. The basic time-step was maintained at one day. The addition of detail to the individual crop models was avoided.

Model integration: combining the IBSNAT models

Combining the IBSNAT models to create CropSys involved four tasks. The simplest was removing redundancies. Some code was repeated in the models and the extra copies had to be removed.

A more complex task was reconciling subprograms designed for the same function but programmed in different ways. The basic model of soil water in SOYNIT was somewhat different than that in the CERES family of models. Either the SOYNIT model or the CERES model could be used in CropSys, but not both. The soil model found in SOYNIT was selected, and as a result, CropSys can produce simulated results for sole crop soybean numerically identical to SOYNIT. For thc cereals, this is strictly true only when there is no moisture or nitrogen stress. Differences in the calculations of water movement in the soil may create small discrepancies between the soil stress factors calculated by CropSys and those calculated by the CERES models.

A third task was adding the submodels for interspecific competition and associated state variables not currently in the IBSNAT models. Dead leaf area, for example, was required for properly simulating relay intercropping but was not present in some of the crop models. The variable was therefore added to the models, along with algorithms for its calculation.

Once the models for interspecific competition were built, the fourth task was to modify each crop model to make them sensitive to competition. The proper linkage points had to be identified and the structure of the crop models changed to accept the information from the competition routines.

Competition for light and photosynthesis

Competition for light and its interface with the photosynthesis routines in the IBSNAT crop models illustrate the last two tasks in model integration.

Two options, varying by level of detail, are provided for simulating competition for light. A simple two-layer canopy model was designed for mixed intercrops. This option requires simulation of plant height for each species – a state variable not present in the sole crop models. Models of plant height were therefore developed and added to each crop model. Leaf area is assumed to have a uniform distribution with height. Shadows are randomly distributed in layers.

The second model of light interception was designed for row and strip intercrops; it follows the individual shadows cast by each row over the course of each day. Basic parameters of canopy architecture – height and lateral extent of foliage – were not present in the IBSNAT models, so code for those state variables was added for each crop. Both light model options provided an estimate of light intercepted by species *i*, noted as I_i. In the second option, information was also provided on the intensity of solar radiation passing through the canopy of the taller species, estimated at the height of the shorter species, as well as the fraction of land surface shaded by the short species that was also shaded by the tall species.

Having assumed there is only one species in the field, the IBSNAT models make direct use of weather-station-measured solar radiation in their photosynthesis calculations. This was not permissible in the intercrop model: the potential for one crop to steal light from the other must be considered. Introducing the competition routine was straight-forward for six of the species (maize, sorghum, millet, barley, rice, and aroids), which had their basic photosynthesis equation, *P*, as a linear function of incoming solar radiation, *b* ($MJ\ d^{-1}$):

$$P = a \cdot b \cdot c \cdot d$$

where *a* is the conversion efficiency (constant; $g\ MJ^{-1}$), *c* is a zero-to-one factor for the fraction of incoming light intercepted (unitless), and *d* is a function of various stress factors (unitless). In these cases, sensitivity to interspecific competition was introduced by replacing the product $b \cdot c$ with I_i, the actual quantity of light intercepted by species *i*, calculated in the CropSys competition routine:

$$P_i = a \cdot I_i \cdot d$$

When species *i* has light stolen from it by a competitor, I_i is reduced resulting in a proportional reduction in photosynthesis, P_i.

Unlike the other CERES models, CERES-Wheat contained a photosynthesis equation that was nonlinear with respect to incoming solar radiation:

$$P = 7.5 \cdot b^{0.6} \cdot c \cdot d$$

As before, in order to introduce sensitivity to competition, *b* had to be removed from the equation and I_i inserted. The necessary form of the equation, $P_i = a\ I_i\ d$, was created by solving for *a*, the conversion efficiency, which became a function of solar radiation instead of a constant. Given the form of the function:

$$a \cdot b = 7.5 \cdot b^{0.6}$$

then

$$a = 7.5 \cdot b^{0.6} \cdot b^{-1} = 7.5 \cdot b^{-0.4}$$

With this function for *a*, the CERES-Wheat photosynthesis equation could be written in the form used by the other CERES models:

$$P = a \cdot b \cdot c \cdot d$$

But, while the product $b \cdot c$ can be replaced with I_i, as done with the other models, the conversion efficiency remains a function of b, solar radiation measured at the weather station. This is physiologically incorrect; photosynthesis should be a function of the plants' local environment, subject to potential shading from a competitor. CropSys therefore calculates a conversion efficiency as a function of s, the simulated light intensity at the top of the wheat canopy when shaded:

$$a_s = 7.5 \cdot s^{-0.4}$$

In strip intercrops, part of the wheat may fall outside the shadow boundary of the competitor. In that case a second conversion coefficient, a_b, is calculated based on full sunlight, b. The average for the entire wheat canopy is estimated based on the fraction of light intercepted under the competitor's shade, f_s, and that in the full sun:

$$a = a_s \cdot f_s + a_b \cdot (1 - f_s)$$

The photosynthesis function can now be written in its standard form, $P_i = a\, I_i\, d$, which is sensitive to competition, both in terms for the amount of light available for photosynthesis and, for wheat, sensitive to the efficiency of photosynthesis in the shade.

There is no direct reference to b in the photosynthesis equation of SOYGRO version 5.41 and SOYNIT:

$$P = P_{max} \cdot c \cdot d$$

P_{max} is a maximum photosynthesis calculated as a quadratic function of incoming solar radiation, with a plateau:

$$P_{max} = f \cdot b + g \cdot b^2, \qquad b < -0.5 \cdot f \cdot g^{-1}$$
$$P_{max} = h, \qquad b \geq -0.5 \cdot f \cdot g^{-1}$$

where f, g, and h are constants. Given the dependence of P_{max} on b, the sole-crop form of the photosynthesis equation was rearranged. Following the same logic as the modifications to CERES-Wheat, two conversion efficiencies were calculated. The conversion efficiency for soybean in the shade of a taller competitor was calculated as:

$$a_s = f + g \cdot b_s, \qquad b_s < -0.5 \cdot f \cdot g^{-1}$$
$$a_s = h \cdot b_s^{-1}, \qquad b_s \geq -0.5 \cdot f \cdot g^{-1}$$

Soybean's conversion efficiency in full sunlight, a_b, was estimated with a similar function, replacing b_s with the b. The overall conversion efficiency was calculated from a_s, a_b, and f_s, and inserted into the final form of the photosynthesis equation, as done with wheat.

In summary, by reorganizing the photosynthesis equations found in the IBSNAT crop models, we were able to introduce sensitivity to interspecific competition. Light stolen from one species by another reduces the energy available for photosynthesis and may influence the efficiency of light utilization. As the equations were reorganized, the integrity of the basic photosynthesis models was maintained: the estimates of sole-crop photosynthesis within CropSys were kept mathematically identical to the IBSNAT counterparts.

Competition for below-ground resources

In order to simulate competition for water and nitrogen, modifications were required at multiple points in the water balance and the nitrogen balance of the IBSNAT models. At each point, the fundamental hypothesis of CropSys – that models of intraspecific competition are transferable to interspecific competition – was used to guide the modeling.

In the IBSNAT models, moisture stress effects are calculated as functions of potential uptake of water, U (which represents the ability of the root/soil system to supply water to the shoot), divided by potential transpiration, E_P (the ability of the atmosphere to remove water from the shoot). The model for E_P depends on leaf area index, L, and the potential evapotranspiration, E_O:

$$E_P = E_O \cdot (1 - e^{-L})$$

(for $L \leq 3.0$). We hypothesized the form of the equation holds in intercrops, permitting a calculation of potential transpiration for the community, E_{Pc}, by using community leaf area index in the function.

The potential transpiration of each species, E_{P1} and E_{P2}, was calculated next, subject to the restriction that $E_{Pc} = E_{P1} + E_{P2}$. Demands for transpiration (E_P expected in the absence of a competitor) were calculated for each species using the sole-crop function. E_{P1} and E_{P2} were then calculated as a fraction of demand, with the degree of the reduction calculated from each species' competitiveness, here assumed to equal the competitiveness for light: the taller species is more successful in intercepting light and is given a correspondingly greater role in transpiration.

In the IBSNAT models, potential (soil moisture limited) root water uptake in layer j, U_j, is calculated according to the radial flow equation:

$$U_j = (0.00267 \cdot e^{62 \cdot W_j})/(6.68 - \ln(R_j))$$

where W_j is the plant extractable water in layer j and R_j is the root length density (parameters here are from CERES-maize, Jones *et al.* 1986).

In CropSys, potential root water uptake for each intercropped species, on a whole-profile basis, was required to calculate moisture deficit factors. The radial flow equation was extended to the intercrop and used to estimate U_{Cj}, the potential uptake of the community in layer j. Because of differences among the sole-crop models, CropSys estimates the community uptake twice by using the community root length density in the algorithm of each sole-crop model; U_{Cj} was then calculated as a weighted average of the two. A demand for water was calculated for each species in each layer using the sole crop algorithms.

Potential root water uptakes, U_{1j} and U_{2j} for species 1 and 2, were then calculated by adjusting the demands so that $U_{Cj} = U_{1j} + U_{2j}$. (U_{1j} and U_{2j} will normally be higher than their respective demands, reflecting the shorter distance water must flow when roots length densities are higher). Once calculated in each layer, estimates of potential uptake are made by summing U_{1j} and U_{2j} across the entire profile.

By providing sensitivity to competition above-ground, in the calculation of each species potential transpiration, and below-ground, in the calculation of each species potential root water uptake, the water model in CropSys produces the following behaviours: intercropping, by raising root length densities, tends to increase the efficiency of root water uptake. This may allow the plants to pick up the moisture they need to avoid stress, but it will also accelerate the use of soil moisture, possibly leading to more intense moisture stress later. A competitor may steal soil moisture, increasing the potential for moisture stress, but the competitor may also provide shade, thereby reducing the energy load on the species and moderating stress conditions.

Competition for nitrogen in CropSys follows the same pattern of calculations used in the soil water model. The actual uptake of N is limited either by the root/soil interface or by the internal requirements of the plant, with the degree of N stress being a function of the two. CropSys version 2.0 contains the soybean model SOYNIT, which allows simulation of within season and between season benefits of biological nitrogen fixation in multiple cropping. Competition for nitrate and ammonium are handled separately.

Links to measured data for non-IBSNAT species

Process-level simulations can provide a researcher with a wealth of information that is difficult, or impossible, to obtain from field measurements. While CropSys version 2.0 is able to simulate seven grain crops and the aroids, problems in agricultural development often involve a much broader set of species. Weeds, ground covers, green manures, trees, and a variety of minor crops are commonly found in multiple cropping systems. Therefore, an option was created in CropSys to link data from the measured species with the simulation. An analysis can then be performed on the resource use of the non-IBSNAT species in the system. CropSys may thereby help researchers better understand their experiments by providing estimates of competition and carryover effects.

When linking to measured data, CropSys replaces what is normally a process-level simulation of plant physiology with simple interpolation routines which update the necessary state variables daily. Knowledge of those state variables permits CropSys to estimate resource use and interspecific competition. When competition is restricted to light, the user need only provide measurements of leaf area and canopy dimensions, with an extinction coefficient. CropSys will then estimate the amount of light the measured species stole from its competitor. In order to estimate competition for water and soil moisture carryover effects, root mass in each layer and specific root length must

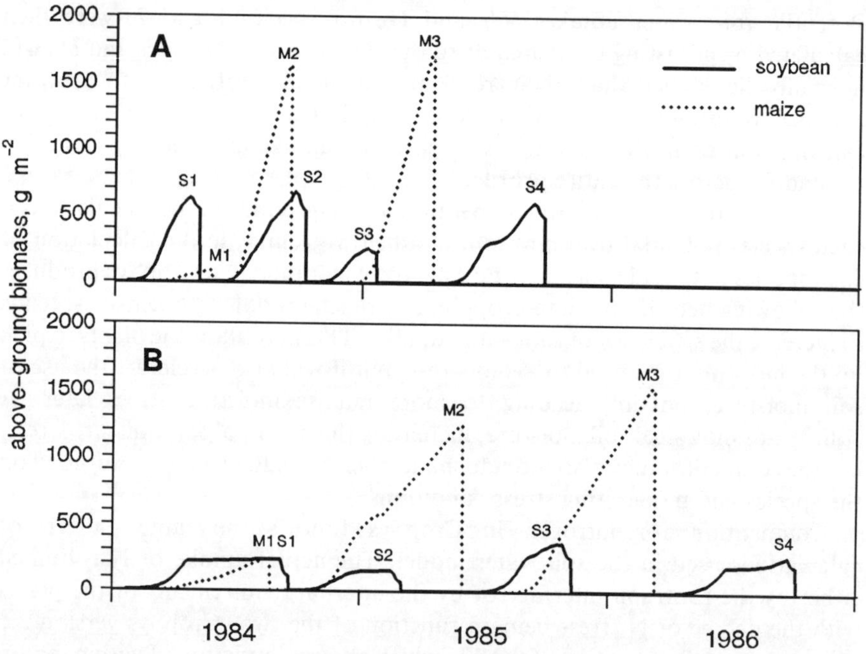

Figure 1. Simulated sequence of crops grown at two elevations on Maui, Hawaii, comparing growth curves at a warm site (A) and a cool site (B). The first cycle was an intercrop of soybean with a low population density of maize (M1 and S1). Cycle 2 was a maize / soybean intercrop with a full density of the maize. In the third cycle, maize (M3) was relay planted into the soybean (S3). Sole crop soybean (S4) finished the sequence.

also be provided. Nitrogen carryover effects and competition for nitrogen can also be analyzed if information is supplied for the mass of each plant part and their critical N concentrations.

Two copies of the measured-data links are included in CropSys version 2.0. Users can use the facility for both components of an experimental intercrop or sequence, or can link to measured data for one species while simulating the other.

Simulation examples

Managing crop sequences can be complex, particularly when the environment is suitable for continuous plantings, as is the case in Hawaii. Figure 1 illustrates the simulation of maize and soybean in four crop cycles at two elevations. Crop development at the cool site is delayed and crops quickly become out of phase with the warm site. Note the depression in soybean growth associated with the winter transition between 1984 and 1985. Those months fall during the third soybean crop (S3) at the warm site but only the second (S2) at the cool site. CropSys allows the user to specify the appropriate conditions for planting (i.e.

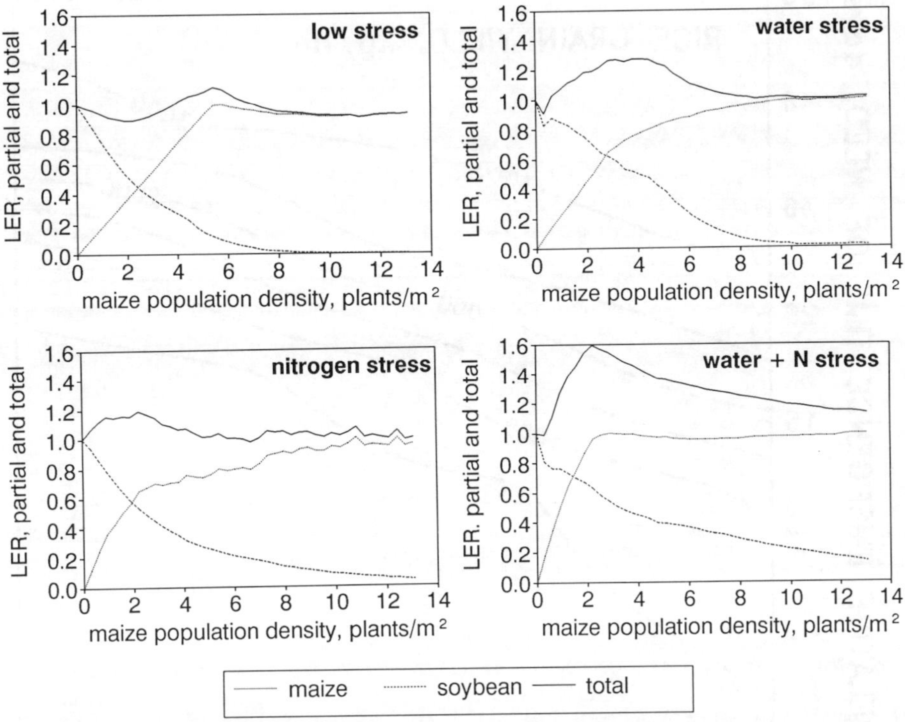

Figure 2. Analysis of simulated land equivalency ratio (LER) for an additive density series of maize (cv. Pioneer X304-C) intercropped with a constant density of soybean (cv. Kahala, 38.5 plants m^{-2}) managed for low and high levels of water and nitrogen stress in Waimanalo, Hawaii, planted June 1, 1990. LER is based on grain yield, with the reference yield for the maize sole crop taken from the optimum density within each management level. Reference yields for maize were 7780, 4220, 3690, and 1830 kg ha^{-1}, and for soybean were 3800, 960, 3490, and 1050 kg ha^{-1} for the low stress, water stress, nitrogen stress, and water plus nitrogen stress cases, respectively.

soil temperature, soil moisture, length of time the field has been fallow, length of time since another crop was planted, calendar dates). Simulated planting is delayed until the conditions are met. This facilitates long-term sequences for which actual planting dates can not be fixed prior to the simulation. Simulated planting also more closely matches farmer practice, and CropSys can evaluate different strategies for planting (i.e. relay intercrop a legume in sorghum only if soil moisture is above a certain threshold).

Population density and proportion are two standard factors used in experiments to characterize plant interaction in mixtures (Chaper 8 Harper 1977; De Wit 1960). Exactly how the factors of density and proportion are used is somewhat problematic: given the practicalities of field experiments, some combinations of density and proportion will not be represented, with a concomitant loss of information. Figure 2 illustrates the use of CropSys in exploring density and proportion relationships in intercrops. The figure represents an additive series of maize and soybean, though the user could just

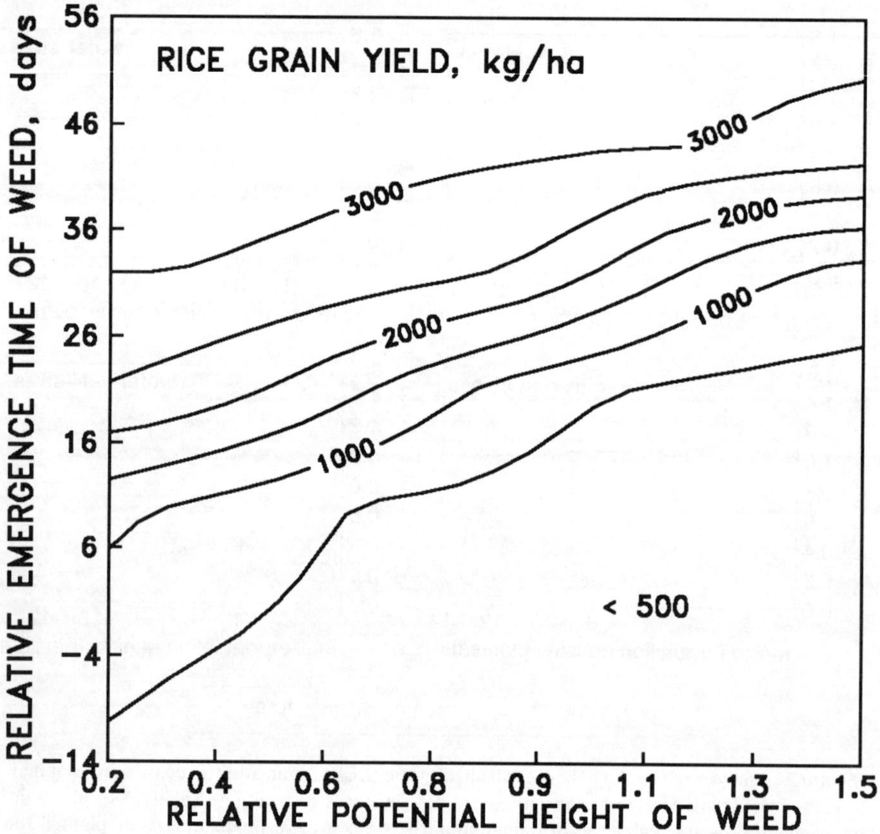

Figure 3. Simulated grain yield of upland rice (cv. RD 7, 100 plants m^{-2}) subject to competition from 'pseudo-weeds' (simulated by CERES-Maize at 6 plants m^{-2}) of different heights (relative to the height of the rice: maximum weed height / maximum rice height), given different emergence dates (date of weed emergence minus date of rice emergence) in Waimanalo, Hawaii, 1990.

as easily generate any series of densities for both crops, including replacement series. CropSys automatically generates the combinations needed in the analysis given the starting and ending densities for both species.

While IBSNAT currently has no simulation model for weeds, crop models within CropSys can be used to act as 'pseudoweeds', providing a means for exploring and teaching basic weed/crop relationships. Figure 3 shows the kind of analysis possible. The simulated yields for upland rice decline as weed height increases and as weeds emerge earlier in the season, illustrating the relative importance of a weed-free period versus plant morphology.

A fundamental motivation for the development of CropSys was to provide risk assessment for multiple cropping. Figure 4 illustrates that capability. In nitrogen limited environments, biological N fixation should reasonably have a benefit on cereal-based cropping systems, and the example shows the improvement in yields when upland rice is preceded by an N-fixing legume. The

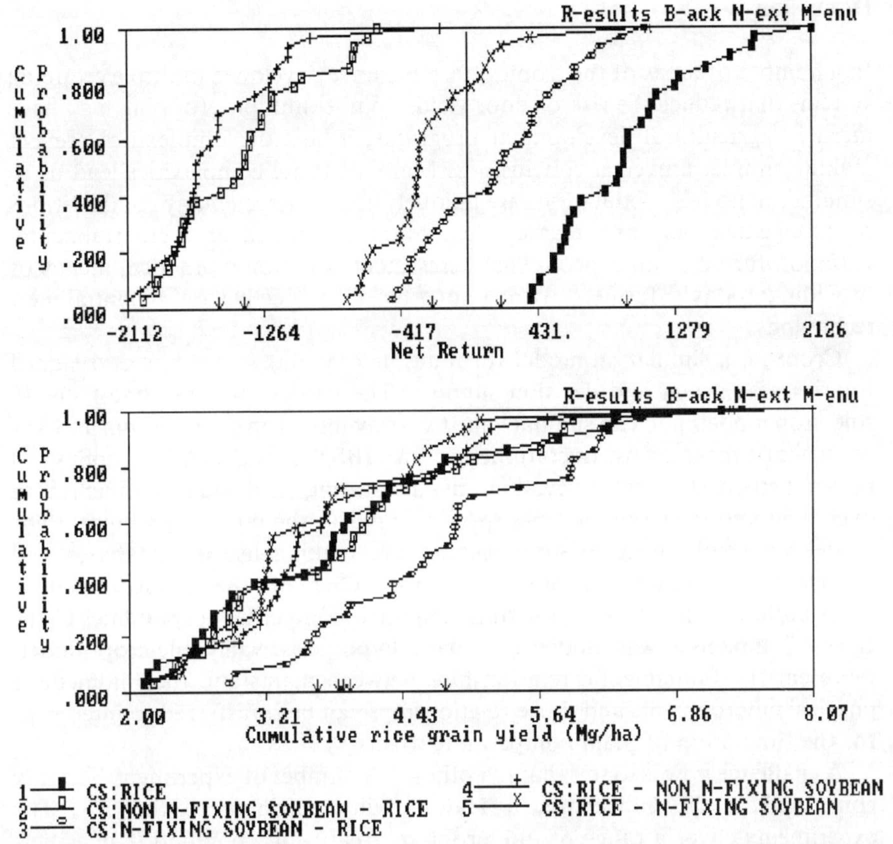

Figure 4. DSSAT presentation of an analysis of net return and rice grain yield of upland rice-based cropping systems in Khon Kaen, Thailand: (a) Rice (cv. RD 7, 100 plants m^{-2}, planted August 3), (b) non-nodulating soybean (a generic early maturing, group 00, cultivar, 30 plants m^{-2}, planted April 30) followed by rice (August 3), (c) nodulating soybean (April 30) followed by rice (August 3), (d) rice (August 3) followed by non-nodulating soybean (December 6), and (e) rice (August 3) followed by nodulating soybean (December 6). Each strategy was integrated over three years. A small amount of N fertilizer was applied before the first rice crop. Grain market price: \$343 t^{-1} rice, \$331 t^{-1} soybean; N \$0.50 kg^{-1} plus \$10 ha^{-1} for application and tillage; cost per 1000 seeds: \$0.40 for rice, \$10.80 for soybean.

improved yields were not enough to offset extra production costs, and a monoculture of rice proved to be the dominant strategy based on the economic analysis. Double-cropping of nodulating soybean before rice gave a yield benefit due to N fixation, but showed a 40% chance of making a negative net return.

Discussion

In a number of areas of the tropics, farmers have developed multiple cropping systems that reduce the risk of poor yields. While multiple cropping may have merit in agricultural development programs, a host of complexities prevent making simple, universal conclusions. Many of those complexities lead us to simulation models. Simulations are uniquely able to screen many combinations of management options and assess them over the broad range of environmental variation farmers can expect. That assessment is the only practical means of providing today's farmers with a long-term perspective on potential new technology.

CropSys, a simulation model for multiple cropping systems, was designed for risk assessment and decision support. The model contains code from 10 sole-crop models and can simulate any two-way intercrop combination and any sequence of those crops. In conjunction with IBSNAT's DSSAT, CropSys can perform stochastic strategy evaluations, integrating yield and economic return over time and over the multiple species found in the cropping systems. The model was built using existing technology, making use of databases and application programs already developed. One of the model's central contributions – the introduction of interspecific competition capabilities to the IBSNAT models – was guided by a basic hypothesis: valid sole-crop models represent the fundamental relationships between plants and their immediate physical environment, and those relationships can be transferred to intercrops for the simulation of plant competition effects.

A challenge now is to test that hypothesis. A number of experiments recently completed at the University of Hawaii will contribute to the test. The experiments cover a range of intercropping treatments: population densities, row widths (including very wide rows) and arrangements, row orientations, relay planting, and irrigation level. Observations from the experiments include yield and basic growth and development data, along with a large number of light interception measurements. Maize, soybean, taro, and upland rice have been evaluated, and additional experiments are planned with wheat and nodulating versus non-nodulating isolines of soybean. The goal of the work is to test and then deliver a simulation of multiple cropping systems that can improve decision making in the process of agricultural development.

Acknowledgements

This research was supported in part by the U.S. Department of Agriculture under CSRS Special Grant No. 87-CRSR-2–3135 and No. 88-34135–3587, managed by the Pacific Basin Advisory Group, and by the Soil Management Collaborative Research Support Program, U.S. Agency for International Development grant no. DAN-1311-G-SS-1083, subgrant no. SM-CRSP-022.

References

Caldwell R M (1989) AGSYSTEM: An object-oriented model of the agricultural systems hierarchy. Page 27 in Proceedings of the workshop on crop simulation, March 28–30, 1989, Biological Systems Simulation Group, Urbana, Illinois, USA.

Caldwell R M (1990a) CropSys: a simulation model for maize / soybean cropping systems. Page 30 in Proceedings of the workshop on crop simulation, March 27–29, 1990, Biological Systems Simulation Group, Lincoln, Nebraska, USA.

Caldwell R M (1990b) Simulation of low input corn / soybean cropping systems. Page 15 in Agronomy abstracts, ASA, Madison, Wisconsin, USA.

Caldwell R M (1990c) Multispecies simulations: problems and approaches. Pages 18–19 in Proceedings of the workshop on modeling pest-crop interactions, January 7–10, 1990, International Benchmark Sites Network for Agrotechnology Transfer (IBSNAT), Dept. of Agronomy and Soil Science, CTAHR, University of Hawaii, Honolulu, USA.

Caldwell R M, Hansen J (1990) User's guide for CropSysms version 1.0: A simulation model for maize/soybean cropping systems. Dept. of Agronomy and Soil Science, Univ. of Hawaii, Honolulu, USA.

Caldwell R M, Hansen J, Fernandez A (1991) Comparison of object-oriented and traditional programming methods for multispecies simulations. Page 7 in Proceedings of the workshop on crop simulation, March 25–27, 1991, Biological Systems Simulation Group, Beltsville, Maryland, USA.

Clements F E, Weaver J E, Hanson H C (1929) Plant competition: an analysis of community functions. Carnegie Inst. Wash. Publ. 398, Washington D.C., USA. 340 p.

De Wit C T (1960) On competition. Versl. Landbouwk. Onderz. 66:1–82. Pudoc, Wageningen, The Netherlands.

Donald C M (1963) Competition among crop and pasture plants. Adv. Agron. 15:1–118.

Francis C A, Ed. (1986) Multiple cropping systems. Macmillan Publishing Company, New York, USA. 383 p.

Grace J B, Tilman D (1990) Perspectives on plant competition. Academic Press, San Diego, USA. 484 p.

Hansen J (1991) An agroecosystem model for simulation of multiple cropping systems using IBSNAT crop submodels and minimum data sets. Page 34 in Proceedings of the workshop on crop simulation, March 25–27, 1991, Biological Systems Simulation Group, Beltsville, Maryland, USA.

Harper J L (1977) Population biology of plants. Academic Press, London, England. 892 p.

International Benchmark Sites Network for Agrotechnology Transfer (1986) Documentation for IBSNAT crop model input and output files, version 1.0, IBSNAT technical report 5. Dept. of Agronomy and Soil Science, CTAHR, University of Hawaii, Honolulu, USA.

Jones C A, Kiniry J R, Ed. (1986) CERES-Maize: A simulation model of maize growth and development. Texas A&M University Press, College Station, Texas, USA. 194 p.

Jones C A, Ritchie J T, Kiniry J R, Godwin D C (1986) Chapter 4. Subroutine structure. Pages 49–111 in Jones C A, Kiniry J R (Ed.) CERES-Maize: A simulation model of maize growth and development. Texas A&M University Press, College Station, Texas, USA. 194 p.

Jones J W, Boote K J, Jagtap S S, Hoogenboom G, Wilkerson G G, Mishoe J W (1988) SOYGRO V5.41: Soybean crop growth and yield model, IBSNAT version, technical documentation, draft, University of Florida, Gainesville, Florida, USA.

Joyce L A, Kickert R N (1987) Applied plant growth models for grazinglands, forests and crops. Pages 17–56 in K. Wisiol and J.D. Hesketh, (Ed.), Plant growth modeling for resource management, Volume I, Current models and methods. CRC Press, Inc., Boca Raton, Florida, USA. 170 p.

Lynam J K, Sanders J H, Mason S C (1986) Economics and risk in multiple cropping. Pages 250–284 in Francis C A (Ed.) Multiple cropping systems. Macmillan Publishing Company, New York, USA. 383 p.

Ritchie J, Sing U, Godwin D, Hunt L (1989) A user's guide to CERES-Maize V2.10, International Fertilizer Development Center, Muscle Shoals, Alabama, USA. 86 p.

Russell J T, Caldwell R M (1989) Effects of component densities and nitrogen fertilization on efficiency and yield of a maize/soyabean intercrop. Expl. Agric. 25:529–540.

Vandermeer J (1989) The ecology of intercropping. Cambridge University Press, Cambridge, England. 237 p.

Willey R W, Heath S B (1969) The quantitative relationship between plant population and crop yield. Adv. Agron. 21:281–321.

Optimization of cropping patterns in tank irrigation systems in Tamil Nadu, India

K. PALANISAMI

Water Technology Center, Tamil Nadu Agricultural University, Coimbatore, Tamil Nadu 641 003, India

Key words: cropping system, financial return, groundwater, India, irrigation, MOTAD, optimization, rice, risk, simulation, TANK IRRIGATION SYSTEM SIMULATION MODEL, water stress, water-limited yield

Abstract

Tank irrigation systems of South India account for about one third of the total rice irrigated area. In the recent past, poor maintenance of the tanks both by government and farmers contributed for the declining tank performance, as indicated by the declining rice area irrigated as well as by reduction in rice yield due to growing water scarcity at the end of the crop period. Since in five out of ten years the tanks received only about only half of their full storage, farmers often incur losses in rice production due to water scarcity. Both the government and farmers are feeling that the cropping pattern in the tank irrigation systems should be changed so that it is possible to stabilize the farmers income by minimizing the risk of crop failure under rice cultivation.

The tank simulation model developed earlier was updated using the coefficients from a system of simultaneous equations. The stress days and yield reduction equations were improved in the tank simulation model. Possible cropping patterns that may suit the tank systems in different years as well as data on net income, input requirement and crop rotation for normal, surplus, deficit and failure tank irrigation periods were used to workout the expected income under different crop mix in the optimization model. The minimization of the total absolute deviation (MOTAD) was used to derive the optimum cropping pattern. The results indicate that optimal cropping pattern for the tank systems consists of 25% rice and 75% non-rice crop.

Introduction

Tank irrigation systems of South India are centuries old and they account for over 30% of the total irrigated area. Tanks are small scale gravity irrigation systems constructed across the gentle slope of the valley to collect and store rainfall run-off for irrigation mainly to irrigate one rice crop during the September – December months. A typical tank system is depicted in Figure 1. Several constraints limit the productivity of these tanks. Tank siltation, foreshore encroachment and poor maintenance of the structures are the major above-outlet problems; absence of water users' organizations, a poor distribution system, and inadequate ground water supplies for supplementation are the major below-outlet problems. The cropping pattern followed by the farmers is mainly one season rice (September–October to November–December) and in most of the years farmers incur crop losses due to water scarcity, particularly at the end of the crop season. Since the filling of tanks

F.W.T. Penning de Vries et al. (eds.), Systems Approaches for Agricultural Development, 413–425.

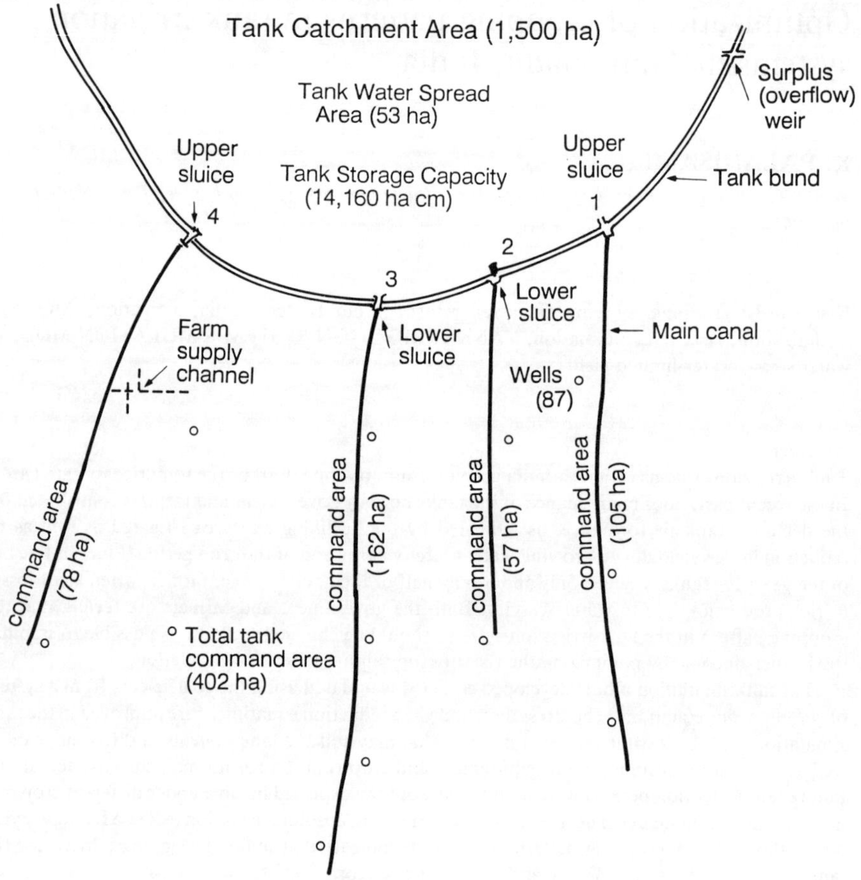

Figure 1. The tank irrigation system.

follow an uncertain pattern due to erratic and uncertain rains, farmers cannot predict the level of tank filling and hence in most cases resort to late rice planting and incur crop losses. Over years, constraints in water use are getting worse, putting even more emphasis on optimum use of water. One of the possibilities of optimum water use will be substituting rice with non-rice crops that need less water such as maize, sorghum, groundnut, pulses, and cotton. Hence optimization of the cropping pattern in the tank irrigation systems is highly important in the context of growing water scarcity.

The main objective of the paper will be to establish a method to optimize the cropping pattern in tank irrigation systems by integrating the simultaneous equations and simulation models developed independently.

Methodology

First stage: application of simultaneous equations model

When the relationship between crop yield and inputs is depicted as a single equation model including, for example, irrigation water, fertilizer, labour, crop management, etc., it is assumed that the only endogenous variable is crop yield and that the explanatory variables are truly exogenous. In practice, however, many of these factors are simultaneously determined: they are part of a general system and are interdependent as opposed to independent. In this case a single equation model will lead to some bias, and the estimated partial regression coefficients will be neither consistent nor efficient. This bias is overcome by the use of simultaneous equation model (Palanisami and Easter 1987; Palanisami and Flinn 1989). A set of seven structural equations constitute the simultaneous-equation model. The model is:

Tank water use:	$Y_1 = f_1(x_1,x_2,x_3,x_4,x_5,y_6)$	(1)
Well water use:	$Y_2 = f_2(Y_1,Y_5,x_3,x_6,x_7,x_8,x_9)$	(2)
Stress days:	$Y_3 = f_3(Y_2,Y_5,x_1,x_3)$	(3)
Nitrogen input:	$Y_4 = f_4(Y_3,Y_5,x_{10},x_{11},x_{12})$	(4)
Labour use:	$Y_5 = f_5(Y_3,Y_4,x_{12},x_{13})$	(5)
Management:	$Y_6 = f_6(Y_3,x_{12},x_{14})$	(6)
Crop yield:	$Y_7 = f_7(Y_3,Y_4,Y_5,Y_6)$	(7)

where:

$Y_1 =$ tank water use (ha·cm)

$Y_2 =$ well water use (ha·cm)

$Y_3 =$ stress days measured (number of days in excess of three when there is no standing water in the field)

$Y_4 =$ quantity of nitrogen applied ($kg\ ha^{-1}$)

$Y_5 =$ quantity of labour used (mandays ha^{-1})

$Y_6 =$ crop management index

$Y_7 =$ crop yield ($t\ ha^{-1}$)

$x_1 =$ farmer expenditure on water management for main and secondary canal maintenance, and labour for water diversion ($Rs\ ha^{-1}$)

$x_2 =$ number of upper sluices

$x_3 =$ number of wells in the command area

$x_4 =$ dummy variable for tank type (1 for systems tanks and 0 for others)

$x_5 =$ deviation of rainfall from mean rainfall (%)

$x_6 =$ dummy variable for well ownership (1 = owned; 0 otherwise)

$x_7 =$ well water/crop price ratio

$x_8 =$ service charges for well water supplies ($Rs\ ha^{-1}$)

$x_9 =$ cost of tertiary canal cleaning ($Rs\ ha^{-1}$)

$x_{10} =$ nitrogen-crop price ratio

$x_{11} =$ value of farm assets (Rs)

$x_{12} =$ credit availability (Rs)

$x_{13} =$ labour-crop price ratio

x_{14} = crop management expenditure (Rs ha^{-1})

Tank water use at the farm level (Y_1), was estimated on a per hectare basis as the product of depth of irrigations and number of irrigations. Irrigation depth was approximated from the distance of the farm from the canal outlet. If the farm was located between 0 and 0.3 km from the outlet, then the assumed depth was set at 3 cm, between 0.3 and 1.0 km at 2 cm, and above 1.0 km at 1 cm; such depths were confirmed with water-depth stakes at different locations. Well water use at the farm level (Y_2) was derived as a product of the number of well irrigations times the depth of irrigation. The depth of each irrigation was set at 2 cm.

The crop management index (Y_6) was based on the timeliness of farming operations and adequacy of input use, when compared to recommended levels. The crop cultural practices: land preparation, transplanting, fertilizing, weed control, tank irrigation, well irrigation, plant protection, and harvest were included to derive the management index (Palanisami and Easter 1983). The results of this model have been used in the yield stress days equation in the simulation model.

Second stage: tank irrigation system simulation model

Model description

A flow chart of a tank irrigation system simulation model is shown in Figure 2. Inputs to the model are of three types. First there are weather and related factors such as run-off, and evaporation. Secondly, there are factors related to tank hydrology such as tank catchment area, seepage, sill level of the surplus weir, full tank capacity and the number of sluices. The third type of input is farmer related such as crop area, water delivery to crops, well output and the number of wells (Palanisami and Flinn 1988).

Srivilliputhur Big Tank (SBT) in Ramanathapuram district was selected to test the model, because: (i) the extensive data basis has already been developed for the tank, it is one of 10 tanks included in an earlier study (Palanisami and Easter 1983); (ii) most of the problems observed in the tank are similar to problems in other tanks; and (iii) the tank command area and storage area are representative of other tanks. The catchment area of SBT is over 1500 ha, and its storage capacity is 14160 ha·cm with a water spread area of 53 ha. The irrigable crop area of the tank is 402 ha.

Water balances

The model, programmed in Fortran IV, operates on a daily basis. Daily tank water balances are computed by adding the inflow from rainfall in the catchment to the previous day's water balance, and subtracting outflows from the tank, and seepage and percolation losses on that day. The elevation of the tank water surface is calculated which permits the estimation of possible outflows from each of the four tank sluices. The model calculates when the tank

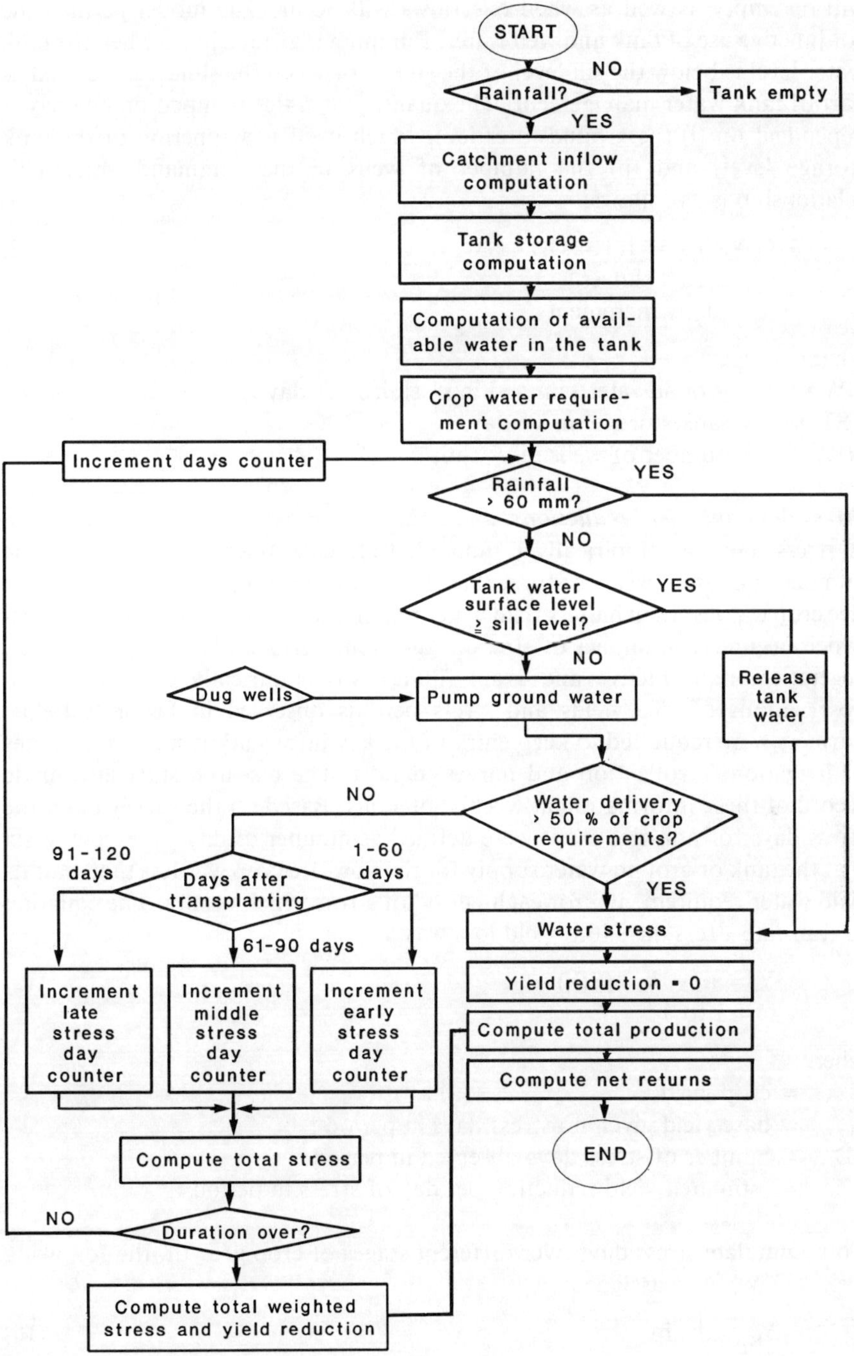

Figure 2. Flow chart of the tank simulation model.

will be empty as well as when overflows will occur. The model permits the conjunctive use of tank and well water. Pumping may take place when the tank water level is below the sill level of the sluice, or when the sluices are closed as part of tank water management. The quantity of water pumped on any day is dependent on (i) the groundwater level which itself is a function of the tank storage level, and (ii) the number of wells in the command area. This relationship is as follows:

$$GWAT_{sk} = f(TSTO_k) \cdot NW_s \quad (8)$$

$s = 1$ to 4 sluices
$k =$ daily increments

where:
$GWAT_{sk}$ = groundwater pumped in s^{th} sluice, k^{th} day.
$TSTO_k$ = tank storage on k^{th} day.
NW_s = number of wells in s^{th} sluice.

Stress days and yield reductions

A stress day was empirically estimated, following Wickham (1983), as the number of days in excess of three that the tank or ground water supply for the rice crop was less than half its field water requirements, for each period of stress which occurs. The impact of stress at different periods of crop growth (i.e., vegetative, reproductive and grain filling) was empirically estimated from observations of rice yields and stress periods observed in farmer's fields. Farmers were requested to keep entry of the key information such as the dates of irrigation, fertilization and harvest details. The research staff also made record of these information at weekly intervals. Based on the survey data, the stress days for non-rice crops were defined as number of days in excess of six that the tank or ground water supply for the non-rice crop was less than half its field water requirements, for each period of stress which occurs. The equation to translate stress days into yield loss was:

$$Y = a_t - b_t \cdot (SD_t) \quad (9)$$

$t = 1$ to 3

where:
Y = crop yield (t ha^{-1})
a_t = base yield given no stress days in period t
SD_t = number of stress days observed in period t
b_t = estimated yield reduction per day of stress in period t

To accumulate stress days over different stages of crop growth, the following weights (W_t) were assigned:

$$W_t = b_t/b_3 \quad (10)$$

$t = 1$ to 3

This permitted the estimation of the weighted stress days (WSD_s) for sluice s from the actual incidence of stress in that sluice and crop growth stages as:

$$WSD_s = SD_{st} \cdot W_t \qquad (11)$$
$$s = 1 \text{ to } 4$$
$$t = 1 \text{ to } 3$$

These weighted stress days were then used in the system of seven simultaneous equations described before to arrive at the actual yield as follows:

$$Y_{sk} = c_k - d_k \cdot WSD_s \qquad (12)$$
$$k = 1 \text{ to } 3$$
$$s = 1 \text{ to } 4$$

where:
Y_{sk} = actual yield in sluice s, location k
c_k = crop yield without moisture stress, location k as derived in the simultaneous equation system,
d_k = estimated coefficient of the yield reduction due to weighted stress days, location k as derived in the simultaneous equation system.

Total yield reduction (YRD_{sk}) within a given location served by a particular sluice is then directly calculated as:

$$YRD_{sk} = d_k \cdot WSD_s \cdot ARE_{sk} \qquad (13)$$
$$s = 1 \text{ to } 4$$
$$k = 1 \text{ to } 3$$

where:
ARE_{sk} = crop area of the group of farms in location k, sluice s (ha)
The total crop production of the group of farms ($TOTPRN_{sk}$) in location k, sluice s is:

$$TOTPRN_{sk} = c_k \cdot ARE_{sk} - YRD_{sk} \qquad (14)$$

Finally, the net return, in Rs ha^{-1}, to the group of farms in location k, sluice s is:

$$NETRN_{sk} = (TOTPRN_{sk} \cdot PR + VBP_{sk})/ARE_{sk} - COSTHA_{sk} \qquad (15)$$

where:
$NETRN_{sk}$ = the net return to the group of farms in location k, sluice s (Rs h^{-1})
PR = crop price (Rs t^{-1})
VBP_{sk} = value of by-product, in location k, sluice s
$COSTHA_{sk}$ = cost of production per ha, in location k, sluice s

Third stage: optimization – application of MOTAD model
Optimization will normally refer to maximization of the net income or minimization of the cost or risk. In the tank irrigation systems, farmers are more interested to stabilize their net income by minimizing the income deviations from year to year. Hence, selection of optimal crop patterns that will minimize their income variability (income risk) is considered more relevant and

Table 1. Alternative cropping patterns in tank irrigated areas.

Cropping pattern	Rice	Non-rice
X_1	100%	–
X_2	75%	25%
X_3	50%	50%
X_4	25%	75%

Table 2. Cropping pattern in tank irrigated areas.

Situation	1st crop (*Samba*) (Jun-Dec)	2nd crop (*Navarai*) (Jan-Apr)	3rd crop (*Sornawari*) (Apr-May)
Tankfed up to Nov-Dec			
a) Light soil	Groundnut	*Ragi*	–
Rice	*Ragi*	–	
Rice	Pulses	–	
b) Medium soil	Rice	*Ragi*	–
Rice	Pulses	–	
Groundnut	*Ragi*	–	
Rice	–	–	
Tankfed up to March-April			
a) Light soil	Rice	Groundnut	–
Rice	*Ragi*	–	
Rice	Pulses	–	
b) Medium soil	Rice	Rice	Rice
Rice	Rice	–	
Rice	*Ragi*	–	
Tankfed, supplemented by wells			
a) Light soil	Rice	Rice	*Ragi*
Rice	Groundnut	Pulses	
Rice	Rice	–	
b) Medium soil	Rice	*Ragi*	–
Rice	Rice	*Ragi*	
Rice	Groundnut	Pulses	
Rice	Rice	Rice	
Rice	Groundnut	Rice	

Ragi = finger millet.

risk programming models will help achieve this goal. One of the commonly used risk programming models is Minimization Of Total Absolute Deviations (MOTAD) which is a linear programming alternative for the quadratic programming (Hazell and Norton 1986). This model is relevant when the variance of the farm income is estimated using time series data. Given the total

Table 3. Frequency of failure of the rice (main) crop in tank irrigation systems, Ramanathapuram, Tamil Nadu, India.

Year type	Probability	Number of farmers planted rice (Total = 602)	Percent failure
Surplus	0.1	594	0
Full	0.2	578	19
Deficit	0.5	556	39
Failure	0.2	234	76

income deviation, $W = \sigma(Z_t^+ + Z_t^-)^2$, since sum of the negative income deviations below the mean, σZ_t^-, must always equal to the sum of the positive deviations, σZ_t^+, it is sufficient to minimize either of these two sums and to multiply the results by 2 to obtain $W^{1/2}$ (Hazell and Norton 1986). The model as such minimizes the total deviation in net income subject to the constraints outlined above. By parametrising the expected income values (λ), the model could be able to select different crop pattern with the corresponding income and standard deviation. The MOTAD model used is as follows:

$$\min 0.5W^{1/2} = \sigma Z_t^- \tag{16}$$
$$\sigma(c_{jt} - c_j)X_j + Z_t^- \geq 0 \text{ for all } t \tag{17}$$

and

$$\sigma c_j X_j = \lambda \tag{18}$$
$$\sigma a_{ij} X_j \leq b_i \text{ for all } i \tag{19}$$
$$X_j, Z_t^- \geq 0 \text{ for all } j, t \tag{20}$$

where:

X_j = level of jth activity
c_j = expected net income of the jth activity
c_{jt} = net income of the jth activity in t^{th} year
λ = expected income constraint (scalar)
a_{ij} = requirement of ith resource for j^{th} activity
b_i = resource availability constraint
Z_t^- = negative deviation in income

Optimization of the cropping pattern

The cropping pattern in the tank irrigated areas is mainly rice and in several cases non-rice crop is grown in the second season (Table 2). About 85% of the tank are non-system tanks, i.e. they depend purely on rainfall run-off in their own catchment areas. In the existing cropping pattern, only rice is grown by all the farmers. The per hectare water use was about 158 ha.cm, labour use was 64 mandays and the capital requirement was Rs. 1900. The total production simulated at the tank level was 11162 t of rough rice with yield reduction of 1516 t.

Table 4. Input use per hectare in different cropping options, Tamil Nadu, India.

Crop pattern	Water used (ha.cm)	Labour used (days)	Income (Rs)
X_1	158	85	1900
X_2	131	64	1721
X_3	103	51	1453
X_4	76	37	1123

1 US \$ = Rs 25.6

Table 5. Net income for the four cropping patterns (Table 1), in 10^6 Rs per tank.

Year	Probability	X_1	X_2	X_3	X_4
Surplus	0.1	.848	.845	.783	.721
Full	0.2	.798	.798	.780	.748
Deficit	0.5	.615	.779	.781	.789
Failure	0.2	.096	.167	.187	.192
Income*	.571	.667	.645	.654	

* Income = expected income, computed as net income x probability of the occurrence of different 'tank filling years'.

Growing rice crop here is highly risky in terms of expected crop failure due to water scarcity. About 39% and 76% of the farmers who have planted rice in the deficit and failure years experienced crop failures (Table 3). Hence optimization of the crop pattern for minimizing the net income deviation (risk) through efficient water use in the tank command areas is the main concern of the farmers.

The input use details and net returns under the suggested cropping pattern are given in Tables 4 and 5.

In the tank simulation model, rainfall data for four years, representing normal (1987), surplus (1981), deficit (1989) and failure (1988) years were used to simulate the tank inflows. Then, both rice and non-rice crop combinations, such as maize and sorghum were introduced in different proportions in all the tank sluice command areas (Table 1). The model simulated stress days, yield reduction and finally net income for different rice and non-rice crop patterns. Then the expected income (which is the net income under different crop patterns times the probability of the occurrence of different tank filling years) was arrived for different crop patterns. Along with the expected income, land, labour, water, capital (income) and crop rotation constraints, deviations of the individual year's net income from the expected income for each crop pattern was included in the MOTAD model as four risk rows (Appendix 1). Then the model was run with and without crop rotation to select the optimal crop pattern for the tank irrigation systems to minimize the deviation in net income.

Table 6. Results of the MOTAD model.

With crop pattern rotation							
λ (Rs)	Change (%)	X_1	X_2 ------ ha ------	X_3 (Rs)	X_4 (Rs)	Total income	SD
671675	–	–	–	200.5	200.5	481601	348197
604507	−10	–	–	180.6	180.6	433829	313658
738842	+10	–	–	200.5	200.5	481601	348197
806010	+20	–	–	200.5	200.5	481601	348197
537340	−20	–	–	160.5	160.5	385626	278807

Without crop pattern rotation							
λ (Rs)	Change (%)	X_1	X_2 ------ ha – -----	X_3 (Rs)	X_4 (Rs)	Total income	SD
671675	–	–	–	–	400.76	480111	347120
604507	−10	–	–	–	360.68	432099	312407
738842	+10	–	–	–	401.00	480398	347327
806010	+20	–	–	–	401.00	480398	347327
537340	−20	–	–	–	320.61	384089	277696

1US $ = Rs 25.6.
λ = expected income in Rs. derived from the ordinary Linear programming solution with maximization of net income.
SD = standard deviation of income = T@
T = total income in Rs and
@ = $2/s\{II/2(s-1)\}^{1/2}$.

The results indicate that the cropping pattern with 25% rice and 75% non-rice crops has entered the optimal solution compared to 100% rice as currently practiced by the farmers (Table 6). When the expected return has increased further, the choice of the crop pattern has not changed from the already selected pattern (X_4). However, when the expected income has been reduced by 10 and 20% from the mean expected income level, the same crop pattern entered the optimal solution, but the area has been reduced. Similarly when the optimal plan was allowed with possible crop rotation between the selected choices (i.e. X_1 to X_4), the crop pattern X_3 and X_4 entered the optimal solution which again demonstrated the importance of non-rice crops in the crop pattern of the tank irrigation systems. It is interesting to note that when crop pattern rotation was allowed, X_3 (50% rice and 50% non-rice) equally shared the area with X_4 indicating the flexibility between 25% and 50% rice area. The standard deviation of the income which is the indicator of the risk also indicated that crop pattern with higher income levels have higher risk. When the prices of fertilizers, labour and crop output prices were increased by 5% and 10% the results did not show any change in the optimum crop pattern selected by the model, indicating the stability of the rice and non-rice crop pattern in the long-run perspective.

Conclusion

Income variability from rice production in the tank irrigation systems over years affected farmers heavily as the farmers used to incur crop losses in five out of ten years when the tank water supply was inadequate during the later part of the crop season. Most of the farmers have expressed their desire to change the cropping patterns so as to minimize their income variability (risk) in crop production. The simulation model has stimulated the stress days, rice yield reduction and net income for different tank filling years. The net income generated in the simulation model was used to derive the net income deviations for different years and were then used as major constraints in the MOTAD model along with other constraints such as land, water, labour, capital (income), crop rotation. The results indicated that the optimal crop pattern was with 25% rice and 75% non-rice and when the rotation was allowed among the different crop patterns, both 25% rice and 75% non-rice crop pattern as well as 50% rice and 50% non-rice crop pattern entered the optimal solution and share the tank command area equally.

The results have the following policy implications:

a. rice will no longer be the major crop in the tank irrigation systems due to increasing water scarcity in the tank systems; a change in the crop pattern is also required to minimize farmers income losses over years.
b. government departments, such as the Agricultural and Irrigation Department should encourage the farmers in sharing the command area for different crops depending upon the water storages in the tank.
c. most of the farmers in the tank systems are small holders, many with less than 1 ha, timely availability of non-rice crop seeds and marketing facilities should be strengthened at the block levels.

Appendix

Appendix 1. MOTAD Model – Initial Matrix.

	X1	X2	X3	X4	Z^{-1}	Z^{-1}	Z^{-1}	Z^{-1}	RHS
Minimize					1	1	1	1	
Exp.income	1636	1663	1671	1676					$=\lambda$
Land	1	1	1	1					<401
Labour	85	64	51	37					<21320
Water	158	131	103	76					<43772
Capital	1900	1721	1453	1123					<601500
Rotation	−1	1	−1	1					=0
Risk rows:									
Surplus	479	444	344	190	1				>0
Full	354	327	337	257		1			>0
Deficit	292	279	339	359			1		>0
Failure	−1322	−1247	−1204	−1152				1	>0

References

Hazell P, Norton R D (1986) Mathematical programming for economic analysis in agriculture, Macmillion Publishing Company, New York, USA.

Palanisami K, Easter K W (1983) Tank irrigation in South India- A potential for future expansion in irrigation, University of Minnesota, St Paul, Economic Report no. 83–4. USA.

Palanisami K, Easter K W (1987) Small-scale surface (tank) irrigation in Asia. Wat. Res. Res. 23:774–80.

Palanisami K, Flinn J C (1988) Evaluating the performance of tank irrigation systems. Agric. Syst. 28:161–77.

Palanisami K, Flinn J C (1989) The impact of inadequate water supplies on input use and yield of rice crops. Ag. Wat. Man. 15:347–59.

Agricultural development in Thailand

N. CHOMCHALOW
FAO Regional Office for Asia and the Pacific,
Bangkok, Thailand

Key words: agricultural development, agroindustry, cassave, (socio-economic) constraint, crop quality, food demand, maize, national development plan, research priority, rice, sugarcane

Abstract
Agriculture has played an important role in Thailand from time immemorial. Although its sectoral share in the Gross Domestic Product (GDP) has gradually declined since the First National Economic and Social Development Plan (NESDP) was launched in 1961, agriculture still accounts for one third of total export revenue and workforce. Three stages of agricultural development can be recognized, viz. (i) Pre-1955 (rice monoculture/natural resources based), (ii) Between 1955–1985 (land based resources/labour-intensive method) and (iii) Post-1985 (structural shift in agricultural production). Thailands agricultural development plan and policy were clearly defined in the NESDP which will be elaborated in the present paper. Among the four major sub-sectors, crop production is by far the most important, accounting for 75% of the total revenue, followed by livestock, fishery and forestry. Problems in agricultural development in Thailand include (i) agricultural production structure, (ii) competition in world market and trade protectionism, (iii) low productivity, (iv) low quality of products, and (v) insufficient utilization of natural resources. At present, Thai agriculture is coming towards a crossroad. Its development can no longer depend on area expansion and burgeoning markets. Future development must be based on innovative technology for the production of non-staple commodities or agro-industrial products.

Introduction

Thailand's economic development

With the adoption of the 'Five-Year National Economic and Social Development Plans' starting in 1961, the overall economic growth during the past 30 years was in a range between 7–8% per year (Table 1). During the Fifth Plan (1982–'86), however, the annual rate of growth receded to 4.4%, mainly due to the world economic recession. Growth has recovered significantly during the Sixth Plan (1987–'91), average around 10%, largely from a significant increase in investments, from both the public and private sectors and a large influx of foreign capital. These high annual growth rates result in steady rise in per capita Gross National Product (GNP) which was US$ 1220 in 1988 (FAO 1991). These high rates of per capita GNP growth can be credited to a successful family planning program that reduced the population growth rate from 2.9% in 1960 to 1.4% in 1990.

F.W.T. Penning de Vries et al. (eds.), Systems Approaches for Agricultural Development, 427–443.

Table 1. Growth rate and agricultural sector share of GDP in the period 1961–'91 (%).

Plan	Annual growth rate			Agricultural sector share of GDP
	Year	Agriculture	Total	
First	1961-66	6.4	8.1	39.7
Second	1967-71	5.9	7.8	33.3
Third	1972-76	6.7	7.1	30.3
Fourth	1977-81	3.5	7.1	26.7
Fifth	1982-86	2.6	4.4	23.7
Sixth	1987	−2.0	8.4	16.1
	1988	8.6	11.0	16.9
	1989	6.3	10.8	15.1
	1990	−2.5	10.0	15.2
	1991	3.3*	7.0-7.5*	–

Data from:
(1) Office of the National Economic and Social Development Board, Office of the Prime Minister;
(2) Centre for Agricultural Statistics, Office of Agricultural Economics, Ministry of Agriculture and Cooperatives;
(3) Bank of Thailand.
* = Estimated.

Contribution of agricultural sector

Thailand's development has always been generally based on agricultural production. Her agricultural sector, mainly supported by smallholders, continues to be the basis of the livelihood of the majority of the population of 58 million, of which about one-third is presently employed in agriculture. Although the sector has declined as a proportion of Gross Domestic Product (GDP), it still accounts for 32% of total exports by value, and agricultural imports remain very small. At least 38% of the country's total land area (51.3 Mha) is now used for agricultural activities. Of this, some 60% is paddy field, 22.8% is under field crops and 9.4% is occupied by fruit and other perennial crops.

Planted area has increased significantly during the last two decades, from only 9.7 Mha in 1963 to 19.4 Mha in 1988. This two-fold increase in planted area has produced most of the growth in agricultural production in this period.

Productivities of most economic crops have not yet been improved satisfactorily; yields remain virtually constant, and some even exhibit downward trends. Since the early 1960s, there has been a steady development of irrigation facilities; increased water storage capacity has facilitated the expansion of dry season cropping. Up to 1986, irrigated area covered about 19% of the country's total cultivated land. Thus, agricultural production relies mainly on the weather; and the instability of the monsoonal climate is a major factor causing fluctuation in agricultural production and prices.

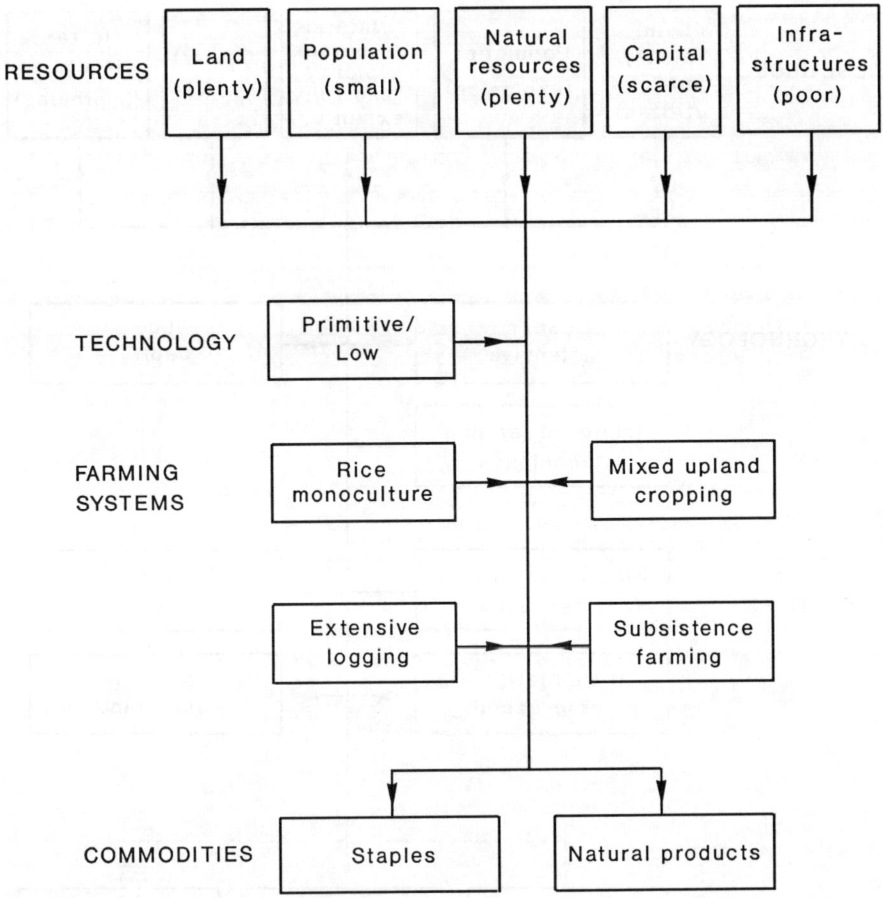

Figure 1. First stage of agricultural development. Pre–1955 (rice monoculture/natural resources–based)

Stages of agricultural development in Thailand

Based on (i) the availability of resources (ii) the level of technology, (iii) the systems of farming practices, and (iv) the types of commodities produced, agricultural development in Thailand can be separated into three, somewhat distinct stages; Pre–1955, Between 1955–'85, and Post–1985.

Pre–1955 (rice monoculture/natural resources–based)

Although agriculture has been the mainstay of Thai economy from time immemorial, it was mainly for internal markets until after World War II when some agricultural development took place. Prior to 1955, land was available in

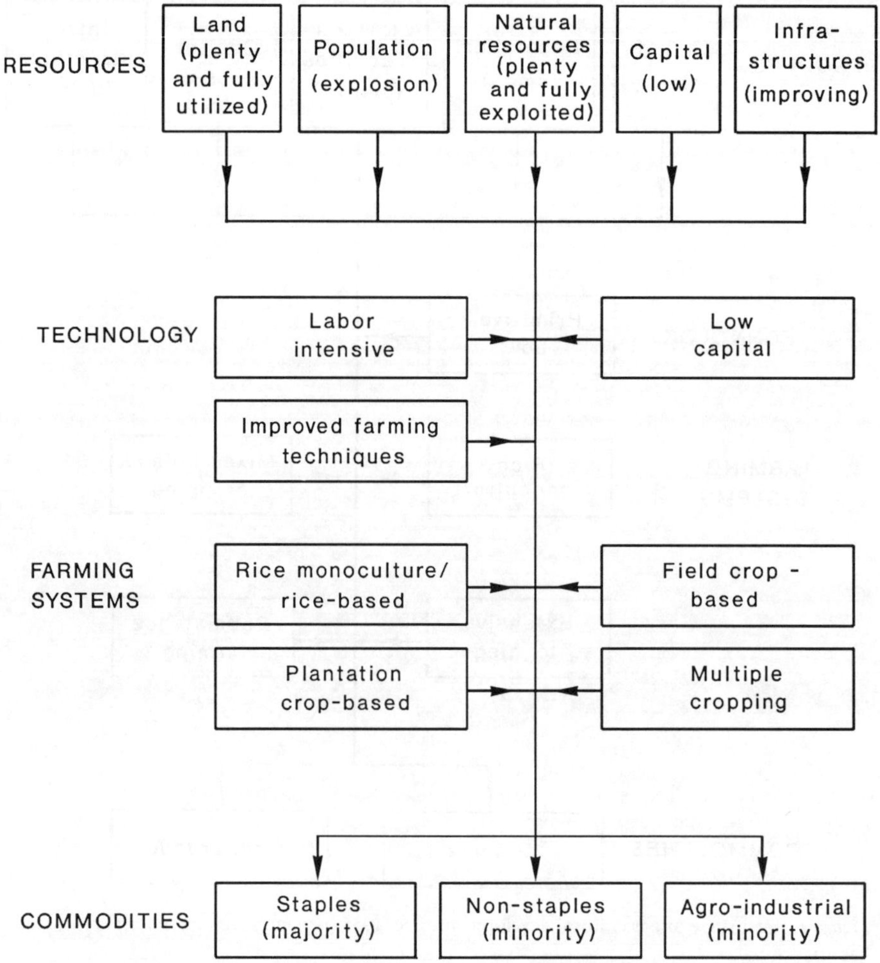

Figure 2. Second stage of agricultural development. Between 1955-1985 (land-based resource/labour intensive)

abundance and agricultural practices were extensive in nature. This limited the use of modern technology. In the rainfed areas farmers were reluctant to risk heavy borrowing for agricultural investment. Low prices of most agricultural commodities were a further constraint on investment. Small farmers with no capital, and with only a small piece of land using simple low-cost technology were the major producers (see Figure 1).

During this period which started even before World War II, rice monoculture and exploitation of forests for timber were the predominant agricultural activities, and both rice and timber were the major export commodities of the country.

With low capital investment and inputs, poor infrastructures, particularly

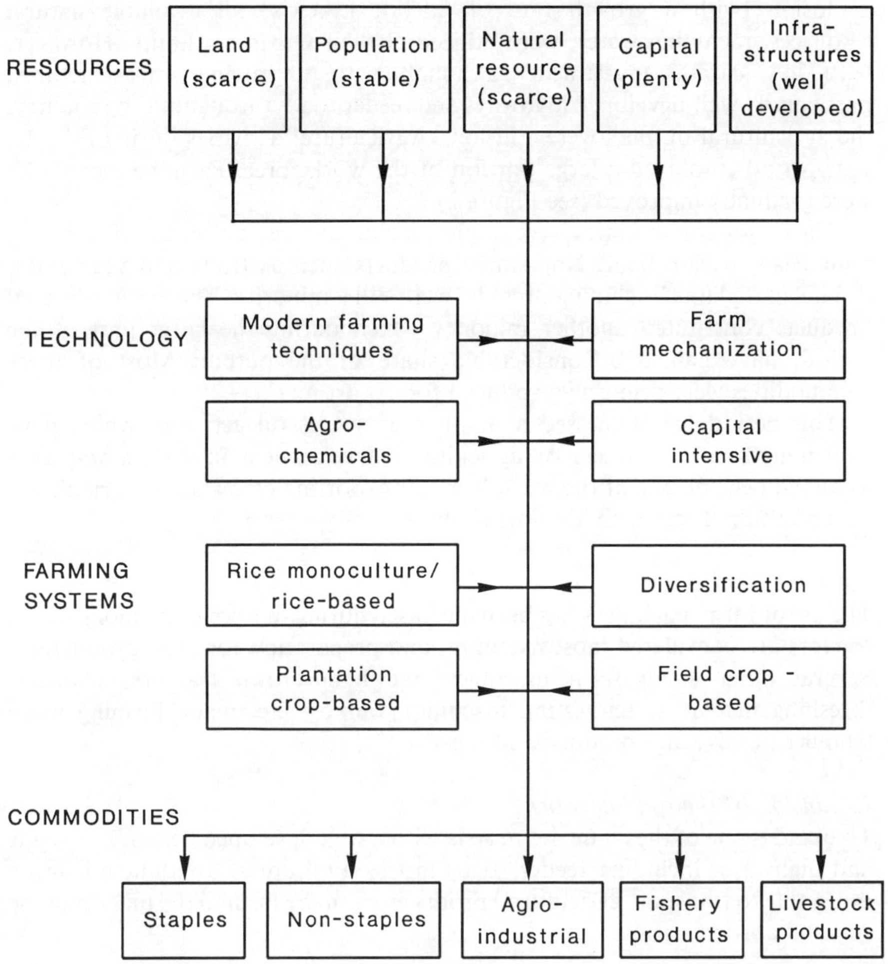

Figure 3. Third stage on agricultural development. Post–1985 (structural shift in agricultural production)

irrigation and transportation facilities, and with primitive or low-cost technology, agricultural production was quite low, yet agricultural products constituted a major element of the GDP.

Between 1955–'85 (land-based resources/labour-intensive method)

This period of slow and intermittent agricultural development which spanned 30 years can be characterized as extensification through an increased use of production area expansion and labour-intensive farming practices. All available lands, mostly acquired through forest clearing, were over-exploited legally and illegally. This was due to the pressure caused by the population explosion (with a growth rate of 2.9%). Likewise all available natural resources, including water, were utilized at their maximum limits. However, with low capital investment, although with improving infrastructures, particularly well-developed highways and feeder roads throughout the country, the agricultural outputs were still low. Agricultural activities during this 30–year period employed a large portion of the work-force. Farming techniques were gradually improved (see Figure 2).

The major commodities were staples, including rice, maize, kenaf, tapioca, mungbean, para-rubber. Non-staple products such as fruits and vegetables, ornamental plants, including orchids, were still a minority. The agro-industrial products constituted another minority which during the latter part of the period, have gained a considerable share of the output. Most of these commodities were primarily produced for export markets.

This period has witnessed a number of successful activities which have contributed to the success of agricultural development in such a way that Thailand became one of the world's major exporting countries of agricultural commodities. These activities included:

Farm mechanization
The use of farm machinery has been increased during this period. Among these, tractors are considered most useful in land preparation for crop production. Several other small farm machines like hand-drawn tractors, planters, threshing machines, harvesting machines, pumps, etc. made farming more labour extensive and productive.

Expansion of the road network
Thailand is one of the countries in Asia with a well-developed network of roads and highways, including feeder roads in the rural areas. This made it more profitable to transport agricultural produces from the farm to the market in the city and abroad.

Viable credit system
Through government and private financial institutions, agricultural credit was made available to advanced farmers who used this capital investment to improve their production capacity.

Post–1985 (structural shift in agricultural production)

In the previous two stages, the abundance of land limited the application of modern farming technology, encouraged labour-intensive methods in agriculture and supported the sector's absorption of a major share of the work-force. Since the mid 1980s, however, this 'land-based' economic growth has become increasingly unsustainable owing to the shrinking availability of new lands. As a result, the development is moving toward a more traditional pattern of growth, with evidence of a rising reliance on non-labour and technology intensive inputs in agriculture, and expanding roles for the industrial and services sectors. This was a shift in employment away from agriculture (see Figure 3).

Although recently initiated, this structural shift in agricultural production has great impact on future economy of the country. The factors which were responsible for this shift are the following.

Innovative technology

The application of innovative technology has been one of the key factors in the success of recent agricultural development. More and more advanced technology has now been incorporated in agricultural production, thus increasing efficiency of the operation and reduction in labour and in land.

Large-scale agro-industrial enterprises

Another important decision has been made by the government to improve agricultural productivity and encourage exports during the decade of 1980s. Large-scale agro-industrial enterprises have received the Board of Investment's support in order to offer commercial cropping opportunities for the smallholders. Such enterprises are also contributing more to agricultural production in their own right.

Disappearance of smallholders

The latter part of the 1980s has been a golden period in Thailand's economic development. Foreign investment has been pouring in from various countries which results in structural shift in agricultural development. One of the obvious shifts is the disappearance on smallholders who prefer to sell their lands to large-scale investors from the city who seek land for plantations, ranching, or to make golf courses and resorts.

Agricultural development plan and policy

The national economic and social development plan

Thailand is one of the first countries in Asia which has made use of the 'national development plan' concept. Initiated in 1961, the five–year 'National Economic and Social Development Plans' have contributed significantly and consistently to the growth rate mentioned earlier. Prior to 1961, agricultural development planning was quite weak, and policy was rather inefficient. Starting from the First Plan onwards, agricultural development has been attaining a top priority in the national development plan because of its importance in the country's economy. During the First Plan period (1961–'66), the agricultural sector experienced a relatively high growth rate of 6.4% per annum with a sectoral share of the GDP of 39.7%, due to the relatively high growth in all sub-sectors, particularly fishery (Table 1). During the Second Plan period (1967–'71), the average growth rate and sectoral share slightly decreased to 5.9% and 33.3% respectively, due to a drop in the livestock sub-sector. In the Third Plan period (1972–'76), the growth rate jumped to 6.7%, but its sectoral share of GDP slightly reduced to 30.3%. The high average growth rate during this period was the result of high world prices. During the Fourth Plan period (1977–'81), as the result of the world economic crisis, the growth rate drastically dropped to 3.5% while sectoral share also dropped to 26.7%. Such depression continued into the Fifth Plan period (1982–'86) with 2.6% growth rate and 23.7% sectoral share. Judging from the sectoral share of the GDP, which had been declining continuously from 39.7% in the First Plan to 23.7% in the Fifth Plan, one would assume a decreasing importance of agriculture and the increasing role of non-agricultural sectors, especially tourism and industry.

In the first year of the Sixth Plan (1987–'91), for the first time in history, the growth rate in the agricultural sector declined by 2%, while the total growth rate increased to 8.4%. Its sectoral share of the GDP also reduced to 16.1%.

The year 1988 was good for the agricultural sector due to high prices of several commodities together with favourable weather conditions which gave rise to the growth of 8.6% and sectoral share of 16.9%.

In 1989, the growth rate dropped to 6.3% and sectoral share to 15.1%. The slower growth rate was a result of delayed and sporadic rainfalls as well as the damage caused by a severe typhoon in the South.

In 1990, agricultural production declined again by 2.5% (Bank of Thailand 1991). Of all the sub-sectors, crop production was badly affected, falling by 4.7% as the result of severe drought at the beginning and the flood caused by a typhoon at the end of the cropping season. Most of the major crops, viz. rice, maize, sugarcane, and cassava were hard hit by these unfavourable climatic conditions while rubber, oil palm and coffee were less affected. The output of other agricultural sub-sectors were similarly affected except for livestock production which continued to expand at the same rate as last year due to increased production of swine and raw milk. Such adverse weather also caused severe damage to baby chicken resulting in drastic reduction in broiler and egg production. Fishery production also declined resulting from (i) damage caused by the typhoon which reduced the availability of catch in the fishing grounds, and (ii) high price of diesel oil. The output of forestry sub-sector continued to decline as expected.

In the last year of the Sixth Plan (1991), the Bank of Thailand (1991) expected that the growth rate will be about 7.0–7.5% which is considered satisfactory in line with the moderating down towards a more sustainable level. The agricultural sector is expected to increase by 3.3% as overall crop production is likely to be nursed by favourable climatic condition. However, rice productions of both the dry and the wet seasons are expected to drop, due to a reduced planting area caused by shortage of water in the main reservoirs and pest damage, respectively.

On the other hand, livestock sub-sector is expected to expand at a high rate as the production of chicken and swine pick-up. The small decline in fishery sub-sector is forecasted as the result of the increase in price of diesel oil since September 1990. As for forestry sub-sector, as there remain stocks of logs fallen during the typhoon disaster of November 1989 available for logging, the forestry production will decline at a less rapid rate than in the previous year.

The National Agricultural Research Policy

The first National Agricultural Research Policy was announced in the Fourth Plan (1977–'81). This included the following issues:

- Research into urgent and important problems confronting farmers and agriculture;
- Production research on quality and yield improvement in agriculture and other important sources of food and forestry;
- Studies on agricultural systems to optimize the use of existing materials and facilities and to conserve the environment for agriculture, forestry and biodiversity;
- Research on post-harvest technology and processing of agricultural products based on suitable raw materials for use in food industry or for other industrial uses;
- Research and utilization of agricultural products to maximum benefit; and
- Studies on basic biological processes as the basis for applied research.

The National Agricultural Research Plan

The National Agricultural Research Plan (NARP) was set up in 1982 in accordance with the Fifth Plan (1982–'86) to elaborate on the guidelines for the Fourth Plan, including irrigation and water management, land use and technology transfer. As for crop production, the new structure is aimed at producing suitable location-specific technology. In the past, research programs have mostly been made at the headquarters in Bangkok because there were only a few researchers in the regional offices. As a result, the research plans formulated were sometimes unrealistic and did not serve the farmer's needs or could not solve the farmers' problems. The NARP is meant to assist the Department of Agriculture, which is responsible for crop research, to foster decentralization of operations to the regional centres, both with regard to formulating research programs and with regard to budget, in order to solve the farmers' problems.

Guidelines and programs for agricultural development

The problems of agricultural development, particularly the productivity and income enhancement by farmers, together with the financial and resource constraints, have forced the government to issue the policy guidelines for agricultural development under the Sixth Plan (1987–'91), for which a short summary of guidelines and programs will be given in the following paragraphs.

Agricultural production development

The commodities selected for promotion were (i) import substitutes, (ii) export oriented, and (iii) agro-industrial products. Import substitute commodities include cotton, soybean, dairy products, wheat, and paper pulp; those with export potential were vegetables, tropical fruits, forest products, cutflowers, meat and marine products; and those agro-industrial products include processed food, dairy products, and paper pulp. It is fair to say that most of the guidelines have been followed and the outcome was encouraging as is evident from the rise in production of these commodities during the Sixth Plan period, which was terminated in September 1991.

Natural resources development

As natural resources are vital to agricultural production, there is an urgent need to protect the environment as well as to improve the utilization efficiency. Deteriorating and scarce land and water resources must receive priority attention. Similarly development and rehabilitation of marine habitats must be stressed. More emphasis must be put on reforestation in parallel with watershed rehabilitation.

Development of farmers' institutions

At present only about 2.2 million households or 40% of agricultural households are participating in farmers' cooperatives. Improvement of efficiency in business management and operating skills through a systematic coordination of production and marketing must receive priority attention. Information on marketing and price movements of agricultural products should be disseminated to the farmers.

Agricultural development policy under the Seventh Plan

In accordance with the policy and plan of agricultural development as formulated in the Seventh Plan (1992–'96), which has been effective from 1 October 1991, the Ministry of Agriculture and Cooperatives issued three principal policies as follows (Sarikaputi 1990).

Increasing and distributing the farmers' income

The main objectives of this policy are to increase the farmers' income and eliminate disparity of incomes of the farmers' and other economic careers. At the same time, attempts will also be made to reduce the risk in agricultural operation through the improvement of production systems and management. Through this approach the prices of agricultural commodities are to be stabilized. Five plans are to be initiated, namely:

- Production distribution at farm level;
- Accelerating increases in production efficiency;
- Promoting processing of agricultural products at farm level;
- Supporting agro-industrial development; and
- Location-specific development.

Conservation and development of natural resources

This policy is aimed at conserving natural environment for sustainable development. It consists of plans to conserve and develop: (i) the forest, (ii) water resources, (iii) land, and (iv) fishery resources.

Development of the quality of life

The policy formulated under this heading aims at raising the standard of living and the quality of life of the farmers and the society through (i) improving the standard of living of the farmers, and (ii) reducing pollution and solving environmental problems.

Major sub-sectors in agricultural production

Agricultural production in Thailand can be classified into crop, livestock, fishery and forestry sub-sectors. Crop production is by far the most important sub-sector, accounting for 75% of the total revenue, on the average, followed by livestock, fishery and forestry sub-sectors.

Crop production

Crop production is the major agricultural activity in Thailand. The last 30–35 years have seen a steady expansion in production of most staple crops, namely, rice, rubber, maize, kenaf, sugarcane, cassava, mungbean, soybean and oil palm. A brief account of each of the major crops is presented below.

Rice

Rice is the most important crop in Thailand. In fact it has long been synonymous with the name of the country. Despite the extensive diversification of agricultural economy since 1960, it is still both the staple food crop for domestic consumption and a major export item of Thailand. In terms of area planted, rice covers more than half of the total arable land. It is grown under irrigated conditions in most of the Central Plain and in the basins of northern part, but it is a rainfed crop in other parts of the country. Production has grown steadily since 1960 as a result of increases in planted area and yield stabilization through irrigation, but yields on a national basis still remain low (2 t ha^{-1}) compared with other countries (3.6 t ha^{-1} average of developing countries in Asia-Pacific Region) (FAO 1991).

Although the percentage share of agricultural GDP originated from rice fell from 42.3% in the First Plan to 29.2% in the Fifth Plan, and down further to about 15.8% in the Sixth Plan, it still remains Thailand's most valuable agricultural export commodity. The already high figure of 4.8 Mt of rice exported in 1988 (up by almost 10% from 1987), was once more surpassed in 1989 with a record of 6 Mt (BAAC 1991). This figure represented almost half of the world trade in rice.

Rubber

Rubber is the second most important foreign exchange earner for Thailand. It has achieved a steady increase in production over the last 25 years as the result of an extensive replanting scheme, using high-yielding stock. This scheme has reached over half of the planted area; thus the potential for further increases in production is high. In 1990, Thailand produced 1.1 Mt of rubber, rank third (after Malaysia and Indonesia) as the world largest rubber producer (FAO 1991). Traditionally, rubber is grown in the South, but is also experimentally tried in the North East with a good chance of success, judging from preliminary results obtained.

Maize

Maize was the first major crop in agricultural diversification drive which began in the second stage of agricultural development. It is now second only to rice in terms of area planted. The crop is cultivated mainly in the Central Plain. In recent years, its expansion into upland regions has had disastrous environmental consequences.

Yields have originally been quite low, but the introduction of new, drought resistant varieties led to record production in 1983/84. In 1990, Thailand produced 3.7 Mt, a drastic drop from 4.7 and 4.4 Mt in 1988 and 1989. The national average yield is 2.1 t ha^{-1} vs. 3.1 t ha^{-1} average from other developing countries in Asia-Pacific region (FAO 1991). Originally grown for export to other Asian countries for cattle feed, it is increasingly being used domestically in animal feed industry. This has reduced export levels since the mid–1980s, and consequently, its importance as export earner has been reduced by half.

Cassava

Cassava products (tapioca) are mainly exported to the EC as animal feed. As its price was quite low compared to other feed ingredients available in the EC markets, there was a sharp increase in demand in the 1970s. This had led to a marked expansion of cultivation in the North East, mainly through substitution of kenaf and sugarcane. Total production is around 20 Mt (of tubers) per annum with the average yield of 14 t ha^{-1} (FAO 1991). Thailand has always been the world's largest producer and exporter of cassava and tapioca from the beginning of tapioca trading with the EC.

Exports of tapioca pellets in 1989 amounted to over 9 Mt, worth US$ 940 millions, an increase of 12% in volume and 8% in value over 1988. However, further expansion is limited by the imposition of an EC quota which will allow Thailand to export 12 Mt during the period 1991–'94. Attempts have been made by the government and the private sector to seek other markets and to diversify its uses within the country. At present production is higher than demand and another effort is being made, with the EC funding, to encourage the farmers to switch to other crops.

Sugarcane

During the 1970s, there was a world-wide shortage of sugar. Taking advantage of supply problems in the world market, Thailand increased sugarcane production drastically until the demands of the world market has been over-satisfied, after which the production is decreasing. Production in 1989 was 36.7 Mt, with the average yield of 55.6 t ha^{-1} (FAO 1991). Thailand is the world's fourth largest sugarcane producing country (after India, China and Pakistan).

Other crops

Thailand is an important producer of several other crops. Kenaf, a fibre crop, expanded rapidly in the North East during the 1960–'70s in response to market opportunity, only to be replaced by cassava which invaded the North East in the mid–1970s, resulting in drastic reduction in production since it is easier to grow cassava and obtain a better income. Several legumes, particularly soybean, mungbean, blackgram and peanut have an important place in many parts of the country, while pineapples are widely cultivated around several new canneries in many parts of the country, but more concentrated in Prachuap Khirikhan area. In 1989, Thailand's canned pineapple exports totalled 350000 t (the world's largest exporter), worth US$ 180 millions, an increase of 2.9% in volume as compared to 1988 (BAAC 1991).

Although not traditionally considered as a crop, silk production should also be included in this sub-sector. Sericulture is regarded as one of the major occupations of Thai farmers, particularly in the North East, who spend most of their leisure time growing mulberry trees, raising silkworms, reeling cocoons, dying yarns, and weaving fabrics, traditionally for their own consumption, but recently for extra-income earning. In term of raw silk, Thailand's production (of about 1300 t per annum) as compared to the total world output (of about 60000 t), is relatively small, but in term of value of fabrics, it is quite large. Moreover, the name 'Thai silk' is very well-known throughout the world for its unique quality. About 80% of silk fabrics produced are domestically sold to the tourists.

Livestock production

Traditionally the livestock sub-sector in Thailand has been the subsidiary element of the subsistence economy with buffaloes used for draught purposes and pigs and poultry for consumption. More recently, the use of draught animals has declined and the raising of small livestock to meet growing urban and export demand has become a feature of the sub-sector. Its percentage share of the agricultural GDP was about 12–13% during the First to the Fifth Plans. By 1988 exports of frozen chicken had reached almost 100000 t, worth US$ 198 millions (BAAC 1991). Another new development is the rise of domestic dairy farming which is vigorously promoted by the government as well as by the private sector.

Fishery production

Thailand is the third largest marine fishing country in Asia, after Japan and China. The sub-sector comprises some 350000 fishermen with 40000 vessels, of which about half are deep sea trawlers, signifying a substantial modernization of the industry which was taken place over the last 20 years.

The total production of the fishery sub-sector expanded at a high rate during the First and Second Plan periods due to the expansion of both fresh water and marine production. From the Third to the Fifth Plan, production fluctuated greatly due to uncertainty of catches, restrictions in off-shore fishing areas of neighbouring countries, and deterioration of water resources.

Since 1985 there has been a rapid expansion of brackish water prawn culture, and after a short period, Thailand has become the most important black tiger prawn producer due to the high export price in 1986. By 1988, this sub-sector alone was exporting 50000 t annually, worth US$ 400 millions. The output of the industry continued to grow in 1989, reflecting earlier coastal farm expansion, which in another way, has led to the widespread destruction of mangrove forests by large-scale operations having dangerous social and environmental consequences. Recently, however, black tiger prawn trade was depressed as the result of competition from other producers and hence a decrease in Japan's imports from Thailand. At the same time, production cost rose significantly primarily because of the shortage of raw materials to produce animal feeds. Yet, Thailand could export a total of 72820 t of prawn products in 1989, worth US$ 634 millions, an increase over the previous year by 14% in volume and 18% in value (BAAC 1991).

Forestry production

The forestry sector constituted about 6.7% of the agricultural GDP during the First Plan. Its share decreased to 3.2% in the Fifth Plan due to a decline in teak production. At present, Thailand's days as a timber producer are officially over. In October 1988, following serious floods in the South, the government banned all loggings. The sub-sector's contribution to the agricultural GDP was around 1–2%. The country's forests, which only 16 years ago covered over 50% of the land area, are now down to as little as 20% as a result of extensive forest clearing for agricultural activities in the 1970s, together with illegal logging which continues upto the present day. Reafforestation efforts so far have fallen short of the target and the government has recently turned to the army and private sector concessionaires for an emergency 'regreening' program for the worst affected Northeastern region.

Problems in agricultural development in Thailand

The growth rate in agricultural development during the First through the Third Plans was consistent but slowed down during the Fourth and Sixth Plan. In addition to low world commodity prices, several internal problems were the major causes of this downward trend in the growth rate (Itharattana 1990). These problems include the followings.

Agricultural production structure

Traditionally, agricultural production in Thailand depends on a few staple crops, namely: rice, rubber, cassava and maize, which are mainly grown in monocultures. These farmers often face high risks from climatic abnormalities, insects and diseases, price fluctuation, and shifts in market demand.

Competition in world markets and trade protectionism

The fluctuation in world market prices and demands have affected Thai agricultural production and prices. Over supply and high competition of the same products from other countries are the main cause of the low price of these commodities. Trade protectionism in developed countries has worsened the situation. This includes the imposition of domestic production subsidies, import quotas, export subsidies and quality regulations. These have resulted in the recent slow growth of Thailand's agricultural production and trade.

Low productivity

Improper farming practices, depleting of land and water on resources, and insufficient adoption of appropriate technology are the main reasons for low productivity, and consequently, high cost of production. These result in the high price of products which cannot compete on the world market.

Low quality of products

As many developed countries which import agricultural products from Thailand are more and more strict in their quality regulations, it is becoming a serious problem for Thailand to produce up-to-date standard products for these countries. Thailand now faces problems of aflatoxin in maize, endrin pesticide residue in mungbeans, biotoxin in frozen seafood, and other toxin residues as well as insects in fruits and vegetables.

Inefficient utilization of natural resources

Soil, water resources and forests are rapidly deteriorating in Thailand as the result of over-exploitation and inefficient utilization of these resources. Such deterioration, in turn, is the main cause of low productivity and high cost of production. In some cases, agricultural production is not feasible.

Summary

Crossroads in Thai agriculture

Thai agriculture is coming towards a crucial point. The high growth rate of the country's economy during the past three decades was, in part, the result of the contribution of agricultural sector which, in turn, was conceivable as the result of planted area expansion (employing extensive work-force and low technology) and of flourishing markets, both of which have now come to a dead end. This situation is worsened by deteriorating natural resources and environment as the results of forest clearing, over-exploitation of natural resources, over-utilization of agro-chemicals, etc. coupled with the frequent occurrence of several natural catastrophies such as typhoons, floods and droughts, and also of abnormal weather conditions such as erratic rainfalls, abnormal high/low temperatures, etc..

Future of Thai agriculture

Based on the present trend of economic development (Table 1), it is anticipated that the agricultural sector share of GDP will be in the range of 15% in 1991 while the percentage of work-force would be around 30%. Such figures will not, and should not, be lower or else, there could be a drastic setback in the economy of the country. In order to maintain these figures, agricultural development in the future must be geared towards producing, in addition to traditional staple commodities for domestic consumption, non-staple commodities of high quality for export markets and for agro-industrial usage, employing non-labour, technology- and capital-intensive agricultural practices. These non-staple commodities include fragrant or special types of rice, tropical fruits, vegetables, ornamental plants, rubber, oil palm, livestock and aquacultural products.

References

BAAC (1991) Annual Report 1989 (1 April 1989–31 March 1990). Bank of Agriculture and Agricultural Cooperatives, Bangkok, Thailand.

Bank of Thailand (1991) Thailand Economic Developments in 1990 and Outlook for 1991 – Special Suppl. Department of Economic Research, Bank of Thailand, Bangkok, Thailand.

FAO (1991) Selected Indicators of Food and Agriculture Development in Asia-Pacific Region, 1980–90. FAO/RAPA, Bangkok, Thailand.

Ithrattana K (1990) Country Paper – Thailand 1. In Proc. Seminar on 'Improving Agricultural Structure in Asia and the Pacific'. Asian Productivity Organization (APO), Tokyo, Japan.

p. 331–358.
Sarikaputi Y (1990) Policy and Plan of the Ministry of Agriculture and Cooperatives and Agricultural Development in the North East. In Proc. Workshop on 'Improvement of Soil and Crops for North East Agricultural Development', Northeast Agricultural Development Centre, Tha Phra, Khon Kaen, Thailand. [in Thai].

A methodological framework to explore long-term options for land use

H. C. VAN LATESTEIJN
Scientific Council for Government Policy,
P.O.Box 20004, 2500 EA The Hague, The Netherlands

Key words: agricultural policy, cropping system, GOAL, land use planning, multiple goal linear programming, nitrogen-limited yield, optimization, simulation, water-limited yield, optimization, simulation, water-limited yield, wheat, WOFOST

Abstract
The Netherlands Scientific Council for Government Policy has developed a methodology that can be used to gather information on options for long-term developments in agriculture in relation to policy objectives. In this paper it is shown how quantitative relations between a number of self-contained technical development processes in agriculture, and socio-economic and environmental policy objectives can be modelled. This model can be used to demonstrate the influence of various policy preferences on future land use changes within the European Community.

A dynamic crop simulation model and a geographical information system that comprise soil characteristics, climatic conditions and crop properties have been used to calculate regional yield potentials for indicator crops. Next a linear programming model that contains several policy derived objective functions is applied to calculate optimal regional allocation of land use. Different sets of restrictions can be put to the objective functions. In this way a number of scenarios is created that reflect different weights given to the policy goals.

In this paper the methodology is described and the potential as well as the flexibility of the approach is illustrated.

Introduction

The Netherlands Scientific Council for Government Policy has carried out a research project to explore possible developments of the rural areas in the European Community (EC) (Netherlands Scientific Council Government Policy 1992). The study provides scenarios that give information on the interactions between a number of more or less self-contained technical development processes in agriculture aimed at productivity gain, and several other 'non-agricultural-production' goals that are to be considered simultaneously. Hence, these scenarios will show the conflicts arising from increasing productivity, market saturation, uneven distribution of production within the EC and increasing concern for regional employment, environment and landscape. The scenarios are used to explore options that emerge when different priorities are given to the goals involved. By demonstrating the consequences of these priorities, valuable information can be gathered to evaluate strategic policy choices that must be dealt with in the current transformation of the Common

F.W.T. Penning de Vries et al. (eds.), Systems Approaches for Agricultural Development, 445–455.

Agricultural Policy of the EC. There is a clear need for a long term agricultural policy that takes major trends and unavoidable changes into account. Rabbinge and Van Latesteijn (1992) deal with these policy implications in more detail. This paper will focus on the methodological aspects of the study.

To construct the scenarios a methodology is developed that uses a systems approach to agriculture to describe possible future changes in land use. In the centre of the methodology a Linear Programming (LP) model is used in conjunction with a procedure called 'Interactive Multiple Goal Programming' (IMGP). A LP-model is generally used to optimize a single objective function. The IMGP procedure makes it possible to optimize a set of objective functions in an iterative process. This reveals the trade-offs between different goals that are modelled by the objective functions.

The IMGP procedure was used by the Council in earlier studies on techno-economic development on a national scale (Netherlands Scientific Council for Government Policy 1983; 1987) and is discussed in detail by Veeneklaas (1990). The present study differs from these earlier studies not only because the topic is quite different, but also because in this case the construction of a useful LP-model requires pre-processing a lot of data.

This paper describes the general features of the methodology.

Why a systems approach

The ongoing productivity rise in agriculture in the EC causes a series of reactions, that lead to problems at EC level. At the level of individual plants and crops agronomic research has brought understanding of the processes involved in this productivity rise. However, policy decisions are taken at much higher levels of aggregation (region, country, EC) and their scientific basis is generally constrained to economic analysis. This presents us with the problem that is illustrated in Figure 1. Research needs are concentrated at intermediate levels. How can we benefit from both the agronomists knowledge at the lower levels and the economists knowledge at the higher levels so as to bridge this gap?

A systems approach might be an answer to that question. Using engineers knowledge, it is possible to construct a model representation of agriculture. Using economists knowledge, it is possible to translate policy goals into quantified objective functions and integrate them in the model. With a model like this it is possible to assess the influence of policy objectives on agriculture and vice versa.

This is the approach adopted in our study. We neither investigate the reaction of farmers or plants to changing conditions, nor the effectiveness of policy instruments on agriculture. Instead we want to assess the flexibility of the agricultural system as a whole, given the fact that various goals are to be fulfilled within this single system. This gives us information on the possibilities within the agricultural system based on the properties of the system itself. It is explicitly not our intention to come up with more or less reliable predictions for

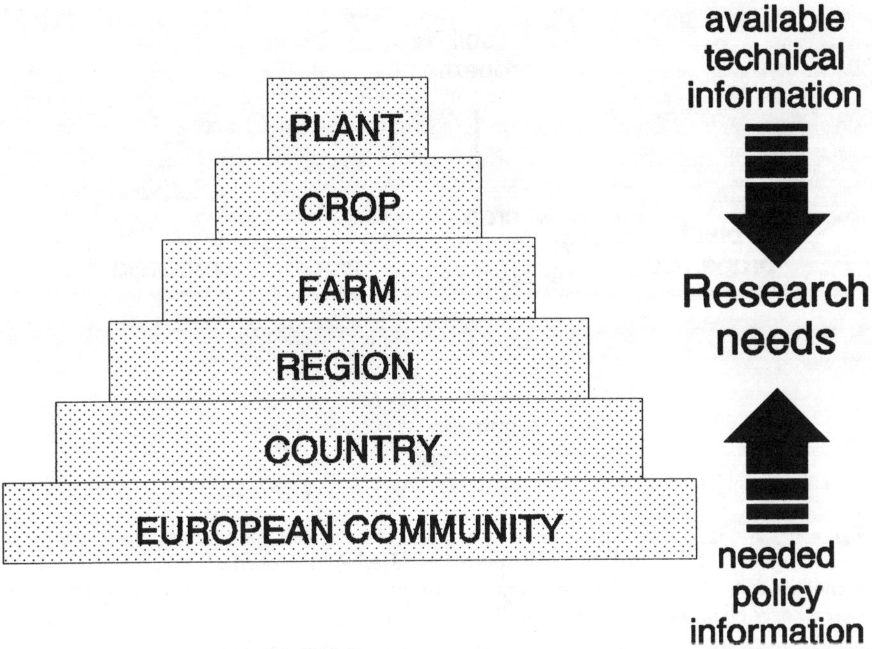

Figure 1. Levels of scale and research needs. Technical information is usually available on plant and crop levels, whilst policy information is needed at regional, national and supra-national level. A systems approach can be used to bridge the gap between these different levels.

the future of agriculture within the EC, but rather to explore the possibilities of the agricultural system. For tactical policy decisions concerning the use of instruments this will not be adequate, but for strategic policy planning purposes this type of analysis is indispensable.

General outline of the methodology

The core of the methodology is formed by a model of (agricultural) land use in the EC which we have baptized 'GOAL' (= General Optimal Allocation of Land use). The model can choose from a limitative set of types of land use to meet an exogenously defined demand for agricultural and forestry products. A number of policy goals that stem from official reports by the EC are coupled to types of land use in the form of objective functions, e.g. maximization of yield per hectare, minimization of regional unemployment in land based agriculture and minimization of the use of pesticides. Distinct policy views can be fed into the model by assigning different preferences to the objectives. Within the GOAL model this is done by restricting the objective functions to a certain domain, e.g.: the total labour force can not be less than six million man year. In this way scenarios can be constructed that show the effects of policy

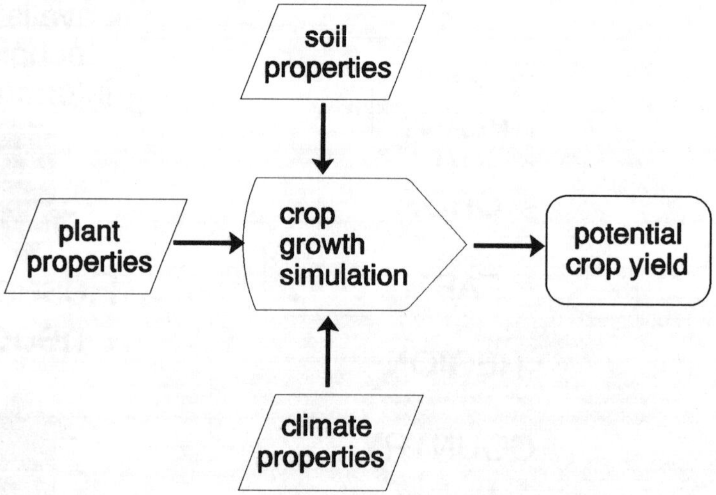

Figure 2. Level I: potential yields of indicator crops are calculated using a crop growth simulation model. Inputs are soil and climate properties and relevant properties of the plant such as phenological development, light interception, assimilation, respiration, partition of dry-matter increase over plant organs and transpiration.

priorities, e.g.: to maintain the labour force the model will have to select types of land use with a relatively high input of labour.

The types of land use that the model can choose from are defined in quantitative terms. Because we want to explore possible long term options, current agricultural practise should not be used as a reference, because it reflects the capabilities and regional differences of this moment, not those of the future. It can be seen that in all areas of the EC agriculture is showing considerable technical and managerial progress. Therefore we must define types of land use that are envisaged over a longer period of time. The concept of 'best technical means' can be used to obtain such types of land use, i.e. agriculture is defined according to the results that are obtained in plant testing stations and experimental farms at this moment. These forerunners are used as a reference for future developments. In that way the results of the model calculations are consistent across all countries. Three levels of analysis were necessary to construct the GOAL model. They are discussed in the next paragraph.

Levels of analysis

Crop level

In Figure 2 the inputs and outputs for the analysis at the individual crop level are visualised. Plant properties, soil properties and climate properties determine the potential crop yield at a given location. To calculate this potential

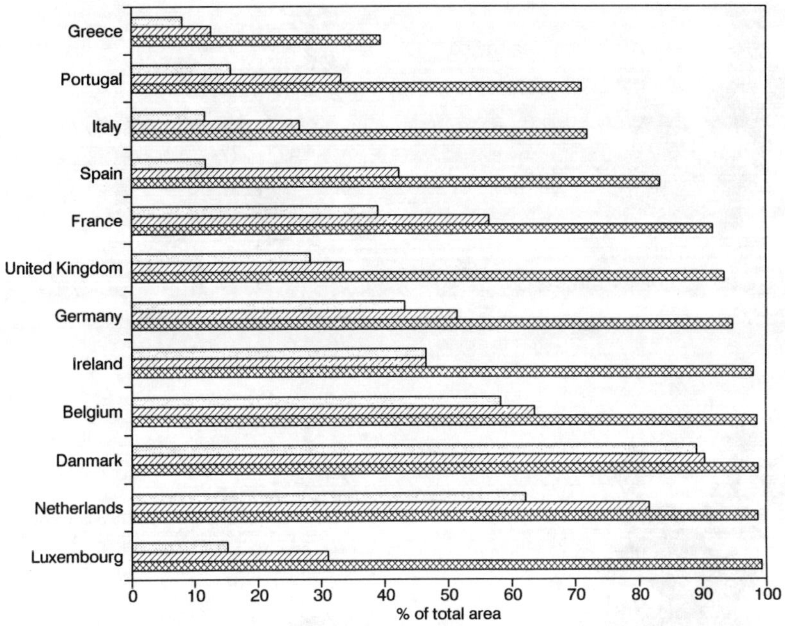

Figure 3. Percentage of area per EC member state suited for grass (largest bar for each state), cereal and root crop (smallest bar) production.

crop yield two steps are necessary. First the suitability of the soil for a certain crop is assessed to exclude all units where that crop can not be grown (e.g. wheat on steep slopes and maize on clay soils). This can be denoted as a qualitative land evaluation. Second, by means of a simulation model, potential yields are calculated for the suitable areas. This can be denoted as a quantitative land evaluation (Van Lanen 1990).

The qualitative land evaluation of the EC is accomplished through the use of a Geographical Information System (GIS) (Van Diepen et al. 1990). The evaluation is executed at the level of Land Evaluation Units (LEUs), a combination of soil and climate conditions that is considered to be homogeneous. For the EC some 22000 units are necessary to cover the total area. By looking at factors like steepness, salinity, and stoniness of the soil the suitability for mechanised farming is assessed. In Figure 3 the total areas suitable for grass, cereals and root crops per EC member state are given. The differences between the member states are obvious. Most of Denmark's area can be used for all three crops, whereas the larger part of Greece is not suitable for arable cropping. In each country the suitable area for grass production exceeds that for cereals, and that for root crops is still smaller.

The quantitative land evaluation is accomplished through the use of the WOFOST crop growth simulation model (Van Keulen and Wolf 1986). For the areas that are suitable the water-limited and potential yields of winter wheat,

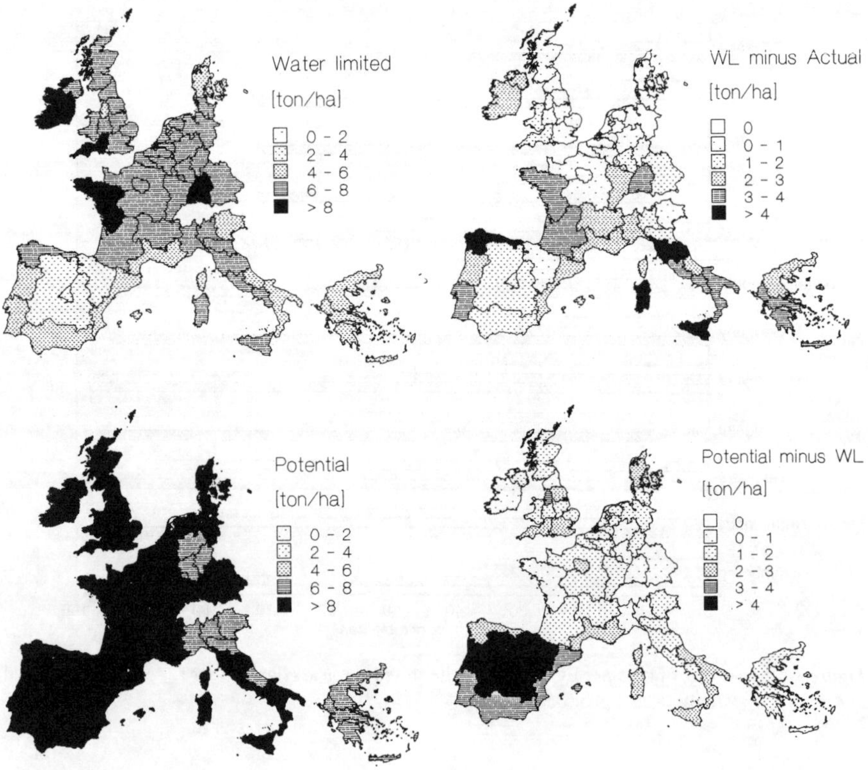

Figure 4. Calculated water limited and potential yield of wheat in the NUTS-1 regions of the EC obtained with the WOFOST crop simulation model. The difference of water limited and actual yield gives an indication of the maximum gain in soil productivity under rainfed conditions. Actual yields are based on data of 1986. The difference between potential and water limited gives an indication of the gain in soil productivity due to irrigation.

maize, sugarbeet, potato, and grass are assessed. The simulation model uses as its inputs: technical information on regional soil (such as water holding capacity) and climate properties and relevant properties of the crop (such as phenological development, light interception, assimilation, respiration, partitioning of biomass increase over plant organs, and transpiration).

Two degrees of water availability are distinguished: rainfed and irrigated. In the rainfed situation maximum yields can be limited by the availability of water at any point during the growing season. In that case the model simulation gives an indication of the attainable yields when no irrigation is applied. This is referred to as water limited yield. In the irrigated situation there are no limitations to crop growth other than those impeded by climate and soil conditions and properties of the crop. In that case the model simulation gives an indication of the maximum attainable yield at a given location. This is referred to as potential yield.

The validation of a simulation model of this type is somewhat problematic.

The simulations are not meant to model actual situations, but give information on production potentials. One way of testing the model is comparing the simulation results with yields that were observed in experimental field situations. This has been done in this study. The assumption made is that in these experimental field situations the production potentials are (nearly) reached by applying state-of-the-art techniques. Although this is not a true validation it is a pragmatic approach to test the simulation model for extreme outcomes.

In Figure 4 the results of the simulations for wheat are given. The simulations are executed at the level of LEUs but the results are averaged at the level of NUTS-1 regions (a classification into broad administrative regions used by Eurostat, the statistical bureau of the EC). If the water limited yield is compared to the actual yield (data of 1986) some conclusions can be drawn. For a number of northern regions the possible rise in yield per unit area appears to be small. This indicates that the limits of soil productivity in these regions are near. In most other regions the simulated water limited yields are much higher (up to 6 t ha^{-1} dry matter) than the actual yields. In those regions soil productivity can still increase, even without irrigation.

In most regions the potential production is much higher. Even in the humid, well developed northern regions irrigation can raise the yield potential by 1–2 t ha^{-1}. It can be concluded from Figure 4 that the difference of what is actually produced and what can be produced according to the simulations is considerable in most of the regions within the EC, especially in the southern regions.

The water-limited and potential yields are used as input at the next level of analysis.

Cropping system level

If one wants to find out land use possibilities in the future, information on individual crops will not be sufficient. All crops are grown in a cropping system that defines all inputs and outputs. Moreover, in most cases monocropping does not provide sustainable agriculture and only a limited number of crop combinations can be used in practical cropping systems. Therefore potential yields of indicator crops are translated into cropping systems that comprise a certain rotation scheme, certain management decisions and a certain use of inputs. In Figure 5 the inputs and outputs at this level of analysis are given. It is striking that at this level, the only viable method is expert judgement. From his experience, both in practise and in experiments, the expert can deduce input and output coefficients of cropping systems. These systems are not commonly used yet, but they might be put into practise within the coming decades. This element in the analysis is crucial yet open to debate due to the subjective choices that are involved. To enable the discussion at this point a full report of all the necessary choices has been published (De Koning et al. 1992).

The following guidelines are used to arrive at an expert judgement on best technical means. It is assumed that cropping systems with excessive input of

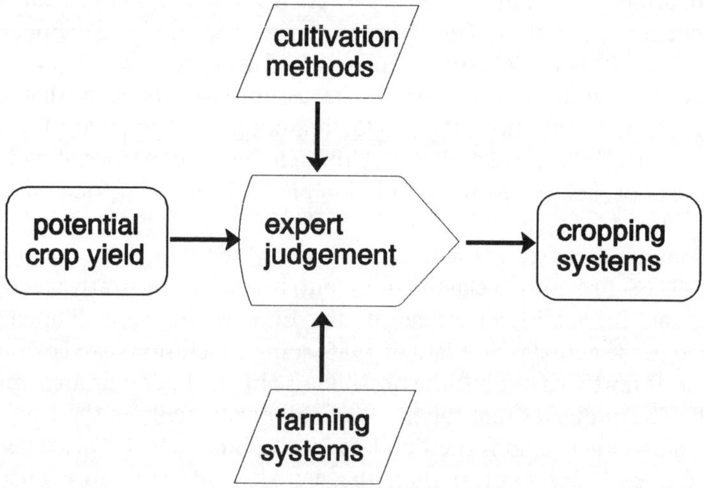

Figure 5. Level II: theoretical cropping systems are defined based on expert judgement. The input consists of the calculated potential yields of indicator crops and information on cultivation methods, farm management, rotation schemes etc. The selection of cropping systems is guided by the principle of 'best technical means', i.e. all inputs are used in an efficient way.

labour are excluded. This means that all cropping systems are mechanized (e.g.: no manual weed control).

For a given level of production (the water limited and potential production levels) the minimal input of resources can be assessed. The theoretical background of this optimization is dealt with extensively by De Wit (1992). He describes this optimum as the situation where each variable production resource is minimized to such a level that all other production resources are used to their maximum. This defines the technical optimum for that particular production situation and will be used as a reference.

To arrive at an economic optimum some substitution of agrochemicals by labour and/or capital is permitted. Expert knowledge is used to define cropping systems that are both economically and agronomically acceptable. We call this set of systems Yield Oriented Agriculture (YOA).

Another deviation from the technical optimum is obtained when more account is taken of environmental hazards related to agriculture. This implies that less environmentally hazardous inputs (such as pesticides and fertilizers) are used, even if this means a slight decrease in yield. Here again the criteria are still rather subjective. We call this set of systems Environment Oriented Agriculture (EOA).

A third deviation from the technical optimum is driven by land use concerns. Under all circumstances it can be foreseen that the agricultural area within the EC will diminish. This can be detrimental to the maintenance of the countryside in some regions. So a set of cropping systems can be defined that deviates from

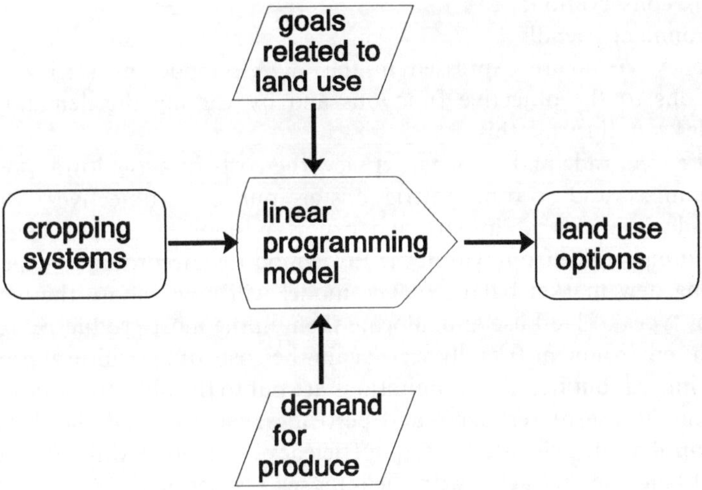

Figure 6. Level III: land use alternatives are calculated using a linear programming model. The model finds an optimal solution to the problem of fulfilling the European demand for agricultural produce while at the same time contributing to the different land use related goals that are incorporated in the model. This can be achieved by choosing between the different cropping systems and locate them in the most appropriate region. The choice is influenced by alternative policy views on developments in agriculture.

the technical optimum and that is characterized by a relatively low soil productivity. We call this set of systems Land use Oriented Agriculture (LOA).

The cropping systems defined for YOA, EOA and LOA are all available to the GOAL model that is used at the next level.

Land use level

At the level of land use possibilities for the EC all information is brought together. Requirements for various goals related to land use together with alternative cropping systems and a demand for agricultural produce are fed into the GOAL model to generate scenarios of different options for land use at the level of NUTS-1 regions within the EC. This is illustrated in Figure 6.

An IMGP procedure is used to optimize a set of objective functions that is incorporated in the model. In this procedure restrictions are put to the objective functions to model preferences in policy goals. Four policy views are used to indicate a desired priority between goals and the levels to which these goals should be fulfilled. The views have been chosen so as to represent a maximum difference between options. They must be regarded as extremes, and their differences give an indication of maximum policy influence. We distinguish:

a free trade and free market;
b regional employment;

c nature conservation;
d environment friendly.

The policy views are expressed in the GOAL model by setting different restrictions to the objective functions and by varying the demand. A few examples can illustrate this.

In the free trade and free market view the costs of agricultural production are minimized and no other restrictions are put to the objectives. Moreover, free trade implies that import and export is allowed, so the demand for agriculture produce from within the EC is modified according to expectations regarding new market balances. The model will now choose the most cost-efficient types of land use and allocate them in the most productive regions.

In the environment friendly view again the costs of agricultural production are minimized, but here strict limitations are put to the objective functions that represent the use of fertilizers and pesticides. Next to that the demand for agricultural produce is fitted to self-sufficiency. The model will now choose for types of land use that agree with the imposed restrictions.

With these data the model creates different scenarios for land use. Policy-makers can now see how their priorities will affect land use and how the effects are distributed over the EC.

However, some requirements cannot be moulded into the rigid outlines of the model. Therefore spatially differentiated claims and demands for nature conservation and development have been assembled in a map (Jongman R, Bischoff N, Dept of Physical Planning and Rural Development, Wageningen University, pers. comm. 1991). This map is matched with the scenarios (=regional allocation of types of land use) to identify potential problematic areas with respect to competing land use.

The study ends with two types of recommendations about the policy requirements that can be derived from the scenarios. First the scenarios, although very different in regional allocation of land use, show common results such as a dramatic decrease of agricultural area from 140 Mha down to as little as 40 Mha, and a 50% decrease in labour in land based agriculture. These results can be looked upon as inevitable and governments might want to mitigate some of the effects. Second the individual scenarios are all extremes and as such not to be pursued, but they indicate directions that might be stimulated with the aid of policy instruments. Here the existing regulatory system of laws, guidelines and subsidies is assessed for its effectiveness on a general level and recommendations are provided with regards to new directions in which the system can be developed.

Conclusion

With the methodological framework described in this paper we have been able to produce scenarios that, given a set of policy goals, describe optimal land use across the EC. These scenarios bridge the gap that was mentioned at the beginning of this paper. Bio-technical and agronomic knowledge together with economic knowledge have been used in a systems approach to agriculture. This synergy of disciplines adds value to the approach. Although the methodology serves a specific aim it can be applied in other situations as well, especially in those situations where integration of bio-technical 'engineers knowledge' and economic 'politicians knowledge' is wanted.

The study does not provide a blueprint for agricultural policy. Instead it presents a set of scenarios that explore possibilities of land based agriculture in the EC. For strategic policy planning this approach can be very useful.

References

De Wit C T (1992) Resource use efficiency in agriculture. Agric. Syst. (in press).

De Koning G H J, Janssen H, Van Keulen H (1992) Input and output coefficients of various cropping and livestock systems in the European Communities. Working Documents W 62, Netherlands Scientific Council for Government Policy, The Hague, The Netherlands. 71 p.

Netherlands Scientific Council for Government Policy (1983) Policy-oriented survey of the future: Towards a broader perspective. The Hague, The Netherlands. 80 p.

Netherlands Scientific Council for Government Policy (1987) Scope for growth: threats to and opportunities for the Dutch economy over the next ten years. The Hague, The Netherlands. 61 p.

Netherlands Scientific Council for Government Policy (1992) Grounds for choice: four perpectives for the rural areas in the European Community. The Hague, The Netherlands. 146 p.

Rabbinge R, Van Latesteijn H C (1992) Long term options for land use in the European Community. Agric. Syst. (in press).

Van Diepen C A, De Koning G H J, Reinds G J, Bulens J D, Van Lanen H A J (1990) Regional analysis of physical potential of crop production in the European Community. Pages 74–79 in Goudriaan J, Van Keulen H., Van Laar H H (Ed.) The greenhouse effect and primary productivity in European agro-ecosystems. Pudoc, Wageningen, The Netherlands.

Van Keulen H, Wolf J (1986) Modelling of agricultural production: weather, soils and crops. Simulation Monographs. Pudoc, Wageningen, The Netherlands. 479 p.

Van Lanen, H A J (1990) Qualitative and quantitative physical land evaluation: an operational approach (thesis), Wageningen Agricultural University, The Netherlands. 195 p.

Veeneklaas F R (1990) Dovetailing technical and economic analysis. (thesis) Erasmus Drukkerij, Rotterdam, The Netherlands. 159 p.

SESSION 6

Education, training and technology transfer

Decision support systems for agricultural development

J. W. JONES
Agricultural Engineering Department, University of Florida, Gainesville, Florida 32611, U.S.A.

Key words: agronomic minimum data set, climate change, data base management, decision support system, DSSAT, Geographical Information System, modeling team, resource use, risk, simulation, wheather data generator (WGEN)

Abstract
Agricultural decision makers at all levels need an increasing amount of information to better understand the possible outcomes of their decisions and to assist them in developing plans and policies that meet their goals. An international team of scientists has recently developed a decision support system for agrotechnology transfer (DSSAT) to estimate production, resource use, and risks associated with different crop production practices. The DSSAT contains crop-soil simulation models, data bases for weather, soil, and crops, and strategy evaluation programs integrated with a user-friendly interface on microcomputers. In this paper, an overview of the DSSAT will be given, followed by example applications and a description of enhancements now being made. Concepts for using components of the DSSAT in new, broader decision support systems for assessing farm and regional outcomes of different policies and practices will be presented. These new tools will help decision makers by reducing the time and human resources required for analyzing complex alternative decisions.

Introduction

A rapidly changing world with high rates of population growth and major changes in political and economic systems has created an urgent need to develop new and revise many existing agricultural systems. During the next decade, changes will be made in infrastructures that support agricultural production, marketing, and distribution, in farms and their enterprises and practices, and in governmental policies and regulations that affect production, natural resources, and the environment. Now, more than ever, decision makers at all levels need an increasing amount of information to help them understand the possible outcomes of their decisions and develop plans and policies for achieving their goals. Decision Support Systems (DSS) are "interactive computer-based systems that help decision makers utilize data and models to solve unstructured problems" (Sprague and Carlson 1982). The goal of such systems is to improve the performance of decision makers while reducing the time and human resources required for analyzing complex decisions. In concept, DSS should support all phases of a decision making process, characterized by Simon (1960) as: (i) searching for conditions calling for a decision by identifying possible problems or opportunities; (ii) creating and analyzing possible courses of action;

F.W.T. Penning de Vries et al. (eds.), Systems Approaches for Agricultural Development, 459–471.

and (iii) suggesting a course or courses of action from those analyzed.

Most of the relatively few agricultural decision aids that have been developed are aimed at the farmer as a decision maker, and attempt to improve operational decisions such as pest control (Hearn 1987; Michalski et al. 1983) fertilizer management (Yost et al. 1988), or a wider set of crop management decisions (Plant 1989; McKinion et al. 1989). The need for agricultural decision support systems extends beyond those for field level operational decisions with the farmer as a decision maker. A different approach was taken by a group of cooperating scientists in the IBSNAT project. The Decision Support System for Agrotechnology Transfer (DSSAT) was designed for users to easily create 'experiments' to simulate, on computers, outcomes of the complex interactions between various agricultural practices, soil, and weather conditions and to suggest appropriate solutions to site specific problems (Jones 1986; Uehara 1989). This system relies heavily on simulation models to predict the performance of crops for making a wide range of decisions. This system has been in use for almost three years. It is time to evaluate this approach and determine priorities for further development of these tools. In this paper, an overview of the DSSAT will be given, including limitations and major enhancements now being made to this field-scale system. Concepts will be presented for broader decision support systems to help identify problems in existing or proposed agricultural systems and to analyze the outcomes of practices and policies at farm and regional scales.

Overview of the DSSAT

The DSSAT was designed to allow users to (i) input, organize, and store data on crops, soils, and weather, (ii) retrieve, analyze and display data, (iii) validate and calibrate crop growth models, and (iv) evaluate different management practices at a site. In adapting and applying the DSSAT to a location, users typically use the following procedures:

1. Conduct field experiments on one or more crops, and collect a minimum data set (MDS) required for validating a crop model. Run the model using the new data to evaluate the ability of the model to predict performance of crops in the region of interest. In many cases, data from previous experiments are used.
2 Enter other soil data for the region and historical weather data for sites in the region. Conduct sensitivity analysis on the crop model(s) to get an overview of the response of the model to alternative practices and weather conditions.
3. Select a set of management practices and simulate each of these over a number of years to predict performance and uncertainty associated with each practice. Compare the alternative practices using means, variances, and cumulative probability distributions of simulated yield, water use, season length, nitrogen uptake, net profit and other responses. Make decisions and recommendations.

Table 1 lists the options that users can choose in the DSSAT to create different

Table 1. Listing of the crop management options to create different strategies, the crop performance variables that can be studied in DSSAT V2.10, and the additional features in V3.0.

Management options, V2.1	Variables available for analysis, V2.1
Crop cultivar	Grain yield
Planting	Pod yield
Plant population	Biomass
Row spacing	Season length
Soil type	Reproductive season length
Irrigation	Seasonal rainfall
Fertilization (nitrogen)	Seasonal evapotranspiration
Initial conditions	Water stress, vegetative
Crop residue management	Water stress, reproductive
	Number of irrigations
	Total amount of irrigation
	Number of nitrogen applications
	Nitrogen applied
	Nitrogen uptake
	Nitrogen leached
	Nitrogen stress, vegetative
	Nitrogen stress, reproductive
	Net returns
Additional options, V3.0	**Additional variables, V3.0**
Crop rotations	Seed used
Harvesting	Runoff
Carbon dioxide	Soil organic C
Pest (damage)	Soil organic N
Water table	Residue applied
	Nitrogen fixation

management strategies, and the simulated performance indicators that can be analyzed. A data base management system provides user-friendly entry and editing of the MDS (IBSNAT 1988). Retrieval programs extract data from the centralized data base and create files for running the crop models. Outputs can be printed or graphically displayed and compared with experimental observations for validating the crop models and conducting sensitivity analyses. Application programs facilitate running crop models for different management practices over several seasons to determine the most promising and least risky combinations of management for various locations and soil types. Graphics programs allow users to easily plot simulated and observed crop and soil data and the results from strategy evaluation analyses.

The programs to perform these functions are written in various computer languages. A shell program provides access to the programs in the DSSAT using pop-up menus, and thus users do not have to worry about which language is being used and how to execute that particular program. Arrow keys are used to select the specific tasks to be performed. A program is included to install DSSAT. This system is available through the IBSNAT office at the University of Hawaii (IBSNAT 1989). Source code for each of the crop models is not included in the DSSAT package, but can be obtained by request from IBSNAT or any of the model authors.

Crop simulation models

The functions of the DSSAT were selected primarily to support the use of crop simulation models in decision making applications. The utility of this system depends on the ability of the crop models to provide realistic estimates of crop performance for a wide range of environment and management conditions and on the availability of the required data. The first release of the DSSAT (V2.1, IBSNAT 1989) contained models of the following four crops: maize (CERES-Maize V2.10; Ritchie et al. 1989), wheat (CERES-Wheat V2.10; Godwin et al. 1989), soybean (SOYGRO V5.42; Jones et al. 1989) and groundnut (PNUTGRO V1.02; Boote et al. 1989). Four additional crop models have since been added: rice (CERES-Rice; Godwin et al. 1990), drybean (BEANGRO V1.01; Hoogenboom et al. 1991), sorghum (CERES-Sorghum), and millet (CERES-Millet). These models are process oriented, and are designed to have global applications; i.e., to be independent of location, season, cultivar, and management system. The models simulate the effects of weather, soil water, cultivar, and nitrogen dynamics in the soil and crop, on crop growth and yield for well drained soils. On a personal computer with a math co-processor they each require less than 1 minute to simulate one growing season.

Genetic coefficients are required by each crop model to simulate the differences in crop performance among varieties. Examples of genetic coefficients are the thermal or photo-thermal time required by a crop to reach a particular growth stage such as flowering or physiological maturity, sensitivity to photoperiod, and maximum number of seed per shell. A library of genetic coefficients is available for many varieties of the crops currently included in the DSSAT. Procedures have been developed to obtain these coefficients for cultivars that are currently not in the DSSAT. Inputs and outputs to the crop models have been standardized and documented (IBSNAT 1986), to increase the efficiency of sharing data, to allow the introduction of other crop models, and to allow application programs in the DSSAT to be used with any of the crop models.

Data Base Management System (DBMS)

For validation, crop models are used to simulate crop responses under specific experimental conditions for which observed data are available. The MDS for validation consists of: (i) weather for the growing season during which the experiment was conducted, (ii) soil properties, and (iii) crop management and experimental data (IBSNAT 1988). Crop management data include planting date, dates when soil conditions were measured prior to planting, planting density, row spacing, planting depth, crop variety, irrigation and fertilizer practices. Programs link weather and experimental data with the crop models by creating crop model input files. The minimum required weather data includes latitude and longitude of the weather station and daily values of incoming solar radiation in MJ m^{-2} d^{-1}, maximum and minimum air temperature in °C, and rainfall in mm. Optional data include dry and wet bulb temperatures and wind speed.

Soil data in DSSAT include pedon data and profile descriptions. Some of the key data include soil classification, surface slope, colour, permeability and drainage class. Soil horizon data include upper and lower horizon depth, sand, silt, clay contents, bulk density, organic carbon, pH, aluminum saturation, and root abundance information. A program uses these data to estimate parameters required for the soil water submodel (albedo, runoff curve number, upper flux limit of the first stage of soil evaporation, drainage coefficient and layer parameters of lower soil water limit for plant growth, drained upper soil water limit, saturated soil water content, and relative root growth distribution). If data are not available in the data base for a particular soil, they can be manually entered by the user through an interactive program.

The standard outputs from the DBMS include: (i) chronological listing of activities and events for an experiment including experimental operations, phenological events, sampling dates, and history of special events; (ii) summary of preplant soil fertility and preplant soil water content for each layer, and (iii) graphs of maximum and minimum temperatures, rainfall, radiation, and degree-days on a 10-day or monthly basis.

Policy and strategy applications

The real power of the DSSAT for decision making lies in its ability to analyze many different management strategies. When a user is convinced that the model can accurately simulate local results, a more comprehensive analysis of crop performance can be conducted for different soil types, cultivars, planting dates, planting densities, and irrigation and fertilizer strategies to determine those practices that are most promising and least risky. The strategy generator and evaluation program in the DSSAT establishes the desired combinations of management practices, links the models to historical weather data or generates multiple years of weather data for the location, runs the model(s), and analyses and presents results to the user. It assists users in evaluating the relative merits of the simulated strategies with respect to any of the variables listed in Table 1 and identifies the best strategy. Cumulative probability functions are presented to help users select a strategy with preferred mean and variability characteristics.

Weather generators, WGEN and WMAKER, developed by Richardson and Wright (1984) and Keller (1982), respectively, are included in the DSSAT. Each generator has two programs; one program to compute coefficients from historical weather data and the second program to generate weather data using these coefficients. The WGEN requires daily long-term historical data on maximum and minimum temperatures, solar radiation, and precipitation, while WMAKER uses monthly means and standard deviations of potential evapotranspiration, average temperature, precipitation, and wet days.

There are many examples in which scientists have used the DSSAT or its components to provide information for decisions ranging from those that deal with field-scale crop management options that improve crop production and minimize risk, to those that identify sites in a region where production of a particular crop could be expanded, to those that estimate changes in crop performance due to the 'greenhouse' effect. Most of these applications have involved scientists at national and international research organizations and universities and have focused on possible solutions to problems of local or regional interest. In the 'greenhouse' effects studies, scientists from nineteen countries used the DSSAT to estimate the effects of possible changes in climate on crop production for use by national and international policy makers. Thus, the direct users of DSSAT have mostly been agriculturalists to whom policy makers, farmers, extension workers and private businesses and others would typically go for recommendations, assessments, advice, and other information for making decisions. Its success is due to the impact that it is having as a tool for those who provide information for decisions. The crop models have limitations, and interpretation of results is made by these users. The DSSAT facilitates their analysis and interpretation of results from specific experiments and extension of those results for decision making under a wider set of conditions.

Modifications in progress

The most important limitations in DSSAT V2.1 are related to limitations in the crop models. Models for only a few crops are contained in the system, and those models do not respond to all environment and management factors. Missing are components to predict the effects of tillage, pests, intercropping, excess soil water and other factors on crop performance. Their value to date has been due in large part to the major regions of the world where weather, water, and nitrogen are the major factors that affect crop performance. Performance of the models may not be good under conditions of severe environmental stresses. Other limitations exist in the design and implementation of the data base system and other components of the current DSSAT. Although most functions are easily accessed by users with minimal computer expertise, some functions have been criticized as being too cumbersome, confusing, or limiting. A final limitation is that the DSSAT is restricted to field scale analyses. It does not include spatial landscape features, capabilities to simulate farm scale productivity and constraints, nor capabilities for aggregation of regional results.

Some rather major modifications are currently being made to the DSSAT

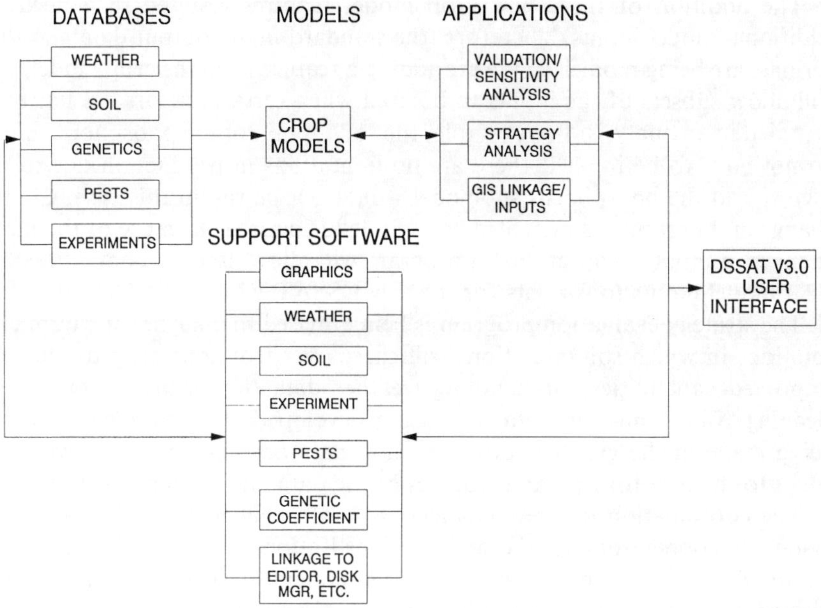

Figure 1. Schematic of the main components of DSSAT V3.0.

and its components to overcome these limitations. Models are being added for potato, barley, aroid, and cassava, and existing crop models are being improved in several ways. A new soil water model is being completed, and it will contain improved infiltration, re-distribution, and root water uptake calculations. Restrictions to percolation will be included in soil inputs so that perched water tables can be simulated along with oxygen stress effects on root and crop growth processes. An option is being added to compute potential evapotranspiration using the Penman equation, which uses humidity and wind speed if those data are available. All of the crops will be sensitive to carbon dioxide concentrations so that climate change studies can be made. A soil phosphorus model is being developed. Capabilities for automatic planting, harvesting, residue management, and fertilizer applications will be included in all crop models, and the DSSAT will be able to sequentially operate the models to simulate crop rotations and the long-term effects of cropping systems on soil N, organic matter, and P availabilities. In addition, the effects of pest damage to the crops can be simulated for various types of pest damage (e.g. insect feeding on different plant parts and disease destroying tissue or whole plants). A program will be included to assist users in estimating genetic coefficients for each crop using data from variety trials and other field experiments. The new DSSAT will contain support software to enable users to create, maintain, search, edit, and display data required for the models and application programs. A schematic of the main components of DSSAT V3.0 is given in Figure 1.

The addition of these new crop model features resulted in a need for additional model inputs. Therefore, the standard input/output data and their formats are being modified. We are adding a feature in all the crop models that will allow subsets of the inputs to be used when certain factors are not to be included in an analysis. For example, the soil inputs for the phosphorus model do not have to be input if there are no limitations in production due to this factor, and the phosphorus submodel would not be run in this case. Climate change and pest inputs may also be excluded if desired. Because of the major changes in inputs and outputs, a program will allow users to convert DSSAT V2.1 model inputs to formats for the new DSSAT V3.0.

The strategy evaluation program is being revised to analyze long term crop rotations in which soil conditions will carry over from one crop to the next. Improved capabilities for handling weather data (importing, reformatting, cleaning, filling missing data) are being developed. Improvements are also being made in the graphics programs and data base management system to adapt to the new formats and features of the data and models, respectively.

The coordination of these changes is indeed challenging. They are being made by cooperators in six different institutions in what is perhaps an unprecedented attempt to integrate component agricultural models and supporting software into one user-oriented package.

Spatial scale decision support systems

Many of the problems faced by agricultural decision makers extend beyond the boundaries of individual fields. Decisions that farmers make about which crop to plant in a particular field and the practices to use depend on the availability of land, variations in soils, previous history of each field, preferences and other information on economics and resource availability (Thornton 1991). Policy makers and agribusiness decision makers may need to know the aggregate agricultural productivity and resource demands over a region under various plans or policies. The field scale DSSAT can be used to organize weather and soil data for many sites on a farm or over a region, and it can simulate various alternatives for each of the sites, which could then be analyzed to estimate farm or regional performance. However, information not contained in the DSSAT would be required, such as the spatial variability of current land use, weather, and soils, and the proposed alternative plans, or arrangements over space, of crops and their management practices.

A higher level of data base organization, definition of strategies, and analysis capabilities are needed for spatial scale decision support systems. As a contrast between field and regional scale decision support systems, consider the form of alternative production plans. At a field scale, the production plan may consist of crop and cultivar choice, planting date,cultural practices, and water, nutrient, and pest management operations. At a farm scale, the production plan would consist of field scale plans for all fields that meet the goals of the

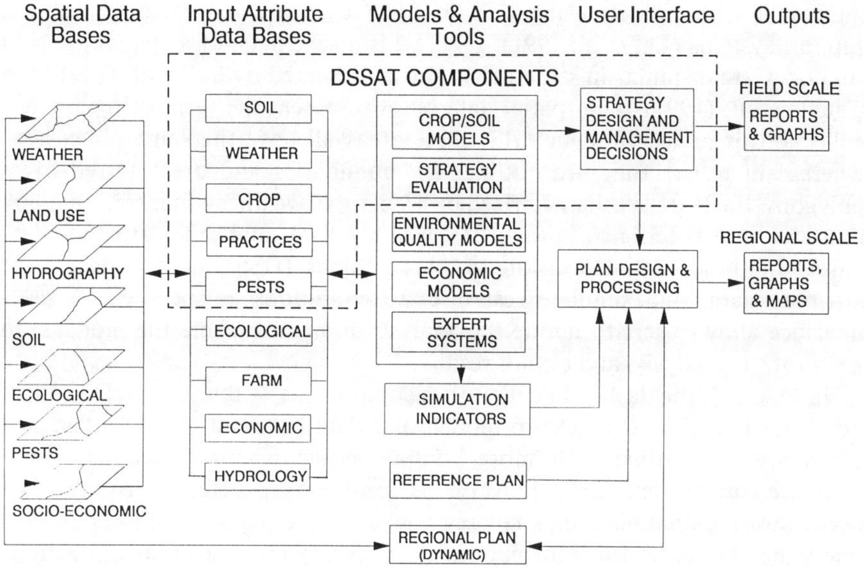

Figure 2. Schematic of a regional decision support system in which components of the DSSAT are used to simulate the performance of agricultural production system across soils and weather regimes in the region.

farmer and can be performed within the constraints of the farm resources. At a regional scale, agricultural production alternatives would involve the combination of land use, soil types, weather, and field-scale management for each unit of land in the region. Performance at the field scale includes crop yield, net returns and other indicators (Table 1), whereas at the farm scale these indicators and others, such as resource use, would be aggregated to provide information for farmers to choose the plan that best meets their goals. At the regional scale, aggregate yields of each crop, and fertilizer, seed, and other resource requirements are needed by decision makers. Policy makers may wish to select policies that encourage the regional plan that results in the desired outcomes whereas agribusiness decision makers may adjust the production or supply of resources to meet expected regional demand.

The field-scale DSSAT provides users with an analysis of the uncertainty in crop performance due to year-to-year weather variability, for each defined management strategy. At farm and regional scales, there will be additional uncertainties, such as spatial variabilities of soil and weather and uncertainties in the selection of crops and practices over space.

Geographic Information Systems (GIS) provide data base, graphics, and other tools for representing and analyzing spatial information. GIS link data to maps and can graphically present characteristics of the region as colours on a map. They also contain programming capabilities that can be used to develop user-friendly access to analysis and reporting functions for decision making. Regional data bases on soil, land use, weather, and other characteristics can be integrated with models using GIS to form user-oriented regional decision support systems (Lal et al. 1991). Figure 2 is a schematic of a spatial decision support system similar in structure to those presented by Lal et al. (1991) and Calixte et al. (1991). The spatial data bases represent the farm or region as a collection of areas. In some GIS, these areas all have the same dimensions whereas in others they are outlined by boundaries and are referred to as polygons. Each area is characterized by a soil type, weather zone, etc., that link that area into specific soil properties, weather data, and other attribute data bases. In these spatial decision support systems (DSS), models access the attribute data, and simulate responses to conditions as specified. A user-interface allows users to define the analysis domain, execute the model(s) to perform the analysis, and obtain results.

In Figure 2, the dashed line identifies components of the current DSSAT to point out that the existing crop models and data bases on soil, weather, and crops are compatible with more comprehensive, spatial DSS. The user-interface for the field scale, however, is considerably different from what is needed for regional planning. The regional scale user interface allows users to query the data bases for information, or to specify proposed plans, practices, and restrictions for simulation of regional responses. The Regional Plan is a data file to store all of the information on the combinations of land use, crops, and practices for the region being analyzed by a user.

The regional DSS can be used in various ways (Thornton 1991):

1. As a store of information on farming systems that can be updated continually, mainly by survey, to produce timely statistics of current land use patterns and production levels and how these are changing over time.
2. As a short-term policy tool, in relation to forecasting for the coming year or the current season. Here, regional simulations of yield and resource use would give an estimate of requirements for imports of agricultural inputs and exports of commodities, any aid requirements in response to a bad season, and likely international loan requirements, for example.
3. As a long-term policy tool, to investigate the effects of change on regional production and resource requirements. Such changes might be economic, technological, or climatic, or they might be substantial policy or trade changes to the economic environment within which farmers operate.

These regional systems allow decision makers to focus on the problems at the scale at which they make decisions. We have already developed a prototype of a regional DSS in which IBSNAT crop models and data bases were linked to PC ARC/INFO (ESRI 1988) in a prototype study to evaluate the potential for bean and rice production in three regions in Puerto Rico (Calixte et al. 1991).

One of the difficult requirements of farm or regional decision support systems is that for the selection of farm or regional plan alternatives. At the field scale, users define specific crop × cultivar × planting date × fertilizer management × irrigation management ×... combinations to evaluate for specific sites. This format provides a high level of flexibility, but requires time to formulate a wide range of alternatives. At a field scale, this has not been a major problem, but when this must be done for a region or even a farm, the time required increases rapidly. For example, to simulate agricultural production over a farm or region, every parcel of land must have a defined crop, cultivar, and all other management practices defined. This results in a large number of combinations. Additional tools are needed to help users select combinations in an efficient way. Possibilities include the use of the farm or regional plan from the previous year with users interactively changing practices for selected areas, expert system diagnosis of farm or regional data and development of a set of plans, and the use of optimization techniques to modify plans followed by simulation of farm or regional performance.

Conclusions

The need for information for agricultural decision making at all levels is increasing rapidly. The generation of new data and its publication is not sufficient to meet these increasing needs. Unless it is put into a format that is easily and quickly accessible, new data and research findings may not be used effectively. Model-based decision support systems are needed to increase the performance of agricultural decision makers while reducing the time and human resources for analyzing complex decisions. The possibility and value of such systems has been demonstrated. A field scale, model-based decision support system has been developed through the close cooperation of a group of interdisciplinary scientists at different institutions. None of the institutions had the resources nor the expertise to develop this system alone, and it is a notable achievement that such cooperation led to this integrated system.

In view of these needs and what has been accomplished to date, it is clear that agricultural decision support systems are still in an early stage of development. Problems faced by decision makers are for more comprehensive than those that can be addressed by the current DSSAT. Further developments are needed in three basic directions. First, the crop models should be improved to account for factors not yet included. This is already being done by the developers of the models in the DSSAT and by other crop modeling groups. These efforts need to be focused to produce fewer crop models with broader capabilities and increased accuracy. Secondly, models of other crops and of other agricultural enterprises should be developed with capabilities to link into the same data bases. Modeling efforts exist for animal, aquatic, and forest systems, and efforts are needed to integrate them for describing the biophysical performances of each of those production systems. Finally, farm and regional decision support systems need to be designed that integrate a broader range of models and analysis capabilities to provide decision makers with predictions of farm and regional performance and resource requirements, and to assess the variability and sustainability of agricultural systems. This will require an exceptional effort by scientists from various disciplines on an international scale.

References

Boote K J, Jones J W, Hoogenboom G, Wilkerson G G, Jagtap S S (1989) PNUTGRO V1.02 Peanut Crop Growth Simulation Model: User's Guide. Florida Agricultural Experiment Station Journal No. 8420, Agricultural Engineering Department and Agronomy Department, University of Florida, Gainesville, Florida 32611, USA.

Calixte J P, Lal H, Jones J W, Beinroth F W and Perez-Alegria L R (1991) A model-based decision support system for regional agricultural planning. ASAE Paper. American Soc. Agric. Engr., St. Joseph, Michigan 49085, USA.

ESRI (1988) PC ARC/INFO Starter Kit. Environmental Systems Research Institute, Inc. Redlands, California, USA.

Godwin D, Ritchie J T, Singh U and Hunt L (1989) A user's guide to CERES-Wheat V2.10. International Fertilizer Development Center. Muscle Shoals, Alabama 35662, USA.

Godwin D C, Singh U, Buresh R J and De Datta S K (1990) Modeling of nitrogen dynamics in relation to rice growth and yield. Trans. 14th Int. Congress Soil Sci. Kyoto, Japan. Vol. IV:320–325.

Hearn A B (1987) SIRATAC: A decision support system for cotton management. Rev. of Marketing and Agric. Econ. 55:170–173.

Hoogenboom G, White J W, Jones J W and Boote K J (1991) BEANGRO V1.01: Dry bean crop growth simulation model. Florida Agric. Exp. Sta. Journal No. N-00379. University of Florida, Gainesville, Florida 32611, USA.

IBSNAT (1986) Technical Report 5. Decision Support System for Agrotechnology Transfer (DSSAT). Documentation for the IBSNAT Crop Model Input and Output Files. Version 1.0. Dept.of Agronomy and Soil Sci., College of Tropical Agric. and Human Resources. University of Hawaii, Honolulu, Hawaii 96822, USA.

IBSNAT (1988) Technical Report 1. Experimental Design and Data Collection Procedures for IBSNAT, 3rd ed., Revised. Dept. of Agronomy and Soil Sci., College of Tropical Agric. and Human Resources. University of Hawaii, Honolulu, Hawaii 96822, USA.

IBSNAT (1989) Decision Support System for Agricultural Transfer V2.10 (DSSAT V2.10). Dept. Agron. and Soil Science, University of Hawaii, Honolulu, Hawaii 96822, USA.

Jones J W (1986) Decision support system for agrotechnology transfer. Agrotechnology Transfer 2:1–5. Dept. of Agronomy and Soil Science, University of Hawaii, Honolulu, Hawaii 96822, USA.

Jones J W, Boote K J, Hoogenboom G, Jagtap S S, Wilkerson G G (1989) SOYGRO V5.42 Soybean Crop Growth Simulation Model: Users Guide. Florida Agricultural Experiment Station Journal No. 8304. Agricultural Engineering Department and Agronomy Department, University of Florida, Gainesville, Florida 32611, USA.

Keller, A A (1982) Development and analysis of an irrigation scheduling program with emphasis on forecasting consumptive use. MSc Thesis. Agricultural and Irrigation Engineering Department, Utah State University. Logan, Utah, USA.

Lal, H, Jones J W, Fonyo C, Boggess W G and Campbell, K L (1991) Lake Okeechobee Agricultural Decision Support System (LOADSS). ASAE Paper 91-2623, American Society of Agricultural Engineers, St. Joseph, MICHIGAN, USA.

McKinion, J M, Baker D N, Whisler F D, Lambert J R, Landivar J A and Mullendor G P (1989) Application of the GOSSYM/COMAX system to cotton crop management. Agric. Syst. 31:55–65.

Michalski R S, Davis J H, Bisht V S and Sinclair J B (1983) A computer-based advisory system for diagnosing soybean diseases in Illinois. Plant Disease 4:459–463.

Plant R E (1989) An integrated expert system decision support system for agricultural management. Agric. Syst. 29:49–66.

Richardson C W and Wright D A (1984) WGEN: A model for generating daily weather variables. U.S. Department of Agriculture, Agricultural Res. Service, ARS–8, 80 p.

Ritchie J T, Singh U, Godwin D and Hunt L (1989) A user's guide to CERES–Maize V2.10.

International Fertilizer Development Center. Muscle Shoals, Alabama 35662, USA.

Simon, H (1960) The New Science of Management Decisions. Harper & Row, Publishers. New York, USA.

Sprague R H Jr and Carlson E H (1982) Building Effective Decision Support Systems. Prentice-Hall, Inc., Englewood Cliffs, New Jersey, USA.

Thornton P (1991) Application of crop simulation models in agricultural research and development in the Tropics. International Fertilizer Development Center. Muscle Shoals, Alabama 35662, USA.

Uehara G (1989) Technology Transfer in the Tropics. Outlook Agric. 18:38–42.

Yost R S, Uehara G, Wade M, Sudjadi M, Widjaja-Adhi I P G and Zhi-Cheng Li (1988) Expert systems in agriculture: Determining the lime recommendations for soils of the humid tropics. Res. Ext. Series 089–03.88. College of Tropical Agriculture and Human Resources. University of Hawaii, Honolulu, Hawaii, USA.

Constraints in technology transfer: a users' perspective with a focus on IPM, Philippines

T. H. STUART

Institute of Development Communication, University of the Philippines at Los Baños (UPLB), Philippines

Key words: extension, farmer participation, Integrated Pest Management, Philippines, research priority, resource use, risk, rural development, socio-economic constraint, technology transfer, women

Abstract

The Philippines has a strong and dynamic national agricultural research and development system coordinated by the Philippine Council for Agriculture, Forestry and Natural Resources Research and Development (PCARRD) under the Department of Science and Technology (DOST). Since it was established in 1972, PCARRD has generated significant technologies and information in agriculture, forestry and natural resources through the National Agriculture and Resources Research & Development Network (NARRDN). The system espouses the cyclic, sequential technology-oriented stages in R&D activities consisting of technology generation, technology adaptation, technology verification and technology dissemination and utilization.

It is within PCARRD's R&D context that the author classifies the many factors which may become constraints to technology transfer in the farmer-extension worker-researcher linkage. Applying a user-oriented technology transfer model, the author traces factors in the decision-making process of a technology information user.

The paper presents some of the author's experiences in research communication as coordinator of PCARRD's multi-agency regional applied communication program in the 13 regions of the country, as well as her project team's technology transfer experiences in a multi-disciplinary Project on Integrated Nutrient and Pest Management Extension among men and women farmers in rice and vegetable farming systems in Laguna, Philippines.

Introduction

The Philippine Council for Agriculture, Forestry and Natural Resources Research and Development (PCARRD), one of the five councils under the Department of Science and Technology (DOST) is mandated to develop, coordinate and monitor the national research and development (R&D) program in agriculture and natural resources in the country. The bottom line of this national R&D program is the generation of information and technologies and their effective communication to the farmer and his family to ultimately improve their farm productivity, income and general welfare. The prime mover of R&D in the PCARRD system is the National Agriculture and Resources Research and Development Network (NARRDN). PCARRD established this network in 1975. To date, the network has over a hundred active R&D nodes consisting of research centres, government agencies and state colleges and universities.

PCARRD has fine-tuned the NARRDN by introducing the R&D consortium

F.W.T. Penning de Vries et al. (eds.), Systems Approaches for Agricultural Development, 473–484.

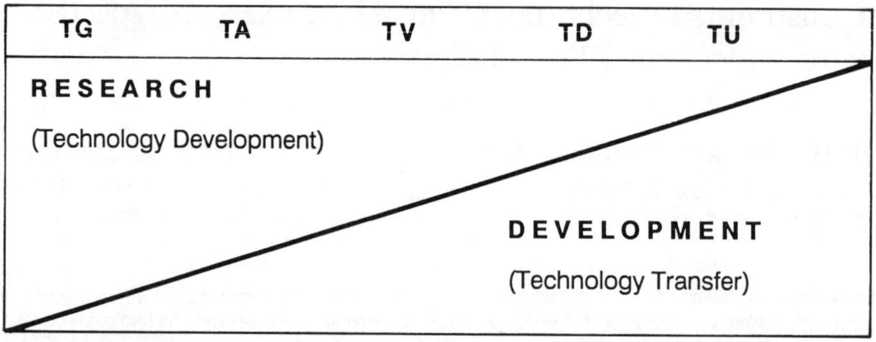

Figure 1. A technology transfer paradigm showing the process and stages in technology research and development (Stuart 1990).

concept, an organizational arrangement which enables the regional R&D agencies to manage and maximize the use of their limited resources through interagency cooperation.

A consortium consists of a number of satellite institutions in a region ranging from eight to twenty-one. There are at present 13 R&D consortia and two research centres in the 13 geopolitical regions of the country.

A base agency is selected to lead the consortium based on its better capability, resources and facilities for R&D than other members. Through the NARRDN, the agriculture R&D sector continuously conducts need-based and problem-oriented research to address farmers' needs and problems on 32 commodity areas.

As the network generates new or better technology, it becomes increasingly necessary to communicate these new findings to all sectors and to train extension workers who must in turn train farmers in the proper use of these technologies. Correspondingly, it becomes necessary to continuously monitor constraints in the system to identify the directions in which technologies must be modified to better serve farmers. Technology must therefore be understood in the context of the physical, biological, social and political environment of the farmer (Librero 1991).

As Castillo (1986) puts it, technology options must be understood "to improve the goodness of fit between what exists and what is possible and feasible". Technologies do not transfer by themselves. Hence, there is a need to address constraints in technology transfer.

Paradigm

A model for analyzing the technology transfer process is presented in Figure 1. The process is viewed as subsumed within the larger process of technology R&D. In this schema, R&D is managed as a single system in a cyclic relationship with sequential technology-oriented stages of generation (TG), adaptation (TA), verification (TV) and utilization (TU) (Stuart 1990).

Technology Generation (TG)

Technology generation (TG) is fundamental research that focuses on the development of production systems that improve the efficiency and economic viability of using agriculture and natural resources.

Technology Adaptation (TA)

Technology Adaptation (TA) is followed by the research designed to evaluate the stability and replicability of the performance of the component technologies over space and time. TA is conducted in research stations or farmers' fields under the supervision of the researcher.

Technology Verification (TV)

Research then moves towards Technology Verification (TV) or research on comparative performance of improved packages of technologies (POT) and farmers' practices under existing farm environments and in the context of relevant farming systems. It is conducted by farmers in their own farms under the supervision and assistance of researchers and extension workers.

Technology Dissemination (TD)

A technology which has a consistently outstanding performance in TG, TA or TV may be considered a 'mature technology' ready for Technology Dissemination (TD) if it is technically feasible, economically viable, socially accepted and environmentally sound. TD includes sequential activities of piloting, packaging and extension.

Technology pilot testing
Technology pilot testing is the process of establishing community-level linkages with local farmer-cooperator groups to confirm and demonstrate the technical feasibility, economic viability, social acceptability, environmental soundness, in other words, the relative advantage of the technology or POT over existing ones. Through their participation in piloting activities, farmers can demonstrate to co-farmers how the technology is managed, thus initiating a first level 'multiplier' or 'spill-over' effect of the improved technology.

Technology information packaging
Technology information packaging involves the translation of technical research reports into popular layman's terms and into the local dialect by communication specialists. It is the process of transforming the various

technology components into extension-communication materials in the level of understanding of and in formats preferred by the extension workers, the farmers and the laymen.

Technology extension

Technology extension is the participatory process whereby government and non-government field technicians carry out a non-formal education process using interpersonal group approaches supported by appropriate communication materials in helping farm families identify their production needs, improve their knowledge and predisposition toward increasing their productivity and net income. These activities are conducted through the provision of Development Support Communication (DSC) along with other necessary support services, with the ultimate aim of attaining food security, quality nutrition and generating savings for reinvestments into complementary income-generating projects (Stuart 1990).

Technology Utilization (TU)

The technology transfer process takes one complete turn when the technology is effectively communicated and ultimately adopted by users in a sustained manner, until new problems for R&D emerge.

In the paradigm, the interface is viewed as dynamic and may be explained literally using the biblical quotation: "He must increase even as I must decrease" (John 3:30); that is, as the R&D process moves from TG to TU, research activities decrease while development activities increase (Creencia and Burgos 1990). In reality, the process is cyclic and iterative.

The three sub-processes subsumed under the technology dissemination (TD) process are recognized as vital in providing the necessary conditions for TU. TD links the R&D continuum through pilot-testing of technologies and of DSC-technology packaging, and training on technologies and DSC by researchers, communication specialists, extensionists and farmers. These activities likewise operationalize the interface between research and extension toward sustainable development. It is at this level that interdisciplinary and multidisciplinary cooperation and coordination are crucial.

A holistic cycle of technology transfer as depicted in the paradigm should supplant the one-way, bullet-theory view associated with the word 'transfer'. Rather, it should be seen as a dynamic, multidirectional, interactive and user-oriented framework that does not only focus on technology per se but on the processes by which people and institutions share and use technology for the benefit of the greater number. In this framework, the management of multidisciplinarity becomes institutionalized in the process. This implies that the scientists (technical and social scientists), the communication specialists, the extensionists, and the farmers – all who represent the human dimension of technology transfer must be regarded as central to any analysis of the process itself (Bonifacio 1990).

Table 1. Factors in the technology transfer process that impinge on the farmer's adoption decision (Stuart 1991). (Table 1 is continued on the next page.)

For whom?	What?	By whom?
Research institutions	*Technology/information*	*Development institutions*
Mission, mandate	Environmental	Policies on t-transfer
Policies	– Ecosystem	
Structure	– Climate	Structure of extension
Programs	– Topography	communication system
Commitment to users'	Biological	
welfare in R&D	– variety	
Credibility	– pests &	Programs for t-transfer
Multidisciplinarity	diseases	and technical assistance
Operational	– weeds	
linkages with GO &	– soil fertility	
NGO extension	– water availability	Incentive system
and communication		
systems	Economic	Resource availability
	– economic incentives	
	i.e., subsidy etc.	Operational linkages of
	– cost and returns	extension with research
	– input availability	and user systems
	– support services	
	– post harvest facilities	
	– marketing infrastructure	
	– market information	

Given the above framework then, the following schema (Table 1) further provides a classification of factors that demystifies constraints analysis in technology transfer (IRRI 1979; Stuart 1991).

When technical scientists are confronted with non-adoption or low adoption rates of new technology, they turn the question over to the social scientists who must provide scientific explanations as well. The search for answers to the questions (From Whom? What? By Whom? How? For Whom? and With What Consequences?) on users' technology adoption behaviour thus becomes systematic and holistic, when all concerned look into the process together, and not piecemeal.

The IPM Project

Since 1988, a multidisciplinary team of five women scientists from the University of the Philippines at Los Baños (UPLB), has been implementing a project on Integrated Pest Management (IPM) Extension and Women in selected farming villages. With funding from the International Development Research Centre (IDRC) in Ottawa, Canada, the project is currently in its

Table 1. Continued.

How?	For whom?	With what consequences
Technology transfer strategies	*Farmers and other users process*	*Effect & impact of*
Institutional/group approach	*Institutional*	*Functional*
	– organizational development	short term
– cooperatives, farmers'/		long term
womens'/youth	– organizational membership	Dysfunctional
associations/		short term
organizations	– organizational participation	cumulative
– trainings, seminars,		long term
meetings	– leadership patterns	Direct effect
– demonstrations		Indirect effect
– field days, tours	Socio-Cultural	
	– education	
	– social networking	
	– relevant experience	
Individual approach	– access and exposure	
– social networking for	to information	
information and	Psychological	
input exchange	– risk-taking	
– consultations	– attitude & outlooks	
Mass media	– aspirations	
– publications	Economic	
– radio	– resource availability	
– audio-visuals		
Indigenous/folk media	– income	
	– farm size	
Local information	– tenure	
centers & systems	– other occupations	
Interagency communication		
approaches		
Service delivery arrangements		
Technical assistance		
Marketing Information		
dissemination		

second phase. It is a research and development project designed to document the dynamics of rice and vegetable IPM technology generation, verification, dissemination and utilization right at the farmers' fields. While the project works closely with more male farmers, it puts special focus on the role and participation of women at different levels, i.e., as researcher, as extension worker, as farmer, in the different stages of technology transfer. From January 1988 to June 1990, Phase I of the Project was implemented in five rice and vegetable growing villages in Calamba, Laguna. For Phase I, IPM was posed to the farming families as an alternative approach to the farmer's sole dependence on synthetic pesticides.

For Phase II, August 1990 to July 1993, the project expanded its research and extension activities to three villages in Majayjay, Laguna, and now covers integrated nutrient management. Rapid composting technology is being demonstrated, using a compost-fungus activator (CFA) known as Trichoderma, as an alternative to commercial fertilizer. With its participatory methodology, the project has enabled the project leaders and its research staff to gain a more holistic understanding of the processes and constraints in technology development and technology transfer on appropriate integrated nutrient and pest management (INPM).

Constraints to technology transfer in the IPM project

Challenges in operationalizing multidisciplinarity

The multidisciplinary approach to research and development ranks high in the scale of values in conducting R&D, as the process moves toward the development side of the spectrum (Figure 1). However, its actual implementation has its challenges and problems particularly for the coordinator. This is expected, considering human nature, and the high level of expertise from varying orientations, expectations, commitments and perceptions in dealing with a common mission. The challenge is heightened when operational linkages among team members (research side), the extension system and local non-governmental organizations (NGO's) (development side) and the clients (user side) must be instituted with regular communication activities (see Table 1) in sustaining efforts toward the participatory methodology.

Multi-faceted nature of IPM technology

Any technology must be dealt with in terms of its ecological, biological, technical, economic, social-psychological and even its political dimensions. Aside from its inherent complexity in the technical sense, the IPM technology must be communicated to extension workers and farmers within the context of other equally complex dimensions as stated earlier.

Role of extension and communication systems

A program for technology transfer, or the sharing of technical, material and information inputs among the various participants in the R&D process must be constantly fine-tuned to the needs of end-users. In the planning and design of R&D programs, resources for technology transfer are usually inadequate if deliberately included at all, ex ante. (This is a general observation and not reflective of the IPM project).

It is necessary to point out this constraint because of the need for R&D

program planners to recognize the role of and the need to involve extension communication and other social scientists as well as representatives of the user system in the planning stage. This should assume that the technology transfer strategy is built-in with the required resources that will realistically operationalize the interfaces in each stage of the R&D spectrum.

Farmers' aversion to risk

Understandably, when it involves their livelihood, and a whole cropping season at that, small farmers are not that keen to participate in research that may pose risks in their productivity. A farmer cooperator for example sprayed his vegetable crop every week, despite the total absence of or signs of infestation. Asked why he did it, he simply answered "For insurance". Even if farmers themselves observed an absence of pest attack, some of them still sprayed as they had been used to do.

Unlearning old practices to learn better ones

Farmers in the Philippines have developed a dependence on synthetic pesticides for crop protection, a practice they adopted following the effective technology transfer program of Masagana 99 ('Bountiful harvest of 99 cavans per hectare'), a national rice production program launched in the mid-70's. With the introduction of IPM, and INPM, it has now become another challenge for development communicators to help farmers unlearn the Masagana 99 recommendations and learn the new IPM recommendations.

Resource-intensive technology

Farmers, through their long years in farming, are able, though crudely, to weigh costs and benefits of technology being introduced to them. It is for this reason that many of our farmer-cooperators often altered IPM, water and fertilizer management recommendations. It is also why many rejected the rapid composting technology. They perceive the technology as costly in terms of their time and energy and additional hired labour which they cannot afford or are not willing to invest in.

Another case in point is the IPM recommendation to manually control golden snail, a pest that feeds on young rice plants. Farmers complained that manual, or mechanical control was too laborious, time-consuming and at times ineffective. The latter is because snails from neighbouring farms tend to spread through the irrigation water. Hence, some farmers, against the project's advice, insisted on increasing the concentration of pesticides against the golden snails. Pest monitoring to determine whether an economic threshold level (ETL) was exceeded was found to be the least acceptable of IPM practices because of the additional time, labour and technical skill it entailed.

Credit-marketing tie-up with traders

Attached to the farmers' unwillingness to take risks is the pressure to harvest quality produce at any cost due to prior credit-marketing arrangements with traders. This is more common among small vegetable farmers than the rice farmers in the project. Such an arrangement has implications in the selection of cooperators for farm-level participatory research activities.

Lack of credit

Farmers were observed to use fertilizer at lower than recommended levels. The most important constraint cited by both cooperators and non-cooperators for their low use of fertilizers in all seasons was lack of funds. Credit was unavailable to many, or if available, was perceived as unaffordable, or many were unwilling to take the risk associated with borrowing money.

Perceived relative advantage of indigenous over recommended technology

A farmer in the Calamba IPM project pointed out that the rapid composting technology required too much labour, and that results were not always encouraging. He would rather stick to his old method of composting which involves plowing the grass and stalks directly into his field.

The project experiences in Calamba point at the reality that IPM is a people-oriented technology. The IPM concept by itself is a complex one, posing a continuing challenge for the project team because it is difficult to comprehend as it is difficult to teach. Moreover, the adoption of the technology depends on the perception of the clientele of its comparative benefits in relation to their traditional pest control practices (Adalla et al. 1990; 1991).

IDRC-PCARRD research on technology assessment: constraints to technology adoption

In another IDRC-funded research project which was coordinated by PCARRD, entitled Technology Assessment for Agriculture in the Philippines, factors affecting technology adoption were studied across various commodities throughout the Philippines (Librero 1990). Table 2 shows the factors that affect farmers' decisions to adopt technologies related to six commodities, namely cabbage, strawberry, coffee, cacao, rice-based farming systems and mango. For cabbage, the factors that affected adoption were farm area, number of parcels of land, education, membership in a farmer's organization, contact with technicians, and financial assistance.

Adoption of technologies related to strawberry production was linked to farm area, number of family members, membership in farmer's organizations

Table 2. Factors affecting technology adoption.

Factors	Cabbage	Strawberry	Coffee	Cacao	Rice-based farming system	Mango
Farm area	x	x				
Number of parcels	x					
Education of farmer	x		x		x	
No. of family members		x		x		
Age of farmer					x	
Tribe						
Religion						
Cosmopoliteness			x			
Membership in farmer's organization	x	x	x			x
Soil type		x				x
Contact with technicians	x					x
Financial assistance	x					
Tenure					x	
Income						x

and soil type. For coffee, adoption decisions were related to education, cosmopoliteness, or the farmer's degree of exposure to new information from the metropolis, and his membership in farmer's organizations. For cacao farmers, the number of family members was the only factor found to affect adoption, while for rice-based farming systems, the factors were education, farmer's age, and tenure. For mango, adoption of technologies was related to membership in organizations, contact with technicians and income.

Farmers in the same IDRC-PCARRD study gave several reasons for not adopting pest control technologies. The crops covered were coconut, coffee on the main islands of Luzon and Mindanao, cacao and rice. Table 3 shows that for coconut, farmers cited the following constraints: financial, use of indigenous technology, laborious, not familiar with pesticides, lack of knowledge, others do not adopt, and ineffective inputs.

Coffee farmers in Luzon who did not practice pest control gave the following reasons: financial constraint, laborious techniques, lack of knowledge, lack of water source and facilities, small farm area, inaccessibility from road, and they did not perceive pests as a problem. Among Mindanao farmers, the reasons were that it was expensive while funds were inadequate.

Among cacao farmers, four reasons were cited for non-adoption: expensive, financial constraint, laborious techique, and pests were not perceived as a problem.

Rice farmers listed eight reasons for non-adoption: expensive, financial constraints, not familiar with pesticide, unavailability of input, others do not adopt, negative effect of chemicals to men and crop, and pests were not perceived as a problem.

Table 3. Farmer's reasons for not adopting IPM technologies.

Reasons	Coconut	Coffee		Cacao	Rice
		Luzon	Mindanao		
Expensive			x	x	x
Financial constraint	x	x	x	x	x
Utilized indigenous sources	x				
Laborious	x	x		x	
Not familiar with pesticides	x				x
Lack of knowledge	x	x			
Lack of water source and facilities		x			
Small farm area		x			
Inaccessibility from road		x			
Unavailability of input					x
Others do not adopt	x				x
Ineffective inputs	x				x
Effect of chemicals to men and crop					x
Not perceived as a problem/ not necessary	x	x		x	x
Peace and order situation					

In sum, non-adoption could be explained in terms of economic, technical and social reasons. Inputs were either not available or were expensive. Or, farmers were financially constrained. Other reasons commonly cited were lack of knowledge about the technology and the view that the technology was not necessary (Librero 1990).

Conclusions

Actual experience and empirical data from R&D projects in the Philippines, notably those discussed in this paper, point at the social, psychological, economic and technical factors that impinge on the users' decisions (Table 1) in the technology transfer process. The paradigm for technology transfer as presented in Figure 1 shows its increasing role as the R&D process moves toward technology verification, dissemination and utilization. It is in these stages that social science, i.e., people-oriented research methodologies should become built into an interdisciplinary R&D scheme that goes beyond the technical, i.e., technology-oriented research methodologies. Technology transfer, in the sense of a bottom-up, two-way flow of information, then becomes truly relevant toward understanding users' constraints and hence, toward improving the usefulness, efficiency and socio-economic viability of agricultural and natural resources technology and information.

References

Adalla C B, Hoque M M, Rola A C, Stuart T H, Sumayao B R (1990) Terminal Report, IPM Extension and Women Philippines, Phase I (1988–1990). IDRC-UPLB, College, Laguna, Philippines.

Adalla C B et al. (1991) Progress Report, Integrated Nutrient and Pest Management (INPM) Extension and Women, Philippines, Phase II, August 1990-June 1991. IDRC-UPLB, College, Laguna, Philippines.

Bonifacio M F (1990) Policy Analysis of Technology Transfer – A Synthesis. In: Analysis of Policy Issues Associated with Technology Transfer in Rural Development. PCARRD Book Series No. 87/1990. Los Baños, Philippines.

Castillo G T (1986) Trip report. Women in Rice Farming Systems Research/Action Research Program. IRRI, Los Baños, Philippines. Mimeograph.

Creencia R D, Burgos B M (1990) Technologies in Agriculture, Forestry and Natural Resources. In Technology Transfer for Sustainable Development, PCARRD Book Series No. 101/1990. PCARRD, Los Baños, Philippines.

IRRI (1979) Farm Level Constraints to High Rice Yields in Asia: 1974–1977. International Rice Research Institute, Los Baños, Philippines.

Librero A R (1990) Technology Assessment for Agriculture in the Philippines. In PCARRD Book Series 92/1990. Los Baños, Philippines.

Librero A R (1991) Estimating the Ceiling Level of Technology Adoption: Some Theoretical Considerations. Paper presented at the Validation Workshop on Estimated Parameters of Probability of Research Success, Spill-over Effects, Technology Adoption and Elasticities for Selected Commodities. February 28-March 1, 1991. PCARRD, Los Baños, Philippines.

Stuart T H (1990) The System of Technology Verification, Packaging and Dissemination in the Agricultural Research Network. Paper presented at the Consultative Workshop on The Role of the National Agricultural Education System (NAES), Jan. 28-Feb. 1, 1990. UPLB, College, Philippines.

Stuart T H (1991) Issues in Development Message Transfer Through Popular Culture and Media in the Philippines. Paper presented at the Seminar-Workshop on Using Entertainment and Advertising for Development Communication, January 21–24, 1991, AMIC, Singapore.

Postgraduate education in Agricultural Systems: the AIT experience

J.A. GARTNER
Asian Institute of Technology, G.P.O. Box 2754, Bangkok 10501, Thailand

Key words: Agricultural Systems, Asian Institute of Technology, curriculum development, extension, Farming Systems Research, IBSNAT, integrated systems, post-graduate education, SARP, team teaching, Yin and Yang

Abstract
For the past 20 years, seasoned agricultural professionals with a holistic view of the world have recognized the need to add 'a systems approach' to the conventional paradigm in agricultural education, research and development. To maintain and strengthen people and institutions evolving with this point of view, young professionals need the chance to undertake a formal programme of education steeped in the philosophy, theory and practice of systems thinking.

One university responding to this need was the Asian Institute of Technology, an international postgraduate institution in Bangkok, Thailand. There, over the period 1986-'91, Agricultural Systems has evolved as a recognized field of study with a programme of coursework and research leading to the degree of Master of Science.

The experience gained from this initiative suggests that the systems dimension can be added to a student's thinking within the 20-month period for the degree. Nevertheless, this remains superficial unless students have a powerful grounding in the provisional facts of natural science, a physical awareness of farming as a human activity system, and a practical grasp of the social, economic, political and other forces that enhance or inhibit change in the progress or regress of agricultural systems.

Introduction

The troubling ambiguity in the world food problem (Boerma 1975) is this: our knowledge base in agriculture is now so great that mere manipulation of the margin between costs of inputs and prices of outputs can produce either huge surpluses of agricultural products, as exemplified by the Common Agricultural Policy of the European Economic Community, or tragic deficits, as exemplified by the socially disturbed and overly controlled economies of some countries. What this really means is that the output of agricultural products is powerfully influenced by the rewards farmers receive for the efforts they make and the risks they take. This simple fact is sobering. Professional scientists need to keep their feet on the ground and not become unduly impressed with the technical advances that they can achieve, even though new technology is a vital part of agricultural development. The potential impact of new technology on farming systems is dependent upon a wide range of social, industrial, economic, political and military factors and forces, which interact and dominate at different levels in the hierarchy of systems. Figure 1 suggests this in terms of

F.W.T. Penning de Vries et al. (eds.), Systems Approaches for Agricultural Development, 485–504.

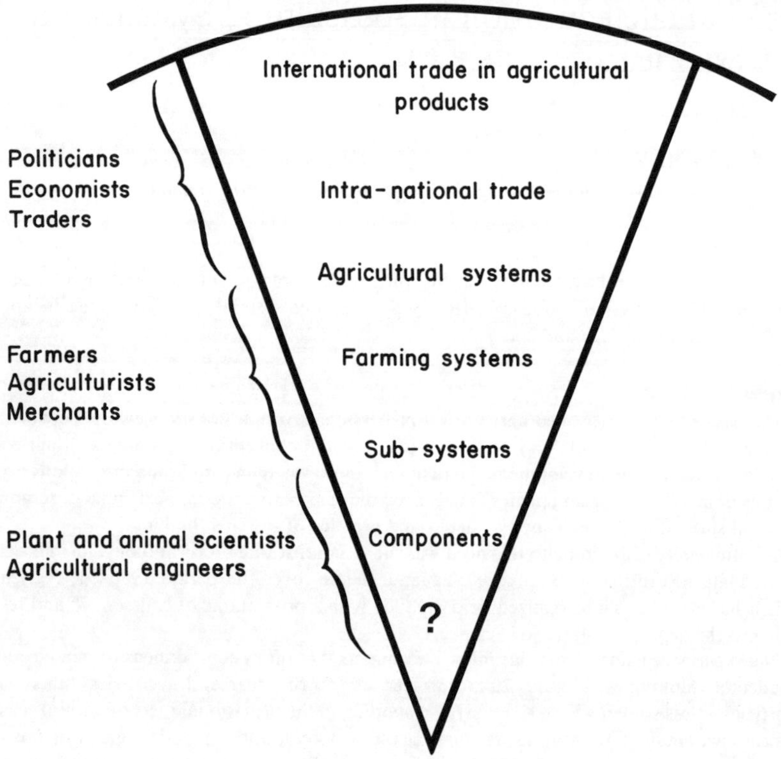

Figure 1. Levels of focus in agricultural development and some of the people involved (adapted from Gartner 1984). Factors and forces in the external environment of a farming system may have greater potential for improvement than those in its internal environment. The term 'farming system' may then inhibit the teaching and understanding of the large picture. To overcome this the term 'agricultural system' is used to encompass the delivery systems for providing essential materials and services to farmers and for getting products to consumers. The question mark indicates further possible levels of focus in this hierarchy of systems.

some of the people involved.

Whether we want to solve the world food problem depends upon the political will of all nations that subscribe to the FAO mandate to relieve hunger and poverty (Boerma 1975) which are inextricably linked. This will require new insights, and changes, not only at the scientific and technical level, but at the non-science level as well.

Wants and needs, haves and have nots

A problem has been defined as a state of unsatisfied need. A situation becomes a problem when the operation of a system does not provide what we want. If what we want equals what we have then what we need equals zero and there is no problem. In political terms this is conservatism i.e. no change or inertia.

Therefore, any problem situation must lie within what we need. If we want to solve the problem, or improve the problem situation, this will require change or motion; this is radicalism. Conflict arises out of these two positions in deciding what we want. These key questions can be formed into the following equation:

$$\text{WHAT WE WANT} = \text{WHAT WE HAVE} + \text{WHAT WE NEED}$$

The point is, that in the trinity of subjects dealing with human affairs–sociology, politics and economics–we talk about wants and needs on their own and without constraint. We never seem to set them in the context of what we have. What we have is the equivalent of the initial state in systems theory.

Why bother with this idea? There are those who want to solve the world food problem and there are those for whom it is convenient to keep. Within the set who want to solve the problem there are many different ways proposed for how it should be done. While the other set retains power, a few will continue to have much and many will continue to have not; for the latter, hunger for freedom may be more important than freedom from hunger. It seems worthwhile then to have some basis upon which to argue for what we want, and to understand the consequences of change or no change in agricultural systems.

A new dimension

Since the 1960's, seasoned agricultural professionals with a holistic view of the world, as well as considerable subject knowledge, have argued the need to add 'a systems approach' to the traditional science and technology paradigm in agricultural development. They advocated that some people need to step outside the confines of the subject in which they specialize in order to understand the linkages and interactions between all the factors and forces that enhance or inhibit change in agricultural systems. But agriculture continued to be taught at university level in terms of scientific subjects and disciplines.

To maintain and strengthen people and institutions evolving with the systems point of view, young professionals need the chance to undertake a formal program of education and training steeped in the philosophy, theory and practice of systems thinking in the subject, agriculture and in the activity, farming.

A question of purpose

In this paper I use the word development to imply change to an existing situation, but decisions as to whether it is good, bad or in the right direction depend on one's view of the world.

On what basis then, can we begin discussion to reconcile the different points of view on what we want, which range from the inertia of no change to those concerned with what may constitute motion towards desirable change? Spedding (1979) opens his book "An Introduction to Agricultural Systems"

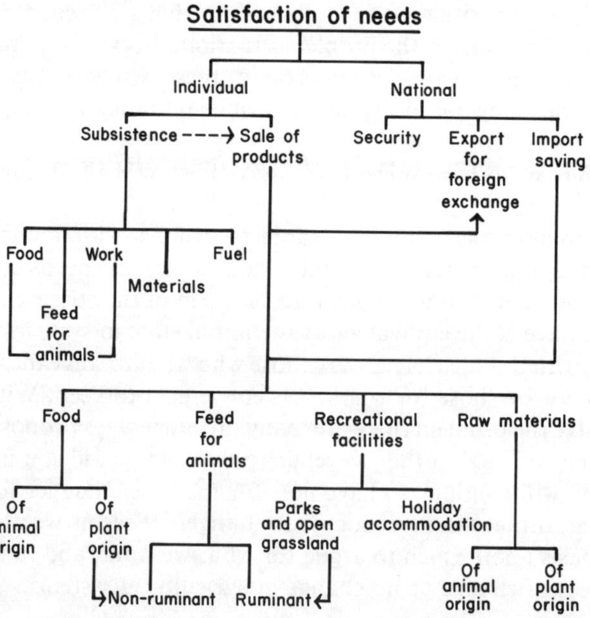

Figure 2. The purposes of agriculture (from Spedding, 1979).

with a chapter called "The Purposes of Agriculture" which contains a similarly-named diagram (Figure 2) headed "Satisfaction of needs". The needs range from individual to national. Discussions on agricultural development need to start with questions of purpose.

Within that context, the purpose of education should be to enable us to understand natural phenomena, explain the past, live in the present, and prepare for change in the future; postgraduate education should be the acme of the formal, institutional part of that cumulative process.

Systems theory, systems practice

On the question of how a program of education in agricultural systems might be delivered, Bawden (1990a) makes a subtle distinction between learning *systems* and *learning* systems. He pursues the point in his plenary paper to this symposium (Bawden 1992). In it he argues that through a shift in systemicity from inquiry into agricultural systems to inquiry through agricultural systems, we may be able to transcend the old trinity of education, extension and research, and place them into a higher-order category of critical human inquiry. Such an inquiry would insist that we put behind us such artificial constructs as: First, Second and Third Worlds; North and South; least developed, developing and developed countries.

In the systems idea of hierarchy, there is one world with many countries and multitudes of individuals. In the systems idea of emergence, these individuals are irreducibly free to make a choice to take a chance on a change, or to accept things as they are, and do nothing; they are born into families which combine together into communities (so vividly portrayed by Steichen 1955, as "The Family of Man") which enter into societies which form sovereign states to make up a family of nations (so optimistically named the United Nations in 1945).

Against this background of ideas, postgraduate education in agricultural systems at the Asian Institute of Technology (AIT) was conceived, born and nurtured in its infancy. The story of that experience, as perceived and recounted by the author, is deeply rooted in the dilemma created (Grene 1974) when scientific reductionist thinking is confronted by the counter positions of holism and emergence, the hallmarks of systems thinking (Checkland 1984).

Genesis 1980–1985

The genesis of an activity depends on the chance congruence of events, the people associated with them, and the choices they make. In systems theory something comes from somewhere, is processed within a physical or abstract structure, and goes somewhere; in practice the process is best described by Bawden (1990b) as "a complex mess of mutual influences". In 'The AIT Experience', these influences covered the full range from the conservative to the radical, with all the conflict inherent therein.

The opportunity to establish in Asia a postgraduate centre for the education of young agricultural professionals in a systems approach arose out of a pair of project ideas submitted to FAO Headquarters in Rome from the Regional Office in Bangkok in 1980. One idea was for the promotion of integrated farming systems. The other arose out of increasing interest in the dilemmas of risk-prone, rainfed farming systems.

A kaleidoscope of influences

There was much confusion about the use of the word 'systems' in the material that supported these project ideas. Nevertheless the two ideas were subsumed in one project proposal which the United Nations Development Programme (UNDP) approved for preparatory assistance; the primary output was to be a full-scale project proposal. From commencement in January 1982 it ran into a complex mess of competing influences.

In the 1970's multi- and inter-disciplinarity became by-words in development activities and the 'bottom-up' approach arose to challenge the traditional 'top-down' approach in the generation and transfer of agricultural technology (Gartner 1990). One response was Farming Systems Research (Hildebrand 1990) known as FSR, with its variants associated with extension (FSR&E) and development (FSD), each with their own advocates. FSR became firmly rooted in Asia under the sponsorship of the International Rice Research Institute and the International Development Research Centre of Canada following the formation of the Asian Cropping Systems Network in 1975 (Zandstra et al. 1981). Another approach was Agroecosystems Analysis (AEA) (Conway 1983). It became well established in Khon Kaen and Chiang Mai Universities in Thailand; later it was institutionalised in the Southeast Asian Universities Agroecosystems Network. In a wider context, Integrated Rural Development (IRD) preceded it (FAO 1975). In 1983, an FAO Expert Consultation (FAO 1984) was held in Bangkok to bring representatives of these approaches together to establish the common ground of 'a systems approach' to agricultural development. Abstract research, involving mathematical modeling and systems simulation, had hardly entered the spectrum of methods and approaches then in vogue; the IBSNAT project began in 1983 (Uehara and Tsuji 1992) and the SARP project in 1984 (Ten Berge 1992).

As this kaleidoscope of influences unfolded, it became clear that what was missing from 'the systems movement' in Asia were substantive, formal education and training programs in the philosophy and theory of a systems approach, and how it might be applied in practice to agricultural development. This conclusion was based on the argument (and experience) that it is not possible to pluck people from the traditional science and technology paradigm in education, research and development and drop them into the maelstrom of the new systems paradigm without prior formal exposure.

Short-term training courses had been developed and delivered to meet the demand for people to staff the many projects proliferating under the banner of FSR/E/D, AEA and IRD. Out of necessity these courses dealt with methods, which can have only limited impact on the way people think. It was asserted that future leaders, who can respond with imagination and innovative flair to rapidly changing circumstances in agriculture, would come from rigourous education programs in agricultural systems at the postgraduate level where education and research merge in a symbiotic relationship.

Differences and similarities

This situation was quite different from that which gave rise to The Hawkesbury Experience (Hawkesbury Agricultural College 1985) in Australia (Bawden et al. 1985) from which I have borrowed my sub-title. My intent goes deeper than mere borrowing, however, because the idea of experiential learning (the equivalent of empiricism in science) is deeply ingrained in systems thinking; this means that those actually involved in the education process are also learning.

The quality that the two situations had in common was complexity and uncertainty in the face of diversity: the diversity of changes taking place rapidly in the external environment of farms in Australia, and the diversity of farming systems in Asia confronting those who would pretend to 'improve' them. What constitutes and improvement, of course, is always open to debate (Spedding 1984).

A new project

Two view points were clearly emerging–they fit Bawden's distinction between learning *systems* and *learning* systems–and headed for conflict, as the preparatory assistance proceeded through its activities, consultation and review. Eventually, a full-scale regional project was agreed with the title Farming Systems Development in Asia: Crop/Livestock/Fish Integration in Rainfed Areas. Advance authorization for FAO to proceed was given by UNDP in January 1985.

Structure and process

The project involved the establishment of a network of six universities and six line agencies in six countries – China, Indonesia, Philippines, Sri Lanka, Thailand and Vietnam – with AIT as the Regional Lead Centre (RLC). AIT is located on the outskirts of Bangkok in Thailand. It is an international, postgraduate university with a regional mandate for education and research.

The primary activity in the first phase was education and training. In theory, and mostly in practice, this existed on three interacting levels. First, one well-qualified student from the counterpart university or line agency in each of the six countries was to come to AIT in each of two years to study for an M.Sc. in Agricultural Systems over 20 months. Second, six students, mainly drawn from line agencies, and who had completed an undergraduate degree, were to commence studies at each of the national universities, for either a postgraduate diploma of 12 months or for a longer M.Sc. program; this last depended on each university's regulations, and its confidence and capacity to develop and deliver a suitable programme. Third, students were to do practical work and research studies in villages with farmers interested in the integrated farming of crops, livestock and fish; part of that practical work was to be interactive training programmes with the students exchanging knowledge for farmer experience. The potential synergy in this structure was reinforced by exchange between students on study tours during vacation time.

Why AIT as the RLC?

There is a great deal to be said for students to be educated within the physical and cultural environments of their own country and region. However, it is essential that this education process does not occur in isolation from the influences and realities that lie beyond the boundaries of those environments. With its international faculty, staff and student body, AIT straddles the two 'worlds' of East and West. Thus it provided a unique opportunity to satisfy both these criteria for an effective education in agricultural systems for Asian and other students. Moreover, AIT is distinct from national educational institutions, and of a higher order in the hierarchy of education systems: it is funded by many governments but politically independent; it is not affected by national education policies; and no one culture is dominant. All teaching is done in an international language, English. The academic standards set result in a high-status degree which carries considerable influence in the region for those who obtain it.

Clearly the advantages of AIT were many, yet the equivocal attitude of some concerned people towards the proposal that AIT should be invited to be the RLC was curious to say the least; even more curious was a reluctance to accept the fact once project implementation began. It was almost as if AIT was some incomprehensible alien institution instead of being a unique member of the family of international institutions. It is not the purpose of this paper to speculate why, but I suspect the attitude is immanent in the plaint of Boerma (1975) in his paper Political Will and the World Food Problem: "what basically disturbs governments (institutions, individuals – my insert) is any proposal that means their relinquishing some measure of individual national control".

AIT accepted the invitation to be the RLC because it is donor dependent and thus receptive to externally funded programs that contribute to Asian development. A Letter of Agreement between FAO and AIT was signed in September 1985. At that time the first M.Sc. program was due to begin in January 1986 with six students: three from Thailand, two from Philippines and one from Sri Lanka.

Evolution 1986–1991

The type of person we want to enter the M.Sc. program in Agricultural Systems is one who is intelligent rather than informed, primarily a biologist but who is unafraid of either economics or mathematics. Getting dirty hands will not be an obstacle for this person nor will communicating with farmers and others concerned with agricultural development.

Such a person will have, or develop, the breadth of vision to encompass the concerns of rural development and the need to maintain the physical environment in a productive and unpolluted state.

Most importantly this person will become an opportunity maker and taker and a problem solver in the community he is meant to serve, as well as one of its highest paid public servants.

The first part of the description of the type of person required comes from Spedding (1970) at the time when ecology had entered the thinking of some agricultural scientists and the first round of mathematical modeling and systems simulation in agriculture had reached its zenith. The next part indicates the need for skills to win respect from, and gain the confidence of, the non-scientific community.

Agricultural development can generate the primary wealth that leads to rural development. However, the people involved must be careful that the changes wrought do not destroy the physical and social landscapes in which development takes place.

Finally, many people are versed in the litany of problems confronting us. One human tendency is to see them as a burden to be shouldered tragically; another is to look for opportunities for improvement and to know the excitement of solving the problems that stand in the way of their achievement. Snow (1964) provides an interesting account of the problems of communication between people with such tendencies.

The curriculum

To establish a new paradigm of study in an existing institution, there are two options: either create a revolution, which means changing everything, or work within the existing paradigm until an identifiable new field of study emerges. The choice depends on circumstances and people. A revolution took place at Hawkesbury which had a long history of traditional vocational teaching in agriculture. At AIT the second option was more appropriate.

Dilemmas

AIT began as the SEATO Post-graduate School of Engineering in 1959; a Division of Agricultural and Food Engineering (AFE) was set up in 1977. Systems thinking was concentrated mainly in the engineering disciplines of management science, operational research and systems analysis. A proposal to establish a field of study based on the philosophy and natural science strands of systems thinking was like an alien from outer space. It was either resisted energetically or embraced without understanding what it really was: a program to deal with agriculture and farming as 'human activity systems' in contrast to 'designed physical systems'. These last are the province of engineers who, at times, regard even scientists with suspicion!

This difference in point of view, incomprehensible at first, continues to be a major hurdle that needs to be overcome in the process of grafting a new curriculum onto the traditional teaching paradigm as well as the existing systems view. But the conclusion from the conflict is positive from one point of view: the search for a solution to the problem of communication between the protagonists led to Peter Checkland's book, Systems Thinking, Systems Practice, first published in 1981 and revised in 1984. His account of the evolution of the systems movement laid bare the bones of the dilemma. The flashpoint of insight and understanding comes not from consensus but at the point of contention between two internally consistent arguments. So where was the problem?

Initial debates on the curriculum were inspired by the work and teaching of Colin Spedding and his colleagues at the University of Reading in England, which enriched the author's own experience in the science and practice of agricultural development since 1960. Until his retirement in 1990, Colin Spedding was Professor of Agricultural Systems. His work falls mainly within the natural science strand of systems thinking. It includes ideas from the sociology and philosophy strands but there is little in it which indicates the engineering strand.

From the outset, it was considered that systems philosophy, theory and practice should be at the core of the curriculum, with subject matter and methods on the periphery. The argument for this was based on several premisses in addition to 'a question of purpose' already discussed above. First, the questions, what system is to be improved, what constitutes an improvement, and from whose point of view, must be decided at the start of any research and development program designed to make changes in an agricultural system that purport to lead to its improvement. Second, a systems approach is holistic. Third, a change introduced in one part of a system will induce a change in its other parts. Fourth, whether this change will cause an unimportant ripple or an irreversible wave in the system is difficult to predict, but of great consequence. And fifth, farming is a way of life and a business. These premisses, it was assumed, would create a demand for relevant knowledge from traditional subjects and disciplines and for appropriate methods.

At this stage, the word 'integrated' in the project title turned into an albatross around the project's neck. One wrong reason for AIT accepting the new program was that it would integrate the disparate faculty and fields of study in AFE. As perceived by the engineers, this meant making a 'collection' of existing courses to form the curriculum of the new program. This perception is not unique for it has been observed within renowned agricultural institutions! It was, and must be, resisted strenuously. The right reason for accepting the program was that it provided an opportunity to integrate, and give system relevance to, the knowledge being taught, and not the people who taught it, in the four existing fields of study viz. agricultural machinery and management, agricultural soil and water engineering, aquaculture, and post-harvest technology.

Linkages

The opportunity to link with aquaculture was very strong, both in teaching and in its embryo research program, both on-campus and off-campus/on-farm, involving the integrated farming of crops, livestock and fish, which was the primary focus of the UNDP/FAO project. This led to the design and construction of a 2.5 ha teaching farm on campus operated under rainfed conditions.

The relationships in time and space between the components and resources of small-scale integrated crop/livestock/fish farming systems in rainfed areas are highly complex. It was believed that if students gained knowledge and experience within the framework of these systems, they would develop the confidence to tackle any problem situation in agricultural or aquacultural development with imagination and ingenuity. However, it was clearly understood that this focus provided the vehicle to carry the M.Sc. program. In no way did it preclude gaining experience in other systems that have evolved in response to different sets of physical, economic and social conditions; for this, students went off campus.

It was concluded that the curriculum should be set within the context of people managing and operating a complex, dynamic, economic, biological system which had to adapt to an ever-changing external environment. Nevertheless, such a curriculum should not neglect the power of science, and technology derived from it, to change systems for defined purposes. Furthermore, any educational program in agricultural systems should now extend Spedding's diagram (Figure 2) to include the satisfaction of international needs; modern concerns with pollution and degradation of the physical environment demand this due to undesirable outputs from production systems, the effects of which can transcend political boundaries.

Structure, process and content

The AIT Masters program comprises a total of five terms, each of about 12 weeks, spread over 20 months. A minimum of 30 units of lectures and practica are taken over three terms and a research thesis for 25 units of credit must be presented. One unit of credit is equivalent to 12 hours of contact in lectures or 36 hours of practica; the norm for a one-term course is three units.

Initially, three core courses dealing with systems issues were created specifically for students in the program. In this arrangement the practica (six hours per week) were in tandem with the lectures (one hour per week) which were spread over three terms. Delivery was by 'team teaching'. This allowed 'systems faculty' to maintain contact with the students throughout the year and provided the opportunity to set coursework against the background of the ever-changing cycle of the seasons as reflected by the teaching farm. Later, these advantages had to be given up to allow other students to take the courses; timetabling and class size precluded this without some modification of delivery.

Team teaching was not appreciated by students because different styles of delivery and points of view caused confusion. It was modified first and then dropped completely. It was very difficult to coordinate due to the idiosyncratic behaviour of faculty, and courses lacked internal consistency. It is better to have one faculty responsible for a whole course in order to maintain continuity of delivery and the internal consistency of an argument, even though there is a loss in the linkages between courses and their relationship to the physical environment. New solutions must be sought for this Gordian knot.

The lectures and practica were re-arranged into sets dealing with: systems thinking; individual farm design, operation and improvement; methodology of farming systems research and development; and factors and forces in the external environment that affect farming and agriculture. These core courses are presently entitled:

- A Systems Approach to Agricultural Development (lead course)
- Farming as a Human Activity System
- Farming Systems Research and Development
- Agriculture Sector Analysis.

Wrapped around this core are courses entitled Crop Production Systems, Livestock Production Systems and Aquaculture Systems. These provide the essential biology involved in the breeding, feeding, health and husbandry of crops, livestock and fish all set within a systems context. The links to 'business' and 'systems involving people' are provided by the courses Farm Management Economics and Agricultural Development and Planning. The links to technological 'hardware' and 'software' are provided by the courses Agricultural Mechanization and Management and Agricultural Systems Analysis. In this group of courses, some of which are electives, the program begins to merge with the existing courses in AFE and AIT.

In the beginning

Applicants with a teaching, research and/or development background in agriculture, agricultural economics, agricultural engineering, aquaculture, veterinary science, and related disciplines have been accepted for the M.Sc. program. However, in 1991, the first break with tradition occurred when a student qualified in law and public administration was accepted; he had a responsible position in the making of agricultural policy.

Students are screened on academic merit and present expertise, followed by years of relevant experience, institutional background and country of origin. The final selection tries to balance as many subject areas with as many countries as possible without sacrificing academic merit. The number of countries represented has been as high as ten in the maximum class size of fourteen.

Sixty three students from 175 candidates have entered the program. Not more than 20 percent of students were the children of farmers. The scientific knowledge base of most was not as strong as expected; few had ever proposed and conducted an experiment using modern methods of scientific investigation. Some had exposure to FSR/E/D and AEA activities. Most were well-versed in the language of popular development literature but few were confident of expressing their own views. Twenty five percent of students were women; in the 1992 intake this figure is expected to be 50%, without bias due to current fashion!

In the middle

In addition to abstract concepts, the education process in agricultural systems requires qualitative and quantitative exploration of fundamental relationships between component parts of a system which operate together for a common purpose, and which react as a whole to stimuli from an ever changing external environment. The objective of this is to inculcate students with an ability to assess the feasibility of change for defined purposes, and to establish what is possible over a given time horizon. Once this foundation is built, students need to be confronted with physical reality and given problem-solving or situation-improving tasks. One consequence of this process is that one can become uncomfortably isolated from the security of one's previous learning and experience (Gartner 1982).

Most students suffered discomfort very early on when confronted with systems thinking. Many had difficulty in accepting the concept of a hierarchy of systems, and that a component of one system could be a system itself with its own boundary and properties. Once this was grasped, it was easy to accept, with some relief, that component research was just as valid and important as 'whole' systems research provided that the problem or situation being studied was relative to the system to which the solution or improvement would apply. Some students felt a loss of self-esteem when competing (in courses that were new to them) with students who had not left their home ground. Slowly, as their minds began to bend with the construction of a systems framework, they realised that they were becoming expert in their own right in placing subject knowledge within that framework, and viewing systems as a whole, rather than in terms of their separate parts; this enhanced their results and their self-esteem.

One of the most difficult questions in systems research is how to start. This distressed all students, but in particular, those who elected to investigate their own vague problem areas in their thesis research. This period requires close attention and encouragement by faculty without yielding to a student's difficulty; some simple advice is 'start!' by asking the tough questions: what do I want? what do I have? what do I need?

Language and communication

Another difficulty that is unique to AIT is the problem of a common, primary language of communication. First there is the difficulty of translating between cultures, particularly when it comes to abstract concepts. Second there is the practical difficulty of conversing with farmers and others when faculty and students carry out off-campus and on-farm field work; having Thai students and staff helps but this removes a student one step from the action. This problem disappears with the establishment of national education programs, but the problem of how to communicate in the 'secondary language' of the different types of people concerned with agriculture does not. If we are to avoid intellectual incest, systems education needs to establish ways and means of finding something in common (Gartner 1990) with non-scientists whose value-sets are driven by different imperatives (Prentice 1976) over different periods of time.

Coincident with the language dilemma is that national institutions can zero in on substantive issues in national agriculture already within a student's purview. Nevertheless, an international program, by bringing together students of different countries, drives home the point that agricultural development issues at the international level can dramatically affect those at the national and individual farm level. In this lies the uniqueness and strength of AIT.

At the end

Of the 63 students who entered the agricultural systems program, 49 have graduated, one was dismissed, and 13 have just completed their first year.

I am sure some students went down from the program none the wiser about the philosophy and theory that underlay 'a systems approach' but the experience did give them two things: a great deal of confidence and an ability to 'make connections', as one visiting faculty put it. They also learnt something of the provisional facts of science and the method by which they are verified or refuted. For other students, the setting up of a philosophical and intellectual framework in the beginning enabled them to fuse the abstract worlds of East and West into an emergent whole. With this framework they were able to evaluate the facts of science and the artefacts of technology, together with a knowledge of the human factors and forces that enhance or inhibit change in the progress or regress of agricultural systems. This, plus their 'AIT experience', reinforced their new-found confidence.

Where are they now?

No formal tracer study has been undertaken yet. Informal feedback on the whereabouts and activities of graduates indicates that most have returned to their former roles in education, research and extension. Three have gone on to do Ph.D. studies, one has entered an international organization, and one is in agricultural banking. Several have chosen not to return to their countries due to political and social disturbances. One interesting aspect is that a number have won short-term training jobs in the aquaculture program ahead of subject matter specialists.

What else?

AIT took responsibility for the program when UNDP/FAO project funding and execution concluded in August 1988. Agricultural Systems was ratified by its Senate as a new field of study in November 1988. The fourth intake of students commenced in January 1989 with the minimum number of three faculty required for the program. These positions were funded by the Australian International Development Assistance Bureau (AIDAB), the German Agency for Technical Cooperation and AIT. Support to AIT by AIDAB, and consequently the faculty position, will cease in 1992. AIT has established a second position but external funding will be needed to sustain the program and to expand it to an effective group of six full-time professionals. One position in such a group would be for someone competent in modeling and systems simulation to facilitate abstract research in association with physical research and farming systems development.

Laundering experience

One of George Bernard Shaw's famous aphorisms was that "those who can, do, those who can't, teach". In systems, just as in science, those who can, should teach, to gain new insight into what they have been doing. This exposes new areas of uncertainty and ambiguity which generates the momentum for a new wave of understanding in the relationship between wants and needs, haves and have nots. A Chinese proverb expresses the situation very well:

What we hear, we forget;
What we see, we remember;
What we do, we understand.

The AIT Experience was like an intellectual journey leavened with a touch of the irrationality and extra-rationality that goes with adventure; there were good times and not so good times as well as chance events and unexpected outcomes. As the journey progressed the links began to fall into place in a chain that led all the way back in time for 2,500 years to an obscure statement of the Greek, Heraclitus, which first set me on the systems journey about 30 years ago.

Continuous change

Heraclitus stated that it is not possible to throw a stone into the same river twice. Bawden (1990, pers. comm.) has countered with an alternative translation: it is not possible to step into the same river twice. At first glance the two translations are the same but in fact they are not. They represent the positions of the observer and the participant in agricultural systems: the observer is on the bank of the river, the participant is in the water! The views are different: the first leads to analysis *ad nauseam* while the second is restricted by its individuality. There needs to be a balance between the people in the two positions, and they need to change places occasionally so that each will understand the other.

The analogies of Heraclitus coincide with those suggesting a relationship between structure and process in the enigmatic poems of the Chinese, Lao-Tse who lived at the same time. These poems are regarded as central to Taoism; a prevalent theme throughout is the dynamic interdependence of the passive and active principles of yin and yang (Campbell 1985). These two forces (Figure 3) cannot be viewed in isolation since neither can exist except in relation to each other. In development, the yielding yin is essential for progress and the unyielding yang is the stabilising force which is necessary to sustain it; a sustained and balanced development can only occur if the 'changeable' and 'unchangeable' are both given equal priority (Lee 1989).

Since the time of Heraclitus and Lao-Tse, concepts of change have been in conflict with the human need for certainty and the rigid structures within which social and religious life have been conducted. They also represent the conflict between the individual and the group all the way up to the state, and the state as a unity in the family of states.

Education: liberal or vocational?

This leads to the question that has raged (Allen 1988) since the time of Lao-Tse and his contemporary, Confucius, and a trio of Greeks–Socrates, Plato and Aristotle–who entered the fray of designing abstract systems of thought to guide human conduct sometime after Heraclitus. Extended accounts of the debate are in Knowles (1978). In the language of our times the question is: should education be liberal or vocational and for whom, an individual or the state?

Confucius and Plato favoured vocational education for the harmonious integration of an individual in a state where everyone knows their place and there will be no change in roles or the relationships between those roles; these rigidities and 'certainties' have influenced Eastern and Western education systems, and thus social systems, ever since. In contrast, Lao-Tse and Aristotle emphasized the cultivation of the individual; learning was for the sake of understanding. Aristotle, rejecting the teaching of his master, saw a liberal education as the means to prepare the individual for the active enjoyment of leisure which to him meant the disinterested search for truth! The advent of technology derived from science, the principles of which were first set out by Aristotle, have yielded untold leisure of a different sort in modern times not only for the 'freemen' of Greece but for many of the 'slaves' as well, who are now free to do something more than 'labour'. Conflict and harmony are inherent in the relationship between these points of view; they are the yin and yang of the education debate.

Paradoxically, it is a liberal education which now provides stability and changes slowly as it is informed by vocational needs. Vocational education must change rapidly to adapt to changes in knowledge and technology in order

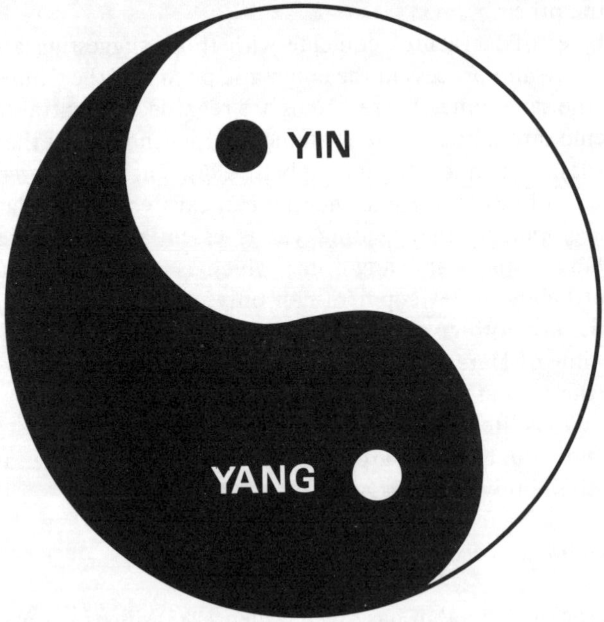

Figure 3. The dynamic interplay between the yielding force – the yin – and the unyielding force– the yang. The dot in each half of the circle represents the seed from which an opposite force will grow when a situation reaches its climax (from Lee 1989).

for systems involving people to survive and evolve. The paradox disappears when one imagines Figure 3 in its dynamic form, for the seed of the yang is perpetually germinating in the yin, and vice versa, in an endless round of becoming. Any social system that neglects to seek a balance between both, at each level in the hierarchy of education systems it constructs for its own purposes, is doomed to sterility, eventually to disappear as the forces of involution lead to decline and decay.

Research: physical or abstract?

Another point of departure between Aristotle and his teacher Plato is highly relevant to the context of this symposium. When faced with the complexity of the phenomena he observed in marine life, Aristotle despaired of finding explanations in Platonic mathematical abstractions. He concluded that perfect mathematical forms were not a useful model for plants, animals, and human beings (Checkland 1984)! Nevertheless, Plato's enthusiasm for mathematical certainty helped to establish the belief that to express the workings of the world mathematically is to take an important and useful step beyond mere qualitative description. This led the way for abstract modeling of 'worlds within worlds' once the technology had been invented.

These views represent the yin and yang of modern agricultural research with potential for conflict or harmony. Abstract modeling does help in understanding complex situations and in the testing and generation of complex hypotheses for change, but if it is not informed by physical experimentation and experience in the field, then it will be barren and a waste of money and effort. Arguments about this have raged in the agricultural systems and science fraternities since modeling and systems analysis first began in the 1960's (Dent and Anderson 1971; Morley 1968; Williams 1967).

When funds and faculty allow, Agricultural Systems, as a field of study, can combine abstract concepts and modeling of systems with physical experimentation and experience in field situations. Most importantly, however, students should be imbued with the confidence and courage to 'have a go', with the knowledge they have already, to make the connections which stimulate change. When funds are in short supply, the order of priority in accomplishment should be from this last to the first, from the 'matter' of a subject to the 'method' by which its knowledge may be increased, understood and used.

Final words

Why this concern with people who lived so long ago? First, it was an extraordinary time when men started to ask serious questions about human life and society, and how it should be lived, after about 20,000 years of human evolution (Wells 1922). Second, their ideas and those of others who lived at the same time, continue to influence our lives. Third, the issues do not seem to have changed much. What has changed is technology derived from scientific knowledge in association with human inventiveness. The power now at our control, from the wealth associated with this technology, to shape, create or destroy landscapes, environments or civilizations is extraordinary. And yet there are still millions of farmers using techniques and implements which had been devised, if not widely adopted (Grigg 1974), at the time when the thinkers mentioned above were alive. Not only that, in reaction to the technological power now available, we are re-inventing the techniques and implements of the past. This is incongruous and verges on the absurd.

We need to explore new ways of looking at agricultural systems and their relationship to: the social systems that subsume them; the designed physical systems and abstract systems that support them; and the natural systems within which their activities take place. In the Introduction I suggested that it was time to dispense with the categories of First, Second and Third Worlds and other artificial constructs. In their place I want to suggest that we use the notional classification of the Western philosopher Karl Popper (Magee 1973): World 1, the objective world of material things; World 2, the subjective world of minds; and World 3, the world of objective structures which are the products, not necessarily intentional, of minds or living creatures, but which, once produced, exist independently of them. If we continue our exploration of systems thinking in the context of this classification, we may be better placed to contend with change in the future.

The most difficult change to manage will be the expectation of more and more people to have a fairer share of the world's resources, and their derived products, the most important of which is food.

Acknowledgements

During genesis and after, the strength of Peter Myers in FAO was unfailing. At implementation the articulate encouragement of David Thorup in UNDP was invaluable. Peter Chudleigh, John Petheram and Christy Peacock made significant contributions to teaching in the early years and argued constructively about how it should be done. Throughout 'the experience' Peter Edwards was always there. He provided helpful criticism of the first draft of this paper; other colleagues refined it. The resistance of all those who found it difficult to accept change helped clarify the argument for it.

References

Allen M (1988) The Goals of Universities. The Society for Research into Higher Education and Open University Press, Milton Keynes, UK.

Bawden R J (1990a) Of agricultural systems and systems agriculture: systems methodologies in

agricultural education. Pages 305–323 in Systems Theory Applied to Agriculture and the Food Chain. Jones J G W, Street P R (eds.) Elsevier Applied Science, London, UK.
Bawden R J (1990b) Systems thinking and practice in agriculture. Paper presented to a Conference of the American Dairy Science Association's Extension and Education Computer Committee, Raleigh, North Carolina, USA.
Bawden R J (1992) Systems approaches to agricultural development: The Hawkesbury experience. Agric. Syst. (submitted)
Bawden R J, Ison R L, Macadam R D, Packham R G, Valentine I (1985) A research paradigm for systems agriculture. Pages 31–42 in Remenyi J V (Ed.) Agricultural Systems Research for Developing Countries. ACIAR, Canberra, Australia.
Boerma A H (1975) Political will and the world food problem. Coromandel Lecture, New Delhi, India. FAO, Rome, Italy.
Campbell R (1985) Fisherman's Guide: A Systems Approach to Creativity and Organization. New Science Library, Boston, USA.
Checkland P (1984) Systems Thinking, Systems Practice. John Wiley & Sons, Chichester, UK.
Conway G R (1983) Agroecosystem analysis. ICCET Series E No. 1. Imperial College of Science and Technology, University of London.
Dent J B and Anderson J R (1971) Systems Analysis in Agricultural Management. John Wiley & Sons, Sydney, Australia.
FAO (1975) Integrated Rural Development. FAO, Rome, Italy.
FAO (1984) Report of the FAO Expert Consultation on Improving the Efficiency of Small-scale Livestock Production in Asia: A Systems Approach. FAO, Rome, Italy.
Gartner J A (1982) Replacement policy in dairy herds on farms where heifers compete with cows for grassland–Part 3: A revised hypothesis. Agric. Syst. 8:249–272.
Gartner J A (1984) About the consultation: its purpose, structure and output. Pages 1–9 in Gartner J A (Ed.) Proceedings of the FAO Expert Consultation on Improving the Efficiency of Small-scale Livestock Production in Asia: A Systems Approach. FAO, Rome, Italy.
Gartner J A (1990) Extension education: top(s) down, bottom(s) up and other things. Pages 325–350 in Jones J W G, Street P R (Eds.) Systems Theory Applied to Agriculture and the Food Chain. Elsevier Applied Science, London, UK.
Grene M (1974) The Understanding of Nature. Dordrecht: Reidel D, The Netherlands.
Grigg D B (1974) The Agricultural Systems of the World. Cambridge University Press, Cambridge, UK.
Hawkesbury Agricultural College (1985) The Hawkesbury Experience. Bawden R J (Ed.) Hawkesbury Agricultural College, Richmond, New South Wales, Australia.
Hildebrand P E (1990) Farming systems research-extension. Pages 131–143 in Systems Theory Applied to Agriculture and the Food Chain. Jones J G W, Street P R (Eds.) Elsevier Applied Science, London, UK.
Knowles A S (Ed.) (1978) The International Encyclopaedia of Higher Education. Jersey-Bass, London, UK.
Lee C (1989) Rice-fish extension in Tung Kula Ronghai. Paper presented at the 2nd Rice-Fish Farming Systems Research and Development Workshop, Muñoz, Philippines.
Magee B (1973) Popper. Fontana/Collins, London, UK.
Morley F H W (1968) Computers and designs, calories and decisions. Aust. J. Sci. 30:405–409.
Prentice W (1976) Wisdom – not knowledge. New Scientist, November 1976:270–71.
Snow C P (1964) The Two Cultures: and a Second Look. Cambridge University Press, Cambridge, UK.
Spedding C R W (1970) The relative complexity of grassland systems. Pages A126–A131 in Norman M J T (Ed.) Proceedings of the XI International Grassland Congress, Surfers Paradise. University of Queensland Press. Queensland, Australia.
Spedding C R W (1979) An Introduction to Agricultural Systems. Applied Science Publishers, London, UK.
Spedding C R W (1984) The systems approach in agriculture: principles and problems in its

application. Pages 10–17 in Gartner J A (Ed.) Proceedings of the FAO Expert Consultation on Improving the Efficiency of Small-scale Livestock Production in Asia: A Systems Approach. FAO, Rome, Italy.

Steichen E (1955) The Family of Man. The Museum of Modern Art, New York, USA.

Ten Berge H F M (1992) Building capacity for systems research at national agricultural research institutes: SARP's experience. Pages 517–540 in Penning de Vries F W T, Teng P S, Metselaar K (Eds.) Systems Approaches for Agricultural Development, Proceedings of the International Symposium on Systems Approaches for Agricultural Development, 2–6 December 1991, Bangkok, Thailand (this volume).

Uehara G and Tsuji G Y (1992) The IBSNAT project. Pages 507–515 in Penning de Vries F W T, Teng P S, Metselaar K (Eds.) Systems Approaches for Agricultural Development, Proceedings of the International Symposium on Systems Approaches for Agricultural Development, 2–6 December 1991, Bangkok, Thailand (this volume).

Wells H G (1922) A Short History of the World (Revised 1976; Pelican Edition, 1982). Penguin books, Harmondsworth, England.

Williams W T (1967) The computer botanist. Aust. J. Sci. 29:266–270.

Zandstra H G, Price E C, Litsinger J A and Morris R A (1981) A Methodology for On-farm Cropping Systems Research. IRRI, Los Baños, Philippines.

The IBSNAT project

G. UEHARA and G. Y. TSUJI
Department of Agronomy and Soil Science, University of Hawaii,
1910 East-West Road, Honolulu, Hawaii 96822 U.S.A.

Key words: agronomic minimum data set, decision support system, technology transfer, IBSNAT, modeler, research priority, agricultural development, simulation, collaborative research

Abstract
The purpose of the International Benchmark Sites Network for Agrotechnology Transfer (IBSNAT) project is to assemble and distribute a portable, user-friendly, computerized decision support system which enables users to match the biological requirements of crops to the physical characteristics of land to attain objectives specified by the user. The decision support software consists of (i) a data base management system to enter, store and retrieve a minimum set of soil, crop, weather, site and management data to validate and apply the software, (ii) crop models capable of simulating genotype × environment × management interactions, and (iii) application programs that enable a user to analyze and display results of multi-year agronomic experiments conducted in the computer.

The project is based on a systems analysis and simulation approach, and is guided by the premise that it is more convenient to use models to find alternative ways to improve agroecosystem performances than to experiment with the system itself.

The main incentive to participate is the project's capacity to enable the participants to achieve greater research efficiency through better integration of effort. The key to project success has been to foster and transfer leadership to the participating scientists and to sustain an environment of trust that allows them to do what is right and necessary. The IBSNAT experience clearly demonstrates that international cooperation is possible, worthwhile, and necessary to deal with local and global issues of food security, environmental quality and sustainable development.

The IBSNAT project

A systems approach to agricultural development represents a major departure from the way we now perceive and study agricultural systems. This approach, is based on the premise that improvements to one component of a system cannot be assumed to lead to an improvement in the performance of the whole system without an understanding of how system components interact. It makes the bold assumption that it is more convenient to use models to study a system than to conduct experiments on the system itself. But deeply held scepticism (Passioura 1963) about whether biological processes as complex as those found in agricultural systems can be adequately understood and modelled remains a barrier to wider adoption of this approach. The strong, visual appeal of seeing treatment difference in field experiments adds to the reluctance of researchers to abandon an institutionalized and entrenched methodology in favour of a new and uncertain one requiring unrewarding, interdisciplinary team work.

F.W.T. Penning de Vries et al. (eds.), Systems Approaches for Agricultural Development, 505–513.

A small but growing number of people are now convinced that a systems approach is not only desirable, but necessary to deal with the complex problems that arise from interactions between humans and the environment. But the systems approach can make a difference only if it is widely accepted by the scientific community, and it in turn, can assemble products which can be used by a large number of farmers and policy makers. The policies formulated by governments, and practices selected and implemented by millions of farmers within the next few decades will have profound impacts on the quality of life for billions of people throughout the next century. If the premise on which this symposium is based is sound, how do we persuade and mobilize the critical number of scientists, development agencies, government offices and users of land, air, water and biotic resources to adopt and apply systems thinking?

What may be helpful is a compendium of experiences and lessons learned by those who have participated in systems-based programs. With this purpose in mind, this paper attempts to describe the participants and accomplishments of the International Benchmark Sites Network for Agrotechnology Transfer (IBSNAT) project, and the lessons learned from its efforts to establish and implement a systems-based research project.

Establishing a systems-based research program

As in any purpose-oriented endeavour, the quality of people involved in a program determines the quality of its efforts. In the implementation of systems research, competent individuals from many diverse disciplines must work in close, self-imposed coordination to attain a common goal. There are many competent scientists, but not all of them are able or willing to work in a team effort. Part of the reluctance to work in interdisciplinary teams stems from reward systems that favour individual effort over group effort. It is much easier to measure output from an individual research project than to weigh the contributions of each member in a team effort. A common complaint among members of model building research teams, for example, is that the modelers take an unfair share of the credit at the expense of the data generators. This problem continues to plague model building teams and is an issue that remains largely unresolved.

Securing the commitment of internationally recognized scientists is difficult because most of them are already fully occupied and often overworked. Forcing them to join interdisciplinary teams by administrative edict or even by changing the reward systems, rarely produces the desired result. Most who are forced to join interdisciplinary teams manage to do what they have always done previously. Thus, membership into an interdisciplinary team must be voluntary. Individuals who join teams do so to accomplish tasks they know require the combined skills and effort of a large and diversified group. To work effectively, team members must share a common vision of what is wrong with the existing situation and what steps must be taken to rectify it. They must also believe, and be prepared to participate in, interdisciplinary team efforts.

In systems research, each individual is responsible for at least one component of the system under investigation. In such groups individuals interact in much the same way as components interact in the system. Our experience with the IBSNAT project suggests that most research organizations, however large or comprehensive, do not have the number of scientists with the required attributes to field a large interdisciplinary team. While there may be exceptions to the rule, it is easier to form interdisciplinary teams by identifying and inviting outstanding researchers from a large number of institutions than to field a team from a single institution. The scarcity of systems-oriented scientists compels systems researchers to look beyond their own institution to find like-thinking individuals. Worldwide, such individuals form a rich and ready source of talent needed to build interdisciplinary teams.

Networking

Networking is necessary because it is no longer possible for individual scientists or single institutions or even nations however large or wealthy, to deal with complex global issues and problems. Owing to the scarcity of system-oriented scientists willing and able to participate in interdisciplinary efforts, the entire international scientific community must now become the source for scientific talent. The use of interdisciplinary teams is essential because virtually all of the critical environmental problems caused by agricultural practices and policies are systems problems–not disciplinary ones, and the way research institutions are now organized and administered does not lend itself to interdisciplinary efforts. It is, therefore, not likely that individual institutions operating alone can mobilize effective teams to deal with large systems problems. It is much easier for an external development agency to bring these individuals together to form the effective problem solving teams the world now so urgently needs.

The IBSNAT project

The IBSNAT project has often been referred to as a modeling project. Sometimes this has been interpreted to mean that models are developed as ends in themselves. This is unfortunate, but probably unavoidable, since the project publications tend to focus on models. Systems scientists, however, use models as a means to capture, condense and organize knowledge. Models are the means by which knowledge about systems and their performance is made portable and accessible to users whose livelihood and welfare depend on systems performance.

When the IBSNAT project was first established, a number of individuals and groups already had functional, dynamic, process- based crop simulation models ready for global testing. Several within this group believed that field testing of the models, preferably in an international network of benchmark research sites, would expose imperfections and allow the models to be refined. Thus the principal intent of the IBSNAT project was not to develop models, but to enable existing modeling groups to demonstrate the utility of their models by demonstrating their capability to simulate outcomes of alternative crop production strategies anywhere in the world. As products of systems synthesis and analysis, functional models are important only insofar as they enable decision makers to generate the desired information with which to support decision making.

One of the great tragedies of science and technology is that the knowledge necessary to support decision making is not accessible to those who need it most. Farming systems research and development programs have demonstrated repeatedly that accessibility to promising new crops, crop cultivars, products or practices does not guarantee their adoption by client groups. What farmers need in addition to the innovations is convincing evidence that an innovation will improve farm performance in a way specified by them. But conducting on-farm trials to determine the suitability of a particular innovation defeats the purpose of research by reducing it to a trial-and-error exercise.

The principal aim of the IBSNAT project was to enable a broad range of users from farm advisers to policy makers to apply scientific knowledge to assess, in a matter of hours, at no cost or risk to the user, how adoption of a specific technology might affect systems performance. The ex ante analysis would be conducted in response to 'what if' questions posed by the technology adopter. Answers to whole sets of 'what if' questions would be the means by which the appropriateness of an innovation is judged, and the decision to adopt or reject it is ultimately made. Ten years ago the key question was whether answers to 'what if' questions raised by farmers or policy makers might not be generated more efficiently by models than by on-farm or on-station trials.

Evidence for the potential utility of crop simulation models appeared over ten years ago when the United States Agricultural Research Service crop modelling group in Temple, Texas demonstrated that its maize model could predict grain yields of field experiments conducted over several years in Hawaii, Indonesia and the Philippines by researchers of the University of Hawaii Benchmark Soils Project. What was remarkable about the results was the ability of the model to account for, and explain large yield variances associated with differences in climate and rates of nitrogen fertilizer application. What was even more remarkable was the model's ability to predict yields in tropical environments never before visited by the model. But the capacity to make predictions depended not only on accessibility to models, but just as critically, on accessibility to soil and climate data with which to drive the models, and the project purpose was modified to include the number and kinds of data one would need to apply models to answer 'what if' questions. The concept of Minimum Data Set (MDS) for model validation and application became an important guiding principle for IBSNAT from its inception in 1982 to the present.

The Minimum Data Set symposium

To inaugurate the IBSNAT Project, an international symposium on the Minimum Data Set for Agrotechnology Transfer was convened at the International Crop Research Institute for Semi-Arid Tropics (ICRISAT 1984) in Hyderabad, India in March 1983. The main purpose of the meeting was to specify the minimum amount of environmental, crop, and management data a model user would need to validate and apply existing crop models. The concept of a Minimum Data Set (MDS) was borrowed from earlier attempts by Australian scientists to initiate a similar effort. These earlier attempts did not go beyond the conceptual stage for lack of demonstrated utility of crop models and the scarcity of wholly operational crop models (Nix 1984).

The Hyderabad meeting accomplished its two principal objectives. It identified crops for the project to focus and reached agreement on the minimum set of crop, soil, weather and management data with which to simulate growth and development of the recommended crops. The long-term aim of the project was to develop a solid foundation for dealing with the soil-plant- atmosphere continuum so that strong links between the biophysical and socio-economic aspects of farming systems could later be forged. The participants recommended that the scope of work be limited to ten food crops including four cereals (maize, rice, sorghum, and wheat), three grain legumes (dry bean, groundnut, and soybean), and three root crops (aroid, cassava, and potato). Millet and barley were later added to this list. The minimum data set was published as Technical Report No. 1 (IBSNAT 1984) and revised in 1986 (IBSNAT 1986), and again two years later (IBSNAT 1988).

Reaching agreement on the MDS was a major achievement for the project. Fortunately, the state of crop models existing at the time enabled the participants to focus on data needed by the models. By 1983, most crop models were operating on a daily time step and considerable convergence in their data requirements had already occurred. Thus, the MDS was largely predetermined by the models themselves. Even so, it took great restraint on the part of even the modelers to keep the data set at the lowest possible minimum. There was great temptation, for example, to add relative humidity, wind velocity and direction, and pan evaporation to the weather MDS of maximum and minimum air temperature, solar radiation and rainfall. In the beginning what was important was not so much the number and kinds of variables that were included in the MDS, but the simple fact that an MDS, imperfect as it was, existed. That the initial set was flawed is evidenced by the number of revisions it has undergone. The MDS is imperfect now and will always be imperfect because knowledge of the processes scientists are trying to simulate is imperfect. But the simple fact that an MDS existed gave a sense of unity and coherence to a dispersed and decentralized project operating out of more than fifteen countries.

In the final analysis it was the utilitarian goals of the IBSNAT Project that determined the size and nature of the MDS. It was reasoned that a large and complex MDS would merely add to the burden of data acquisition in the developing countries. There was full agreement among all involved that the project would not assemble decision aids, however powerful they might be, which would be rejected by the very clients they were meant for simply because the input data to use them were unavailable or too difficult and costly to obtain.

The participants

A unique feature of the IBSNAT project has been the role played by the participating scientist in the design, management and implementation of project activities. This project quickly evolved into a participatory effort out of practical necessity. Unlike projects that deal with a single crop, or component such as soil, water or climate, IBSNAT, designed as a systems project, was intended to deal with many crops and all relevant biophysical and socioeconomic aspects of agricultural systems. The task was to find suitable individuals to lead sub-programs to make up the whole. The first two participants were selected by the principal investigator, but thereafter, all others were jointly identified and selected by the existing group. This method of inviting new participants into the group had two advantages. First the group as a whole provided a larger base from which to choose competent candidates, and second, the close and intimate working relations required for interdisciplinary work made it imperative that a new member was the choice of the group rather than an individual. One danger of forming research teams in this way is that it tends to bring together individuals who think alike, and run the risk of reaching consensus on a flawed concept. True interdisciplinary teams, however, are composed of individuals with such diverse backgrounds and training that some convergence in thinking, particularly with respect to agreement on project goals, research methodology, and client orientation is not only desirable but necessary. A strong commitment to a client-oriented effort, driven and guided by client-needs, was a key element shared by team members.

A commitment to a client-oriented, problem-solving project has probably been the single characteristic that has united and sustained the team members. They had long realized that the problems faced by clients ranging from farmers to policy makers were not disciplinary problems but systems problems that could only be diagnosed and solved through interdisciplinary efforts. Another characteristic of the participants was their willingness to adjust their on-going programs to accommodate project goals and objective, thereby greatly leveraging the overall efforts. The incentive to operate in this way came from knowledge that the value and quality of the product assembled by the team would far exceed anything a single member working alone could produce. The participants of the IBSNAT project were, therefore, not individuals who needed additional resources to do research, but leaders of on-going research programs who were themselves looking for opportunities to work in a more integrated manner with workers in other disciplines. The IBSNAT project provided them with this opportunity.

Accomplishments

The IBSNAT project focused on the creation of a single integrated product (Jones et al. 1989) called the Decision Support System for Agrotechnology Transfer (DSSAT). DSSAT is a computer software that is designed to enable a user to match the biological requirements of crops to the physical characteristic of land to attain objectives specified by the user. The system consists of (i) a data base management system to enter, store and retrieve the minimum set of soil, crop, weather and management data to validate and apply crop simulation models, (ii) a set of validated crop models to simulate outcomes of genotype by environment by management interactions, and (iii) application programs that enable the user to conduct agronomic experiments in the computer. A single experiment may contain as many as 15 treatment combinations, and may be simulated for up to 50 consecutive years using historical or estimated weather data to assess long-term performances in a way that cannot be done within one or two years of on-farm or on-station trials. Whole probability distribution of outcome of alternative agronomic, economic and environmental strategies are analyzed and displayed in tables and graphs for easy comparison.

Since agriculture deals with a problem domain that is nearly infinite, a DSSAT that attempts to address all problems for all users will not be practical, possible or even desirable. On the other hand, DSSAT should be able to accommodate the range of problems and conditions encountered on a single farm and allow the farmer to assess alternative combinations of management practices for the crops, soils and weather patterns of a particular farm. DSSAT also uses the same minimum set of soil, crop and weather data to evaluate alternative management practices, irrespective of agroecological zone. The MDS also serves the important role of helping to guide policy makers and planning agencies collect and store the proper kinds of data. Lastly, DSSAT can also be used as an effective training aid for university students, agricultural advisers and policy makers.

Lessons learned

One lesson learned is that better integration of effort offers the easiest and most cost-effective way to increase research efficiency at this time. Unfortunately most research institutions are organized and administered in a way that fosters continued reliance on disciplinary research for prestige and scholarly excellence. The reward systems, and the research and publication standards set by disciplinary societies also contribute to perpetuation of the existing situation.

There exists, however, in nearly every institution, pockets of systems-oriented researchers who are ready and eager to join interdisciplinary teams. While they represent a small fraction of the total scientific community, and occur in numbers too small to field interdisciplinary teams within any single institution, they constitute a large, under-utilized resource internationally. These individuals have the following characteristics:

- They are mission- and goal-oriented.
- They are committed to systems-based interdisciplinary research.
- Their research priorities are set by client needs.
- They respond to client needs by producing user-friendly decision aids designed to enable clients to diagnose and solve problems on their own.
- They are product-oriented.
- They are process-oriented and know the value of basic research.
- They tend to think alike and share a common vision of the purpose of research.
- They are eager to form networks that enable them to attain higher goals which are otherwise unattainable.

The IBSNAT experience demonstrates that establishment of multi-disciplinary, international, collaborative research networks composed of individuals with the above characteristics is not only possible and worthwhile, but essential for dealing with systems problems.

Future needs

Decision support systems, however powerful or reliable, serve no useful purpose unless they are used to improve decision making. Although much needs to be done to improve existing models and other decision aids, two other factors prevent their widespread use. First, the minimum data set for soils, crops and weather needed to apply the models and decision aids is not generally available to users. A major task for the future is to compile and organize existing data for each country and region, and to design, plan and implement a global program to fill missing data gaps. Because development of data bases is not as exciting as model development, researchers and policy makers must not make the error of neglecting the former.

The second need is to train a critical number of farm advisers, researchers, educators, administrators and policy makers to use the new decision aids. To make a difference, a critical number of users in every part of the world must choose to replace unsound practices and policies with a new, more productive and environmentally sustainable set. Fortunately the rate of increase in computer literacy remains high even in the developing countries and accessibility to computers is rapidly becoming a non-problem. The time is ripe to develop a global plan to train people to use the knowledge captured in decision aids to make better choices now, to ensure a better future.

Conclusion

While there is widespread agreement that a systems approach based on interdisciplinary effort is needed to address agricultural and environmental issues, large scale implementation of this approach is constrained by the inability of the development community to field the necessary interdisciplinary teams. This situation largely stems from the disciplinary and commodity orientation of most research organizations. In spite of this, most institutions have a small number of researchers who are strongly inclined towards systems-based research. On an international scale, this group represents a large, underutilized resource which can be mobilized to respond to complex tasks that require interdisciplinary action. The IBSNAT experience indicates that the fielding of international interdisciplinary teams is not only possible, but highly productive and cost-effective.

References

International Benchmark Sites Network for Agrotechnology Transfer (1984) Experiment Design and Data Collection Procedure for IBSNAT: The minimum data set for systems analysis and crop simulation. University of Hawaii, Honolulu, USA.

International Benchmark Sites Network for Agrotechnology Transfer (1986) Experimental Design and Data Collection Procedure for IBSNAT: The minimum data set for system analysis and crop simulation. 2nd edition. Technical Report 1. Department of Agronomy and Soil Science, University of Hawaii, Honolulu, USA.

International Benchmark Sites Network for Agrotechnology Transfer (1988) Experimental Design and Data Collection Procedure for IBSNAT: The minimum data set for systems analysis and crop simulation. 3rd edition. Technical Report 1. Department of Agronomy and Soil Science, University of Hawaii, Honolulu, USA.

International Crops Research Institute for the Semi-Arid Tropics (1984) Proceedings of the International Symposium on Agrotechnology Transfer. Kimble V (Ed.) ICRISAT, Patancheru, A.P. 502 324, India.

Jones J W, Jagtap S S, Hoogenboom G, Tsuji G Y (1989) The structure and function of DSSAT. Pages 1–14 in Proceedings of the IBSNAT Symposium on Decision Support System for Agrotechnology Transfer. 81st Annual Meeting of the American Society of Agronomy, Las Vegas, Nevada. Department of Agronomy and Soil Science, University of Hawaii, Honolulu, USA.

Nix H A (1984) Minimum data sets for Agrotechnology Transfer. Pages 181–188 in Kimble V (Ed.) Proceedings of the International Symposium for Agrotechnology Transfer. ICRISAT, Patancheru, India.

Passioura J B (1973) Sense and nonsense in crop simulation. J. Aust. Inst. Agric. Sci. 39:181–183.

Building capacity for systems research at national agricultural research centres: SARP's experience

H.F.M. TEN BERGE
Center for Agrobiological Research (CABO-DLO), P.O. Box 14, 6700 AA Wageningen, The Netherlands

Key words: collaborative research, International Agricultural Research Center, IRRI, National Research Center Agricultural, research priority, research team, SARP, simulation, socio-economic constraints, systems analysis, technology transfer, training course

Abstract
The SARP project (Simulation and Systems Analysis for Rice Production) was established to build research capacity in systems analysis and simulation modelling in developing countries. It has developed into an informal network consisting of national agricultural research centres and universities. Nine countries represented by 16 national agricultural research centres (NARCs) constitute the network, supported by the International Rice Research Institute (The Philippines), the Centre for Agrobiological Research and the Wageningen Agricultural University (The Netherlands).

This paper presents an overview of the main activities and developments during the course of SARP. The first section describes the aims and structure of the project, the participants, the key elements of the training program, and the research topics and themes that have evolved.

The second part discusses the potential role of the systems approach in agricultural research, and indicates how the techniques disseminated by SARP may contribute to the functioning of NARCs in the region. Finally, limitations are identified and tasks for international centres are suggested to help NARCs overcome some of the difficulties experienced.

Overview of the SARP project

Goals and institutional structure

SARP (Simulation and Systems Analysis for Rice Production) is a project of collaboration among a number of agricultural research centres in Asia and in The Netherlands. Its immediate goal is to build research capacity in the field of systems analysis and crop simulation (SAS) at national research centres and universities in Asian countries – in this paper collectively named NARCs – and at the International Rice Research Institute (IRRI) in The Philippines.

The project is founded on the idea that research efficiency and efficacy can be much enhanced by widespread application of modern systems research techniques. This applies both to national and international research systems, and to their mutual linkage. On a longer time scale, SARP's goals are to contribute to the understanding of rice ecosystems and to compile available knowledge, both essential prerequisites for further enhancement of sustainable productivity in rice-based cropping systems.

F.W.T. Penning de Vries et al. (eds.), Systems Approaches for Agricultural Development, 515–538.

SARP was initiated by the DLO–Centre for Agrobiological Research (CABO–DLO) and the Department of Theoretical Production Ecology of the Wageningen Agricultural University (TPE–WAU), in collaboration with IRRI. Currently, most of the activities are coordinated by IRRI. Sixteen NARCs participate in the project, not yet counting the centres that have recently become involved through national training programs.

Financial support was obtained from the Directorate for International Cooperation (DGIS) of the Dutch Ministry of Foreign Affairs. Considerable inputs in the form of research and support staff have been contributed in all phases of the project by each of the participating and organizing institutes.

History

The project started in 1984 with an emphasis on training. At that time, suitable personal computers (IBM PC–AT) had become available at affordable cost. These were the first PCs powerful enough to handle full scale crop models which had, so far, been developed on Mainframe computers. Crop models – and, in more general terms, dynamic simulation and other techniques of systems analysis – had developed over the past decades into practical tools for agricultural research (e.g. Jones and Street 1990; Rabbinge et al. 1990; Penning de Vries et al. 1988; Penning de Vries et al. 1989). Hardware limitations were no longer hampering the widespread application of SAS techniques in agricultural research. However, the shortage of scientists trained in this field remained a bottleneck.

Courseware had earlier been developed for international postgraduate courses, designed to train crop researchers of various disciplines (Penning de Vries and Van Laar 1982; Rabbinge et al. 1989). Such short courses were first given in Wageningen, but later also in a number of other countries including Belgium, Spain, Venezuela, Bulgaria, Finland, and Turkey. For the purpose of SARP, the simulation language CSMP (Continuous System Modelling Program, IBM 1975) was converted to a PC version (PCSMP). This version has since spread to modelling groups throughout the world, including some beyond the direct SARP environment. Simulation modules and other training materials available at TPE–WAU were adapted to suit the particular needs associated with the rice environment (Penning de Vries et al. 1989).

An important lesson taken from earlier short training courses was that results – in terms of effective transfer – were generally poor. This was partly attributed to a lack of continued support and follow-up activities, but also to the relative isolation in which researchers pursued their modelling work after returning to their home base. This appeared to be a general problem, irrespective of country or scientific discipline. Simulation capacity, it seemed, is gained only gradually and through continued effort by groups rather than individuals. In addition, guidance in applying the new techniques to solve practical problems turned out to be an essential element.

Apart from the preparation of courseware, the first phase of the project

(1984–1987) included the training of eight teams of scientists. It was immediately followed in 1987 by a new phase to repeat the international training program twice. This second phase was concluded in 1991 with the international symposium on Systems Approaches for Agricultural Development. In a third and final phase, that started in 1992, coordination activities will be transferred from the Dutch institutes to the NARCs and to IRRI.

The participants

Specific NARCs were invited to participate in the project, on the basis of criteria such as research achievements, opportunity for impact, and the availability of basic data. Of equal importance was the commitment of the institute to allow team members to spend half of their time on simulation and related activities, throughout the one-year formal training program. NARCs were asked to select researchers with field experience and holding an MSc or PhD degree.

SARP staff visited all the NARCs prior to the training, to meet participants, obtain an impression of current research, and to explain the principles of the training program and the associated requirements. Teams were also instructed to prepare basic data sets on the local environment, including weather, crops and soils. These would later be used for exercises and in case studies.

Methods

Methods used to transfer SAS techniques included formal training courses and, at a later stage, collaborative research on topics common to most of the participating institutions.

Training methods

In order to improve the impact of training beyond what had been achieved with other trainees in earlier short courses, SARP featured the following particular characteristics:

- each of the national centres participated by a multidisciplinary team of four researchers, not by individuals; this was to establish a critical mass within each of the centres; such teams included at least one crop physiologist or plant physiologist or agronomist, to ensure a focus on the crop itself; the other team members were specialists from various fields such as soil science, phytopathology, entomology, plant breeding, etc., in different combinations; teams were headed by a team leader, responsible for contacts with other teams and with SARP staff;
- research leaders at the NARCs were involved as team supervisors; they participated only in certain sections of the training program; the supervisor's role was to ensure that topics addressed by his team were sufficiently central to the home institute's workplan, and that the team's needs (time, equip-

ment, fields) were supported by the administration of the institute;
- support was given to the teams during the training program in the form of consulting visits at the team's home base; these visits were continued after completion of the training program;
- tools and training materials, including a PC–AT computer, software, books etc. were donated to each team.
- simulation models were 'open', i.e. they were presented as a starting point or hypothesis to be tested and adapted by users, not as a finished end-product; modelling as an approach to studying real world problems was the central issue, not learning to work with a particular package of truth 'wrapped' elsewhere.

The training program was conducted three times: 1986–1987, 1988–1989, and 1990–1991. A detailed script on the organizational aspects was recently compiled by Jansen (1991). The training consisted of the following components:
- a basic course on simulation and systems analysis; this course extended over eight weeks and included the handling of PCs and basic software (operating system, word processing), principles of dynamic simulation, the simulation language, hands-on use of simulation modules, input preparation (weather, crops, soils), and output interpretation; an important final chapter was the formulation of one or more case studies per team;
- case studies conducted at the team's home institute; the issues addressed were selected by the participants in accordance with ongoing research at their institutes; teams conducted their case studies during eight months, combining simulation with field or laboratory experiments; SARP staff visited each of the teams for about one week during this period;
- a training workshop for teams to present the results of the case studies; these were compiled by Penning de Vries et al. (1987), Van Laar et al. (1989) and Van Laar et al. (1991).

Collaborative research

All NARCs wished to remain involved in SARP activities after completion of the training program. As a natural follow-up on the earlier case studies, plans for continuation of systems research at the NARCs gradually evolved. To ensure that future work would be well embedded within the general research planning at the NARCs, team supervisors from the NARCs participated in drafting new research proposals during the final workshop closing the formal training program (Penning de Vries 1989; Kropff et al. 1991).

Four themes emerged as a framework for research coordination: (i) 'Agroecosystems' dealing with agroecological zonation, and timing of crops and crop sequences; (ii) 'Potential Production', including crop responses to light and temperature, and morphogenesis; (iii) 'Crop and Soil Management', focussing on water and nitrogen uptake in lowland rice, and the dynamics of both in soil and crop; and (iv) 'Crop Protection', with emphasis on damage mechanisms by pests and diseases.

Table 1. Number of scientists that received SARP training, per year, country and institute. Acronyms are explained in Appendix 1.

	1986	1988	1990	Total
Bangladesh (BRRI)	0	4	4	8
China				11
CNRRI		4	3	
ZAU		4		
India				23
GPUAT	4		2	
CRRI	4		2	
TNAU-TNRRI		4	2	
TNAU-WTC			2	
IARI-WTC		2	1	
Indonesia				12
SARIF	4			
BORIF		2	2	
SURIF		2	2	
Malaysia (UPM)	4		2	6
Philippines (UPLB)	4		2	6
Sri Lanka (RARCB)	4		2	6
South Korea (CES-ASI)		4		4
Thailand (KKU)	4		3	7
IRRI	4		4	8
Total	32	26	33	91

Apart from the training workshops mentioned above, other meetings centred around these themes were organized to promote disciplinary interaction among NARCs. Activities regarding each of the four themes were coordinated by a SARP staff member ('theme coordinator').

A research network was initiated as a basis for coordinating the continued research activities. A SARP newsletter was issued regularly to disseminate general information. Participants remained hesitant, however, to use it as a forum to present results and ideas.

Results

Training results

Table 1 gives an overview of the three consecutive batches of NARCs of which a team was trained. The geographic distribution over nine countries of these 16 institutes is depicted in Figure 1. Table 2 lists the disciplines of the scientists. The topics addressed in the case studies are given in Table 3.

While entirely fresh teams joined in the first two training programs, only additional members to existing teams were trained in the third one. This was to reinforce the teams in each of the NARCs, extending team size to six scientists. It also opened the possibility that team members instruct their new colleagues on basic issues of computing, simulation etc., thus allowing a reduction of the

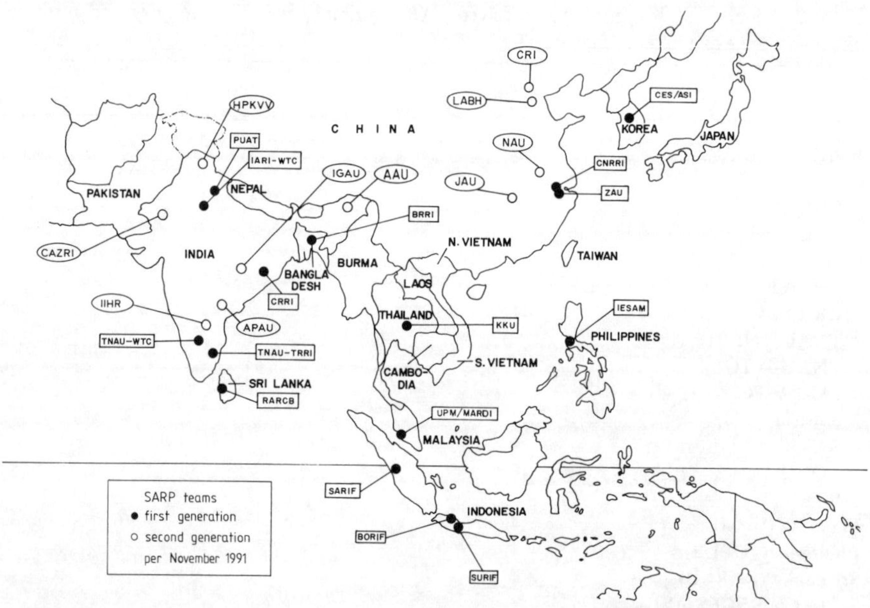

Figure 1. Geographic distribution of NARCs that participated in SARP training programs.

Table 2. Number of scientists that received SARP training, per year, and discipline.

	1986	1988	1990	Total
Agronomy	8	4	5	17
Plant/crop physiology	3	7	4	14
Entomology	5	3	5	13
Plant Breeding	2	4	2	8
Soil Chemistry/fertility	3	4	1	8
Plant pathology	3	1	3	7
Soil physics	2	2	4	8
Statistics/computer	3	–	3	6
(Micro)meteorology	2	1	1	4
Forestry	–	–	1	1
Irrigation/water management	–	–	1	1
Food engineering/aquaculture	–	–	1	1
Weed science	–	–	1	1
Horticulture	–	–	1	1
Nematology	1	–	–	1
Total	32	26	33	91

formal course to six weeks. An additional reason for not adopting new teams was the danger of too thin support, in view of the limited capacity of SARP staff.

IRRI participated with a team of trainees in 1986–1987 only, as a start to build its own capacity.

Table 3. Case studies conducted during training program.

I. Cropping systems	
– rice peanut cropping system	KKU'86
– potential production late planted aman	BRRI'88
– potential production under direct seeding	CES-ASI'88
– potential production of rice and optimization of cropping systems	CNRRI'88
– effects of plant population and planting date on yield	TNAU-TNRRI'88
– potential production in rice-rice-wheat/barley cropping system	ZAU'88
– productivity and stability of rice-wheat rotation in India	IARI-WTC'88
– evaluation of tank irrigation systems	TNAU-WTC'90
– rainfed peanut cultivation in NE Thailand	KKU'90
– potential production of peanut at four sites in the Philippines	UPLB-IESAM'90
– potential production in wheat-Paulownia intercropping	CAF'90
II. Potential production	
– daylength sensitivity and potential production of late japonica rice in East china	ZAU'88
– rice production under low light conditions	CRRI'86
– assimilate partitioning and planting density	UPM'86
– potential production of rice under low temperature conditions	CES-ASI'88
– leaf senescence, leaf nitrogen content and leaf photosynthesis	KKU'90
– potential production of rice different cultivars	GPUAT'86
– simulation potential production	IRRI'86
– potential production	RARCB'86
– potential production different varieties at two locations in Indonesia	SURIF-BORIF'88
– potential production of early rice varieties in different regions in south china	CNRRI'88
– potential production of short duration varieties at Coimbatore	TNAU-TNRRI'88
– potential production of six rice varieties in West Java	BORIF-SURIF'90
– potential production of IR50	TNAU-TNRRI'90
– potential production of early maturing japonica rice	ZAU'90
– potential production of rice variety KDML105	KKU'90
III. Crop and soil management	
– effect of water shortage in lowland rice	CRRI'90
– soil water balance in rainfed rice	TNAU-TNRRI'88
– effects of soil physical properties on root growth	SURIF-BORIF'88
– wheat growth under waterlogged conditions	ZAU'88
– effect of impeded drainage on root growth and yield in rice	TNAU-WTC'90
– water limited production lowland rice	UPLB'86,IRRI'86
– water limited production upland rice	IRRI'86
– water balance and crop productivity in rainfed sorjan systems	BRRI'90
– water uptake distribution as a function of root distribution	IARI-WTC'90
– effect on nitrogen on rice production	CRRI'90, CNRRI'88, GPUAT'86,BRRI'90
– determination of optimum nitrogen application in different rice varieties and soils	CES-ASI'88
– nitrogen requirement and recovery on two soils in Java	BORIF-SURIF'90
IV. Crop protection (all in rice)	
– damage by bacterial leaf blight	CRRI'86, GPUAT'86 TNAU-TNRRI'90
– damage by sheath blight	GPUAT'90,RARCB'90
– epidemiology of leaf blast	SARIF'86

Table 3. Continued.

– leaf blast damage	CNRRI'90
– damage by sheath rot and ufra	BRRI,'90
– stemborer damage	IRRI'86,ZAU'88,GPUAT'90
– leaffolder damage	TNAU-TNRRI'88, CNRRI'90
– Malayan blackbug damage	UPM-MARDI'90
– damage by weeds in directseeded rice	UPM-MARDI'90
– effects of frequency and timing of weeding	UPLB'86

Table 4. Topics of basic and adaptive research currently addressed by the NARCs participating in SARP.

Potential production,	
A. Irrigated rice	
sink formation (tillering)	UPM
late grain filling	UPM
maintenance respiration	UPM, CES
photosynthesis: varietal differences	CES
low temperature effects	CES
low light tolerance and stem reserves	CRRI
parameterization varieties	CNRRI, SURIF
phenology leaf area	CNRRI
ideotypes new vars	CNRRI
ratooning	CNRRI
photoperiod sensitivity	TNRRI
B. Irrigated upland crops	
sink limitation wheat	IARI-WTC
C-partitioning wheat	IARI-WTC
senescene wheat	IARI-WTC
varietal characterization soybean	BORIF/SURIF
Crop and soil management	
A. Irrigated rice	
response to nitrogen and uptake efficiency	ZAU, BORIF, CRRI, GPUAT
timing of nitrogen appilication (relation to temperature)	CES
rooting	CES
balance of nitrogen to other nutrients	CNRRI
groundwater pollution by nitrogen	TNAU-WTC
B. Irrigated upland crops	
nitrogen uptake in wheat	IARI-WTC
C. Rainfed lowland rice	
drought response	CRRI, BORIF/SURIF, TNRRI
interaction nitrogen-water	CRRI, BRRI
response to nitrogen	KKU, BORIF/SURIF, CRRI, BRRI, TNRRI
iron toxicity	RARCB

Table 4. Continued.

D. Rainfed upland crops	
response to nitrogen in wheat, barley	ZAU, IARI-WTC, BRRI
nitrogen from green manures	ZAU
drought response wheat	IARI-WTC, BRRI, CAS
drought response maize	UPM
drought response peanut and boybean	KKU
waterlogging wheat	ZAU
waterlogging pulses	KKU
varietal characterization soybean	BORIF/SURIF
phosphorus uptake in various crops	BORIF/SURIF, UPLB
Crop protection	
A. Irrigated rice	
competition with weeds	UPM, TNRRI
blast epidemiology vs micro-climate	ZAU
damage relations stemborer	UPM, CNRRI
damage relations leaf folder	UPM, CNRRI, TNRRI
damage relations blast	CRRI, CNRRI, TNRRI
damage relations stemborer	ZAU, BORIF/SURIF SURIF, CNRRI, GPUAT,
damage relations sheath blight	GPUAT
damage relations bacterial leaf blight	GPUAT, TNRRI
damage relations gall midge	TNRRI
damage relations brown plant hopper	TNRRI
B. Irrigated upland crops	
various damage relations in wheat	IARI-WTC
C. Rainfed lowland rice	
competition with weeds	TNRRI
blast damage relations	CRRI
damage relations stemborer	KKU, BORIF/SURIF, RARCB
damage relations rice tungro, blast	CRRI
damage relations ufra, sheath rot	BRRI

Results of collaborative research

Presently, all except two NARCs maintain the structure and function of their teams, and team members are actively applying crop modelling in their work. Collaborative research became a method as well as a result of the project. Two things can be inferred from the NARCs' continued interest in SARP: (i) a longer period of close cooperation than the one-year training program is required to fully implement simulation skills; (ii) even while no substantial material compensation was offered, there is a scope for research collaboration with NARCs on systems analysis and simulation.

If activities are not sufficiently focussed, research progress is slow and interaction is ineffective. A selection of topics for continued SARP support was made, therefore, in consultation with the teams. Criteria were that topics

Table 5. Topics of strategic and applied research currently addressed by the NARCs participating in SARP.

Regional zonation studies for potential production and/or water-limited production; rice and other crops	UPLB, UPM, IARI-WTC, BRRI, KKU, GPUAT, TNRRI, RARCB, BORIF, SURIF
Optimization cropping systems including identification of alternative crops; long term stability; plant spacing; crop timing; sorjan; nitrogen and drought contingency; tank irrigation	BRRI, TNRRI, ZAU, BORIF/SURIF, CES, UPLB, CAS, IARI-WTC, TNAU-WTC
Assessment of effects of global climate change	UPM, CES, IARI-WTC, UPLB
Screening of new varieties and design of ideotypes	TNRRI, RARCB, CNRRI

addressed in a network context should be relevant to more than only a few NARCs, and that the application of systems analysis techniques to these subjects can be expected to significantly enhance research output. The subjects currently addressed by the teams are listed in Tables 4 and 5. These cover a wide spectrum from basic and strategic to applied and adaptive research topics.

Forty research papers presented at thematic workshops have been compiled in a book prepared at the occasion of the SAAD symposium (Penning de Vries et al. 1991).

'Second generation' teams

Some of the NARCs have proceeded to organize their own national training programs for sister institutes. The team of Zhejiang Agricultural University (China) gave a course in 1989, adding four more institutes; the Indian Council for Agricultural Research initiated six new teams in 1991, five of them at NARCs that had not been involved so far. Figure 1 includes these 'second generation' NARCs as well.

To support such national initiatives, training materials were made available. The national courses were based on the blueprint of the international SARP courses (Jansen 1991). Although SARP staff was invited to lecture on specified topics, both the Indian and Chinese courses were organized and conducted chiefly by former trainees. Additional coursewares such as self-instructive simulation packages and slide-tape modules (IRRI 1991) are being developed and disseminated for this particular purpose. Three textbooks on simulation have been translated into Chinese. The Chinese groups also started issuing their own newsletter in Chinese for communication among sister institutions.

External review

An external review team assessed the SARP project and its results (Mc William et al. 1990). In brief, the recommendations are that in a follow-up:

- emphasis should be on consolidating the SAS research capacity at the NARCs initiated in the earlier phases of the project; the network should not be further expanded;
- teams should be supported particularly in demonstrating the usefulness of crop simulation modeling at their home institutes;
- a better linkage with IRRI's research program should be developed, based on effective use of simulation modelling;
- collaborative research projects involving IRRI scientists and other SARP participants should address high-priority factors affecting productivity of crops in rainfed environments; in this context, models should be further developed.

Wider perspective

Introduction

SARP builds on the viewpoint that agricultural research by the national and international centres can be much improved by a more extensive application of systems research methods. For several reasons it is not possible to unambiguously prove that this view is correct. Firstly, the claim is not precise: which activities pertain to the domain of systems research, and which do not? Secondly, such an attempt would require the differential assessment (ex ante, ex post) of institutional performance (research quality, research impact). Even where changes in NARS and their outputs can be documented, identifying the causes of change and the contributions of different agencies and programs to performance is difficult and no methods are available to do this (Horton 1991).

Nevertheless, a serious attempt should be made to demonstrate the potential that the systems approach holds particularly for the NARCs. National and international research systems have been distinguished (e.g. Horton 1990) in terms of their principal outputs and impacts. Research and development (R&D) technology is the principal output of international centres, whose impact should be evaluated at the institutional level (NARC's). Production technology, in contrast, is the output of national systems and its impact is to be assessed in the production environment. This dichotomy is not practical in developing products and technology associated with systems research. Although such products can with little doubt be identified as R&D instruments, an active role of NARCs in co-developing the technology itself is required, as a complement to international activities. This opens new ways, but also leads to new problems at the level of NARCs. In the following sections these issues will be discussed, using the SARP experience as an example.

Systems research in agriculture

What is systems research?

'The systems approach is a way of doing things and thinking about things. It can be applied to many different subjects, and its characteristics tend to include methods and techniques' (Spedding 1979). Such phrases point at the difficulty of identifying what is so particular about the systems view. Yet, a number of characteristic concepts can be listed.

Central is the idea that one must identify, delineate and understand a system in order to be able to influence it in a predictable manner (Spedding 1990). An important consequence is that one has to deal with system boundaries rather than disciplinary boundaries. In our abstracted view, such delimited parts of the real world convert external stimuli into responses or behaviour. The systems researcher tries to understand this behaviour as the result of a set of processes, which are to be made explicit in the form of relations between system variables. Among them are measurable, inherent properties or system characteristics. Computer programs can assist this understanding, especially when dynamic behaviour is studied.

Agricultural systems

Agricultural systems are governed by two 'layers' of processes: (i) biological, chemical and physical processes, that can be described in terms of causal relations between variables, and (ii) a group of decision processes leading to management actions (Knol et al. 1987). The latter are then imposed on the reduced system, i.e. that part of the system governed by the ecophysiological processes (i). Human behaviour, playing a prominent role in the second 'layer', is generally so complex that it should be normative rather than be dynamically modelled (Penning de Vries 1990; Van Dusseldorp 1991). Although capacity building in the SARP project has focussed on the first group of processes, it must be mentioned that much efforts are currently made to forge systematic links between the above two layers (e.g. Van Keulen 1990, 1992; Veeneklaas 1990).

Production systems are often called 'agricultural ecosystems' or 'agroecosystems' to underline that certain governing ecological principles were retained when the natural ecosystem was transformed by the advent of cultivation, control, subsidy, harvest, etc. (Conway 1990). Systems research at this broader level distinguishes such characteristics as productivity, stability, sustainability and equitability among human beneficiaries (Conway 1987).

Tools and methods

Systems research has now grown into many branches. In each of them, research tools have been developed along the way. Together they constitute a 'systems tool-box' with pieces of software, such as dynamic simulation languages, simulation models, data bases, calibration packages, geographic information systems, linear optimization programs, expert systems, etc.; and the associated hardware. Some of the tools are used to gain understanding of underlying processes, others to synthesize information towards higher aggregation levels and study the behaviour of these larger systems' descriptions. Most of the software tools consist of a (scientific) knowledge component and a framework to carry that information. For the different instruments, these two take different forms.

Less visible but certainly no less important is another element of systems research: the way of looking at problems, the mutual reinforcement of inventory, analysis and synthesis, the awareness that practical problems can hardly be solved in disciplinary isolation, and hands-on skills to sensibly apply the available technology. This 'attitude' is an essential complement to the scientific information about systems, the software to manipulate it, and the hardware tools.

Systems Analysis and Simulation (SAS)

The SARP project tries to establish a firm foundation in systems analysis and simulation first. Systems analysis is a method of investigating systems in terms of particular types of variables (rates, states, inherent system properties, boundary conditions, etc.) and their mutual relations; simulation is a technique to assemble the collected information into dynamic models (abstractions) of the real system. Important is that only essential information is retained, irrelevant details are discarded, and research is prevented from drifting into a quest for unnecessary understanding.

Along with the skills to build and use simulation models, participants develop a new, process oriented view on research. Model building and application enhance systems thinking, and are intricately linked with this gradual process. For most SARP teams, interests are at the crop and cropping systems level, with photosynthesis, respiration, growth and development as the principal processes, closely followed by groups of processes responsible for biotic and abiotic stresses (Table 4).

Modelling and experimentation in the four research strata

In the remainder of this contribution, the term 'systems research' will refer to the development and application of SAS techniques as defined above. Their potential benefits at different research strata can now be inspected. The following levels are distinguished (CGIAR 1981): *basic research* to generate new understanding; *strategic research* for the solution of specific research problems; *applied research* to create new technology; and *adaptive research* to adjust technology to specific needs of a particular set of environmental conditions. SARP activities include applications of SAS techniques in each of these four strata.

In basic research, the role of the systems approach is to summarize, delineate and express existing knowledge in quantitative terms. The resulting explicit hypotheses enable a sharper research focus, leading to better experimentation and an enhanced exchange of insights. Thus, models become a 'vehicle' to enhance thinking. Within SARP, basic research concerns the crop and crop organ level: effects of diseases on photosynthesis, of temperature on nutrient uptake, of plant spacing on assimilate partitioning, of drought and waterlogging on transpiration, etc. Examples in this category are listed in Tables 3 and 4.

For research of strategic character, experimentation is often impractical. Four SARP teams are involved with IRRI in a project funded by USA's Evironmental Protection Agency to investigate the likely effects of climate change on rice production. Agroecological zonation, too, receives much attention from the SARP teams. Table 5 includes the topics of strategic research addressed in the project.

The design of ideotypes suited to a particular environment with the help of models is an example of applied research. Several SARP teams are investigating alternative cropping systems with the help of models. They are listed in Table 5. Some aspects of cropping systems, such as the agronomic consequences of direct seeding in rice, are addressed by teams listed in Table 4.

Adaptive research is about extrapolating experiences from one environment to another. Here, the main forte of modelling is in identifying the scope for transferring production technology to other environments. This was early recognized by the IBSNAT modelling group, founded upon the Benchmark Soils Project, which explicitly aimed at supporting the transfer of production technology among different sites (Plucknett et al. 1990). When used to narrow down the likely causes of a particular problem associated with a given environment, modelling for adaptive purposes resembles its role in basic research. For the SARP teams listed in Table 4, adaptive and basic research can therefore often not clearly be separated.

On all topics summarized in Table 4, teams merged simulation with experimental work. In basic and adaptive research, modelling cannot replace experimentation; it does change, however, the role of experimentation into one of validating assumptions and collecting a new kind of information. This often has a more general validity and therefore greater 'transferability' than empirical information about production technologies (ISNAR 1984).

In strategic and applied research, models are viewed as accepted knowledge and their role is more explorative, assessing potentials of crop types, environments, impacts of changes, etc. Here, cautious application of models can replace experimental work, or fill the gap where experiments are not possible.

Across the whole spectrum of basic, strategic, applied and adaptive research, a systems view increases the consistency among research components.

What SAS can do for NARCs

General tasks of NARCs

Potential applications of SAS can also be reviewed in the light of the national centres' specific tasks. It has been argued (ISNAR 1984) that national public research institutes, for their successful operation, need to ensure the mutual reinforcement of three groups of variables : (i) those that determine the policy environment; (ii) a set of operational processes common to all research systems; (iii) organizational structure. Although processes in groups (i) and (iii) will be affected in the long run by the spreading of SAS techniques, the discussion will be limited to group (ii), because the SARP project aims primarily at these processes. One aspect of (iii), however, is also of direct relevance and will be discussed below: the ensuing need for multidisciplinary action.

ISNAR (1984) expresses the operational processes in terms of seven activities:

- directing activities towards the country's priorities;
- developing and maintaining a physical infrastructure that responds to agroecological characteristics and economic potential;
- mobilizing and effectively utilizing financial resources;
- developing and maintaining a critical mass of qualified scientific personnel;
- taking advantage of scientific capabilities at national and international levels;
- assuring the flow of information between research and extension workers, farmers, policy makers, and the public;
- monitoring and evaluating program implementation.

Priorities and financial resources

Agroecological zonation is now widely seen as an activity, essential for the identification of research priorities, regional allocation of resources, and for tailor-made introduction of new production technologies. According to Oram (1990), "decentralization of research regionally and to the farm level, based on agroecological characterization and on an interdisciplinary systems approach, offers the most effective solution to diagnosis of the agricultural potential and provides the essential feed-in and feedback mechanism to upstream researchers and to policymakers". IRRI has recently reorganized its research structure on the basis of a "rice ecology" classification (IRRI 1989). Systems research plays a key role in zonation studies. Simulation models serve to functionally integrate different types of data (weather, soil, crops), and are a key instrument to assess yield potentials for various crops and input levels. These assessments, in turn, are the starting point for yield gap analysis in the adaptive research stratum. SAS induces the development of structured and accessible databases, and points at gaps in essential information. Global radiation and groundwater depth are obvious examples of such gaps, as experienced in SARP. While radiation is important for any agroecological study, groundwater depth is of particular relevance to rainfed rice potentials.

Virtually all SARP teams (Table 5) are engaged in agroecological zonation with the help of crop models and databases. The scope of SAS techniques can be further enhanced by coupling simulation models to geographic information systems. Whereas zonation is a strategic activity that helps in regional priority setting, SAS in its role of focussing process research is an instrument to identify priorities also in basic and adaptive research.

SARP does not explicitly address the effective use of financial resources available at the level of NARCs, nor does it contribute any significant funding to NARCs. In view of the high growth rate – relative to real research expenses – of scientific personnel employed in public agricultural research (Pardey and Roseboom 1990), increased research efficiency is likely to hold more promise than a mere quantitative expansion of research capacity at the NARCs. Increasing this efficiency through the introduction of SAS techniques is the main aim of the SARP project.

Development and mobilization of scientific capacity
Scientific skills are sharpened by SAS because it forces the explicit and quantitative formulation of assumptions. One consequence of this systematic compilation of scientific information in models is that models are efficient tools for instruction and communication. Models are also an easy starting point for self-instructive training materials, emphasizing the basic biological and physical sciences that are essential to solve the complex environmental problems agriculture is facing. As set forth in the first half of this paper, training of scientific personnel is a major component of the SARP project.

The use of models facilitates the mobilization of scientific capacities available elsewhere, as a consequence of the uniformity in 'scientific language'. Models with a clear modular structure allow a quick exchange of ideas and theories developed by the various participants in a collaboration scheme. In SARP, such exchange has been effectuated by combining the experience in crop modelling gained at the Dutch institutes with the broad base of knowledge on rice production available at NARCs and at IRRI. All participants can potentially benefit from such collaboration, but the various roles of each of the institutions has to be given further thought.

Information flow
The impact of the systems approach will be felt mainly by those directly involved in research, either in the role of scientist or research manager. Research managers – and to some extent policy makers – have been involved in SARP workshops on research planning, and also in the SAAD symposium. Thus it was tried to incorporate modelling in five-year workplans of institutes. This has been successful at many of the NARCs, and notably also at IRRI. Some of the products through which the NARCs communicate with the outside world – policy makers, extension service, farmers – can now be developed with less effort and more tailored to the user's demands. Zonation maps are one such example, directed chiefly at policy makers. Optimization schemes for regional planning are another example (see Van Keulen and Veeneklaas, 1922). Data bases and advisory systems will, in the long run, support extension services and maybe farmers. Computer-based advisory systems represent concise and versatile expertise. Whereas the SARP project made no direct attempts to affect information flow at the various in- and outlets of the national systems, many of

its participants have explicit tasks in planning, extension and education; spin-off in these directions is therefore expected. Much of the required software is still to be developed, by the NARCs themselves and by their collaborators.

Organizational aspects at NARCs

The promotion of systems research may have consequences for institutional organization. In a number of NARCs participating in SARP, the multidisciplinary teams work effectively across the department boundaries. Researchers originally trained in the fields of soil science (soil fertility, soil physics) and crop protection (entomology, phytopathology) are now actively paying attention to crop processes, both in their thinking and in experimental work. Experiments on soil conditions and pests are conducted in close collaboration with crop physiologists or agronomists. Equipment is frequently exchanged among departments, scientists assist each other, and departments become more flexible in sharing 'their' experimental fields. It is too early to assess whether this trend leads to a more ecosystem-oriented organization.

Other direct consequences follow for the organization of NARCs. The new approach implies a shift from the empirical testing of a given production technology (new varieties, soil management, pest management, etc.) at many locations and over several years, to a more detailed investigation of causal relationships that determine the success of the particular technology. (One example in SARP is the testing of new varieties at Tamil Nadu's TNRRI: Palanisamy et al. 1992) This can lead to a gross reduction in the number of field trials. It is not known to the author whether the number of 'traditional' trials has already been reduced at any of the participating NARCs, as a result of SARP.

Participants have much commented on the other side of the coin: increased emphasis on causal relations has led to higher costs per experiment, both in terms of material and labour investment. This is partly the result of the need to control or quantify experimental conditions. The constant flow of new genotypes – traditionally screened in multi-year-site trials – now calls for continued efforts in varietal parameterization. New demands are also made upon field assistants; and often more chemical or physical analyses of plant and soil materials have to be performed. Studying a crop as a dynamic system also implies that some observations are to be repeated at different development stages. These new demands call for renewed consideration of the various tasks attributed to national, international and upstream research systems, and their modes of interaction.

The role of International Agricultural Research Centers (IARCs) and networks

Development, transfer and application of SAS

To alleviate these new pressures, what help should NARCs expect from the international research community? To answer this question, it must be assessed how each of the elements of the systems approach is best developed and applied, and which of these elements are transferable.

No real problems, other than cost, are associated with the hardware component. All required items are commercially available in most countries, and international research systems have no specific roles to play other than, possibly, providing training courses to teach the operation of equipment. Such international courses are available, for example at the Asian Institute of Technology in Bangkok. To a large extent the same is true for software packages designed to carry or manipulate information (simulation languages, data base management systems, including those for geographic information), and optimization tools. They can be purchased and the technical operation of these packages can be learned from available courseware. It is not true, in contrast, for simulation models.

Attitude ('systems thinking') and the body of knowledge about real agricultural systems can less easily be transferred. This problem is particularly apparent in the case of simulation models for adaptive research. For this purpose, models must be adapted or developed locally, not be fully developed elsewhere for subsequent dissemination. This requires the local generation and modelling of scientific knowledge on processes underlying crop production. Application of models by 'operators', rather than scientists aware of assumptions and associated limitations, constitutes a real threat to research quality at the NARCs. Our current understanding of systems is often not good enough, and generalized models lack the specificity required to study and solve particular local problems. (For strategic applications – e.g. the use of potential production models in zonation studies – this is different because the general principles are sufficiently understood.)

Thus, a paradox seems to arise. Faced with resource limitations and with the need to quickly answer practical questions, NARCs should also investigate knowledge gaps in terms of underlying processes with relevance to practical problems.

International coordination

Concerted international efforts are a possible way out. These should be focussed on sufficiently common problems, and can be conducted within a network structure or in the form of research consortia. Parallels can be found in other fields of agricultural research. Biotechnology provides useful tools to meet national demands, but some form of international support is required for local adaptation and application. Persley (1990) highlighted the scope and associated requirements for cooperation in this field, in terms of research organization.

Modelling, as it builds on basic biological and physical principles valid over a wide range of species and environmental conditions, serves as an international 'language'. Modelling in a network format can be a basis for collaborative research among NARCs; and between NARCs, IARCs and other centres. SARP teams have made the first steps during the past two years, notably in fertilizer management and crop protection. This has resulted in the joint planning of experiments, the exchange and comparison of data and model sections, and a dividing of tasks among research groups.

IARCs should support such trends by developing and maintaining organizational structures for international collaboration, and by providing training facilities to effectively transfer SAS skills. They should also take up those aspects of process research that require expensive experimentation or highly specialized personnel. The results of such work must be presented in the form of relations that can further be validated in the various participating countries. One obvious task for IARCs in this direction is the characterization of released germplasm in terms of physiological parameters. Not only does this require particular equipment often not available at NARCs, repeating such assessments of inherent varietal properties in the various countries is also a waste of human and material resources.

There is another issue where some kind of centralization would be helpful. Current simulation models are often presented as 'knowledge packages', not as frameworks to be loaded with information by the user. Apart from the commercially available simulation languages proper, the model structure itself is a carrier of scientific information. Such structures should be developed by specialized groups as user friendly frameworks, that can be 'plugged' with particular pieces of information. This enables a quick exchange and evaluation of hypotheses and established knowledge, and allows easy linkage with crop, soil and weather datasets.

While forms of interaction and support are still developing in SARP, it is useful to consider the prerequisites for successful cooperation in networks. According to Plucknett et al. (1990) and Smith et al. (1990), there are many such requirements. They list: identification of a widely shared problem; self-interest to motivate participants; involvement of participants in planning and management of the network; clear definition of the problem or focus of the network; a baseline study to produce an authoritative founding document; a realistic research agenda; flexible research and management; constant infusion of new ideas and technologies; regular workshops to provide opportunities to assess progress and discuss issues; collaborators to contribute resources; outside funding to facilitate travel, training and meetings; collaborators with sufficient training and expertise to contribute effectively; relatively stable network membership; efficient and enlightened leadership. Several of these requirements were met, albeit to varying extent, in the phases of SARP now completed; others deserve more attention in future work. These are listed below, along with other bottlenecks experienced by the various parties.

Prospects and limitations

There is no central scientific problem common to all participants. The project's chief aim remains the promotion of the research methodology, not the study of a particular scientific issue. Nevertheless, it is necessary to reduce the number of problems under study in order to make sufficient impact. Consolidation of teams and full acceptation of the approach by the NARCs can only be reached through clear demonstrations of the potentials of SAS in solving actual problems. In the next phase, therefore, SARP will focus more than before on a few elements of common interest. In each of the four existing themes, two types of activities are planned: process research on the most relevant gaps in current understanding of production systems, and the development of tools that can find direct application in executing the NARCs' specific tasks. These tools can take different forms, in view of the diversity of tasks listed above. Such application products will be developed as much as possible by the NARCs themselves.

Among the problems voiced by the participating scientists during the past phase were scepsis and limited recognition by colleagues not involved in modelling, and the difficulty to allocate time to simulation work. The production of tangible outputs is the most valid answer to these problems. If outputs are well chosen, they will help provide quick answers to practical questions raised by extension services. Interference of these pressing demands with modelling work, another complaint often heard at this stage among participants, will thus partly disappear in the long run.

To ensure full involvement of participants, both the central research themes and the desired application outputs are to be identified and indicated by the NARCs. A founding workshop for this purpose will be held as the start of the next phase (1992). Yearly workshops on each of the themes are planned, to increase interaction among NARCs, improve research planning, and adjust outputs developed in the course of the project.

A lack of specific experimental equipment, and difficulties with repair and maintenance of both experimental and computing equipment, have been mentioned as bottlenecks, too. Limited funding will be made available in the next phase to cover some of the costs made at NARCs in direct relation with the project. The principle remains, however, that offering considerable compensations would obscure the approach's viability, at the level of NARCs, beyond the current project.

The comment most repeated by participants is that skills in handling the SAS tools are still limited. More project staff will therefore be involved to ensure closer guidance through visits and workshops, and no fresh teams will be included to avoid reduction of support.

The SARP network is informal and the stability of teams is difficult to judge. No formal commitments are made with the NARCs on the production of outputs, nor on the maintenance of a team. Although team stability has not yet become an obvious problem, it is noticed at some of the centres that a team of 4–6 individuals is small. In particular, as many team members are relatively young and look for professional opportunities, more fellowships for short and long term visits abroad will be offered. This includes a number of PhD fellowships.

From a project management viewpoint, SARP faces other challenges, too.

Communication among all participants is a logistic problem in every network. Mail is often lost. Recent developments opened new ways for electronic mail through PC's, without the need for mainframes at either the sending or receiving station. The development of an electronic network will have a high priority.

While team visits by SARP staff so far aimed at solving problems in programming and computing, this task may be alleviated by electronic mail. Visits can then be devoted more to joint planning and interpretation of results. One point not yet really solved is how to effectively support teams, all of which merge various disciplines, with the help of contact persons (visiting SARP staff) who do not fully cover all the disciplines simultaneously?

Another difficulty, in some cases, is competition from other agencies and programs offering substantial financial compensations. In the end, of course, it is the NARCs that have to choose from realistic opportunities.

In the near future many practical bottlenecks will have to be resolved. It is expected that the development of central (international) services as outlined in the previous section provides substantial help, but this will take time.

The last phase of the SARP project includes the transfer of coordination activities to a number of national centres and to IRRI. A sustained collaboration between NARCs, IARCs and specialized centres is envisaged after this project, leading to effective application of systems techniques in solving the wide spectrum of problems threatening sustainable development. No good examples are available to show how a transfer of network coordination can best be implemented. Neither is clear whether some of the NARCs will really be interested to coordinate thematic activities throughout the network as it exists now. Within the next few years, the SARP project nevertheless anticipates the transfer of theme coordinators' activities to team members and/or IRRI staff. During 18 months, each of the four research themes will be guided by two coordinators, one SARP staff and one member of a national centre. It is expected that, after this period, NARCs will completely take over the full responsibility of thematic coordination. Practical and formal problems related to institutional responsibilities will have to be solved in this process.

In a possible future follow-up of SARP, techniques designed to go beyond simulation and crop level studies are to be disseminated, too. The limited mandates of NARCs, and the lack of expertise – at commodity oriented institutes – on other commodities and non-agronomic research fields essential to farming and rural development, may emerge as new bottlenecks.

Appendix

Appendix 1. List of acronyms.

AAU	Assam Agricultural University (*Jorhat, India*)
APAU	Andhra Pradesh Agricultural University (*Hyderabad, India*)
BRRI	Bangladesh Rice Research Institute (*Joydebpur, Bangladesh*)

CABO-DLO	DLO-Center for Agrobiological Research (*Wageningen, The Netherlands*)
CAZRI	Central Arid Zone Research Institute (*Jodhpur, India*)
CES	Crop Experimental Station (*Suweon, South Korea*)
CNRRI	China National Rice Research Institute (*Hangzhou, China*)
CRI	Cotton Research Institute (*Anyang, China*)
CRIFC-BORIF	Central Research Institute for Food Crops, Bogor Research Institute for Food Crops (*Bogor, Indonesia*)
CRIFC-SURIF	Central Research Institute for Food Crops, Sukamandi Research Institute for Food Crops (*Sukamandi, Indonesia*)
CRRI	Central Rice Research Institute (*Cuttack, India*)
DGIS	Directorate General for International Cooperation (*The Hague, The Netherlands*)
DLO	Agricultural Research Department, Ministry of Agriculture, Nature Conservation and Fisheries (*Wageningen, The Netherlands*)
EPA	Environmental Protection Agency (*USA*)
GIS	Geographic Information System
GPUAT	Pantnagar University of Agriculture and Technology (*Pantnagar, India*)
HPKVV	Himachal Pradesh Krishi Vishwa Vidyalaya (*Palampur, India*)
IARC	International Agricultural Research Centre
IARI-WTC	Indian Agricultural Research Institute, Water Technology Centre *(New Delhi, India)*
IBSNAT	International Benchmark Sites Network for Agrotechnology Transfer
IGAU	Indira Gandhi Agricultural University (*Raipur, India*)
IIHR	Indian Institute for Horticultural Research (*Bangalore, India*)
IRRI	International Rice Research Institute (*Los Baños, Philippines*)
ISNAR	International Service for National Agricultural Research (*The Hague, The Netherlands*)
JAU	Jiangxi Agricultural University (*China*)
KKU	Khon Kaen University (*Khon Kaen, Thailand*)
LAB	Langfang Agricultural Bureau (*China*)
MARDI	Malaysian Agricultural Research and Development Institute (*Serdang, Malaysia*)
NARC	national agricultural research centre
NARS	national agricultural research system
NAU	Nanjing Agricultural University (*Nanjing, China*)
RARCB	Regional Agricultural Research Center (*Bombuwela, Sri Lanka*)
SAAD	Systems Approaches for Agricultural Development
SARP	Simulation and systems Analysis for Rice Production
SAS	simulation and systems analysis
TNAU-TNRRI	Tamil Nadu Agricultural University-Tamil Nadu Rice Research Institute (*Aduthurai, India*)
TNAU-WTC	Tamil Nadu Agricultural University- Water Technology Center (*Coimbatore, India*)
TPE	Department of Theoretical Production Ecology (WAU) (*Wageningen, The Netherlands*)
UPM	Universiti Pertanian Malaysia (*Serdang, Malaysia*)
UPLB	University of the Philippines in Los Baños (*Los Baños, Philippines*)
WAU	Wageningen Agricultural University (*Wageningen, The Netherlands*)
ZAU	Zhejiang Agricultural University (*Hangzhou, China*)

References

Consultative Group of the International Agricultural Research (1981) Second review of the Consultative Group on International Agricultural Research, CGIAR, Washington.

Conway G R (1987) The properties of agroecosystems. Agric. Syst. 24:95–117.

Conway G R (1990) Agroecosystems. In Jones J G W, Street P R (Eds.) Systems theory applied to agriculture and the food chain. Elsevier, New York, USA.

Horton D E (1990) Assessing the impact of international research: concepts and challenges. In Methods for diagnosing research system constraints and assessing the impact of agricultural research, Proceedings ISNAR/Rutgers Workshop on Agricultural Technology Management, 6–8 July 1988, Rutgers University, New Jersey, USA.

Forrester J W (1970) Collected Papers. Wright-Allen Press, New York, USA.

Horton D E (1991) Two institutional aspects of impact assessment: evaluation of and in national agricultural research systems. Paper presented at the workshop on Assessment of International Agricultural Research Impact, Cornell International Institute for Food, Agriculture and Development, Ithaca, June 16–19. New York, USA.

International Business Machines (IBM) (1975) Continuous Systems Modelling Program III.(CSMP III). Program reference manual SH19–7001–2, Program Number 5734–XS9. IBM Data processing division, White Plains, New York, USA.

International Rice Research Institute (1989) Implementing the strategy: work plan 1990–1994. International Rice Research Institute, Manila, Philippines.

ISNAR (1984) Considerations for the development of national agricultural research capacities in support of agricultural development. International Service for National Agricultural Research. The Hague, The Netherlands.

Jansen D M (1991) Aspects of organizing training programs in simulation. Internal Report CABO–TPE–IRRI. Los Baños, Philippines.

Jones J G W, Street P R (Eds.) (1990) Agroecosystems. Elsevier, New York, USA.

Knol O M, Nijland G O, Verberne F C M, Wierenga K (1987) Systeemanalyse van de Nederlandse Landbouw. Interim rapport eerste fase. Werkgroep Systeemanalyse Landbouw, Wageningen, The Netherlands.

Kropff M J, Bindraban P S, Van Laar H H (Eds.) (1991) Crop simulation research in the SARP network: research themes and plans for 1991–1996. Internal Report CABO–TPE–IRRI. Los Baños, Philippines.

McWilliam J R, Collinson M P, Van Dusseldorp D B W (1990) Report of the evaluation of the SARP project. Report to the Dept. of Research and Technology of the Dutch Ministry of Foreign Affairs, The Hague, The Netherlands.

Oram P (1990) Agricultural productivity growth and the structure and organization of agricultural research. In Technology policy for sustainable agricultural growth, IFPRI Policy Briefs, Washington, USA.

Palanisamy S, Mohandass S, Thiyagarajan T M, Abdul Kareem A, Penning de Vries F W T (1992) Simulation in pre-testing of rice genetypes in Tamil Nadu. Pages 63–75 in Penning de Vries F W T, Teng P S, Metselaar K (Eds.) (1992) Systems Approaches for Agricultural Development, Proceedings of the International Symposium on Systems Approaches for Agricultural Development, 2–6 December 1991, Bangkok, Thailand (this volume).

Pardey P G, Roseboom J (1990) Development of national agricultural research systems in an international quantitative perspective. In Technology policy for sustainable agricultural growth, IFPRI Policy Briefs, Washington, USA.

Penning de Vries F W T (Ed.) (1989) Crop simulation research in the SARP network: plans, themes and views. Internal Report CABO–TPE–IRRI. Los Baños, Philippines.

Penning de Vries F W T (1990) Can crop models contain economic factors? Pages 89–103 in Rabbinge R, Goudriaan J, Van Keulen H, Penning de Vries F W T, Van Laar H H (Eds.) Theoretical production ecology. PUDOC, Simulation Monographs 34. Wageningen, The Netherlands.

Penning de Vries F W T, Van Laar H H (Eds.) (1982) Simulation of plant growth and crop production. Simulation Monographs, PUDOC, Wageningen, The Netherlands.

Penning de Vries F W T, Jansen D M, Bakema A H, Rabbinge R (Eds.) (1987). Case studies in crop growth simulation. Internal Report CABO–TPE–IRRI. Los Baños, Philippines.

Penning de Vries F W T, Rabbinge R, Jansen D M, Bakema A (1988) Transfer of systems analysis and simulation in agriculture to developing countries. Agricultural Administration and Extension 29:85–96.

Penning de Vries F W T, Jansen D M, Ten Berge H F M, Bakema A (1989) Simulation of ecophysiological processes of growth in several annual crops. Simulation Monographs 29, PUDOC, Wageningen, The Netherlands.

Penning de Vries F W T, Kropff M J, Van Laar H H (Eds) (1991) Simulation and systems analysis for rice production (SARP). PUDOC, Wageningen, The Netherlands.

Persley G J (Ed) (1990) Agricultural Biotechnology: Opportunities for International Development. Biotechnology in Agriculture Series no 2., CAB International. London, UK.

Plucknett D L, Smith N J H, Ozgediz S (1990) Networking in international agricultural research. Cornell University Press, Ithaca, New York, USA.

Rabbinge R, Ward S A, Van Laar H H (Eds.) (1989) Simulation and systems management in crop protection. Simulation Monographs 32, PUDOC, Wageningen, The Netherlands.

Rabbinge R, Goudriaan J, Van Keulen H, Penning de Vries F W T, Van Laar H H (Eds) (1990) Theoretical production ecology. Simulation Monographs 34, PUDOC, Wageningen, The Netherlands.

Smith N J H, Plucknett D L, Ozgediz S (1990) Effective networking in international agricultural research. In ICRISAT (International crops research institute for the semi-arid tropics). Chickpea in the Nineties: proceedings of the Second International Workshop on Chickpea Improvement, 4–8 December 1989, ICRISAT Centre, India. ICRISAT, Patancheru, A.P. 502 324, India.

Spedding C R W (1979) An introduction to agricultural systems. Applied Science Publishers, London, UK.

Spedding C R W (1990) Agricultural production systems. Pages 239–248 in Rabbinge R, Goudriaan J, Van Keulen H, Penning de Vries F W T, Van Laar H H (Eds.) Theoretical production ecology. Simulation Monographs 34, PUDOC, Wageningen, The Netherlands.

Van Dusseldorp D B W (1991) Integrated rural development and interdisciplinary research: a link often missing. (in prep)

Van Keulen H (1990) A multiple goal programming basis for analysing agricultural research and development. Pages 165–276 in Rabbinge R, Goudriaan J, Van Keulen H, Penning de Vries F W T, Van Laar H H (Eds.) Theoretical production ecology. Simulation Monographs 34. PUDOC, Wageningen, The Netherlands.

Van Laar H H, Ten Berge H F M, Jansen D M, De Jong P D, Penning de Vries F W T (Eds.) (1989) Case studies in crop growth simulation. Internal Report CABO–TPE–IRRI. Los Baños, Philippines.

Van Laar H H, Bindraban P S, Kropff M J, Ten Berge H F M, Jansen D M, De Jong P D, Penning de Vries F W T (Eds.) (1991) Case studies in crop growth simulation. Internal Report CABO–TPE–IRRI. Los Baños, Philippines.

Veeneklaas F R (1990) Dovetailing technical and economic analysis. Doctoral dissertation, Erasmus University, Rotterdam. ISBN 90–9003379–3.

INDEX

Systems Approaches for Sustainable Agricultural Development

1. Th. Alberda, H. van Keulen, N.G. Seligman and C.T. de Wit (eds.): *Food from Dry Lands.* An Integrated Approach to Planning of Agricultural Development. 1992 ISBN 0-7923-1877-3
2. F.W.T. Penning de Vries, P.S. Teng and K. Metselaar (eds.): *Systems Approaches for Agricultural Development.* Proceedings of the International Symposium (Bangkok, Thailand, December 1991). 1993 ISBN 0-7923-1880-3; Pb 0-7923-1881-1

KLUWER ACADEMIC PUBLISHERS – DORDRECHT / BOSTON / LONDON